NUREG-1757
Vol. 2, Rev. 1

Consolidated Decommissioning Guidance

Characterization, Survey, and Determination of Radiological Criteria

Final Report

Manuscript Completed: September 2006
Date Published: September 2006

Prepared by
D.W. Schmidt, K.L. Banovac, J.T. Buckley,
D.W. Esh, R.L. Johnson, J.J. Kottan,
C.A. McKenney, T.G. McLaughlin,
S. Schneider

Division of Waste Management and Environmental Protection
Office of Nuclear Material Safety and Safeguards
U.S. Nuclear Regulatory Commission
Washington, DC 20555-0001

ABSTRACT

As part of its redesign of the materials license program, the U.S. Nuclear Regulatory Commission (NRC), Office of Nuclear Material Safety and Safeguards (NMSS) has consolidated and updated numerous decommissioning guidance documents into a three-volume NUREG. Specifically, the three volumes address the following topics:

(1) "Decommissioning Process for Materials Licensees";

(2) "Characterization, Survey, and Determination of Radiological Criteria"; and

(3) "Financial Assurance, Recordkeeping, and Timeliness."

This three-volume NUREG series replaces NUREG-1727 (NMSS Decommissioning Standard Review Plan) and NUREG/BR-0241 (NMSS Handbook for Decommissioning Fuel Cycle and Materials Licensees). This NUREG series is intended for use by NRC staff, licensees, and others.

Volume 2 of the NUREG series, entitled, "Consolidated Decommissioning Guidance: Characterization, Survey, and Determination of Radiological Criteria," provides guidance on compliance with the radiological criteria for license termination (License Termination Rule (LTR)) in 10 CFR Part 20, Subpart E. This guidance takes a risk-informed, performance-based approach to the demonstration of compliance. The approaches to license termination described in this guidance will help to identify the information (subject matter and level of detail) needed to terminate a license by considering the specific circumstances of the wide range of NRC licensees. Licensees should use this guidance in preparing decommissioning plans, license termination plans, final status surveys, and other technical decommissioning reports for NRC submittal. NRC staff will use the guidance in reviewing these documents and related license amendment requests.

Volume 2 is applicable to all licensees subject to the LTR.

PAPERWORK REDUCTION ACT STATEMENT

The information collections contained in this NUREG are covered by the requirements of 10 CFR Parts 19, 20, 30, 33, 34, 35, 36, 39, 40, 51, 60, 61, 63, 70, 72, and 150 which were approved by the Office of Management and Budget, approval numbers 3150–0044, 0014, 0017, 0015, 0007, 0010, 0158, 0130, 0020, 0021, 0127, 0135, 0199, 0009, 0132, and 0032.

PUBLIC PROTECTION NOTIFICATION

The NRC may not conduct or sponsor, and a person is not required to respond to, a request for information or an information collection requirement unless the requesting document displays a currently valid OMB control number.

CONTENTS

CONTENTS

FIGURES

CONTENTS

TABLES

CONTENTS

APPENDICES

CONTENTS

FOREWORD

> **NRC staff suggests that licensees contact NRC or the appropriate Agreement State authority to assure understanding of what actions should be taken to initiate and complete decommissioning at facilities.**

In September 2003, U.S. Nuclear Regulatory Commission (NRC) staff in the Office of Nuclear Material Safety and Safeguards (NMSS)[1] consolidated and updated the policies and guidance of its decommissioning program in a three-volume NUREG series, NUREG-1757, "Consolidated NMSS Decommissioning Guidance." This NUREG series provides guidance on: planning and implementing license termination under the NRC's License Termination Rule (LTR), in the Code of Federal Regulations (CFR), Title 10, Part 20, Subpart E; complying with the radiological criteria for license termination; and complying with the requirements for financial assurance and recordkeeping for decommissioning and timeliness in decommissioning of materials facilities. The staff periodically updates NUREG-1757, so that it reflects current NRC decommissioning policy.

In September 2005, the staff issued, for public comment, draft Supplement 1 to NUREG-1757, which contained proposed updates to the three volumes of NUREG-1757. Draft Supplement 1 included new and revised decommissioning guidance that addresses some of the LTR implementation issues, which were analyzed by the staff in two Commission papers (SECY-03-0069, Results of the LTR Analysis; and SECY-04-0035, Results of the LTR Analysis of the Use of Intentional Mixing of Contaminated Soil). These issues include restricted use and institutional controls, onsite disposal of radioactive materials under 10 CFR 20.2002, selection and justification of exposure scenarios based on reasonably foreseeable future land use (realistic scenarios), intentional mixing of contaminated soil, and removal of material after license termination. The staff also developed new and revised guidance on other issues, including engineered barriers.

The staff received stakeholder comments on Draft Supplement 1 and prepared responses to these comments. The stakeholder comments are located on NRC's decommissioning Web site, at http://www.nrc.gov/what-we-do/regulatory/decommissioning/reg-guides-comm.html, and the NRC staff responses are located on the same Web site and also in the Agencywide Documents Access and Management System at ML062370521. Supplement 1 has not been finalized as a separate document; instead, updated sections from Supplement 1 have been placed into the appropriate locations in revisions of Volumes 1 and 2 of NUREG-1757. The staff plans to revise Volume 3 of this NUREG series at a later date, and that revision will incorporate the Supplement 1 guidance that is related to Volume 3.

---

[1] As of September 2006, NRC is planning to reorganize NMSS and the Office of State and Tribal Programs (STP) to create two new offices: the Office of Federal and State Materials and Environmental Management Programs, which will focus on materials programs; and the new NMSS, which will focus on fuel cycle programs. This reorganization is scheduled to take effect on October 1, 2006. This document contains references to NMSS and STP. These references will be updated in future revisions of this document.

NRC is currently moving toward increasing the use of risk information in its regulation of nuclear materials and nuclear waste management, including the decommissioning of nuclear facilities. NRC's risk-informed regulatory approach to the decommissioning of nuclear facilities represents a philosophy whereby risk insights are considered together with other factors to better focus the attention and resources of both the licensee and NRC on the more risk-significant aspects of the decommissioning process and on the elements of the facility and the site that will most affect risk to members of the public following decommissioning. This results in a more effective and efficient regulatory process.

The term "risk-informed," as used here, refers to the results and findings that come from risk assessments. A risk assessment is a systematic method for addressing risk. The end results of such assessments (e.g., the calculation of predicted doses from decommissioned sites) may relate directly or indirectly to public health effects. NRC staff has developed this guidance to implement the risk-informed approach and intends that the guidance be implemented in a risk-informed manner.

The primary decommissioning guidance documents used by licensees and NRC staff are NUREG-1757 and NUREG-1700, "Standard Review Plan for Evaluating Nuclear Power Reactor License Termination Plans." Table 1 below describes the general applicability of these documents. NUREG-1537, "Guidelines for Preparing and Reviewing Applications for the Licensing of Non-Power Reactors," contains guidance for non-power reactor licensees and NRC staff, which includes a section on decommissioning and license termination for non-power reactors.

Table 1. Contents and Applicability of Key Decommissioning Guidance Documents

Volume and Status [1]	Title	Licensees to Which the Guidance Applies
NUREG-1757, Vol. 1, Rev. 2; September 2006	Consolidated Decommissioning Guidance: Decommissioning Process for Materials Licensees	Fuel cycle, fuel storage, and materials licensees.[2] Limited applicability to reactor licensees (see text below).
NUREG-1757, Vol. 2, Rev. 1; September 2006	Consolidated Decommissioning Guidance: Characterization, Survey, and Determination of Radiological Criteria	All licensees that are subject to the LTR (fuel cycle, fuel storage, materials, and reactor licensees).
NUREG-1757, Vol. 3; September 2003	Consolidated NMSS Decommissioning Guidance: Financial Assurance, Recordkeeping, and Timeliness	Fuel cycle, fuel storage, and materials licensees.[2]
NUREG-1700, Rev. 1; April 2003	Standard Review Plan for Evaluating Nuclear Power Reactor License Termination Plans	Power reactor licensees.

1 Versions listed are current as of September 2006. Please refer to the NRC's Public Electronic Reading Room at http://www.nrc.gov/reading-rm/doc-collections/nuregs to obtain the most up-to-date version.

2 Licensees regulated under 10 CFR Parts 30, 40, 60, 61, 63, 70, and 72 (for 10 CFR Parts 60, 61, and 63, only the ancillary surface facilities that support radioactive waste disposal activities). Because uranium recovery facilities are not subject to 10 CFR Part 20, Subpart E, refer to NUREG-1620, Rev. 1, Section 5, for decommissioning guidance for uranium recovery facilities that are subject to 10 CFR 40, Appendix A.

The current document, NUREG-1757, Volume 2, Revision 1, is applicable to all licensees that are subject to the LTR. As mentioned above, this document incorporates changes based on finalizing the guidance of draft Supplement 1. Table 2 describes the most significant changes to the guidance in this volume.

Table 2. Summary of Major Changes to Volume 2, Revision 1

Subject	Affected Sections
Engineered Barriers	Section 3.5 Appendix P
Scenario Justification Based on Reasonably Foreseeable Land Use	Chapter 5 Section I.3 Appendix M
Removal of Material after License Termination	Section G.1.1 Section G.3
Other Issues and Changes	Section 2.1 NEW Section 2.8 Section 4.0 Appendix D Appendix E Section I.2

NUREG-1757 is intended for use by applicants, licensees, NRC license reviewers, and other NRC personnel. It is also available to Agreement States and the public.

This NUREG is not a substitute for NRC regulations, and compliance with it is not required. The NUREG describes approaches that are acceptable to NRC staff. However, methods and solutions different than those in this NUREG will be acceptable, if they provide a basis for concluding that the decommissioning actions are in compliance with NRC regulations.

Larry W. Camper, Director
Division of Waste Management and Environmental Protection
Office of Nuclear Material Safety and Safeguards

ACKNOWLEDGMENTS

The writing team thanks the individuals listed below for assisting in the development and review of this revision of the report. All participants provided valuable insights, observations, and recommendations.

The team thanks Justine Cowan, Loleta Dixon, and Agi Seaton of Computer Sciences Corporation. The team also thanks Terry L. Johnson, consultant to NRC, for assistance regarding engineered barriers.

The Participants

Nuclear Regulatory Commission Staff

Cameron, Francis X.
Cameron, Jamnes L.
Flanders, Scott C.
Gillen, Daniel M.
Hull, John T.
Isaac, Patrick J.
Jensen, E. Neil
Leslie, Bret W.
McConnell, Keith I.
Nicholson, Thomas J.
Orlando, Dominick A.
Ott, William R.
Persinko, Andrew
Philip, Jacob
Smith, Brooke G.
Spitzberg, Blair B.
Treby, Stuart A.

Organization of Agreement States

Cortez, Ruben (Texas Department of State Health Services)
Galloway, Gwyn (Utah Department of Environmental Quality)
Helmer, Stephen (Ohio Department of Health)
Young, Robert N. (Tennessee Department of Environment and Conservation)

Conference of Radiation Control Program Directors

Hsu, Stephen (California Department of Health Services)

ABBREVIATIONS

The following terms are defined for the purposes of this three-volume NUREG report.

ACAP	Alternative Cover Assessment Program
ADAMS	Agencywide Documents Access and Management System
AEA	Atomic Energy Act (of 1954, as amended)
AEC	U.S. Atomic Energy Commission (became Energy Resource Development Agency and Nuclear Regulatory Commission)
ALARA	As low as is reasonably achievable
ALCD	Alternative Landfill Cover Demonstration
ANSI	American National Standards Institute
APF	Assigned Protection Factors
ASME	American Society of Mechanical Engineers
ASTM	American Society for Testing and Materials
Bq	becquerel
BRT	Bankruptcy Review Team
BTP	Branch Technical Position
CAM	Continuous Air Monitor
CATX	Categorical Exclusion
CEDE	Committed Effective Dose Equivalent
CERCLA	Comprehensive Environmental Response, Compensation, and Liability Act
CEQ	Council on Environmental Quality
CFR	Code of Federal Regulations
Ci	curie
cpm	counts per minute
DCD	Decommissioning Directorate (Nuclear Regulatory Commission)
DCGLs	Derived Concentration Guideline Levels
DFP	Decommissioning Funding Plan
DOE	U.S. Department of Energy
DOT	U.S. Department of Transportation

ABBREVIATIONS

DP	Decommissioning Plan
dpm	disintegrations per minute
DQA	Data Quality Assessment
DQO	Data Quality Objective
DWMEP	Division of Waste Management and Environmental Protection (Nuclear Regulatory Commission)
EA	Environmental Assessment
Eh	Redox potential
EIS	Environmental Impact Statement
EMC	Elevated Measurement Comparison
EML	DOE Environmental Measurements Laboratory (formerly the Health and Safety Laboratory)
EPA	U.S. Environmental Protection Agency
EPAD	Environmental and Performance Assessment Directorate (Nuclear Regulatory Commission)
EPA/NRC MOU	Memorandum of Understanding between the Environmental Protection Agency and the Nuclear Regulatory Commission dated October 9, 2002
ER	Environmental Report
FEP	Feature, Event, and/or Process
FFIEC	Federal Financial Institutions Examination Council
FHLM	Federal Home Loan Mortgage Corporation
FNMA	Federal National Mortgage Association
FONSI	Finding of No Significant Impact
FR	*Federal Register*
FSS	Final Status Survey
FSSP	Final Status Survey Plan
FSSR	Final Status Survey Report
FUSRAP	Formerly Utilized Sites Remedial Action Program
GEIS	Generic Environmental Impact Statement
GNMA	Government National Mortgage Association
GPO	Government Printing Office

HEPA	High-efficiency particulate air
HSA	Historical Site Assessment
IC	Institutional Control
ICRP	International Commission on Radiological Protection
IMC	Inspection Manual Chapter
IMNS	Division of Industrial and Medical Nuclear Safety (Nuclear Regulatory Commission)
IP	Inspection Procedure
IROFS	Items Relied on for Safety
ISA	Integrated Safety Analysis
ISCORS	Interagency Steering Committee on Radiation Standards
ISFSI	Independent Spent Fuel Storage Installation
ISO	International Organization for Standardization
LA/RC	Legal agreement and restrictive covenant
LA	License Amendment
LBGR	Lower Bound [of the] Gray Region
LLD	Lower limit of detection
LPDR	Local Public Document Room
LTC	Long-term control
LTP	License Termination Plan
LTR	License Termination Rule
MARLAP	Multi-Agency Radiological Laboratory Analytical Protocols Manual
MARRSIM	Multi-Agency Radiological Survey and Site Investigation Manual (NUREG-1575)
mCi	millicurie
MCL	Maximum Contaminant Level
MDA	Minimum Detectable Activity
MDC	Minimum Detectable Concentration
MIP	Master Inspection Plan
MOU	Memorandum of Understanding

ABBREVIATIONS

mrem	millirem
mSv	millisievert
NAIC	National Association of Insurance Commissioners
NAS	National Academy of Sciences
NCRP	National Council on Radiation Protection and Measurements
NCS	Nuclear Criticality Safety
NCSA	Nuclear Criticality Safety Analysis
NEPA	National Environmental Policy Act
NIST	National Institute of Standards and Technology
NMMSS	Nuclear Materials Management and Safeguards System
NMSS	Office of Nuclear Material Safety and Safeguards (Nuclear Regulatory Commission)[2]
NOAA	National Oceanic and Atmospheric Administration
NORM	Naturally Occurring Radioactive Material
NRC	U.S. Nuclear Regulatory Commission
OC	Office of Controller
OCC	Office of the Comptroller of the Currency
OE	Office of Enforcement (Nuclear Regulatory Commission)
OGC	Office of General Counsel (Nuclear Regulatory Commission)
OSHA	U.S. Occupational Safety and Health Administration
PCBs	Polychlorinated Biphenyls
pCi	picocurie
PDF	Probability Density Function
PDR	Public Document Room
P&GD	Policy and Guidance Directive
pH	Hydrogen power

---

[2] As of September 2006, NRC is planning to reorganize NMSS and STP to create two new offices: the Office of Federal and State Materials and Environmental Management Programs, which will focus on materials programs; and the new NMSS, which will focus on fuel cycle programs. This reorganization is scheduled to take effect on October 1, 2006. This document contains references to NMSS and STP. These references will be updated in future revisions of this document.

PM	Project Manager
PMF	Probable maximum flood
PMP	Probable maximum precipitation
PPE	Personal protective equipment
PSR	Partial Site Release
QA	Quality Assurance
QAPP	Quality Assurance Project Plan
QA/QC	Quality Assurance and Quality Control
RAI	Request for Additional Information
RCRA	Resource Conservation and Recovery Act
REMP	Radiological Environmental Monitoring Program
RF	Resuspension Factor
RG	Regulatory Guide (also known as Reg Guide)
RIS	Regulatory Issue Summary
ROD	Record of Decision
RSO	Radiation Safety Officer
RSSI	Radiation Site Survey and Investigation [Process]
RWP	Radiation Work Permit
SCP	Site Characterization Plan
SCR	Site Characterization Report
SDMP	Site Decommissioning Management Plan
SDWA	Safe Drinking Water Act
SER	Safety Evaluation Report
SOPs	Standard Operating Procedures
SRP	[NMSS Decommissioning] Standard Review Plan (NUREG–1727)
SSAB	Site-specific advisory board

ABBREVIATIONS

STP	[Office of] State and Tribal Programs (Nuclear Regulatory Commission)[3]
Sv	sievert
TAR	Technical Assistance Request
TDS	Total Dissolved Solids
TEDE	Total Effective Dose Equivalent
TENORM	Technologically Enhanced Naturally Occurring Radioactive Material
TI	Transport Index
TLD	Thermoluminescent Dosimeter
TOC	Total Organic Carbon
TODE	Total Organ Dose Equivalent
TRU	Transuranic(s) [radionuclides]
UECA	Uniform Environmental Covenants Act
UMTRA	Uranium Mill Tailings Remedial Action
UMTRCA	Uranium Mill Tailings Radiation Control Act
USDA	U.S. Department of Agriculture
USACE	U.S. Army Corps of Engineers
U.S.C.	U.S. Code
USGS	U.S. Geological Survey
WAC	Waste acceptance criteria
WRS	Wilcoxon Rank Sum [test]

---

[3] As of September 2006, NRC is planning to reorganize NMSS and STP to create two new offices: the Office of Federal and State Materials and Environmental Management Programs, which will focus on materials programs; and the new NMSS, which will focus on fuel cycle programs. This reorganization is scheduled to take effect on October 1, 2006. This document contains references to NMSS and STP. These references will be updated in future revisions of this document.

GLOSSARY

The following terms are defined for the purposes of this three-volume NUREG report.

Affected parties. Representatives of a broad cross-section of individuals and institutions in the community or vicinity of a site that may be affected by the decommissioning of the site.

Acceptance Review. The evaluation the NRC staff performs upon receipt of a license amendment request to determine if the information provided in the document is sufficient to begin the technical review.

Activity. The rate of disintegration (transformation) or decay of radioactive material. The units of activity are the curie (Ci) and the becquerel (Bq) (see 10 CFR 20.1003).

ALARA. Acronym for "as low as is reasonably achievable," which means making every reasonable effort to maintain exposures to radiation as far below the dose limits as is practical, consistent with the purpose for which the licensed activity is undertaken, and taking into account the state of technology, the economics of improvements in relation to the state of technology, the economics of improvements in relation to the benefits to the public health and safety, and other societal and socioeconomic considerations, and in relation to utilization of nuclear energy and licensed materials in the public interest (see 10 CFR 20.1003).

Alternate Criteria. Dose criteria for residual radioactivity that are greater than the dose criteria described in 10 CFR 20.1402 and 20.1403, as allowed in 10 CFR 20.1404. Alternate criteria must be approved by the Commission.

Aquifer. A geologic formation, group of formations, or part of a formation capable of yielding a significant amount of ground water to wells or springs.

Background Radiation. Radiation from cosmic sources, naturally occurring radioactive material, including radon (except as a decay product of source or special nuclear material) and global fallout as it exists in the environment from the testing of nuclear explosive devices or from past nuclear accidents such as Chernobyl that contribute to background radiation and are not under the control of the licensee. Background radiation does not include radiation from source, byproduct, or special nuclear materials regulated by NRC (see 10 CFR 20.1003).

Broad Scope Licenses. A type of specific license authorizing receipt, acquisition, ownership, possession, use, and transfer of any chemical or physical form of the byproduct material specified in the license, but not exceeding quantities specified in the license. The requirements for specific domestic licenses of broad scope for byproduct material are found in 10 CFR Part 33. Examples of broad scope licensees are facilities such as large universities and large research and development facilities.

Byproduct Material. (1) Any radioactive material (except special nuclear material) yielded in, or made radioactive by, exposure to the radiation incident to the process of producing or utilizing special nuclear material; and (2) the tailings or wastes produced by the extraction or concentration of uranium or thorium from ore processed primarily for its source material content, including discrete surface wastes resulting from uranium solution extraction processes (see 10 CFR 20.1003).

Categorical Exclusion (CATX). A category of regulatory actions which do not individually or cumulatively have a significant effect on the human environment and which the Commission has found to have no such effect in accordance with procedures set out in 10 CFR 51.22 and for which, therefore, neither an environmental assessment nor an environmental impact statement is required (see 10 CFR 51.14(a)).

Certification Amount of Financial Assurance. See *prescribed amount of financial assurance.*

Certification of Financial Assurance. The document submitted to certify that financial assurance has been provided as required by regulation.

Characterization survey. A type of survey that includes facility or site sampling, monitoring, and analysis activities to determine the extent and nature of residual radioactivity. Characterization surveys provide the basis for acquiring necessary technical information to develop, analyze, and select appropriate cleanup techniques.

Cleanup. See *decontamination.*

Closeout Inspection. An inspection performed by NRC, or its contractor, to determine if a licensee has adequately decommissioned its facility. Typically, a closeout inspection is performed after the licensee has demonstrated that its facility is suitable for release in accordance with NRC requirements.

Confirmatory Survey. A survey conducted by NRC, or its contractor, to verify the results of the licensee's final status survey. Typically, confirmatory surveys consist of measurements at a fraction of the locations previously surveyed by the licensee, to determine whether the licensee's results are valid and reproducible.

Critical Group. The group of individuals reasonably expected to receive the greatest exposure to residual radioactivity for any applicable set of circumstances (see 10 CFR 20.1003).

DandD code. The Decontamination and Decommissioning (DandD) software package, developed by NRC, that addresses compliance with the dose criteria of 10 CFR Part 20, Subpart E. Specifically, DandD embodies NRC's guidance on screening dose assessments to allow licensees to perform simple estimates of the annual dose from residual radioactivity in soils and on building surfaces.

Decommission. To remove a facility or site safely from service and reduce residual radioactivity to a level that permits (1) release of the property for unrestricted use and termination of the license or (2) release of the property under restricted conditions and termination of the license (see 10 CFR 20.1003).

Decommission Funding Plan (DFP). A document that contains a site-specific cost estimate for decommissioning, describes the method for assuring funds for decommissioning, describes the means for adjusting both the cost estimate and funding level over the life of the facility, and contains the certification of financial assurance and the signed originals of the financial instruments provided as financial assurance.

Decommissioning Groups. For the purposes of this guidance document, the categories of decommissioning activities that depend on the type of operation and the residual radioactivity.

Decommissioning Plan (DP). A detailed description of the activities that the licensee intends to use to assess the radiological status of its facility, to remove radioactivity attributable to licensed operations at its facility to levels that permit release of the site in accordance with NRC's regulations and termination of the license, and to demonstrate that the facility meets NRC's requirements for release. A DP typically consists of several interrelated components, including (1) site characterization information; (2) a remediation plan that has several components, including a description of remediation tasks, a health and safety plan, and a quality assurance plan; (3) site-specific cost estimates for the decommissioning; and (4) a final status survey plan (see 10 CFR 30.36(g)(4).

Decontamination. The removal of undesired residual radioactivity from facilities, soils, or equipment prior to the release of a site or facility and termination of a license. Also known as remediation, remedial action, and cleanup.

Derived Concentration Guideline Levels (DCGLs). Radionuclide-specific concentration limits used by the licensee during decommissioning to achieve the regulatory dose standard that permits the release of the property and termination of the license. The DCGL applicable to the average concentration over a survey unit is called the $DCGL_W$. The DCGL applicable to limited areas of elevated concentrations within a survey unit is called the $DCGL_{EMC}$.

Dose (or *radiation dose*). A generic term that means absorbed dose, dose equivalent, effective dose equivalent, committed dose equivalent, committed effective dose equivalent, or total effective dose equivalent, as defined in other paragraphs of 10 CFR 20.1003 (see 10 CFR 20.1003). In this NUREG report, dose generally refers to *total effective dose equivalent (TEDE)*.

Durable institutional controls. A legally enforceable mechanism for restricting land uses to meet the radiological criteria for license termination (10 CFR 20, Subpart E). Durable institutional controls are reliable and sustainable for the time period needed.

Effluent. Material discharged into the environment from licensed operations.

Environmental Assessment. A concise public document for which the Commission is responsible that serves to (1) briefly provide sufficient evidence and analysis for determining whether to prepare an environmental impact statement or a finding of no significant impact, (2) aid the Commission's compliance with NEPA when no environmental impact statement is necessary, and (3) facilitate preparation of an environmental impact statement when one is necessary (see 10 CFR 51.14(a)).

Environmental Impact Statement. A detailed written document that ensures the policies and goals defined in the NEPA are considered in the actions of the Federal government. It discusses significant impacts and reasonable alternatives to the proposed action.

Environmental Monitoring. The process of sampling and analyzing environmental media in and around a facility (1) to confirm compliance with performance objectives and (2) to detect radioactive material entering the environment to facilitate timely remedial action.

Environmental Report (ER). A document submitted to the NRC by an applicant for a license amendment request (see 10 CFR 51.14(a)). The ER is used by NRC staff to prepare environmental assessments and environmental impact statements. The requirements for ERs are specified in 10 CFR 51.45–51.69.

Exposure Pathway. The route by which radioactivity travels through the environment to eventually cause radiation exposure to a person or group.

Exposure Scenario. A description of the future land uses, human activities, and behavior of the natural system as related to a future human receptor's interaction with (and therefore exposure to) residual radioactivity. In particular, the exposure scenario describes where humans may be exposed to residual radioactivity in the environment, what exposure group habits determine exposure, and how residual radioactivity moves through the environment.

External Dose. That portion of the dose equivalent received from radiation sources outside the body (see 10 CFR 20.1003).

Final Status Survey (FSS). Measurements and sampling to describe the radiological conditions of a site or facility, following completion of decontamination activities (if any) and in preparation for release of the site or facility.

Final Status Survey Plan (FSSP). The description of the final status survey design.

Final Status Survey Report (FSSR). The results of the final status survey conducted by a licensee to demonstrate the radiological status of its facility. The FSSR is submitted to NRC for review and approval.

Financial Assurance. A guarantee or other financial arrangement provided by a licensee that funds for decommissioning will be available when needed. This is in addition to the licensee's regulatory obligation to decommission its facilities.

Financial Assurance Mechanism. Financial instruments used to provide financial assurance for decommissioning.

Floodplain. The lowland and relatively flat areas adjoining inland and coastal waters including flood-prone areas of offshore islands. Areas subject to a one percent or greater chance of flooding in any given year are included (see 10 CFR 72.3).

Footprint. The portion of a site undergoing decommissioning, which is comprised of all of the areas of soil containing residual radioactivity, where intentional mixing is proposed to meet the release criteria.

General Licenses. Licenses that are effective without the filing of applications with NRC or the issuance of licensing documents to particular persons. The requirements for general licenses are found in 10 CFR Parts 30 and 31. Examples of items for which general licenses are issued are gauges and smoke detectors.

Ground Water. Water contained in pores or fractures in either the unsaturated or saturated zones below ground level.

Historical Site Assessment (HSA). The identification of potential, likely, or known sources of radioactive material and radioactive contamination based on existing or derived information for the purpose of classifying a facility or site, or parts thereof, as impacted or non-impacted (see 10 CFR 50.2).

Hydraulic Conductivity. The volume of water that will move through a medium in a unit of time under a unit hydraulic gradient through a unit area measured perpendicular to the direction of flow.

Hydrology. Study of the properties, distribution, and circulation of water on the surface of the land, in the soil and underlying rocks, and in the atmosphere.

Impact. The positive or negative effect of an action (past, present, or future) on the natural environment (land use, air quality, water resources, geological resources, ecological resources, aesthetic and scenic resources) and the human environment (infrastructure, economics, social, and cultural).

Impacted Areas. The areas with some reasonable potential for residual radioactivity in excess of natural background or fallout levels (see 10 CFR 50.2).

Inactive Outdoor Area. The outdoor portion of a site not used for licensed activities or materials for 24 months or more.

Infiltration. The process of water entering the soil at the ground surface. Infiltration becomes percolation when water has moved below the depth at which it can be removed (to return to the atmosphere) by evaporation or transpiration.

Institutional Controls. Measures to control access to a site and minimize disturbances to engineered measures established by the licensee to control the residual radioactivity. Institutional controls include administrative mechanisms (e.g., land use restrictions) and may include, but are not limited to, physical controls (e.g., signs, markers, landscaping, and fences).

Karst. A type of topography that is formed over limestone, dolomite, or gypsum by dissolution, characterized by sinkholes, caves, and underground drainage.

Leak Test. A test for leakage of radioactivity from sealed radioactive sources. These tests are made when the sealed source is received and on a regular schedule thereafter. The frequency is usually specified in the sealed source and device registration certificate and/or license.

Legacy site. An existing decommissioning site that is complex and difficult to decommission for a variety of financial, technical, or programmatic reasons.

License Termination Plan (LTP). A detailed description of the activities a reactor licensee intends to use to assess the radiological status of its facility, to remove radioactivity attributable to licensed operations at its facility to levels that permit release of the site in accordance with NRC's regulations and termination of the license, and to demonstrate that the facility meets NRC's requirements for release. An LTP consists of several interrelated components including: (1) a site characterization; (2) identification of remaining dismantlement activities; (3) plans for site remediation; (4) detailed plans for the final radiation survey; (5) a description of the end use of the facility, if restricted; (6) an updated site-specific estimate of remaining decommissioning costs; and (7) a supplement to the environmental report, pursuant to 10 CFR 51.33, describing any new information or significant environmental change associated with the licensee's proposed termination activities (see 10 CFR 50.82).

License Termination Rule (LTR). The License Termination Rule refers to the final rule on "Radiological Criteria for License Termination," published by NRC as Subpart E to 10 CFR Part 20 on July 21, 1997 (62 FR 39058).

Licensee. A person who possesses a license, or a person who possesses licensable material, who NRC could require to obtain a license.

MARSSIM. The *Multi-Agency Radiation Site Survey and Investigation Manual (NUREG–1575)* is a multi-agency consensus manual that provides information on planning, conducting, evaluating, and documenting building surface and surface soil final status radiological surveys for demonstrating compliance with dose- or risk-based regulations or standards.

Model. A simplified representation of an object or natural phenomenon. The model can be in many possible forms, such as a set of equations or a physical, miniature version of an object or system constructed to allow estimates of the behavior of the actual object or phenomenon when the values of certain variables are changed. Important environmental models include those estimating the transport, dispersion, and fate of chemicals in the environment.

Monitoring. Monitoring (radiation monitoring, radiation protection monitoring) is the measurement of radiation levels, concentrations, surface area concentrations, or quantities of radioactive material and the use of the results of these measurements to evaluate potential exposures and doses (see 10 CFR 20.1003).

mrem/y (millirem per year). One one-thousandth (0.001) of a rem per year. (See also *sievert.*)

National Environmental Policy Act (NEPA). The National Environmental Policy Act of 1969, which requires Federal agencies, as part of their decision-making process, to consider the environmental impacts of actions under their jurisdiction. Both the Council on Environmental Quality (CEQ) and NRC have promulgated regulations to implement NEPA requirements. CEQ regulations are contained in 40 CFR Parts 1500 to 1508, and NRC requirements are provided in 10 CFR Part 51.

Naturally Occurring Radioactive Material (NORM). The natural radioactivity in rocks, soils, air and water. NORM generally refers to materials in which the radionuclide concentrations have not been enhanced by or as a result of human practices. NORM does not include uranium or thorium in source material.

Non-impacted Areas. The areas with no reasonable potential for residual radioactivity in excess of natural background or fallout levels (see 10 CFR 50.2).

Pathway. See *exposure pathway.*

Performance-Based Approach. Regulatory DECISION-MAKING that relies upon measurable or calculable outcomes (i.e., performance results) to be met, but provides more flexibility to the licensee as to the means of meeting those outcomes.

Permeability. The ability of a material to transmit fluid through its pores when subjected to a difference in head (pressure gradient). Permeability depends on the substance transmitted (oil, air, water, and so forth) and on the size and shape of the pores, joints, and fractures in the medium and the manner in which they are interconnected.

Porosity. The ratio of openings, or voids, to the total volume of a soil or rock expressed as a decimal fraction or as a percentage.

Potentiometric Surface. The two-dimensional surface that describes the elevation of the water table. In an unconfined aquifer, the potentiometric surface is at the top of the water level. In a confined aquifer, the potentiometric surface is above the top of the water level because the water is under confining pressure.

Prescribed Amount of Financial Assurance. An amount of financial assurance based on the authorized possession limits of the NRC license, as specified in 10 CFR 30.35(d), 40.36(b), or 70.25(d).

Principal Activities. Activities authorized by the license which are essential to achieving the purpose(s) for which the license was issued or amended. Storage during which no licensed material is accessed for use or disposal and activities incidental to decontamination or decommissioning are not principal activities (see 10 CFR 30.4).

Probabilistic. Refers to computer codes or analyses that use a random sampling method to select parameter values from a distribution. Results of the calculations are also in the form of a distribution of values. The results of the calculation do not typically include the probability of the scenario occurring.

Reasonable Alternatives. Those alternatives that are practical or feasible from a technical and economic standpoint.

Reasonably foreseeable land use. Land use scenarios that are likely within 100 years, considering advice from land use planners and stakeholders on land use plans and trends.

rem. The special unit of any of the quantities expressed as dose equivalent. The dose equivalent in rems is equal to the absorbed dose in rads multiplied by the quality factor (1 rem = 0.01 sievert) (see 10 CFR 20.1004).

Remedial Action. See *decontamination.*

Remediation. See *decontamination.*

Residual Radioactivity. Radioactivity in structures, materials, soils, ground water, and other media at a site resulting from activities under the licensee's control. This includes radioactivity from all licensed and unlicensed sources used by the licensee, but excludes background radiation. It also includes radioactive materials remaining at the site as a result of routine or accidental releases of radioactive material at the site and previous burials at the site, even if those burials were made in accordance with the provisions of 10 CFR Part 20 (see 10 CFR 20.1003).

RESRAD Code. A computer code developed by the U.S. Department of Energy and designed to estimate radiation doses and risks from RESidual RADioactive materials in soils.

RESRAD-BUILD Code. A computer code developed by the U.S. Department of Energy and designed to estimate radiation doses and risks from RESidual RADioactive materials in BUILDings.

Restricted Area. Any area to which access is limited by a licensee for the purpose of protecting individuals against undue risks from exposure to radiation and radioactive materials (see 10 CFR 20.1003).

Risk. Defined by the "risk triplet" of a scenario (a combination of events and/or conditions that could occur) or set of scenarios, the probability that the scenario could occur, and the consequence (e.g., dose to an individual) if the scenario were to occur.

Risk-Based Approach. Regulatory DECISION-MAKING that is based solely on the numerical results of a risk assessment. (Note that the Commission does not endorse a risk-based regulatory approach.)

Risk-Informed Approach. Regulatory DECISION-MAKING that represents a philosophy whereby risk insights are considered together with other factors to establish requirements that better focus licensee and regulatory attention on design and operational issues commensurate with their importance to public health and safety.

Risk Insights. Results and findings that come from risk assessments.

Robust engineered barrier. A man-made structure that is designed to mitigate the effect of natural processes or human uses that may initiate or accelerate release of residual radioactivity through environmental pathways. The structure is designed so that the radiological criteria for license termination (10 CFR 20, Subpart E) can be met. Robust engineered barriers are designed to be more substantial, reliable, and sustainable for the time period needed without reliance on active ongoing maintenance.

Safety Evaluation Report. NRC staff's evaluation of the radiological consequences of a licensee's proposed action to determine if that action can be accomplished safely.

Saturated Zone. That part of the earth's crust beneath the regional water table in which all voids, large and small, are ideally filled with water under pressure greater than atmospheric.

Scoping Survey. A type of survey that is conducted to identify (1) radionuclide contaminants, (2) relative radionuclide ratios, and (3) general levels and extent of residual radioactivity.

Screening Approach/Methodology/Process. The use of (1) predetermined building surface concentration and surface soil concentration values, or (2) a predetermined methodology (e.g., use of the DandD code) that meets the radiological decommissioning criteria without further analysis, to simplify decommissioning in cases where low levels of residual radioactivity are achievable.

Sealed Source. Any special nuclear material or byproduct material encased in a capsule designed to prevent leakage or escape of the material.

sievert (Sv). The SI unit of any of the quantities expressed as dose equivalent. The dose equivalent in sieverts is equal to the absorbed dose in grays multiplied by the quality factor (1 sievert = 100 rem) (see 10 CFR 20.1004).

Site. The area of land, along with structures and other facilities, as described in the original NRC license application, plus any property outside the originally licensed boundary added for the purpose of receiving, possessing, or using radioactive material at any time during the term of the license, as well as any property where radioactive material was used or possessed that has been released prior to license termination

Site Characterization. Studies that enable the licensee to sufficiently describe the conditions of the site, separate building, or outdoor area to evaluate the acceptability of the decommissioning plan.

Site Characterization Survey. See *characterization survey.*

Site Decommissioning Management Plan (SDMP). The program established by NRC in March 1990 to help ensure the timely cleanup of sites with limited progress in completing the remediation of the site and the termination of the facility license. SDMP sites typically have buildings, former waste disposal areas, large volumes of tailings, ground-water contamination, and soil contaminated with low levels of uranium or thorium or other radionuclides.

Site-Specific Dose Analysis. Any dose analysis that is done other than by using the default screening tools.

Smear. A radiation survey technique which is used to determine levels of removable surface contamination. A medium (typically filter paper) is rubbed over a surface (typically of area 100 cm^2), followed by a quantification of the activity on the medium. Also known as a swipe.

Source Material. Uranium or thorium, or any combination of uranium and thorium, in any physical or chemical form, or ores that contain by weight one-twentieth of one percent (0.05 %) or more of uranium, thorium, or any combination of uranium and thorium. Source material does not include special nuclear material (see 10 CFR 20.1003).

Source Term. A conceptual representation of the residual radioactivity at a site or facility.

Special Nuclear Material. (1) Plutonium, uranium-233 (U-233), uranium enriched in the isotope 233 or in the isotope 235, and any other material that the Commission, pursuant to the provisions of Section 51 of the Atomic Energy Act, determines to be special nuclear material, but does not include source material; or (2) any material artificially enriched by any of the foregoing but does not include source material (see 10 CFR 20.1003).

Specific Licenses. Licenses issued to a named person who has filed an application for the license under the provisions of 10 CFR Parts 30, 32 through 36, 39, 40, 61, 70 and 72. Examples of specific licenses are industrial radiography, medical use, irradiators, and well logging.

Survey. An evaluation of the radiological conditions and potential hazards incident to the production, use, transfer, release, disposal, or presence of radioactive material or other sources of radiation. When appropriate, such an evaluation includes a physical survey of the location of radioactive material and measurements or calculations of levels of radiation, or concentrations or quantities of radioactive material present (see 10 CFR 20.1003).

Survey Unit. A geographical area consisting of structures or land areas of specified size and shape at a site for which a separate decision will be made as to whether or not the unit attains the site-specific reference-based cleanup standard for the designated pollution parameter. Survey units are generally formed by grouping contiguous site areas with similar use histories and having the same contamination potential (classification). Survey units are established to facilitate the survey process and the statistical analysis of survey data.

Technologically Enhanced Naturally Occurring Radioactive Material (TENORM). Naturally occurring radioactive material with radionuclide concentrations increased by or as a result of past or present human practices. TENORM does not include background radioactive material or the natural radioactivity of rocks and soils. TENORM does not include uranium or thorium in source material.

Timeliness. Specific time periods stated in NRC regulations for decommissioning unused portions of operating nuclear materials facilities and for decommissioning the entire site upon termination of operations.

Total Effective Dose Equivalent (TEDE). The sum of the deep-dose equivalent (for external exposures) and the committed effective dose equivalent (CEDE) (for internal exposures) (see 10 CFR 20.1003).

Transmissivity. The rate of flow of water through a vertical strip of aquifer which is one unit wide and which extends the full saturated depth of the aquifer.

Unrestricted Area. An area, access to which is neither limited nor controlled by the licensee (see 10 CFR 20.1003).

Unsaturated Zone. The subsurface zone in which the geological material contains both water and air in pore spaces. The top of the unsaturated zone typically is at the land surface, otherwise known as the vadose zone.

Vadose Zone. See *unsaturated zone.*

1 PURPOSE, APPLICABILITY, AND ROADMAP

1.1 PURPOSE AND APPLICABILITY OF THIS VOLUME

The purpose of this volume is to

- Provide guidance to NRC licensees for demonstrating compliance with the radiological criteria for license termination. Specifically, provide guidance relevant to demonstrating compliance with 10 CFR Part 20, Subpart E, for materials and reactor licensees.

- Provide guidance to NRC staff on methods and techniques acceptable to NRC staff for compliance with the license termination criteria.

- Maintain a risk-informed, performance-based, and flexible decommissioning approach.

This NUREG provides guidance regarding decommissioning leading to termination of a license. Licensees decommissioning their facilities are required to demonstrate to NRC that their proposed methods will ensure that the decommissioning can be conducted safely and that the facility, at the completion of decommissioning activities, will comply with NRC's requirements for license termination. Licensees who are subject to Subpart E should use the policies and procedures discussed in this volume to develop and implement a decommissioning plan (DP) or license termination plan (LTP) (note that throughout this volume, when the term "DP" is used, it may generally be understood to refer to DPs or LTPs). Uranium recovery facilities may find this information useful, but they are not subject to Subpart E. Licensees of Agreement States should contact the appropriate regulatory authority. In many instances, depending on the State, licensees may use this guidance, with the substitution of "Agreement State Authority" for NRC. This volume is also intended to be used in conjunction with NRC Inspection Manual Chapter 2605, "Decommissioning Inspection Program for Fuel Cycle and Materials Licensees."

This volume of NUREG–1757 is being issued to describe, and make available to licensees and the public, (a) guidance on technical aspects of compliance with specific parts of the Commission's regulations; (b) methods acceptable to NRC staff in implementing these regulations; and (c) some of the techniques and criteria used by NRC staff in evaluating DPs and LTPs. Licensees should use this guidance to prepare DPs, LTPs, final status surveys (FSSes), and other technical decommissioning reports for NRC submittal. NRC staff will use the guidance in reviewing these documents and related license amendment requests. The guidance in this volume is not a substitute for regulations, and compliance with the guidance is not required. Methods and solutions different from those described in this volume will be acceptable, if they provide a sufficient basis for NRC staff to conclude that the licensees' decommissioning actions are in compliance with the Commission's regulations. However, the use of nonstandard methods may require more detailed justification for NRC staff to determine acceptability. In addition, the increased complexity and detail of nonstandard demonstrations may result in increased NRC staff review time and, therefore, cost to the licensee.

Volume 2 Does Not Address

- Financial assurance for decommissioning

- Public notification and participation

- Recordkeeping and timeliness in decommissioning

- Decommissioning of uranium recovery facilities

- Disposition of solid materials from licensee control

1.2   ROADMAP TO THIS VOLUME

NRC regulations require a licensee to submit a DP to support the decommissioning of its facility either (a) when it is required by license condition or (b) when NRC has not approved the procedures and activities necessary to carry out the decommissioning and these procedures could increase the potential health and safety impact to the workers or the public. Chapters 4–6 provide acceptance criteria and evaluation criteria for use in reviewing DPs and other information submitted by licensees to demonstrate that the facility is suitable for release in accordance with NRC requirements.

The approach used in this volume is similar to that in Volume 1 of this NUREG report. Volume 1 of this NUREG described the categorization of facilities into Decommissioning Groups 1–7, based on the amount of residual radioactivity, the location of that material, and the complexity of the activities needed to decommission the site. Table 1.1 provides a summary description and examples of each decommissioning group (see Part I of Volume 1 of this NUREG series for more details). Table 1.2 shows the potential applicability of the guidance in this volume to each of these groups. Therefore, where possible, the guidance in this volume has been categorized by the decommissioning groups. For most topics in this volume, the guidance applies to more than one decommissioning group, as shown in Table 1.2. Licensees are encouraged to consult with the appropriate NRC staff to better determine the applicability of the guidance to their facility.

Table 1.1 Description and Examples of Each Decommissioning Group

Group	General Description	Typical Examples
1	Licensed material was not released into the environment, did not cause the activation of adjacent materials, and did not contaminate work areas.	Licensees who used only sealed sources such as radiographers and irradiators
2	Licensed material was used in a way that resulted in residual radioactivity on building surfaces and/or soils. The licensee is able to demonstrate that the site meets the screening criteria for unrestricted use.	Licensees who used only quantities of loose radioactive material that they routinely cleaned up (e.g., R&D facilities)
3	Licensed material was used in a way that could meet the screening criteria, but the license needs to be amended to modify or add procedures to remediate buildings or sites.	Licensees who may have occasionally released radioactivity within NRC limits (e.g., broad scope)
4	Licensed material was used in a way that resulted in residual radiological contamination of building surfaces or soils, or a combination of both (but not ground water). The licensee demonstrates that the site meets unrestricted use levels derived from site-specific dose modeling.	Licensees whose sites released loose or dissolved radioactive material within NRC limits and may have had some operational occurrences that resulted in releases above NRC limits (e.g., waste processors)
5	Licensed material was used in a way that resulted in residual radiological contamination of building surfaces, soils, or ground water, or a combination of all three. The licensee demonstrates that the site meets unrestricted use levels derived from site-specific dose modeling.	Licensees whose sites released, stored, or disposed of large amounts of loose or dissolved radioactive material onsite (e.g., fuel cycle facilities)
6	Licensed material was used in a way that resulted in residual radiological contamination of building surfaces, and/or soils, and possibly ground water. The licensee demonstrates that the site meets restricted use levels derived from site-specific dose modeling.	Licensees whose sites would cause more health and safety or environmental impact than could be justified when cleaning up to the unrestricted release limit (e.g., facilities where large inadvertent release(s) occurred)
7	Licensed material was used in a way that resulted in residual radiological contamination of building surfaces, and/or soils, and possibly ground water. The licensee demonstrates that the site meets alternate restricted use levels derived from site-specific dose modeling.	Licensees whose sites would cause more health and safety or environmental impact than could be justified when cleaning up to the restricted release limit (e.g., facilities where large inadvertent release(s) occurred)

Table 1.2   Applicability of Volume 2 to Decommissioning Groups

	Group 1	Group 2	Group 3	Group 4	Group 5	Group 6	Group 7
Dose Assessment Method	N/A	Screening criteria (Section 5.1, Appendix H)		Site-specific assessment (Section 5.2, Appendices I and M)		Site-specific assessment (Section 5.3, Appendices I and M)	Site-specific assessment (Section 5.4, Appendices I and M)
Dose Assessment for Partial Site Release	No	Yes, for licensees electing partial site releases (Appendices K and L)					
Site Characterization	No	Yes (Section 4.2, Appendix E)			Yes (Section 4.2, Appendices E, F, and G)		
Remedial Action Support Surveys	No	Yes, if remediation is required (Section 4.3, Appendix E)					
Final Status Survey (FSS)	No	Yes (Sections 4.4 and 4.5, Appendices A, B, D, and E)		Yes (Sections 4.4 and 4.5, Appendices A, D, and E)			
Complex Survey Situations (Not Addressed in MARSSIM)	No		Yes (Section 4.6, Appendix G)				
Ground Water Characterization	No			Yes, surface water only (Appendix F)	Yes (Appendix F)		
ALARA Analysis	No	Yes, good housekeeping only (Section 6.2)	Yes (Chapter 6, Appendix N)				

Because of the variability in the amounts, forms, and types of radioactive material used by each decommissioning group, licensees may need to submit a broad range of information types and details to NRC for approval of decommissioning activities. The types of information required could vary because of the radionuclides involved, whether or not remediation is required, or the complexity of the site. The amount of detail discussed in this volume is based on the needs of complicated sites. The NRC staff does not suggest that all licensees should provide the same level of detail. Rather, the amount of detail provided on a specific issue should be commensurate with the complexity of the issue for the facility. Thus, licensees and NRC reviewers should generally determine the level of detail and appropriate methods based on the complexity of the facility as related to a compliance demonstration. Licensees are encouraged to discuss with their NRC license reviewer the appropriate level of detail to be included in the DP, using the checklists of Appendix D of Volume 1 of this NUREG report.

The technical aspects of sites, as related to decommissioning, are often called either "simple" or "complex." The question becomes what defines the technical aspects as "simple" or "complex." One needs to decide what aspect of the decommissioning one is trying to judge. For example, site characterization may be complex at a site, but the FSS, after remediation, may be simple and straightforward.

Unfortunately, there is no precise definition or list of characteristics that can define the technical aspects as either simple or complex without caveats. That is because simple and complex are not distinct boxes but part of a continuum. For example, sites using screening criteria are relatively "simple," technically, and sites proposing both partial release and restricted release with an engineered barrier design along with institutional controls that rely on active maintenance are relatively "complex," technically. While there can be exceptions to the site complexity characterization illustrated in Figure 1.1, Decommissioning Groups 1–3 generally have mostly simple technical aspects, and Decommissioning Groups 5–7 generally have mostly "complex" technical aspects. Group 4 sites, which are sites without initial ground-water contamination, can be of either complexity.

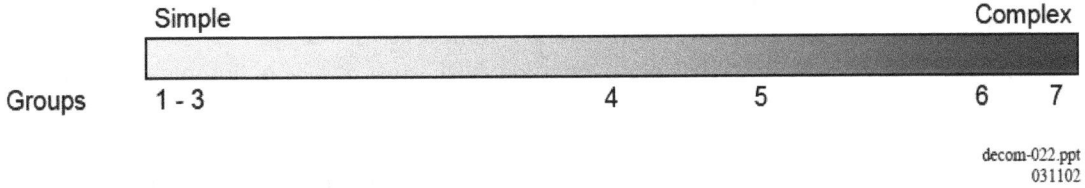

Figure 1.1 **Continuum of Site Complexity.**

Simple sites are generally easy to assess, because site characterization information, survey methods, and models with NRC–reviewed default parameter sets are readily available. These sites have residual radioactivity generally limited to building surfaces or surface soil at a site with simple geological and hydrological characteristics.

Technically complex sites are generally sites with one or more of the following conditions:

- existing ground water or surface water contamination;

- former burials of radioactive material or highly heterogeneous subsurface soil residual radioactivity;

- diversified and extensive surface/subsurface residual radioactivity, that may require data and modeling of these multiple sources at the site, because of the interactions between sources;

- radionuclides that (a) are hard to detect, (b) lack suitable surrogate radionuclides, or (c) have very low effective derived concentration guideline levels (DCGLs);

- current offsite releases such that alternate offsite scenario(s) may be required or use of onsite resident farmer scenario may be inadequate (e.g., sites with multiple receptors);

- planned license termination under restricted conditions (10 CFR 20.1403);

- physical barriers or vaults; or

- unusual physical or lithologic properties, such as a highly fractured formation, karst features, or sinkholes that may significantly impact assumptions of transport models or the overall conceptual model.

These conditions are not rigid definitions as other factors are also important. One such important factor would be the locations where radionuclides are present. For example, a site could be called simple because the predominant radionuclide is a short-lived energetic gamma that is in the surface soil; even if the hydrology at the site is complex, the site would still be called simple because the primary exposure pathway is external exposure, which is an uncomplicated pathway.

Technically complex sites may require more advanced remediation, survey planning, or performance assessment modeling and analysis approaches. Specifically, more advanced approaches may be required to select appropriate models or codes, collect characterization data, justify source-term assumptions, ensure internal consistencies in the associated complex transport models, and design site- or source-specific survey plans. Because of the complex nature of these sites, the scope of NRC staff review will depend on site-specific conditions and on the degree of site complexity. Therefore, a generic NRC staff review of complex sites cannot be articulated in this volume.

Licensees and NRC staff should interact early for information and direction regarding development of a complete DP. Once the decision has been made to decommission, the next step is to determine what information the licensee needs to provide to demonstrate site conditions successfully. If the licensee does not need to submit a DP, the licensee should follow the guidance in Volume 1 of this NUREG report for the appropriate decommissioning group.

If the licensee is required to submit a DP, NRC staff should schedule a meeting with the licensee to discuss both the planned decommissioning and the approach that will be used to evaluate the information submitted to support the decommissioning. NRC staff and the licensee should review the licensed operations, types and quantities of radioactive materials used at the facility, and any other activities (spills, leaks, etc.) that could affect decommissioning operations. NRC staff should also discuss the decommissioning goal envisioned by the licensee (i.e., license termination under unrestricted versus restricted conditions) and the information required to be submitted for the appropriate decommissioning group (described in Chapters 10, 11, 12, 13, or 14 of Volume 1 of this NUREG report). NRC staff should then discuss the acceptance criteria for information to be included in the DP. Finally, NRC staff should prepare a site-specific checklist for evaluating the DP. Appendix D of Volume 1 of this NUREG report provides a generic checklist, that may be used to develop this site-specific checklist. Thus, before the licensee begins to develop its DP, both the NRC staff and the licensee should have a good understanding of the types of information that should be included in the DP, as well as the criteria that NRC staff will use to evaluate the information submitted to support the decommissioning. This should help minimize the need for requests for additional information.

1.3 ROADMAP FOR GUIDANCE ON RESTRICTED USE, ALTERNATE CRITERIA, AND USE OF ENGINEERED BARRIERS

The focus of this volume is on guidance for demonstrating compliance with the dose criteria from 10 CFR Part 20, Subpart E. However, there are additional criteria in Subpart E related to license termination under restricted conditions and the use of alternate criteria for license termination. In addition, some licensees may wish to use engineered barriers as part of the compliance strategy. This section describes where guidance on these subjects may be found in this NUREG (Volumes 1 and 2).

Table 1.3 provides cross-references to sections of Volume 1 and this volume for guidance on aspects of restricted use, use of alternate criteria, and use of engineered barriers.

Table 1.3   Cross-References for Restricted Use, Alternate Criteria, and Use of Engineered Barriers

Issue	Applicable Sections of this Report	
	Volume 1	**Volume 2**
Initial eligibility demonstration for restricted use	17.7.2	n/a
Institutional controls	17.7.3	n/a
Site maintenance and long-term monitoring	17.7.4	n/a
Obtaining public advice	17.7.5 and Appendix M	n/a
Dose modeling for restricted use	17.7.6	5.3
ALARA analysis for restricted use	17.7.3.5	6
Use of alternate criteria	17.8	n/a
Dose modeling for alternate criteria	17.8	5.4
Use of engineered barriers	17.7.3	3.5 and Appendix P
Note: Volume 3 has no applicable sections on engineered barriers.		

1.4   ITERATIVE NATURE OF THE COMPLIANCE DEMONSTRATION PROCESS: A DECISIONMAKING FRAMEWORK

NRC staff developed an overall framework for dose assessment and decisionmaking at sites where the licensee has decided to begin the decommissioning and license termination process. The framework can be used by licensees throughout the decommissioning and license termination process for sites ranging from simple sites to the more complex or contaminated sites. Information is summarized here for using the framework to step through the decommissioning and license termination process; a detailed description is provided in NUREG–1549. This framework was developed for demonstrating compliance using the characterization and dose assessment approach (see Section 2.5), but the concepts may be extended for use in the DCGL development and the FSS approach.

This framework is designed to assist the licensee, NRC staff, and other stakeholders in making decommissioning decisions. By doing so, the process allows the licensee to:

- coordinate its planning efforts with NRC staff input and conduct dose assessments and site characterization activities that are directly related to regulatory decisions;

- optimize cost decisions related to site characterization, remediation, and land-use restrictions;

- integrate analyses for ALARA requirements; and

- elicit other stakeholders' input at crucial points.

The framework is designed to allow the licensee flexibility in the decisionmaking process for demonstrating compliance. As such, the framework provides one method that may be useful for licensees in developing the compliance strategy.

The steps and decision points of the decision framework support assessment of the entire range of dose modeling options from which a licensee may choose, whether it involves using generic screening parameters, changing parameters, or modifying pathways or models. The decision framework, including its steps and decision points, is illustrated in Figure 1.2 (modified from NUREG–1549).

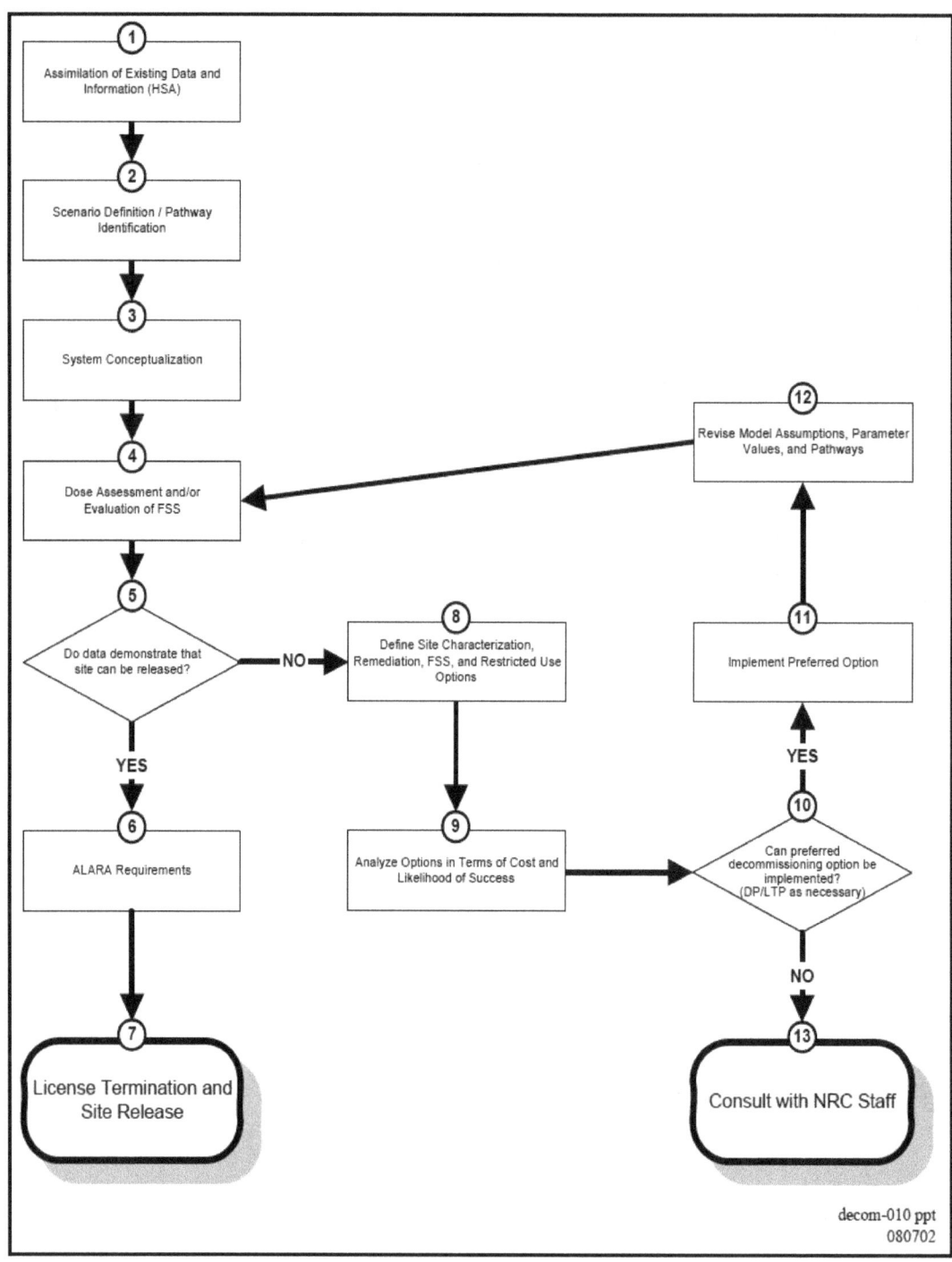

Figure 1.2   Decommissioning and License Termination Decision Framework (modified from NUREG–1549).

1.4.1 CONTENTS AND GENERAL CONCEPTS OF THE ITERATIVE APPROACH IN USING THE DECISION FRAMEWORK

To facilitate the preparation and evaluation of the dose assessments, this framework describes an iterative approach to decisionmaking for license termination. An iterative approach is helpful because of the very wide range of levels of residual radioactivity, complexity of analysis, and potential remediation necessary at NRC–licensed sites. The iterative approach consists of using existing information for generic screening and using site-specific information as appropriate. This approach provides assurance that obtaining additional site-specific information is worthwhile because it ensures that a more realistic dose assessment will generally result in an estimated dose no greater than that estimated using screening. These two phases of the compliance assessment are summarized in broad terms below (more details are provided in NUREG–1549):

1. <u>Generic screening</u>: In this iteration, licensees would demonstrate compliance with the dose criteria of the LTR by using predefined models and generic screening parameters.

2. <u>Use of site-specific information as appropriate</u>: If compliance cannot be demonstrated using generic screening, then licensees should proceed to the next iteration of analysis in which defensible site-specific values are obtained and applied.

The following general concepts apply to using the iterative approach with the decision framework shown in Figure 1.2:

- The approach provides a process for screening sites and for directing additional data collection efforts where necessary or where most helpful toward demonstrating compliance.

- The framework is designed such that the level of complexity and rigor of analysis conducted for a given site should be commensurate with the level of risk that the site poses.

- The licensee would not need to start the process with generic screening but may move directly to use of site-specific information, as appropriate.

- For the process to work efficiently, the licensee is encouraged to involve NRC staff from the very first step through the end of the decisionmaking process.

The framework provides the licensee with a variety of options for performing dose assessments from simple screening to more detailed site-specific analyses. Use of the framework would normally encompass Steps 1–7; however, the amount of work that goes into each of these steps should be based on the expected levels of residual radioactivity and the health risks they pose. Note that in this framework, while all sites may start at the same level of very simple analyses (not a requirement for successful implementation), it is expected that only certain sites would progress to very complex dose assessment and options analyses. Some sites may not need to conduct any options analyses as described in Step 8, and some sites may need to evaluate a limited set of relatively simple and inexpensive options. For example, a site with a contained source of residual radioactivity that is obviously simple to remove would not spend time analyzing large suites of alternative data collection and remediation options. On the other hand, a site with high levels of widely distributed residual radioactivity may use this process to analyze a variety of simple and complex options to define the best decontamination and decommissioning strategy.

Therefore, this approach ensures that the licensee's efforts and expenses will be commensurate with the level of risk posed by the site.

1.4.2   STEPS OF THE DECISION FRAMEWORK

NUREG–1549 provides three separate discussions to illustrate the iterative nature of assessments as site complexity increases. The following is both a summary of the steps of the decision framework and a set of examples to help users understand most of the features of dose modeling in the context of the decision framework. This discussion has been modified slightly from that in NUREG–1549 to make it applicable to a broader range of compliance demonstrations. A number of the examples refer to the use of the DandD and RESRAD dose assessment codes. See Appendices H and I of this volume for details about dose modeling codes with specifics regarding these two dose assessment codes. Licensees desiring further details should refer to NUREG–1549. Refer to Figure 1.2 (modified from NUREG–1549) while reviewing the following steps of the dose modeling framework:

1.   The first step in a compliance assessment involves gathering and evaluating existing data and information about the site, historical site assessment (HSA), including the nature and extent of residual radioactivity at the site. Often, minimal information is all that is needed for an initial screening analysis (e.g., a simple representation of the source of residual radioactivity). Specifically, information is needed to support the decision that the site is "simple" and is qualified for screening analysis. However, licensees should use all information about the site that is readily available. This step also includes the definition of the performance objectives for compliance with decommissioning criteria.

2.   This step involves defining the scenarios and pathways that are important and relevant for the site dose assessment. This step also includes the preliminary determination of whether the licensee plans to adopt an unrestricted use or restricted use option provided for in the LTR. For all assessments using screening concentration tables or DandD, NRC has already defined the generic scenarios and pathways for screening. For site-specific analysis, DandD and RESRAD/RESRAD–BUILD codes may be used, in addition to other codes. The codes used should allow the licensee to select and deselect exposure pathways as appropriate for the site-specific conditions.

3.   Once scenarios are defined and exposure pathways identified, a basic conceptual understanding of the system is developed, often based on simplifying assumptions regarding the nature and behavior of the natural systems. System conceptualization includes conceptual and mathematical model development and assessment of parameter uncertainty. Using DandD for generic screening (and as the basis for screening

concentration tables), NRC has predefined conceptual models for the scenarios along with default parameter distributions (based on NUREG/CR–5512, Volumes 1 and 3). For site-specific analysis, the DandD and/or RESRAD/RESRAD BUILD conceptual model can be used after verification that the site conceptual model is compatible with the conceptual model of the code used.

4. This step involves the dose assessment or consequence analysis, based on the defined scenario(s), exposure pathways, models, and parameter distributions. This step may also involve the evaluation of FSS results. For generic screening, reviewers can accept lookup tables and use the generic models and default parameter probability density functions (PDFs) by running DandD with the appropriate site-specific source term and leaving all other information in the software unchanged. Site-specific assessments allow the licensee to use other codes and change pathways and parameter distributions based on site-specific data and information. Based on Monte Carlo sampling of the input distributions, DandD and RESRAD/RESRAD–BUILD provide various plots and reports of the dose distribution.

5. This is the first major decision point in the license termination decision process. It involves answering the question of whether the dose assessment results and/or FSS results demonstrate compliance with the dose criterion in 10 CFR Part 20, Subpart E. If the results demonstrate compliance, the licensee proceeds with Steps 6 and 7 to demonstrate that the ALARA requirements in Subpart E have been met. If the results are ambiguous or clearly exceed the performance objective, then the licensee proceeds to Steps 8 and 9 for the next iteration of the decisionmaking process.

6. In this step, the licensee can proceed to satisfy ALARA criterion of 10 CFR Part 20, Subpart E, if it is not already addressed. If the ALARA requirements are satisfied, then the licensee initiates the license termination. Note that the DandD or RESRAD codes do not involve or automate these steps.

7. This step includes the administrative and other actions necessary to terminate the license and release the site. See Volume 1 of this NUREG for more details on the specific actions to terminate the license and release the site.

8. Full application of the decision framework involves defining all possible options the licensee might address to defend a final set of actions needed to demonstrate compliance with license termination criteria. Options may include (a) acquiring more data and information about the site and source(s) of residual radioactivity to reduce uncertainty about the pathways, models, and parameters, and thus reduce the calculated dose; (b) reducing actual contamination through remediation actions; (c) reducing exposure to radionuclides through implementation of land-use restrictions; (d) performing an FSS; or (e) some combination of these options.

9. All the options identified in Step 8 are analyzed and compared in order to optimize selection of a preferred set of options. This options analysis may consider the cost of

implementation, the likelihood of success (and the expected costs associated with success or failure to achieve the desired results when the option is implemented), the timing considerations and constraints, and other quantitative and/or qualitative selection criteria.

10. The activities in Steps 8 and 9 provide information for licensees to choose the preferred decommissioning option based on considerations of cost, the likelihood of success, timeliness, and other considerations. Based on the results of the DandD and RESRAD/RESRAD–BUILD sensitivity analysis, for example, a licensee may identify one or more parameters that may be modified, based on the acquisition of site-specific information and data. If new data can reduce the uncertainty associated with sensitive parameters, then the licensee may be able to defend a new calculated dose that meets the license termination criteria. This step may include submission to NRC of a DP, if such submittal is necessary to proceed with the preferred option. If the licensee believes that no viable options exist at this time, the licensee should confer with NRC staff (see also Step 13).

11. Under this step, the preferred option is implemented. The licensee obtains the information necessary to support revisions to the parameters identified in Steps 8 and 9 or performs an FSS.

12. Once data are successfully obtained, the affected parameters for the predefined models are revised as appropriate. Also, data may support elimination of one or more of the exposure pathways in the predefined scenarios. DandD and RESRAD/RESRAD–BUILD codes provide very simple and straightforward modification of the pathways and parameters of interest.

Once the pathways and parameters are revised, the licensee would revisit Steps 4 and 5 to determine the impact of the revisions on demonstrating compliance with the performance objectives. If met, the licensee proceeds to Steps 6 and 7. If the performance objective is still exceeded, the licensee returns to Steps 8 and 9 to analyze remaining options to proceed.

13. In certain limited circumstances, terminating the license may not be feasible. The licensee should consult with NRC staff for case-specific guidance and for the regulatory approvals that may be necessary to maintain, rather than terminate the license.

1.5   BIBLIOGRAPHY AND SUPERSEDED DOCUMENTS

This section provides the reference list for this volume, categorized in the following subsections by type of reference document. Chapter 4 of Volume 1 of this NUREG report provided a more general list of decommissioning references which included statutes, decommissioning regulations, decommissioning inspection manual chapters, and decommissioning inspection procedures.

Use of References Cited in this Volume

This volume refers to a number of other documents for guidance. In some cases, this volume will state that the referenced guidance is approved by NRC staff. However, in some cases, the documents are only referenced for information. In these cases, the specific applicability to a facility should be determined by the licensee, in consultation with NRC staff, as appropriate.

1.5.1 NRC DECOMMISSIONING DOCUMENTS REFERENCED IN THIS VOLUME

- Nuclear Regulatory Commission (U.S.) (NRC), Washington, DC. "Decommissioning Criteria for the West Valley Demonstration Project (M–32) at the West Valley Site: Final Policy Statement." *Federal Register*: Vol. 67, No. 22. pp. 5003–5012. February 2, 2002.

- —————. "Draft Branch Technical Position on Site Characterization for Decommissioning." NRC: Washington, DC. November 1994.

- —————. "Draft Staff Guidance for Dose Modeling of Proposed Partial Site Releases." Memorandum from John T. Greeves to John A. Zwolinski. NRC: Washington, DC. September 28, 2001.

- —————. Inspection Manual Chapter 2605, "Decommissioning Procedures for Fuel Cycle and Materials Licensees." NRC: Washington, DC. November 1996.

- —————. NRC 2003. SECY-03-0069, "Results of the License Termination Rule Analysis," May 2, 2003.

- —————. NRC 2003. SRM-SECY-03-0069, "Staff Requirements - SECY-03-0069 - Results of the License Termination Rule Analysis," November 17, 2003.

- —————. NRC 2004. SECY-04-0035, "Results of the License Termination Rule Anlysis of the Use of Intentional Mixing of Contaminated Soil," March 1, 2004.

- —————. NRC 2004. SRM-SECY-04-0035, "Staff Requirements - SECY-04-0035 - Results of the License Termination Rule Anlysis of the Use of Intentional Mixing of Contaminated Soil," May 11, 2004.

- —————. NRC 2004. Regulatory Issue Summary 2004-08, "Results of the License Termination Rule Analysis," May 28, 2004.

- —————. NRC 2006. SECY-06-0143, "Stakeholder comments and Path Forward on Decommissioning Guidance to Address License Termination Rule Analysis Issues," July 5, 2006.

- —————. NRC 2006. SRM-SECY-06-0143, "Staff Requirements - SECY-06-0143 - Stakeholder Comments and Path Forward on Decommissioning Guidance to Address License Termination Rule Analysis Issues," September 19, 2006.

- —————. NUREG–1496, "Generic Environmental Impact Statement in Support of Rulemaking on Radiological Criteria for License Termination of NRC–Licensed Nuclear Facilities." NRC: Washington, DC. July 1997.

- —————. NUREG–1500, "Working Draft Regulatory Guide on Release Criteria for Decommissioning: NRC Staff's Draft for Comment." NRC: Washington, DC. August 1994.

- —————. NUREG–1501, "Background as a Residual Radioactivity Criterion for Decommissioning–Draft Report." NRC: Washington, DC. August 1994.

- —————. NUREG–1505, Rev. 1, "A Proposed Nonparametric Statistical Methodology for the Design and Analysis of Final Status Decommissioning Surveys: Interim Draft Report for Comment and Use." NRC: Washington, DC. June 1998.

- —————. NUREG–1507, "Minimum Detectable Concentrations with Typical Radiation Survey Instruments for Various Contaminants and Field Conditions." NRC: Washington, DC. June 1998.

- —————. NUREG–1549, "Decision Methods for Dose Assessment to Comply with Radiological Criteria for License Termination, Draft Report for Comment." NRC: Washington, DC. July 1998.

- —————. NUREG–1575, Rev. 1, "Multi-Agency Radiation Survey and Site Investigation Manual (MARSSIM)." EPA 402–R–97–016, Rev. 1, DOE/EH–0624, Rev. 1. U.S. Department of Defense, U.S. Department of Energy, U.S. Environmental Protection Agency, and NRC: Washington, DC. August 2000. Corrected pages for MARSSIM, Revision 1 (August 2000) with the June 2001 updates, are available at the EPA Web site: http://www.epa.gov/radiation/marssim.

- —————. NUREG–1727, "NMSS Decommissioning Standard Review Plan." NRC: Washington, DC. September 2000.

- —————. NUREG/BR–0241, "NMSS Handbook for Decommissioning Fuel Cycle and Materials Licensees." NRC: Washington, DC. March 1997.

- —————. NUREG/CR–5512, Vol. 1, "Residual Radioactive Contamination From Decommissioning: Technical Basis for Translating Contamination Levels to Annual Total Effective Dose Equivalent." NRC: Washington, DC. October 1992.

- —————. NUREG/CR–5512, Vol. 2, "Residual Radioactive Contamination from Decommissioning, User's Manual, Draft Report." NRC: Washington, DC. May 1999.

- —————. NUREG/CR–5512, Vol. 3, "Residual Radioactive Contamination from Decommissioning, Parameter Analysis, Draft Report for Comment." NRC: Washington, DC. October 1999.

- —————. NUREG/CR–5849, "Manual for Conducting Radiological Surveys in Support of License Termination, Draft Report for Comment." NRC: Washington, DC. June 1992.

1.5.2 OTHER NRC DOCUMENTS REFERENCED IN THIS VOLUME

- *Code of Federal Regulations.* 10 CFR Part 20, "Standards for Protection Against Radiation." Sections 10 CFR 20.1001–2402.

- —————. 10 CFR Part 30, "Rules of General Applicability To Domestic Licensing of Byproduct Material." Sections 10 CFR 30.1–72.

- —————. 10 CFR Part 40, "Domestic Licensing of Source Material." Sections 10 CFR 40.1–82.

- —————. 10 CFR Part 50, "Domestic Licensing of Production And Utilization Facilities." Sections 10 CFR 50.1–120.

- —————. 10 CFR Part 70, "Domestic Licensing of Special Nuclear Material." Sections 10 CFR 70.1–92.

- —————. 10 CFR Part 72, "Licensing Requirements for the Independent Storage of Spent Nuclear Fuel And High-Level Radioactive Waste." Sections 10 CFR 72.1–248.

- Nuclear Regulatory Commission (U.S.) (NRC). NUREG–1200, Rev. 3, "Standard Review Plan for the review of a license application for a Low-Level Radioactive Waste Disposal Facility." NRC: Washington, DC. April 1994.

- —————. NUREG–1573, "A Performance Assessment Method for Low-Level Waste Disposal Facilities: Recommendations of NRC's Performance Assessment Working Group." NRC: Washington, DC. October 2000.

- —————. NUREG–1620, Rev. 1, "Standard Review Plan for the Review of a Reclamation Plan for Mill Tailings Sites Under Title II of the Uranium Mill Tailings Radiation Control Act, Draft." NRC: Washington, DC. January 2002.

- —————. NUREG–1623, "Design of Erosion Protection for Long-Term Stabilization, Draft Report for Comment." NRC: Washington, DC. February 1999.

- —————. NUREG–1748, "Environmental Review Guidance for Licensing Actions Associated with NMSS Programs." NRC: Washington, DC. September 2001.

- —————. Staff Requirements Memorandum, SECY–98–144, "White Paper on Risk-Informed and Performance-Based Regulation." NRC: Washington, DC. March 1999.

- U.S. Nuclear Regulatory Commission, "Technical Position on Waste Form (Revision 1)," Division of Low-Level Waste Management and Decommissioning, Washington, DC, January 1991.

1.5.3   OTHER DOCUMENTS REFERENCED IN THIS VOLUME

- Albright, W., Et al., "Field Water Balance of Landfill Final Covers," Journal of Environmental Quality, 33(6), 2317–2332, 2004.

- Department of Energy (U.S.) (DOE). DOE/EM–0142P, "Decommissioning Handbook." DOE: Washington, DC. March 1994.

- Department of Energy (U.S.) (DOE). "Plant Encroachment on the Burrell, Pennsylvania, Disposal Cell: Evaluation of Long-Term Performance and Risk," DOE, Grand Junction Office, Grand Junction, CO, 1999.

- Environmental Protection Agency (U.S.) (EPA), Washington, DC. "Federal Radiation Protection Draft Guidance for Exposure of the General Public." *Federal Register*: Vol. 59, p. 66414. December 23, 1994.

- —————. EPA 402–R–93–081, "External Exposure to Radionuclides in Air, Water, and Soil: Federal Guidance Report No. 12." EPA: Washington, DC. September 1993.

- —————. EPA 520/1–88–020, "Limiting Values of Radionuclide Intake and Air Concentration and Dose Conversion Factors for Inhalation, Submersion, and Ingestion: Federal Guidance Report No. 11." EPA: Washington, DC. September 1988.

- —————. EPA/540/G–89/004, "Guidance for Conducting Remedial Investigations and Feasibility Studies Under CERCLA." EPA: Washington, DC. October 1988(b).

- —————. EPA/600/R–96/055, "Guidance for the Data Quality Objectives Process, EPA QA/G–4." EPA: Washington, DC. August 2000.

- —————. OSWER Directive 9360.0–03B, "Superfund Removal Procedures." EPA: Washington, DC. 1988(c).

- James, A., et al., "Field Performance of GCL under ion exchange conditions," Journal of Geotechnical and Geoenvironmental Engineering, ASCE, vol. 13, no. 10, pp. 897-901, 1997.

- Lin, L., and C.H. Benson, "Effect of Wet-Dry Cycling on Swelling and Hydraulic Conductivity of GCLs," Journal of Geotechnical and Geoenvironmental Engineering, pp. 40-49, 2000.

- Melchoir, S., "In-situ studies on the performance of landfill caps," In Proceedings of the International Containment Technology Conference, pp. 365-373, U.S. Department of Energy, Germantown, MD, 1997.

- Waugh, W.J., et al., "Ecology, Design, and Long-Term Performance of Surface Barriers: Applications at a Uranium Mill Tailings Site," Presented in "Barrier Technologies for Environmental Management," Summary of a Workshop, National Research Council, National Academy of Sciences, 1997.

1.5.4 DOCUMENTS SUPERSEDED BY THIS VOLUME

This volume supersedes the guidance documents listed in Table 1.4, and the superseded documents should no longer be used.

Table 1.4   Documents Superseded by this Report

Document	Title	Date
NRC memorandum	Draft Staff Guidance for Dose Modeling of Proposed Partial Site Releases	09/28/2001
Branch Technical Position	Draft Branch Technical Position on Site Characterization for Decommissioning	11/1994
NUREG–1500	Working Draft Regulatory Guide on Release Criteria for Decommissioning: NRC Staff's Draft for Comment	08/1994
NUREG/CR-5849	Manual for Conducting Radiological Surveys in Support of License Termination	06/1992
Branch Technical Position	Draft Branch Technical Position: Screening Methodology for Assessing Prior Land Burials of Radioactive Waste Authorized Under Former 10 CFR 20.304 and 20.302	10/1996
Branch Technical Position	Disposal or Onsite Storage of Thorium or Uranium Wastes from Past Operations	10/1981

This Volume 2 of this NUREG report also incorporates and updates numerous portions of the SRP, specifically Chapters 5, 7, and 14; and Appendices C, D, and E. The chapters and appendices that have been incorporated into this NUREG are superseded. This three-volume NUREG series supersedes both NUREG/BR–0241 and NUREG–1727 in their entirety.

1.5.5   REQUEST COPIES OF DOCUMENTS

To request single copies of NRC documents from NRC's Offices, see Table 1.5 for addresses and telephone numbers.

Table 1.5   NRC Offices

Location	Address	Phone Number(s)
Headquarters	Washington, DC. 20555–0001	301–415–7000, 1–800–368–5642
Region I	475 Allendale Road King of Prussia, PA 19406–1415	610–337–5000, 1–800–432–1156
Region II	61 Forsyth Street, SW, Suite 23T85 Atlanta, GA 30303	404–562–4400, 1–800–577–8510
Region III	2443 Warrenville Road Lisle, IL 60532–4352	630–829–9500, 1–800–522–3025
Region IV	611 Ryan Plaza Drive, Suite 400 Arlington, TX 76011–4005	817–860–8100, 1–800–952–9677

Note that NRC publishes amendments to its regulations in the *Federal Register*. Documents may be obtained by contacting NRC's Public Document Room (PDR), through the following methods:

Telephone: 1–800–397–4209 or 301–415–4737
TDD (for the hearing impaired): 1–800–635–4512
Facsimile: 301–415–3548
U.S. Mail: U. S. NRC, PDR, O1F13, Washington, DC 20555
Onsite visit to the PDR: One White Flint North, 11555 Rockville Pike (first floor), Rockville, Maryland 20852 (opposite the White Flint Metro Station on the Red Line)

In an effort to make NRC documents and information readily available to licensees and the general public, NRC is placing documents and information on its Internet Web site. Many of the reference sections of this volume refer to a World Wide Web address on the Internet (e.g., http://www.nrc.gov). Applicants and licensees who have Internet access may use the referenced address to find more information on a topic, the referenced document, or information on obtaining the referenced document.

2 FLEXIBILITY IN DEMONSTRATING COMPLIANCE WITH 10 CFR PART 20, SUBPART E

NRC and its licensees share a common responsibility to protect public health and safety. Federal regulations and the NRC regulatory program are important elements in the protection of the public; however, NRC licensees are primarily responsible for safely using nuclear materials. NRC's safety philosophy explains that "although NRC develops and enforces the standards governing the use of nuclear installations and materials, it is the licensee who bears the primary responsibility for conducting those activities safely." This philosophy applies to the decommissioning of licensed facilities. Thus, the licensee has the primary responsibility for compliance with the license termination criteria. The responsibility of NRC staff is to oversee the process and to make a conclusion that there is reasonable assurance that the criteria have been or will be met and then to terminate or amend licensees, as appropriate.

The dose criteria of 10 CFR Part 20, Subpart E, are performance criteria. In this volume, NRC staff has taken a risk-informed, performance-based approach to demonstrations of compliance with the license termination criteria. Thus, there are different methods available to licensees to demonstrate compliance with the criteria. Regardless of the specific method used by a licensee, it is important that the licensee provide sufficient justification for its approach. This chapter discusses some of the aspects of flexibility in methodologies for demonstrating compliance with the license termination criteria. One objective of this chapter is to emphasize the flexibility available in demonstrating compliance with the regulations.

Licensees should consider the flexibility available when demonstrating compliance with the license termination criteria. A licensee may determine that the standard methods are not the best for a given site. The benefit of the performance criteria is the flexibility of approaches allowed to demonstrate compliance.

NRC staff should evaluate any methodology proposed by licensees. However, the use of nonstandard methods may require more detailed justification for NRC staff to determine acceptability. In addition, the increased complexity and detail of nonstandard demonstrations may result in increased NRC staff review time and, therefore, cost to the licensee.

2.1 RISK-INFORMED APPROACH TO COMPLIANCE DEMONSTRATIONS AND REVIEWS

This section provides a summary of the risk-informed approach to regulatory decision-making. Additional details can be found in the NRC Staff Requirements Memorandum, SECY–98–144 (NRC 1999).

NRC has increased the use of risk information and insights in its regulation of nuclear materials and nuclear waste management, including the decommissioning of nuclear facilities. Risk is defined by the "risk triplet" of (1) either a scenario or set of scenarios with a combination of events and/or conditions that could occur, (2) the probability that the scenario(s) could occur, and (3) the consequence (e.g., the dose to an individual) if the scenario(s) were to occur. The term risk insights, as used here, refers to the results and findings that come from risk assessments. The end results of such assessments may relate directly or indirectly to public health effects (e.g., the calculation of predicted doses from decommissioned sites).

A risk-based approach to regulatory decision-making is based solely on the numerical results of a risk assessment. The Commission does not endorse a risk-based regulatory approach but supports a risk-informed approach to regulation. A risk-informed approach to regulatory decision-making represents a philosophy whereby risk insights are considered together with other factors in the regulatory process to better focus licensee and regulatory attention on design and operational issues commensurate with their importance to public health and safety. Explicit consideration of the numerical probability that a scenario would occur (i.e., number 2 of the risk triplet) is not typically used by the NRC staff to determine compliance with the LTR. This is a departure from a purely risk-based approach.

The typical deterministic approach to regulatory decision-making establishes requirements for engineering margin and for quality assurance in design, manufacture, and construction. In addition, it assumes that adverse conditions can exist and establishes a specific set of design basis events (i.e., What can go wrong?). The deterministic approach involves implied, but unquantified, elements of probability in the selection of the specific design basis events to be analyzed. Then, it requires that the design include safety systems capable of preventing and/or mitigating the consequences (i.e., What are the consequences?) of those design basis events in order to protect public health and safety. Thus, a deterministic analysis explicitly addresses only two questions of the risk triplet.

The risk-informed approach has enhanced the deterministic approach by (a) allowing explicit consideration of a broader set of potential challenges to safety; (b) providing a logical means for prioritizing these challenges based on risk significance, operating experience, and/or engineering judgment; (c) facilitating consideration of a broader set of resources to defend against these challenges; (d) explicitly identifying and quantifying sources of uncertainty in the analysis (although such analyses do not necessarily reflect all important sources of uncertainty); and (e) leading to better decision-making by providing a means to test the sensitivity of the results to key assumptions.

Where appropriate, a risk-informed regulatory approach can also be used to reduce unnecessary conservatism in purely deterministic approaches, or can be used to identify areas with insufficient conservatism in deterministic analyses and provide the bases for additional requirements or regulatory actions. Risk-informed approaches lie between the risk-based and purely deterministic approaches (NRC 1999).

NRC's risk-informed regulatory approach to the decommissioning of nuclear facilities is intended to focus the attention and resources of both the licensee and NRC on the more risk-significant aspects of the decommissioning process and on the elements of the facility and the site that will most affect risk to members of the public following decommissioning. While a licensee must comply with all Commission regulations, a licensee whose sites (or aspects of a site) have higher risk significance may need to provide a more rigorous demonstration to support compliance. Furthermore, NRC staff generally will apply more scrutiny to reviews of such sites or situations with higher risk significance. This should result in a more effective and efficient regulatory process. The risk-informed regulatory approach to decommissioning is reflected in this volume, as shown by the following examples:

- NRC has developed and is applying the concept of "decommissioning groups" based on (a) the nature and the extent of the radioactive material present at a site and (b) the complexity of the decommissioning process. The groups are generally related to the potential risks associated with the site, in that the less complex sites with limited distribution of radioactive material may pose lower risks (i.e., manageable risks) to individuals and populations during and following decommissioning (see Section 1.3).

- NRC's framework for decommissioning regulatory decision-making reflects the iterative nature of the compliance demonstration process. The iterative approach to decision-making for license termination provides a process for screening sites and for directing additional data collection effort toward demonstrating compliance. The framework is designed such that the level of complexity and rigor of analysis conducted for a given site should be commensurate with the level of risk posed by the site (see Section 1.5).

- This volume provides two different approaches for demonstrating compliance with the dose-based decommissioning criteria, using either a dose modeling approach or a DCGL approach. The dose modeling approach uses measurements of the actual residual radioactivity at a site after cleanup to more realistically assess the potential dose, and therefore the risk, associated with a decommissioned site. The DCGL approach allows a licensee to calculate, a priori, a concentration limit (DCGL) for each radionuclide based on the dose criteria of the LTR, and to then demonstrate that the residual radionuclide concentrations are below the DCGLs (see Section 2.5).

- This volume provides for demonstrating compliance through either a screening approach or a site-specific approach. The screening approach allows sites that pose lower potential risks to demonstrate compliance through simpler, yet conservative, screening analysis by adopting screening DCGLs developed by NRC (see Sections 2.6 and 5.1 and Appendix H).

- NRC staff recommends using the Data Quality Objectives (DQO) process for establishing criteria for data quality and developing survey designs. The process uses a graded approach to data quality requirements, based on the type of survey being designed and the risk of making a decision error based on the data collected. This process aligns the resources expended to collect and analyze data with the risk-significance of the data (see Section 3.2).

- NRC provides for an approach to dose assessment that accounts for the site-specific risk significance of radionuclides and exposure pathways. NRC staff allows a licensee to identify radionuclides and exposure pathways that may be considered "insignificant" based on their contribution to risk, and remove them from further consideration (see Section 3.3).

- NRC endorses the MARSSIM approach to FSS design and execution. The MARSSIM approach results in a site-specific FSS design that is commensurate with potential risks associated with a site, in terms of the likelihood of exceeding the DCGLs at the site (see Section 4.4).

- NRC staff supports a risk-informed approach to site-specific dose modeling for compliance demonstration in several ways: (a) allowing for site-specific selection of risk-significant exposure scenarios, exposure pathways, and critical groups; (b) expecting selection of conceptual models, numerical models and computer codes that incorporate the more risk-significant elements of a site; (c) expecting site-specific data for the more risk-significant input parameters, and allowing for more generic data for less risk-significant parameters; and (d) encouraging the use of probabilistic techniques to evaluate and quantify the magnitude and effect of uncertainties in the risk assessment, and the sensitivity of the calculated risks to individual parameters and modeling assumptions (see Appendix I).

- NRC allows for early partial release of a portion of a site prior to completion of decommissioning for the entire site, based on the risks associated with the early partial site release (see Appendix K).

- NRC staff supports a risk-informed graded approach for engineered barriers, and this guidance includes an example of how the risk-informed approach is applied to designing erosion protection barriers (see Appendix P). In addition, the staff supports a risk-informed graded approach for selecting institutional controls and for long-term monitoring and maintenance at restricted use sites, which allows licensees to tailor the type of institutional controls and the specific restrictions on future site use based on a risk framework and insights from dose assessments (see Section 17.7 and Appendix M of Volume 1).

2.2   FLEXIBILITY IN SUBMISSIONS

NRC staff expects that certain information will be included in licensees' DPs, including the FSS design (if an FSS will be performed) and a description of the development of DCGLs or the dose assessment, as applicable. Volume 1 of this NUREG provides additional details on the expected content in these submittals. For guidance on lessons learned regarding flexibility related to recently submitted decommissioning plans, refer to Section O.2 from Appendix O of this volume.

Some information is required by regulations (e.g., 10 CFR 30.36(g)(4)) and must be provided in the DP; the DP must include all of the following:

- the conditions of the site, building, or area, are sufficient to evaluate the acceptability of the plan;

- the planned decommissioning activities;

- the methods used to ensure protection of workers and the environment against radiation hazards during decommissioning;

- the planned final radiation survey; and

- an updated cost estimate for decommissioning, comparison with decommissioning funds, and a plan for assuring availability of adequate funds to complete decommissioning.

In addition, DCGLs are typically submitted in the DP. Therefore, the typical approach to supplying information in the DP is for the licensee to obtain all the detailed information needed and to submit the information in the DP. Using the DP checklist (Appendix D of Volume 1 of this NUREG report) as a guide, licensees should coordinate with NRC staff to determine what information should be included in the DP. For example, at a facility for which a MARSSIM final status survey will be performed, the licensee may perform sufficient characterization surveys to determine the appropriate number of samples to obtain for each survey unit for the FSS. In this case, the survey design could be approved by NRC staff with approval of the DP, and the final status survey report (FSSR) may focus primarily on the results of the FSS.

In some cases, all of the desired information will not be available during the DP preparation. For an FSS, the MARSSIM approach requires that certain information needed to develop the final radiological survey be developed as part of the remedial activities at the site; therefore, this information may not be available for the DP. Similarly, some aspects of the DCGL development or dose assessment may not be available before remediation and final surveys are complete.

When some important information is not available at the time of the DP submission, licensees may either (a) make assumptions about the information or (b) commit to following a specific methodology to obtain the information. In the first case, assumptions will be considered by NRC staff to be commitments to ensure and subsequently demonstrate that the assumption is true. The information then would be submitted by the licensee at the completion of remediation, at the completion of FSS design, with the FSSR, or at some other appropriate time. For example, a facility uses the ratio of concentrations of Th-232 to U-238 along with measured concentrations of Th-232 in estimating the concentration of U-238. The licensee may have preliminary information about the ratio of concentrations and, if it is reasonable, may assume that that ratio would be valid for the conditions at the time of the FSS. NRC staff could accept the use of the assumed value for the ratio. The licensee would demonstrate, at a later stage, that the assumed value was valid, perhaps based on measurements made during the FSS.

In the second case, a licensee commits in the DP to following a specific methodology to obtain the information. One such example is a facility for which a MARSSIM final status survey will be performed, where sufficient information may not be available at the time of the DP submission to determine the number of samples to be taken from each survey unit for the FSS. In this case, the licensee may commit to the procedure recommended in MARSSIM for determining the number of samples in a survey unit. This commitment to the MARSSIM methodology would be documented in the DP. The licensee then may determine the number of samples for each survey unit as information is obtained. The FSS design, including the number of samples, could be described as part of an FSSR, which would be evaluated by NRC staff along with the FSS results.

Depending on the circumstances and the type of information that is not specifically included in the DP, NRC staff may consider requiring license conditions to formalize the licensee's commitments. This can be accomplished by specific license condition or by reference to the approved DP (i.e., in the "tie-down" condition). Licensees should consult with NRC staff regarding the details of implementing these types of licensee commitments.

Similar approaches could be taken regarding information needed to complete a dose assessment. One example is a facility for which the fraction of building-surface residual radioactivity that is removable has been determined during scoping surveys, but the licensee does not know whether the fraction will change after remediation activities. In this case, the licensee might assume for its dose assessment that the measured fraction will remain unchanged. NRC staff expects the licensee (a) to make measurements or calculations to demonstrate that the removable fraction was representative of the conditions when remediation is complete and (b) to demonstrate that the dose assessment is representative.

NRC staff normally would not undertake review of DPs or FSSes that use assumptions in lieu of specific information that reasonably could be obtained prior to submission. In general, NRC staff expects that assumptions used in development of DPs submitted for review would be limited to those parameters that could change as a result of the remediation or FSS process itself or to those parameters for which information cannot reasonably be obtained at the time of DP submission. NRC staff should consider other assumptions on a case-by-case basis.

Cautions on Making Assumptions or Committing to a Methodology

Providing all details in the DP may result in more efficient and effective reviews by NRC staff. If a licensee finds it reasonable to use the flexible approaches discussed here, the licensee is cautioned that (a) there may be a more detailed demonstration of compliance necessary and (b) there may be a greater chance that the facility release would not be approved by NRC staff, because some of the overall compliance strategy would be reviewed by NRC staff only at the end of the decommissioning process. In addition, the licensee may be required to resolve the assumptions and commitments to meet license conditions. The licensee should consult with NRC staff regarding details of implementing these flexible approaches.

2.3   USE OF CHARACTERIZATION DATA FOR FINAL STATUS SURVEYS

Although the FSS is generally discussed as if it were an activity performed during a single stage of the Radiation Survey and Site Investigation (RSSI) process (see Chapter 4 and Table 4.1 for more about the RSSI Process), this does not have to be the case. There is no requirement that an FSS be performed at the end of the decommissioning process. Data from other surveys conducted during the RSSI process—such as scoping, characterization, and remedial action support surveys—can provide valuable information for an FSS, provided the data are of sufficient quality.

In some cases, the data obtained from these other surveys may be sufficient to serve as an FSS. Licensees may plan the different phases of the RSSI such that the data obtained will be of sufficient quality to serve as or to supplement the FSS. The DQO process may be applied to all phases of the RSSI, with DQOs developed that will be as robust as those typically developed for the FSS. This approach may result in more costly characterization or remedial action support surveys (to support the more stringent DQOs), which may be balanced against the elimination of a separate FSS. For guidance on lessons learned regarding the use of characterization data for FSSes related to recently submitted decommissioning plans, refer to Section O.2 from Appendix O of this volume.

2.4 CHOICE OF NULL HYPOTHESIS FOR FINAL STATUS SURVEY STATISTICAL ANALYSIS

The default assumption used in the MARSSIM approach to FSSes and followed by NRC staff is that the survey unit is considered contaminated above the limit, unless survey data show otherwise. Thus, the null hypothesis used for the MARSSIM FSS statistical tests is that the concentrations of residual radioactivity exceed the DCGLs. This assumption and null hypothesis is considered Scenario A. In most all cases, NRC staff will consider Scenario A to be the appropriate choice. In some limited cases, a different assumption and null hypothesis, Scenario B, may be appropriate. Scenario B is when the assumption is made that the mean concentrations of contaminants in the survey unit are indistinguishable from those in background. This section provides some guidance on this issue, and more details are provided in NUREG–1505. Table 2.1 provides a summary of the differences between the Scenarios A and B.

FLEXIBILITY IN DEMONSTRATING COMPLIANCE WITH 10 CFR PART 20, SUBPART E

Table 2.1   Comparison of FSS Statistical Test Scenarios

Characteristic	Scenario A	Scenario B
Assumption for statistical test	The survey unit is assumed to fail unless the data show it can be released.[a]	The survey unit is assumed to pass unless the data show that further remediation is necessary.[a]
Null hypothesis	The concentrations of residual radioactivity exceed the DCGLs.	The mean concentrations of residual radioactivity are indistinguishable from those in the background.
Scenario emphasis	Compliance with a dose limit.	Indistinguishable from the background.
What is needed to reject the null hypothesis?	The measured average concentration in the survey unit must be statistically less than the DCGL.	The measured average concentration in the survey unit must be statistically greater than the background.
Rejecting the null hypothesis means	The survey unit passes.[a]	The survey unit fails.[a]
Increasing the number of measurements in a survey unit	Increases the probability that an adequately remediated survey unit will pass.	Increases the probability that an inadequately remediated survey unit will fail.
When should the scenario be used?	Should be used in most cases (i.e., default) when the DCGL is fairly large compared to the measurement variability.	Should be used in special cases (i.e., exception) when the DCGL is small compared to measurement and/or background variability.

Note:

a For both Scenarios A and B, "passing" the FSS means a conclusion that the survey unit may be released, and "failing" means a conclusion that the survey unit may not be released.

Deciding which scenario to use and the process to make that decision are difficult questions. In most cases, when the $DCGL_W$ is fairly large compared to the measurement variability, Scenario A should be chosen. This is because even residual radioactivity below the $DCGL_W$ should be measurable. In some cases, however, it may be more appropriate to demonstrate indistinguishability from the background. When the $DCGL_W$ is small compared to measurement and/or background variability, Scenario B may be appropriate. This is because residual radioactivity below the DCGL$_W$ may be difficult to measure. Background variability may be considered high when differences in estimated mean concentrations measured in potential reference areas are comparable to screening level DCGLs. NUREG–1505 provides an example of the use of Scenario B to demonstrate indistinguishability from the background when the residual radioactivity consists of radionuclides that appear in background, and the variability of the background is relatively high.

As mentioned above, NRC staff's default assumption is that the use of Scenario A is appropriate. The use of Scenario B is expected only for a small number of facilities, and the considerations for any given facility are expected to be site-specific. Therefore, NRC staff recommends that licensees contact NRC early in the licensee's FSS design process to discuss considerations for their situation.

Cautions on the Use of Scenario B for FSS Statistical Tests

- Case-by-case evaluation is required.

- Licensees considering the use of Scenario B for compliance with Subpart E are strongly encouraged to consult with NRC staff early in the planning process.

- Information about the potential use of Scenario B can be found in NUREG–1505, but this should be used cautiously.

2.5 DEMONSTRATING COMPLIANCE USING DOSE ASSESSMENT METHODS VERSUS DERIVED CONCENTRATION GUIDELINE LEVELS AND FINAL STATUS SURVEYS

There is flexibility in the general approach to demonstrating compliance with 10 CFR Part 20, Subpart E, dose criteria. Two major approaches include (a) the dose modeling approach (characterizing the site—after remediation, if necessary—and performing a dose assessment) and (b) the DCGL and FSS approach (developing or using DCGLs and performing an FSS to demonstrate that the DCGLs have been met). Since the second option is commonly the more efficient or simpler method for licensees, most discussions in this NUREG report refer to the use of DCGLs and FSSes as the compliance method. It should be noted that these two approaches are not mutually exclusive; they are just different approaches to show that the dose is acceptable. Table 2.2 shows some advantages and disadvantages of the two approaches.

Table 2.2   Comparison of Dose Modeling to DCGL and FSS Approaches to Compliance

Approach	Advantages	Disadvantages
Dose Modeling	• more realistic • accounts for time of peak dose for mixes of radionuclides • can use additional data collected during decommissioning for site-specific analyses • can guide remediation activities and data collection	• may still need preliminary cleanup goals or DCGLs to design surveys or guide remediation • greater chance of additional iterations of remediation and/or site characterization
DCGLs and FSSes	• simpler to implement • lower chance of not showing compliance with dose criterion after remediation	• using sum of fractions provides level of conservatism for radionuclide mix • additional modeling data (i.e., to modify DCGLs) collected during decommissioning can not be used without license amendment • potential conflict with "peak of the mean" approach

2.5.1   DOSE MODELING APPROACH

Calculating the final dose is the most direct approach to show compliance with the dose criteria in Subpart E. Direct calculation of the total dose—from all radionuclides in a code that correctly accounts for the time of the peak dose for each radionuclide—is a more realistic measure of the potential dose from the site. Another advantage of the dose modeling approach is that a licensee can use dose modeling information during the decommissioning process to guide additional site characterization, remediation, or other decommissioning options. Additional site characterization could be performed to reduce the level of conservatism in the dose model, parameters, or scenario.

An advantage for sites that comply with the Subpart E criteria without any cleanup is that it may be unnecessary to create any DCGLs; however, the quality of the licensee's site characterization data should be sufficient for use as an FSS.

A disadvantage of the dose modeling approach is that changes in the dose modeling, between the approval of the DP and the request for license termination, would result in NRC staff needing to perform a review of the new information before granting approval of license termination. This additional review step could result in further justification, modeling, remediation activities, or site characterization before approval is granted. This additional review step is similar to what can occur for a site that needs no remediation but uses site-specific dose modeling to show compliance as part of the DP.

Another disadvantage of using the dose modeling approach is that cleanup goals or final concentrations may need to be estimated (a) to provide assurance that the approach will result in compliance and (b) to design quality surveys, guide remediation activities, and perform additional site characterization.

2.5.2 DCGL AND FSS APPROACH

For many sites, especially those that need remediation, the DCGL and FSS approach is a simpler system to show compliance with Subpart E. The DCGL and FSS approach is the most commonly used approach for compliance with the license termination rule, and is the approach recommended by the MARSSIM. In the DCGL and FSS approach, the licensee commits to a single concentration value for each radionuclide (i.e., $DCGL_W$) that results in a dose equal to the dose criteria. The $DCGL_W$ derivation can use either generic screening criteria or site-specific analysis. The licensee then uses FSSes to demonstrate that the DCGLs have been met. For sites with multiple radionuclides or sources, a sum of fractions approach is typically used to ensure that the dose from all radionuclides and all sources complies with the Subpart E criteria (see Section 2.7). The DCGLs (and the sum of fractions approach) are usually included in the license. The disadvantages of this approach include the following:

1. The sum of fractions approach (Section 2.7) has an underlying assumption that the peak dose for every radionuclide occurs at the same time. This can result in an additional level of conservatism, depending on the mix of radionuclides.

2. Any changes in the DCGLs (e.g., because of new site information) may require a license amendment and NRC staff review.

3. DCGLs may be difficult to calculate for sites using realistic dose modeling because of potential issues with using "peak of the mean" doses to derive DCGLs.

2.6 MERITS OF SCREENING VERSUS SITE-SPECIFIC DOSE ASSESSMENT

The advantages of selecting a screening dose assessment approach, where it is applicable, are that minimal justification, characterization, and NRC staff review are required. Its disadvantages are that only two potential sources of radiation (i.e., buildings surfaces and soil) are covered and that the results are more conservative than could be arrived at by site-specific modeling. On the other hand, the advantages and disadvantages for site-specific analysis are based on the same principle: flexibility. Site-specific analyses allow a licensee to tailor the analysis to their site conditions, as long as proper justification is available. Site-specific analyses would require the licensee to provide justifications and site-specific information, as necessary, to support changes in parameters or changes of codes/models and default assumptions. Table 2.3 provides a brief summary of attributes and merits of each screening and site-specific analysis approach.

The models, scenarios, and parameters used in screening are intended to be conservative, because the lack of information about a site warrants the use of conservative models and default conditions to ensure that the derived dose is not underestimated. The screening analysis is intended to overestimate the dose, to ensure that, for 90 % of the screening cases, the derived dose is not underestimated. In performing screening analysis, NRC staff should recognize that in the screening analysis, the 90th percentile of the dose distribution is used for calculating compliance, whereas in the site-specific analysis, the "peak of the mean" dose over time (e.g., 1000 years) may be used. As soon as default parameters are changed, source term conditions are modified, or different models or codes are used, a transition from screening to site-specific analysis would be indicated.

Table 2.3   Attributes of Screening and Site-Specific Analysis

Attribute	Screening	Site-Specific
Models/Codes	DandD Version 2 (Others may be accepted.)	Any model/code compatible with the site and approved by NRC staff
Scope of Application	Only for sites qualified for screening	Any site
Parameters	DandD default parameters	Site-specific and/or surrogates with justification
Scenarios/Pathways	DandD default scenarios/pathways	Scenarios/pathways may be modified, based on site condition.
Basis of Dose Selection & Uncertainty	The dose at the 90th percentile of the peak dose distribution within 1000 years	"Peak of the mean" annual doses within 1000 years

NUREG/CR–5512, Volume 1, and the deterministic parameter set from DandD Version 1 have been superseded by NUREG/CR–5512, Volume 3, and DandD Version 2, respectively. Therefore, a licensee should not refer to NUREG/CR–5512, Volume 1, as a primary source for a default deterministic parameter set. Similarly, DandD Version 1, which did not support probabilistic analyses, provided a default deterministic input parameter set. DandD Version 2 has replaced Version 1, and the DandD Version 1 default deterministic parameter set should not be used as a reference data set for any parameters. This is especially important for the Version 1 defaults, as all the defaults in the code were selected by a method that made them highly interdependent. Each single value in the default deterministic data set was selected based on the values of the other parameters. Thus, if a single parameter is changed in DandD Version 1, the appropriateness of every other parameter in the code may be questionable.

2.7 SUM OF FRACTIONS

The sum of fractions is a simple, yet flexible, approach to deal with multiple radionuclides or sources. A source is any discrete material or medium that contains residual radioactivity. For example, a site with residual radioactivity in surface soil, ground water, and in buried piping has at least three sources. The $DCGL_W$ is equivalent to the concentration of a single radionuclide from a single source that would provide 0.25 mSv/y (25 mrem/y) total effective dose equivalent (TEDE). The dose from each radionuclide and source needs to be calculated and then added together. If a licensee only complied with the $DCGL_W$ for each radionuclide in each source, the resulting total dose could be as high as 0.25 mSv/y (25 mrem/y) multiplied by the number of radionuclides multiplied by the number of sources. Unless there was only one source and one radionuclide, the resulting dose would not meet the limits detailed in Subpart E. The dose from all the radionuclides and sources must be equal to or less than the appropriate dose limit in Subpart E.

One simple way to calculate the dose from one radionuclide from one source is to calculate the relative ratio of the residual radioactivity concentration over the $DCGL_W$. Then, the ratio is multiplied by 0.25 mSv/y (25 mrem/y). In fact, for multiple sources or radionuclides, the ratios can be added together and the sum multiplied by the dose limit. Therefore, the sum of the ratios for all the radionuclides and sources may not exceed "1" (i.e., unity). For example, if radionuclides *A* and *B* are present at respective concentrations of *Conc A* and *Conc B*, and if the respective applicable DCGLs are *Limit A* and *Limit B*, then the concentration needs to be limited so that the following relationship exists to meet Subpart E:

$$\frac{Conc\ A}{Limit\ A} + \frac{Conc\ B}{Limit\ B} \le 1 \qquad (2\text{-}1)$$

Similarly, for multiple sources, the sum of the ratios resulting from the sum of the radionuclide contributions may not exceed unity. For example, if the site had a second source, also with radionuclides *A* and *B*, but in concentrations of *Conc A₀* and *Conc B₀*, and DCGLs of *Limit A₀* and *Limit B₀*, the following relationship would need to exist to meet Subpart E:

$$\frac{Conc\ A}{Limit\ A} + \frac{Conc\ B}{Limit\ B} + \frac{Conc\ A_0}{Limit\ A_0} + \frac{Conc\ B_0}{Limit\ B_0} \leq 1 \qquad (2\text{-}2)$$

In the general form, the relationship of the ratios, commonly known as the "sum of the fractions" or the "unity rule," would be for M sources (s) and N radionuclides (r):

$$\sum_{s=1}^{M}\sum_{r=1}^{N} \frac{Conc_{sr}}{Limit_{sr}} \leq 1 \qquad (2\text{-}3)$$

where $Conc_{sr}$ = the concentration of radionuclide r in source s, and
$Limit_{sr}$ = the DCGL$_W$ value for radionuclide r in source s.

For sites with a number of radionuclides and sources, it may be easier to partition the acceptable fraction between various sources or radionuclides. For example, a licensee could commit to keep the ratio from the ground water to less than 25 % of the dose limit. For guidance on lessons learned regarding use of the unity rule related to confirmatory and FSSes, refer to Section O.3.4.2 from Appendix O of this volume.

One major, implicit assumption in using the sum of fractions approach is that peak doses for each radionuclide and source occur simultaneously. Because of the importance of differential radionuclide transport through the environment and because of differing predominant pathways, there are many radionuclides and contaminated media for which peak doses do not occur simultaneously. For example, radionuclides that result in predominantly external dose, such as Co-60, usually have a peak dose right after license termination. For radionuclides that result in peak dose through irrigation or drinking ground water, the peak does not occur until years after license termination. So in a situation like this, when peak doses are from different radionuclides or when sources occur at different times, the sum of fractions approach results in a conservative estimate of the dose at the site. To eliminate this conservatism, the licensee could directly calculate the combined dose using final concentrations from the FSS (see Section 2.5).

2.8 FLEXIBILITY FOR USE OF INSTITUTIONAL CONTROLS AND ENGINEERED BARRIERS AT RESTRICTED USE SITES

The new guidance developed for restricted use sites includes risk-informed and performance-based approaches to institutional controls, engineered barriers, monitoring, and maintenance. These approaches not only enhance the attention to safety by being risk-informed, but also provide flexibility to licensees planning restricted use for a site. The approaches described allow licensees to select the most effective and efficient methods for: restricting site use; designing engineered barriers to mitigate disruptive processes important to compliance; and planning monitoring and maintenance activities that are tailored to the specific site and indicators of potential disruptive processes and engineered barrier performance. These approaches are described in Section 3.5 of this volume and Section 17.7 and Appendix M of Volume 1.

3 CROSS-CUTTING ISSUES

This chapter provides guidance on several cross-cutting issues that relate to multiple aspects of surveys, characterization, and dose modeling. The issues addressed in this chapter include the following:

- transparency and traceability of compliance demonstrations;
- Data Quality Objectives (DQO) process;
- insignificant radionuclides and exposure pathways;
- considerations for other constraints on allowable levels of residual radioactivity; and
- the use of engineered barriers.

Use of the Guidance in this NUREG Report

- The suggestions in this NUREG report are only guidance, not requirements.

- Other methods for demonstrating compliance are acceptable.

- As noted in Section 5.3 of Volume 1 of this NUREG report, licensees are encouraged to have early discussions with NRC staff in developing DPs. This is especially important when NRC guidance is limited on a specific topic. Early discussions can save licensees from following an approach that NRC staff may find unacceptable and can clarify this guidance and identify areas where modification may be helpful for NRC staff's review.

- This volume refers to a number of other documents for guidance. In some cases, this volume states that the referenced guidance is approved by NRC staff. In other cases, the documents are only referenced as potentially relevant information. In these latter cases, specific applicability to a facility should be determined by the licensee in consultation with NRC staff, as appropriate.

3.1 TRANSPARENCY AND TRACEABILITY OF COMPLIANCE DEMONSTRATIONS

Licensees submit various information to justify their conclusions regarding compliance with 10 CFR Part 20, Subpart E. Because of insufficient justification, NRC staff have found a number of licensee submittals to be inadequate to conclude compliance. This section describes some considerations for improving the thoroughness of licensee submittals. Transparency refers to arguments or calculations with descriptions sufficient to replicate the argument or calculation by an independent reviewer. Traceability refers to the sources of information being relatable to the original source. NRC staff encourages licensees to submit compliance demonstrations that are transparent and traceable. This should result in more efficient and effective NRC staff reviews.

To help ensure transparency and traceability, licensees should have the following in their justification:

- Sources of data should be described.

- It may be appropriate only to provide summary data. To the extent that summary data is provided, references to detailed data should be provided and the detailed data should be made available to the NRC staff for review if requested (e.g., in an inspection).

- Data, including units used, should be clearly described in tables and other presentations of data.

- Assumptions should be stated; the difference between assumptions and justified data or parameters should be clear.

- Justifications for parameters or arguments should be provided, especially when nonstandard arguments or nondefault parameters are employed.

- Uncertainties in data and parameters should be described.

3.2   DATA QUALITY OBJECTIVES PROCESS

Compliance demonstration is the process that leads to a decision as to whether or not a survey unit meets the release criteria. For most sites, this decision is supported by statistical tests based on the results of one or more surveys. The initial assumption used by NRC staff is that each survey unit is contaminated above the release criteria until proven otherwise. The surveys are designed to provide the information needed to reject this initial assumption. NRC staff recommends using the Data Life Cycle as a framework for the planning, implementation, assessment, and DECISION-MAKING phases of final surveys. The major activities associated with each phase of the Data Life Cycle are discussed in Section 2.3 of MARSSIM.

One aspect of the planning phase of the Data Life Cycle is the DQO process. The DQO process is a series of planning steps for establishing criteria for data quality and developing survey designs. The DQO process consists of seven steps:

1.   statement of the problem;

2.   identification of the decision;

3.   identification of inputs to the decision;

4.   definition of the study boundaries;

5.   development of a decision rule;

6.   specification of limits on decision errors; and

7.   optimization of the design for obtaining data.

The output from each step influences steps later in the process. Even though the DQO process is depicted as a linear sequence of steps, it is iterative in practice; the outputs of one step may lead to reconsideration of prior steps.

The DQO process uses a graded approach to data quality requirements. This graded approach defines data quality requirements according to (a) the type of survey being designed and (b) the risk of making a decision error based on the data collected. This approach provides a more effective survey design combined with a basis for judging the usability of the data collected. Thus, the DQO process is a flexible planning tool that can be used more or less intensively as the situation requires.

DQOs are qualitative and quantitative statements that satisfy all of the following:

- clarify the study objective;

- define the most appropriate type of data to collect;

- determine the most appropriate conditions for collecting the data; and

- specify limits on decision errors that will be used as the basis for establishing the quantity and quality of data needed to support the decision.

Although the DQO process is generally used for surveys and the steps of an RSSI, the general concepts may also be applied to dose assessments. Licensees are encouraged to apply the general concepts of the DQO process to all applicable parts of their compliance demonstration. The use of the DQO process can help ensure that the type, quantity, and quality of data and calculations used in DECISION-MAKING will be appropriate for the intended application. Additional guidance on the use of DQO process is provided in Section 2.3 and Appendix D of the MARSSIM, and in an EPA guidance report on the DQO process (EPA 2000).

Experience has shown that in developing the final survey design, it is helpful for the licensee to identify all appropriate DQOs in planning and designing the final status survey plan (FSSP). The process of identifying the applicable DQOs ensures that the survey plan requirements, survey results, and data evaluation are of sufficient quality, quantity, and robustness to support the decision on whether the cleanup criteria have been met.

In purpose and scope, the DQO process can include a flexible approach for planning and conducting surveys and for assessing whether survey results support the conclusion that release criteria have been met. The DQO process can be an iterative process that continually reviews and integrates, as needed, new information in DECISION-MAKING and the design of the final survey plan. Finally, the selection and optimization of DQOs should facilitate the later evaluation of survey results and DECISION-MAKING processes during the data quality assessment (DQA) phase. NRC staff has observed that licensees have had difficulties in developing DQOs, especially during the optimization step, and have not taken full advantage of the DQO process. Experience has shown that the process is often rigidly structured by relying too much on characterization data and not readily open to the possibility of incorporating new information as it becomes available. This rigid approach makes implementing any changes difficult and is an inefficient use of resources, since it imposes time delays (e.g., the additional time required to determine how to implement any changes). Refer to Section O.2 from Appendix O of this volume, for guidance on lessons learned regarding use of the DQO process related to recently submitted decommissioning plans.

3.3   INSIGNIFICANT RADIONUCLIDES AND EXPOSURE PATHWAYS

Licensees should note that they are required to comply with the applicable dose criteria; nothing in this discussion should be interpreted to allow licensees to exceed the criteria.

This section provides guidance on conditions under which radionuclides or exposure pathways may be considered insignificant and may be eliminated from further consideration. The dose criteria in 10 CFR Part 20, Subpart E, apply to the total dose from residual radioactivity. Thus, demonstrations of compliance should generally address the dose from all radionuclides and all exposure pathways. However, NRC staff recognizes that there may be large uncertainties associated with survey data and with dose assessment results. In a risk-informed, performance-based paradigm, NRC staff has determined it is reasonable that radionuclides or pathways that are insignificant contributors to dose may be eliminated from further detailed consideration.

NRC staff considers radionuclides and exposure pathways that contribute no greater than 10 % of the dose criteria to be insignificant contributors. Because the dose criteria are performance criteria, this 10 % limit for insignificant contributors is an aggregate limitation only. That is, the sum of the dose contributions from all radionuclides and pathways considered insignificant should be no greater than 10 % of the dose criteria. No limitation on either single radionuclides or pathways is necessary. In cases of restricted release, where two dose criteria apply (one for the possibility of restrictions failing), the 10 % limitation should be met for each dose criterion.

Once a licensee has demonstrated that radionuclides or exposure pathways are insignificant, then (a) the dose from the insignificant radionuclides and pathways must be accounted for in demonstrating compliance, but (b) the insignificant radionuclides and pathways may be eliminated from further detailed evaluations. For example, after sufficient site characterization, suppose a licensee shows that the dose from Sr-90 at the facility is 0.02 mSv/y (2 mrem/y), which is less than 10 % of the dose criterion for unrestricted use. In this case, Sr-90 can be considered insignificant and eliminated from the FSS and from detailed consideration in the dose modeling. However, the dose from Sr-90 has to be considered in demonstrating compliance with the dose criterion.

It is important that the licensee documents the radionuclides and pathways that have been considered insignificant and eliminated from further consideration and that the licensee justifies the decision to consider them insignificant. However, licensees and NRC staff should be aware that remediation techniques (or other activities or processes) may increase concentrations above those previously deemed insignificant. Thus, licensees should also demonstrate that the concentrations deemed insignificant will not increase from other activities. Refer to Section O.1 from Appendix O and Questions 1 and 2, all of this volume, for guidance on which radionuclides can be considered and deselected from further consideration, respectively.

Summary of Determining Insignificant Radionuclides and Exposure Pathways

- Licensees may eliminate insignificant radionuclides and exposure pathways from further detailed consideration. However, the dose from the insignificant radionuclides and pathways must be accounted for in demonstrating compliance with the applicable dose criteria.

- Insignificant means no greater than 10 % of applicable dose criterion.

- Ten percent is an aggregate limit; total dose contributions of all radionuclides and all exposure pathways considered insignificant should not exceed the 10 % limitation.

- No additional limit on single radionuclides or pathways.

- Licensees should also address potential for concentrations to increase during remediation activities.

3.4   CONSIDERATIONS FOR OTHER CONSTRAINTS ON ALLOWABLE RESIDUAL RADIOACTIVITY

There can be situations or standards other than the dose criteria and ALARA requirements of Subpart E that may constrain the final dose below 0.25 mSv/y (25 mrem/y). There are two main causes for constraining the Subpart E dose limit: these causes are (1) partial site release and (2) other standards or regulations.

Partial site release is a situation where a licensee releases a portion of its site for unrestricted use prior to terminating the entire license. While the licensee should demonstrate that the residual radioactivity at the time of unrestricted release of the specific area meets the Subpart E dose limit, the residual radioactivity of the area should also be taken into account during final termination to demonstrate that the entire site met the appropriate release criteria. Dose modeling considerations for partial site release are discussed in Appendix K of this volume. In general, the comments below are also applicable to partial site releases.

Demonstrating compliance with the Subpart E dose limit does not eliminate the licensee's requirement for meeting other applicable Federal, State, or local rules and regulations. These regulations from other governmental agencies may conflict with the requirements of Subpart E, as they may allow higher or lower levels of residual radioactivity on the site or may conflict in other ways, such as limiting decommissioning options or final status. Nevertheless, NRC staff should review a DP for compliance with NRC requirements only, including 10 CFR Part 20,

Subpart D, which incorporates, where applicable, the requirements of 40 CFR Part 190. For example, in reviewing the appropriateness of proposed DCGLs or number of samples per survey unit for an unrestricted site, NRC staff would use the limit of 0.25 mSv/y (25 mrem/y), and not a State's limit of 0.2 mSv/y (20 mrem/y). Thus, any requests for additional information would therefore also be based on compliance with the limit of 0.25 mSv/y (25 mrem/y). Because of differences in scenarios, models, and parameters, licensees should note that the lowest dose standard may not result in the lowest acceptable concentration of residual radioactivity.

3.5   USE OF ENGINEERED BARRIERS

The purpose of this section is to provide guidance to licensees in considering the use of engineered barriers, including a risk-informed graded approach for selecting engineered barriers; engineered barrier analysis process; technical basis for engineered barrier performance; and potential performance and degradation mechanisms. This section also supports Section 17.7.3 of Volume 1 by giving guidance on the information to be submitted in a decommissioning plan for the engineered barrier analysis and technical basis for engineered barrier performance.

In the Commission's view, engineered barriers, are distinct and separate from institutional controls (NRC 2002). Used in the general sense, an engineered barrier could be one of a broad range of barriers with varying degrees of durability, robustness, and isolation capability. Generally, engineered barriers are passive, man-made structures or devices intended to enhance a facility's ability to meet the dose criteria in the LTR. Engineered barriers are usually designed to inhibit water from contacting waste, limit releases of radionuclides (e.g., through groundwater, biointrusion, erosion), or to mitigate doses to inadvertent intruders. Institutional controls are used to limit inadvertent intruder access to, and/or use of, the site to ensure that the exposure from the residual radioactivity does not exceed the established criteria. Institutional controls include legal mechanisms (e.g., land use restrictions) and may include, but are not limited to, physical controls (e.g., signs, markers, landscaping, and fences) to control access to the site and minimize disturbances to engineered barriers.

The functionality and robustness of barriers would be determined using the risk-informed graded approach described in Section 3.5.1 and evaluated on a site-specific basis for each licensee application. However, the general framework that a licensee should consider would not vary from licensee to licensee, only the depth and breadth of information supplied to demonstrate the performance of the engineered barriers may vary. The guidance that follows provides the general framework a licensee should consider for use of engineered barriers in the decommissioning process.

It is expected that engineered barriers will most frequently be used for restricted use sites. However, there may be infrequent cases where engineered barriers are used as one component of a decommissioning approach to achieve unrestricted use of a site. These cases should be infrequent because of the uncertainty associated with the long-term performance of engineered systems without monitoring and maintenance and because the goal should be to achieve unrestricted use without relying on engineered barriers. If an engineered barrier is used at an unrestricted use site, only the passive performance of the barrier to mitigate radiological impacts may be credited (i.e., performance of the barrier without monitoring, inspection, and maintenance) in the dose assessment to demonstrate compliance with the LTR dose criteria. The assessment of performance of the engineered barriers should consider the reasonableness of a breach and the potential degradation of the barriers over time because monitoring and maintenance are assumed to not be active. In addition, other reasonably foreseeable disruptive conditions from humans or natural events and processes should be evaluated, and uncertainty in projecting the passive performance of the barriers must be considered. The dose assessment for demonstrating compliance with the unrestricted use criteria of the LTR, for a site using engineered barriers, should be based on the limiting scenario for the passive performance of the barrier.

Because of the wide range of residual radioactivity encountered at decommissioning sites licensed by NRC, the LTR and NRC's decommissioning guidance are not prescriptive as to the criteria for, or acceptability of, site-specific engineered barriers. Therefore, the licensee has flexibility in the methods used to demonstrate compliance with performance-based criteria of Part 20, Subpart E. Because of this flexibility and because engineered barrier designs are site-specific, it is very important for the licensee to clearly and completely document how the licensee has designed the engineered barriers and the monitoring and maintenance program for the site-specific conditions to maintain performance of the engineered barriers for as long as necessary.

The guidance that follows provides the framework for applying engineered barriers to achieve decommissioning at a site. These five sections are directed towards licensees pursuing restricted use, since it is envisioned that this is the situation where engineered barriers will most frequently be used. The guidance is directed towards designing new engineered barriers; however, it is recognized that some sites may have in-situ engineered features that are part of the existing site being decommissioned, for which the licensee wants to take credit. In these situations, elements of Sections 3.5.2 and 3.5.3 may be most helpful. The guidance is intended to strike a balance between providing adequate direction without being overly prescriptive. Because the application of engineered barriers is site specific as a result of different radiological source terms, different exposure environments, and different natural systems, some elements of the guidance may not be applicable to every site.

In summary, NRC proposes that the following assumptions be used when applying engineered barriers to achieve decommissioning at a site:

- Engineered barriers are distinct and separate from institutional controls.

- Engineered barriers are passive, man-made structures at unrestricted use sites. Engineered barriers may be active (e.g., with monitoring and maintenance) or passive at restricted use sites.

- Engineered barrier evaluation is completed on a case-by-case basis using a risk-informed approach.

3.5.1 RISK-INFORMED GRADED APPROACH TO ENGINEERED BARRIERS

Initially, the general need for an engineered barrier(s) at a specific decommissioning site should be determined by considering if barrier(s) would be needed for compliance with the LTR dose criteria (e.g., mitigates the impact of natural processes such as erosion and infiltration). Once the general need is determined, a risk-informed graded approach should be used for selecting, designing, and providing a technical basis for the engineered barriers at a specific site. The risk-informed graded approach to engineered barriers is similar to the risk-informed graded approach for institutional controls described in Section M.2 of Appendix M of Volume 1 that consists of a general risk framework and associated grades of institutional controls. The same risk framework is used for both institutional controls and engineered barriers and is defined by the hazard level and likelihood of hazard occurrence. The underlying philosophy is that robust engineered barriers and additional basis for engineered barrier performance should be provided for higher risk sites (compared to lower risk sites). As described in Section M.2 of Appendix M of Volume 1, higher risk sites are those sites where the hazards are large (>1.0 mSv/y (100 mrem/y)) when institutional controls are not in place or when the hazard from the material is long-lived (e.g., >100 years), resulting in a longer performance period with higher likelihood of disruption and introducing greater uncertainty in assessing performance over long temporal scales. Generally, "robust" means more substantial, reliable, and sustainable for the time period needed without reliance on ongoing active maintenance. The term "robust" is similar to the term "durable" used for institutional controls at higher risk sites (see Section M.2 of Appendix M of Volume 1). Section 3.5.6 provides a detailed example of the application of the risk-informed graded approach to engineered barriers for erosion protection.

Robust engineered barriers are also needed for sites where the hazards are being significantly mitigated by the functioning of a barrier (s) (i.e., risk reduction). "Significantly mitigated" is site specific and can vary due to differences in the source and other features of the site (see example). In general, an engineered barrier that reduces the hazard more than a factor of 5 would likely be

---

(Example) Site A and Site B both contain soil contaminated with equal concentrations of Sr-90. The primary exposure pathway is from leaching of the contamination to the groundwater pathway. Both sites design an engineered cap to limit infiltration for the next 100 years. Site A has a thick unsaturated zone composed of a clayey soil. The estimated travel time to the groundwater, constrained by observations of past releases of Sr-90, is at least 200 years. Site B has a thin unsaturated zone composed of a sandy soil and the estimated travel time to the groundwater is 60 years. There is no additional data to constrain the numerical estimate of the travel time at Site B.

Conclusion: Site B would need more technical basis for the performance of their engineered cap compared to Site A, because they would have less risk reduction from the natural system and their travel time estimate (which is directly tied to risk via radioactive decay) is more uncertain.

---

considered to be significant. The basis for this factor is that in many circumstances the uncertainty in the risk for restricted use sites can approach or exceed this value.

Although robust engineered barriers would generally be appropriate for higher risk restricted use sites, there is flexibility to also use robust engineered barriers as an approach to not rely on ongoing active maintenance at lower risk restricted use sites. Such an approach might simplify institutional controls and maintenance and thereby reduce costs for a licensee. In some limited cases, a robust engineered barrier also could be used at an unrestricted use site because the robust barrier has been designed so that its performance does not rely on active ongoing maintenance.

The concept of passive performance of an engineered system is not unique to decommissioning of a radiologically contaminated site. There are many long-lived man-made structures that have not benefitted from continual monitoring and maintenance. However, passive performance of an engineered system can not be assumed. As indicated in the sections that follow, passive performance must be justified through design, experimentation, analysis, and support, and uncertainty in projecting performance, particularly over periods of time that exceed the experience base, must be considered.

Specific grading of engineered barriers recognizes that the site-specific factors affecting risk can be highly variable from site to site. As a result, specific grading provides the flexibility to design barriers to achieve the desired functionality and robustness appropriate for a specific site.

Uncertainty related to engineered barrier performance over a long-temporal scale is also contextual in that the length of experience in the use of that type of barrier should be considered relative to how long-lived the contamination is. In general, hundreds to thousands of years would be considered long temporal scales for the application of engineered barriers to decommissioning, because it is almost certain that site-specific experience for the performance of an engineered system does not extend beyond tens of decades. Section 3.5.2 provides guidance on how the barrier analysis process should be used to determine the performance of the engineered barriers. The risk-informed graded approach to engineered barriers is linked to the sections on effective barrier analysis (Section 3.5.2) and the technical basis for engineered barrier performance (Section 3.5.3), as well as Section 17.7.4 of Volume 1, on maintenance and monitoring. For example, where there is a demonstration that the engineered barriers have been designed to be robust and not reliant on ongoing active maintenance and repair, the amount of information and resources needed to support maintenance would be considerably reduced.

The robustness of an engineered barrier and the amount of technical basis provided for an engineered barrier should be consistent with the level of risk for a site (i.e., lower or higher risk) and the amount of risk-reduction provided by the barrier.

More robust engineered barriers and more technical basis would be needed for higher risk sites or where hazards at a site are significantly mitigated by the barrier.

Engineered barriers should be designed with the goal of remaining effective over the time period needed to achieve compliance, especially for long-lived radionuclides. The following items should be considered in designing engineered barriers for decommissioning sites:

- Designs should simplify long-term control and minimize the reliance on active ongoing maintenance and associated costs, especially for sites with long-lived radionuclides.

- Designs should mitigate potential future failures of the engineered barrier over the period needed to achieve compliance and the resulting need for and high cost of major repairs or replacement of major portions of the engineered barrier.

- Designs for sites with long-lived radionuclides should place more reliance on natural materials and less reliance on synthetic materials that are less proven over the long-term. For sites with short-lived radionuclides, synthetic materials may be more advantageous because of less variability in the performance of the materials.

- The determination of adequate financial assurance should consider the cost of monitoring, routine maintenance, and the need for potential major repairs of the engineered barrier over the time period of compliance.

3.5.2   ENGINEERED BARRIER ANALYSIS PROCESS

Note that this section is directed towards licensees pursuing restricted use. Licensees using engineered barriers for unrestricted use may find parts of this section useful, but should only credit the passive performance (without monitoring, maintenance, and inspection) for mitigating radiological impacts.

In order to implement a risk-informed graded approach, an accurate assessment of the performance of the engineered barriers should be provided. To determine compliance with Subpart E for restricted use, a licensee should complete, but not be limited to, an analysis of the following:

1. The contribution of engineered barriers towards compliance with the criteria of 10 CFR 20.1403 with institutional controls in place (including maintenance).

2. The contribution of engineered barriers towards compliance with the criteria of 10 CFR 20.1403 assuming loss of institutional controls (including loss of maintenance) such that the barrier may degrade over time.

The first analysis would typically evaluate a public receptor located at the site boundary prevented from accessing the site by the active institutional controls. The performance of the engineered barriers could include the benefit of active monitoring and maintenance to mitigate degradation while the institutional controls are in place and the dose limit would be 0.25 mSv/y (25 mrem/y).

The second analysis, that assumes loss of institutional controls, would typically include evaluating a reasonably foreseeable inadvertent intruder potentially located on the site. The dose limit applied to this receptor is 1.0 mSv/y (100 mrem/y) [or 5.0 mSv/y (500 mrem/y)]. The performance of the engineered barriers should consider the likelihood of a breach by an inadvertent intruder and the potential degradation of the barriers over time because monitoring and maintenance are assumed to not be active. In addition, other reasonably foreseeable disruptive conditions from humans or natural events and processes should be evaluated. In some cases, the intruder may disrupt the engineered system that could result in higher doses through a different pathway (e.g., an agricultural intruder disrupts an engineered cap with plowing but does not disrupt the contamination directly.) Disruption of the cap could result in increased infiltration and higher groundwater pathway doses.

"Reasonably foreseeable disruptive conditions from humans or natural events and processes" should be interpreted as those processes and events expected to have a probable or likely occurrence over the analysis period. The intent of performance assessments is to provide reasonable assurance, considering uncertainties in engineered and natural systems over long time periods, that the actual performance of the disposal facility will comport with its design. Natural or human processes and events are possible stressors to the engineered barrier design. Unlikely processes or events would not be expected to have a significant influence on risk because of the low probability of occurrence.

The potential impact to engineered barriers from reasonably foreseeable disruptive conditions from humans or natural events and processes can vary by site and by barrier. In some circumstances it may be reasonable to take no credit for the engineered barriers. For example, a hypothetical site installs a geomembrane to control infiltration at a restricted use site. For the analysis of loss of institutional controls, it would be reasonable to assume no performance credit for the geomembrane because it could be easily removed or damaged by near surface processes and this type of barrier is subject to discrete failures. As a second example, a site in a rural setting uses a large, thick earthen cover of low gradation. Complete removal of the earthen cover under these conditions, while possible, would not be reasonably foreseeable considering the cost and limited benefits. Realistic scenarios for future site use for this example (e.g., residential or farming) would dictate the reasonably foreseeable disruptive conditions from humans would commonly be construction of a foundation and a well for a residence that disrupts a portion of the cap, or farming of the site which would result in plowing of the top layers of the cap and resultant potential impacts from the rooting of vegetation and increased infiltration.

In regard to engineered barrier degradation, the licensee should address the two cases where maintenance is either in place or lost, because the assumption for loss of institutional controls includes the loss of maintenance of engineered barriers and physical controls such as fences or signs. It should be noted, however, that for those cases where an erosion control cover is designed in accordance with the uranium mill tailings guidance in NUREG–1623, a case might be made for a durable, long-lived engineered barrier that does not rely upon ongoing active maintenance (i.e., maintenance needed to assure that the design will meet specified longevity requirements) and associated future costs in order to maintain erosional stability of the site. For this case, a licensee may be able to demonstrate that no significant degradation of the erosion control cover is expected.

The barrier analysis process should be used to determine how much performance is being provided by the engineered barrier (i.e., risk reduction). To accomplish this, a licensee should perform analysis with the engineered system present and functioning. However for unrestricted use or when evaluating the loss of institutional controls (active monitoring and maintenance not performed), the barriers may degrade over time. An assumption of instantaneous and complete failure of a barrier is not required. The goal is to clearly identify the expected benefit of the engineered barriers quantitatively in terms of dose reduction. The analysis should also address the uncertainty in the expected performance of the barriers. For example, a comparison could be made of the doses without engineered barriers present to the doses with engineered barriers performing as designed and expected, as well as more pessimistic performance and more optimistic performance. The assumption to model a degraded barrier is generally more realistic than the assumption of the absence of a barrier, and actually may lead to higher doses in some situations; for example, partial failure of a cover or partial failure of a grout wall system can focus water flow and can create a "bathtub" effect. Thus, simple on-off analyses (i.e., where the barrier is either assumed completely present or completely absent) should be used cautiously.

Caution should also be used when analyzing systems with multiple engineered barriers. When multiple barriers are present, the performance of one barrier may mask the potential contribution to performance of another barrier or may degrade its contribution. In these situations, the analysis may need to evaluate various combinations of barriers to determine the individual and cumulative contribution (positive or negative) to performance of the barriers. In particular, characteristics of one barrier that challenge or impair the performance of another barrier should be carefully assessed (e.g. increased infiltration resulting from elimination of vegetation by inert covers or the clogging of a drainage layer with a resultant hydraulic head over the waste). The type of analysis performed should be determined on a case-by-case basis. For the complex and higher risk decommissioning sites and those sites with long-lived radionuclides, the use of probabilistic analysis should be strongly considered, because deterministic analysis may not be able to adequately address the uncertainty in the calculations. However for simpler, lower risk sites and sites with short-lived radionuclides, deterministic analysis with sensitivity analysis may be sufficient. As indicated in Section 3.5.6.1, deterministic methods should be used for selection of the design flood for the development of long-term erosion controls.

In summary, the analysis of engineered barriers should identify and evaluate those conditions or processes that are adverse to performance and result in non-compliance. The following analysis should be provided:

- Analysis with institutional controls taking credit for monitoring and maintenance.

- Analysis assuming loss of institutional controls not taking credit for monitoring and maintenance.

- Analysis assuming loss of institutional controls (or unrestricted use) considering disruptive natural processes as well as reasonably foreseeable human disruptive processes to the barriers.

3.5.3 TECHNICAL BASIS FOR ENGINEERED BARRIER PERFORMANCE

Significant uncertainty exists concerning predicting the service life and long-term degradation rates of most engineered barriers. This section provides guidance on the main elements that should be provided to support the assessment of the performance of the engineered barriers in Section 3.5.2, including:

- The design, features, and functionality of the engineered barriers should be fully-described.

- Technical basis that the barriers will meet the dose criteria, considering the degradation mechanisms, should be provided including consideration of combined and synergistic effects resulting from the real world conditions expected for the barriers (Section 3.5.4).

- Uncertainty in parameters and models used in the assessment of barrier performance and the design of engineered barriers.

- The suitability of numerical models for the estimation of engineered barrier performance should be provided.

- Parametric or component sensitivity analysis should be performed to identify how much degradation of the engineered barrier would result in non-compliance (see Section 3.5.2).

- Model support should be provided for the engineered barrier performance (e.g., analogs, experiments, simple engineering calculations to demonstrate reasonableness of the results).

- Quality assurance and quality control (QA/QC) should be provided for the design and analysis of engineered barriers.

Analysis can be used to understand the impact of uncertainty. For example, 1/4 loss of a cover may not result in non-compliance, but loss of 3/4 of a cover would. Then determine if reasonably foreseeable natural and human processes might cause a loss of 3/4 of a cover. This analysis approach may be an additional way to deal with the uncertainty due to lack of long-term cover degradation data.

Many engineered barriers are not amenable to model validation in the true sense, therefore multiple lines of evidence are recommended. Model support can come in many different forms, including but not limited to: analogs, laboratory experiments, field experiments, formal and informal expert judgement, and engineering calculations to demonstrate reasonableness of the results (e.g., hand calculations when numerical models are used). The level of model support should be commensurate with the risk significance of the engineered barriers to achieving site decommissioning (see Section 3.5.1). If the level of performance needed for an engineered barrier is consistent with past experience at similar sites, and the engineered barriers have similar design and quality assurance, then the model support could be considerably less than for an engineered barrier with performance objectives that significantly exceed engineering experience. When considering engineering experience, care must be taken to ensure that the environmental conditions for the relevant degradation mechanisms are reasonably similar, since many degradation mechanisms can be very sensitive to the environmental exposure conditions.

For engineered barriers that must have very long-term performance (e.g., thousands of years), natural analogs should be considered. The greatest uncertainties stem from extrapolating the results of short-term tests and observations to long-term performance. Standard engineering approaches frequently implicitly assume that the initial conditions persist, however the actual application of a barrier more appropriately could be viewed as a evolving component of a larger, dynamic ecosystem (Waugh, 1997). For some types of engineered barriers, natural analogs might provide information as to the possible long-term changes to an engineered system, and can be thought of as a long-term, uncontrolled experiment. Evidence from natural analogs can help demonstrate that there are real world complements to the postulated numerical predictions. It is important that the functionality of an engineered barrier is considered when developing analogs. The example of earthen mounds constructed by the Native Americans, provided below, provides a reasonable analog for the physical stability of an engineered cap. It does not provide analog information for the ability of engineered caps to limit infiltration or the release of radionuclides. Analog information is uncertain for a variety of reasons, such as unknown past environmental conditions. Therefore, analog information should not be envisioned as providing proof of future engineered system performance, but rather it provides confidence that the engineered system is likely to perform as designed.

Experience is limited for the long-term performance of *some* engineered barriers. However, *other* barriers do have suitable analogs such that the multi-faceted approach to designing, testing, assessing, implementing, supporting, and evaluating engineered barriers provided in this guidance should result in protection of public health and safety. There are a number of analogs to engineered covers that have shown sustained durability, even for thousands of years (discussed in more detail below). In addition, some cementitious materials used by the Romans, for example, are intact after more than a thousand years of exposure to the environment. In the United States, cements in the Erie Canal and used in colonial settlements are two to three hundred years old.

An example of a natural analog, in this case with respect to the durability of earthen covers, but not with respect to limiting infiltration or waste releases in the groundwater pathway, may be the various earthen mounds constructed by Native Americans that have survived for periods of 1000–5000 years. Archaeologists have dated the mounds by excavating bones and artifacts from the mounds and determining the age of the object or the date of its burial. Information on the mounds is readily available by visiting State or national parks associated with the mounds. Also, there is considerable information and reference material available on the internet. Examples of Native American mounds that have survived relatively intact for very long periods of time are shown in Table 3.1.

Table 3.1   Examples of Surviving Native American Mounds

Mound	Location	Approximate Age (years)
Grave Creek Mounds	Moundsville, WV	2500
Hopewell Culture Mounds	Chillicothe, OH	2000
Cahokia Mounds	Collinsville, IL	1000
Poverty Point Mounds	Epps, LA	3500
Watson Break Mounds	Monroe, LA	5500

The mounds vary in size with some being more than 23 m (75 ft) high and more than 100 m (300 ft) long. Mounds of this size reasonably approximate the size of engineered caps that may be installed at decommissioning sites. Therefore, long-term stability of these analog sites provides additional confidence and assurance that wastes may be effectively stabilized for very long periods of time. It should be noted that stability in this context refers to erosional stability and not the ability of the mounds to limit infiltration, which is unknown. More importantly, by examining the causes for failures of any damaged burial mounds (where only portions of the mounds remain and can be examined effectively), guidance for earthen cover designs could be developed. Understanding of the long-term performance of analog systems may allow for additional safety factors to be used in current designs. For example, additional protection, such as rock layers, may be used to further reduce erosion. It should also be noted that the use of archaeological bases for guidance is consistent with previous approaches. In NUREG/CR-2642, "Weathering and Long-Term Protection for Long-Term Stabilization," riprap durability procedures were based in part on examinations of ancient Native American ruins and petroglyphs that were approximately 1000 years old. Riprap durability reviews performed by the staff have included examinations of the weathering of very old grave markers and comparison of photographs taken almost 150 years apart.

When evaluating analogs, it is important to note that the structures that have persisted are most likely the most durable structures. That is why it is important to consider analogs that have persisted as well as those that may have experienced damage or failure. An additional complicating factor is that the initial conditions and past exposure environment for the analogs are not known and may only be estimated. However, developing an understanding of analogs increases the likelihood that a design may be implemented with sustainable long-term performance. It should be reiterated that natural analogs should be only one element of the technical basis for the long-term performance of engineered barriers.

Monitoring and maintenance might be needed in order to verify the effectiveness, durability, and service life of the engineered barriers. This monitoring involves both the environmental system surrounding the engineered facility that could be disruptive and the facility itself (See the risk-informed approach to monitoring in Section 17.7.4 of Volume 1). Non-destructive monitoring technologies that included designed and emplaced sensors are preferred to conventional post-failure monitoring. Novel ideas such as introduction of special dyes and/or tracers within the engineered system may facilitate identification of impending failures. The identification of these and other measurable performance indicators within a monitoring strategy coupled with any needed repair or remediation can be very important in extending the effective service life of these facilities and reducing risk. To the extent practicable, engineered barriers should be designed to support and simplify monitoring and maintenance.

3.5.4 DEGRADATION MECHANISMS AND FUNCTIONALITY OF COMMON ENGINEERED BARRIERS

The purpose of this section is to identify for licensee consideration, the common engineered barriers, the degradation mechanisms for the common barrier types and materials, and typical functionality for these barriers. It is envisioned that this information may help staff or licensees select and design appropriate engineered barriers and understand the considerations that are needed for assessing long-term barrier degradation, so that the overall decommissioning process can be more efficient.

3.5.4.1 Common Barriers

The common barriers provided are those that may be encountered at a decommissioning site. Because technology evolves and a site may have unique considerations, this list should not be viewed as comprehensive. Engineered waste forms are not explicitly listed as a common barrier for decommissioning, because in most instances a decommissioning site is dealing with contamination of environmental media and not explicit engineering of a waste form. The assessment of engineered waste forms has been addressed in low-level waste disposal and that guidance should be considered with respect to waste forms used in decommissioning (NRC, 1991). Lists of common barriers that may be used at decommissioning sites and their primary functionality are:

- Engineered Caps — Multi-layered and composite engineered caps are typically used to limit infiltration, provide for shielding between the contamination and potential receptors, eliminate exposure scenarios, and to limit erosion. Caps for infiltration control are typically either a resistive type or a water balance type.

- Geomembranes — Geomembranes are synthetic materials use primarily to limit infiltration to the contamination.

- Concrete/cement/grout — Engineered cementitious materials are typically used to stabilize contamination, provide a chemically favorable environment for retention of radionuclides, limit water contact, prevent erosion, provide shielding, and limit potential intruder contact with the contamination.

- Vertical barriers — Vertical barriers may be soil-bentonite, soil-cement-bentonite, cement-bentonite, sheet pile (steel or high-density polyethylene), and clay barriers and are primarily used to control the horizontal migration of groundwater.

- Permeable reactive wall — A contaminant plume is channeled between impervious vertical walls, referred to as the funnel, and flows naturally through a permeable reactive barrier gate, where the pollutants are treated in situ during the flow process.

- Interceptor trenches — Used to intercept and collect contaminant releases. Typically only applicable with monitoring and maintenance.

- Chemical barriers — Chemical barriers are used to modify subsurface environmental conditions (e.g., pH, Eh) to limit the solubility of radionuclides or to provide a more favorable geochemical environment for sorption. A good example are engineered cementitious materials (see above). Because of the diverse type of chemical barriers that could be applied, the degradation mechanisms and typical levels of functionality are not provided in the following sections but would need to be evaluated on a case-by-case basis.

3.5.4.2   Degradation Mechanisms

The degradation mechanisms provided may not be comprehensive due to the large variability in conditions and processes from site to site, but they should represent the main degradation mechanisms typically encountered. Degradation mechanisms depend on both the barrier and site-specific conditions. When evaluating degradation mechanisms, careful consideration should be given that the environmental conditions assumed or used in an analysis of long-term performance are representative of the in-situ conditions expected for the engineered barrier.

The main degradation mechanisms are described below for the different engineered barriers.

Degradation of Cement-Based Engineered Barriers

The major environmental degradation processes that affect cement-based engineered barriers are sulfate attack, corrosion of reinforcing steel/carbonation, alkali/aggregate reactions and leaching by acidic subsurface water. Other degradation mechanisms include freeze-thaw deterioration, and microbiological attack. Degradation mechanisms can also be due to poor design and construction of cement-based structures, and can include differential settlement of the structures; stress concentrations; seismic effects; and insufficient structural engineering design. To avoid these latter degradation mechanisms, the structures need to be properly designed, and constructed under strict QA/QC procedures to ensure that their design objectives have been met. Discussions that follow will not address these design and construction-related degradation issues of cement-based materials.

Sulfate Attack

Sulfate attack on concrete can be severe, resulting in cracking of the concrete, and in some cases, its disintegration. Naturally-occurring sulfates of sodium, potassium, calcium and magnesium are sometimes found in subsurface water and soils. Sulfate attack has occurred in several regions of the U.S., particularly in arid regions such as the Northern Plains and the Southwest States. Localized sources of sulfates in groundwater include mine workings, mine tailings, blast furnace slag waste piles, and chemical waste ponds. Water used in irrigation can also be a potential source of sulfate attack due to the gradual accumulation of sulfates in the soils. The main cause of sulfate degradation of cement-based materials is the formation of ettringite in the reaction process. This results in the mechanical expansion and subsequent deterioration of concrete. Another degradation process involves the formation of gypsum which replaces the calcium hydroxide in concrete resulting in expansive stresses and deterioration.

Corrosion of Reinforcing Steel

Cement concrete normally provides a high-alkaline environment which passivates the steel. However the corrosion of steel embedded in concrete has been a serious problem due to the presence of chloride ions. Although chloride ions are common in nature, and small amounts are intentionally added in the mix ingredients of concrete to accelerate set times, the principal sources of chloride ions which cause problems in concrete are from deicing salts, sea water, and chloride ions in surface runoff. Corrosion of reinforcing steel can, in some cases, occur in the absence of chloride ions. This happens in the case of carbonation which results in a reduction of alkalinity of the concrete, thereby depassivating the steel and facilitating corrosion.

Alkali-Aggregate Reactions

Alkali-aggregate reactions are usually internally contained in concrete, and are not dependent on the diffusion of an aggressive solution into the cement-based material. For appreciable amounts of swelling to occur due to this process, a source of water is required. Almost all aggregates react to some extent with alkalies in cement. It is when the reaction results in the formation of expansive products (e.g., gypsum) that serious cracking of the concrete occurs. Expansive alkali-aggregate reactions are known to occur when siliceous (i.e., alkali-silica reactions) and dolomitic (i.e., alkali-carbonate reactions) limestone aggregates are used. In addition, the rate of expansive reaction is also influenced by the size of the aggregates. Alkali-silica reactions are the most common with the majority of the reported instances in the western states. Alkali-carbonate reactions have occurred in some Midwestern and Eastern states.

Leaching

Buried concrete, in contact with percolating subsurface water, can undergo deterioration by the dissolution of the common constituents of cement paste, the alkali salts and sodium hydroxide. Leaching can reduce the pH of the concrete, as well as make it more porous for subsurface water. The rate and extent of the leaching is dependent on the acidity of the water since leaching increases as the pH decreases. The potential leaching capabilities of subsurface water can be related to their Langelier Index. The Langelier Index is related to the total dissolved solids (TDS), total alkalinity, pH and calcium content of the water. A positive index indicates that calcium carbonate will be precipitated, while a negative index indicates lime deficient water capable of dissolving calcium from the hardened cement paste.

Freeze-Thaw Attack

Freezing and thawing damage in cement- based materials occurs when a water-saturated concrete is exposed to prolonged cycles of freezing and thawing. Structures most susceptible to freeze- thaw damage are surfaces of the structures where flowing or ponding water can remain in contact with the concrete structure for extended periods of time. Several precautions should be taken to avoid freeze-thaw damage, including the following: precluding ponded water from the concrete structure, incorporating entrained air, placing properly, consolidating, and curing.

Microbiological Attack

Sulfate-producing bacteria are capable of oxidizing elemental sulfur and sulfides to sulfuric acid under aerobic conditions, which in turn degrade cementitious materials. Some bacteria can attack cement-based materials by transforming ammonia into nitrites or nitrates, or by producing lactic acid or butyric acid. In the normal design life (e.g., 40 years) of conventional cement-based materials, bacterial action does not seem to be a major cause of deterioration. However, for cement-based materials' chemical durability over hundreds of years, the impacts of bacterial activity need to be assessed, though they may be difficult to predict.

Cracking of Concrete

Cracking can originate within concrete due to a number of mechanisms. During placement, if the evaporation rate is great enough, the concrete surface can develop tensile stresses sufficient to crack the concrete. These plastic-shrinkage cracks typically extend through the entire concrete member. Cracking can also be caused by settlement of the concrete member, by flexural stresses, and thermal effects. The continued removal of water by the hydration process will generate a chemical-shrinkage stress that can initiate autogenous shrinkage cracks. Subsequent drying due to ambient conditions will also generate shrinkage stress which generates drying-shrinkage cracks. Absent environmental conditions which may cause concrete-material degradation, cracking can be the most severe degradation mechanism affecting concrete. Transport through cracks in concrete will only be of consequence, if the cracks extend throughout the concrete member. Relatively large amounts of water can be transported through a crack depending on the total potential water-head across the crack, and the crack aperture and density.

Degradation of Engineered Caps

Engineered caps or covers are commonly designed to eliminate or significantly limit infiltration of subsurface water into the waste, but may also be designed to limit gaseous release, provide shielding, or reduce the likelihood of contact with the waste. From an infiltration management perspective, engineered caps are based on either resistive or water balance concepts. Caps based on resistive concepts use impermeable layers to prevent water from contacting the waste. Caps based on water balance concepts attempt to mimic natural systems to evaporate or transpire water from the system.

Resistive engineered cap designs have a wide range of configurations ranging from a one-layer system of vegetated soils, to complex multi-layered designs composed of soils and geosynthetics. Soil materials can include vegetative soils, permeable sand and gravel drainage layers, low hydraulic conductivity clay soils, and filter soils to preclude the migration of fines from soils overlying drainage layers which cause the drainage layers to clog. Geosynthetic materials include geomembranes and geosynthetic clay layers. Composite barriers use both soils and geomembranes/geosynthetic clay layers. Geomembranes are essentially impermeable PVC or polystyrenes layers, while geosynthetic clay layers are composed of a thin layer of bentonite between two geosynthetic textiles which may be used in conjunction with geomembranes. The effectiveness of these engineered caps lies in its ability to significantly limit infiltration of subsurface water into the waste. The ability to drain infiltrating water away from the buried waste, and its ability to release generated gas is a function of the waste stability, the degree of settlement, and slope stability of the cap system. Degradation mechanisms unique to engineered cap materials, other than those specific to inadequate design, construction, QA/QC issues, follow in later sections.

Water balance, or evapotranspiration covers attempt to manipulate the water balance of the source zone by enhancing soil water storage and evapotranspiration. The performance of evapotranspiration covers depends on many factors, especially the climatology, soil hydrology, and plant ecology at a site. Water balance covers may be used in a variety of settings, but may be most effective in arid or semi-arid climates with high potential evapotranspiration.

Compacted Clay Barrier

The function of the clay barrier layer in the engineered cap is to prevent and block infiltration of subsurface water through the cover into the waste. In order to meet this function, the clay must always remain close to saturation. The compacted clay layer is frequently specified to have a saturated hydraulic conductivity of 1E-07 cm/sec or less. The longevity and effectiveness of the engineered cap are influenced by the ability of the clay layer to retain low permeability characteristics. However laboratory and field studies have shown that dessication (severe drying) cracks can form quickly in compacted clay. These cracks, which do not completely self heal, can penetrate the entire thickness of the clay layer, and are not filled by soil from overlying layers in the time frame of available field studies (less than 10 years). The cause for extensive dessication of the clay is thought to be due to vertical water vapor transport. Unanticipated ecological processes (e.g., biointrusion), compaction either dry or wet of optimum, freeze-thaw cracking, and differential settlement can result in the degradation of clay and other resistive barriers.

Drainage and Filter Soil Layers

A common degradation mechanism affecting these drainage and filter components of the engineered cap system is the potential clogging of these materials by finer soil particles from overlying soils including colloidal or biological materials. Clogging of drainage/filter soil layers can greatly reduce the permeability of these materials, and render them unable to perform their intended function to drain subsurface water.

Composite Soil Caps

Composite caps are comprised of a combination of a compacted clay layer, geosynthetic clay layer and geomembrane. In general, composite caps have performed well in the time frame of available field studies (10+years). Geomembranes protect clay barriers/geosynthetic clay layers by eliminating or significantly reducing vertical water vapor migration. Degradation mechanisms of geomembranes include: puncture by granular soils and construction equipment; behavior of "waves" or fabric wrinkles due to temperature and overburden stresses; long-term degradation of geomembranes under the influence of UV light, chemicals and radiation effects; the potential for slippage between geomembranes and adjacent materials; and material embrittlement over time. Some degradation mechanisms, such as exposure to UV light, can be managed with effective QA/QC. Geosynthetic clay layers are sometimes used instead of compacted clay, and seem to perform better with respect to water flow, chemical degradation due to cation exchange, and mass transport (i.e., diffusion and retardation). As with geomembranes however, geosynthetic clay layers have inherent problems due to installation activities (e.g., puncturing and degradation by construction equipment). Geosynthetic clay layers can experience dessication cracking similar to clay layers. In addition, recent research seems to indicate that there exists the potential for cation exchange between commonly available calcium-laden fluids and the sodium in the geosynthetic clay layer bentonite, thus rendering the geosynthetic clay layer incapable of functioning as a low-permeability barrier layer in engineered-cap systems [(James et al., 1997), (Melchoir, 1997), (Lin and Benson, 2000)]. Research is being conducted to more carefully study these degradation mechanisms.

Degradation of Water Balance Caps

The effectiveness of a water balance cap is dependent on the development of a design that is effective over the range of expected natural and ecological conditions. Natural and ecological conditions are inherently variable over the timeframe of most decommissioning analyses. As discussed in Section 3.5.5, water balance caps have been analyzed in great detail at a number of sites. With effective design and development, water balance caps may be very effective, especially in arid and semi-arid climates. In humid climates, water balance caps may be effective at managing a substantial fraction of the infiltration but not achieve design goals.

Infiltration can commonly exceed evapotranspiration in humid climates or in colder climates where a large fraction of infiltration may occur as snowmelt when evapotranspiration is low. Therefore, one of the major lessons learned, albeit not related to physical degradation, is that design of a water balance cap must consider natural and ecological variability over the period of performance.

Degradation mechanisms associated with water balance caps include unanticipated ecological consequences, evolution of soil properties (e.g., soil development), and effects of disturbance (e.g., fire, land use) on plant ecology. An example of unanticipated ecological consequences is the development of deeper rooted plant species that result in pathways for moisture below the design zone for moisture storage and removal.

Degradation of Erosion Control Caps

Degradation of erosion control caps is discussed in Section 3.5.6.4.

Degradation of Permeable Reactive Barriers

Permeable reactive barriers are in-situ constructed walls below the land surface that intercept contaminated groundwater which is funneled through it. Reactive materials in the wall can sorb chemicals and radioactive species on their surfaces and/or precipitate contaminants dissolved in the flowing water. In some cases nutrients and oxygen in a permeable reactive barrier help microbes in the soil to precipitate contaminants and radioactive species. Experience with permeable reactive barriers seem to indicate that not all of them are performing well, often due to poor placement in the ground-water flow field. Material properties, such as grain size of reactive zeolites can be changed in the construction process, thereby reducing the hydraulic conductivity. There seems to be a decrease of barrier performance due to loss of reactivity and permeability over relatively short periods of time (in less than 5 years a number of systems have experienced challenges associated with diversion of water around the permeable reactive barrier due to insufficient permeability).

Degradation of Vertical Barriers

Various types of subsurface vertical barriers are in use. Their primary purpose is to impede or preclude horizontal ground-water flow. These vertical barriers are placed at depths up to 60 m (200 ft), and often vary in thickness from 0.6–1.2 m (2–4 feet). The barriers must extend down to an impermeable natural horizontal barrier such as a clay zone to effectively impede ground-water flow from below. These barriers are often designed as temporary or semi-permanent remediation techniques to isolate contaminated fluids from migrating to uncontaminated surrounding groundwater. Some soil-bentonite mixtures are not able to withstand attack by chemicals such as strong acids, bases, salt solutions and certain organic chemicals. This hastens the deterioration of the barrier. Verification that the vertical wall forms a continuous barrier is critical to the function of this technology. Although it may be difficult to identify flaws in the continuity, and/or gaps in the wall, monitoring is essential to verify their performance. Although vertical walls have been used for decades, the process of designing the proper mix of wall materials to contain specific contaminants is less well developed.

3.5.4.3   Potential Levels of Functionality and Uncertainty

The purpose of this section is to give some general information to help licensees initially consider the use of engineered barriers. As discussed in Sections 3.5.2 and 3.5.3, licensees are responsible for developing acceptable technical bases and conducting analyses of engineered barriers propose for a specific site.

The ranges of functionality or performance for different barriers and the associated uncertainty are based on a broad consideration of observations and analysis throughout the national and international community. Potential ranges of functionality are levels of performance that can likely be supported by technical basis and analyses, not solely based on demonstrated field experience. As the time scales get longer, past direct observation of the performance of engineered barriers that can be cited as basis becomes less likely, and therefore, performance for longer times becomes more uncertain and more based on inference. The functionality provided here can be thought of as the level of performance believed to be reasonably achievable with proper design, analysis (Section 3.5.2), technical basis (Section 3.5.3), and implementation (quality) given current understanding and engineering practice. The ranges provided for potential functionality, help provide direction as to when less technical basis may be needed (assume less than typical performance) compared to more technical basis (credit is taken for more than typical performance). The level of uncertainty should be considered in developing a monitoring and maintenance plan. Barriers with less uncertainty might need less reliance on monitoring and maintenance. In contrast, barriers with higher uncertainty, may need substantial monitoring and maintenance until uncertainties are reduced.

NRC's discussion in this draft section is an initial attempt using readily available information to provide some insights on potential functionality and uncertainty. This section could be expanded based on future studies and inputs from other programs involved with engineered barriers. Therefore, we invite suggestions and information that could further develop this section.

Typical levels of functionality or performance of the main engineered barriers are as follows.

Cement-Based Engineered Barriers

The performance of cement-based materials to isolate radioactive can be divided into two categories:

1. Hydrologic effectiveness or physical containment of the wastes to preclude water contacting the waste.

2. Chemical effectiveness or the ability of the high pH characteristics of the intact and degraded concrete to limit transport of the radionuclides to the accessible environment.

Absent environmental concrete degradation factors such as sulfate attack, chloride corrosion, etc., full depth cracking of the concrete member can be the most severe degradation mechanism causing contact of water with the waste and the subsequent release of radionuclides. Accordingly, the effectiveness of cement-based physical barrier structures need to be monitored for hydrologic effectiveness and the projected service life for the structure should be revised based on analysis of the monitoring data. Assuming adequate design, construction practices and excellent QA/QC followed by a competent monitoring program, service life of tens of years to a few hundred years appears feasible.

In the case of chemical containment of wastes, the duration needed for chemical effectiveness of cement-based materials is dependent on the source term and specific radionuclide time frames (approximately 10 half-life periods). The timeframe of performance is difficult to predict, but could extend a significant period of time beyond the physical containment period (e.g., possibly thousands of years, likely hundreds of years). The longevity of chemical effectiveness is strongly related to the bulk hydraulic properties of the material and the quantity of cement present, because chemical species of the cement matrix are removed from the system by the exchange of pore fluids and also possibly by diffusion.

A cementitious barrier used to limit potential intruder contact with waste, with proper design, construction practices, and QA/QC could be expected to be effective for hundreds of years if it remains unexposed to aggressive environmental conditions (e.g., high sulfate, excessive freeze-thaw cycles). Performance of this type of barrier may be enhanced with appropriate monitoring, repair, and remediation strategies.

Engineered Caps

Based on recent research, extensive dessication of clay barriers in soil caps has compromised the ability of clay barriers to retain low permeability characteristics and preclude infiltration of water through the cover and into the waste (Albright, 2004). Conversely, composite caps composed of a combination of compacted clay buried at sufficient depth, geomembranes and geosynthetic clay liners have performed well in the timeframes of available field studies (10+ years). Monitoring of caps composed of compacted clay barriers and composites can verify the effective lifetimes of these facilities. Given the good performance of composite covers, it is more probable that their functionality will be superior to compacted clay barriers. Current experience provides evidence of hydrologic functionality of tens of years for composite caps appear to be feasible. Longer hydrologic functionality may be feasible with the proper development and implementation of the elements provided in the previous sections, and with continued research. Existing uncertainty in long-term functionality could be reduced by additional technical basis, analyses, testing, and field experience. Geomembranes do not have the experience base of common natural materials or a man-made material like cement, which has been used for hundreds to thousands of years. Therefore, until the experience base is developed, a cautious approach is needed for the long-term performance of novel materials in engineered barriers.

Water balance caps have been studied in the field on large scales, as discussed in Section 3.5.5. When well-designed, constructed, and monitored they have performed well in arid and semi-arid climates. Because they attempt to mimic natural processes, the level of performance achieved over the long-term would be expected to exceed that of resistive infiltration barriers. Whereas a resistive cap is subject to failure by fast pathways (likely to be more discrete, either it is functioning or it is not), a water balance cap is more likely to have different degrees of performance (likely to be more gradual, based on exceeding the design capacity for water management). Water balance caps have been shown in the field to perform well on the order of ten years, and in theory may have some passive performance over the very long term (e.g., thousands of years).

Engineered caps developed for erosion control could have effectiveness that can exceed 1000 years. However, erosion control designs may not be adequate to preclude excessive infiltration. Section 3.5.6 provides a detailed example including some of the design considerations.

Aside from the depth to waste, most engineered caps would not provide a substantial barrier to common practices assumed in intruder analysis (e.g., home construction, well installation).

Permeable Reactive Barriers

There has been limited experience with permeable reactive barriers, approximately 15 years. Effective lifetimes of these barriers, from the literature, appear to be limited to less than 10 years.

Vertical Barriers

As noted previously, it is difficult to identify flaws in the continuity and gaps in constructed vertical walls. In addition, some of these walls are not able to withstand chemicals such as strong acids, bases, and certain organic materials. Moreover, the process of designing a proper mix of wall materials to contain specific contaminants (hazardous chemicals and radionuclides) is less well developed. Accordingly, effective service life of these structures would range in the low single digits of years and performance should be demonstrated by field testing.

3.5.5 SUMMARY OF EXISTING GUIDANCE AND REFERENCE INFORMATION

Table 3.2 provides a summary of existing guidance and reference information that may have some relevance to the application of engineered barriers at decommissioning sites. Early contact with NRC staff is encouraged to discuss which portions of these referenced reports may be appropriate for the site and for the intended purpose of the engineered barriers. Guidance for design of engineered disposal cells for uranium mill tailings sites is provided in NUREG-1620 and NUREG-1623. For sites considering engineered disposal cells for long-term stability (e.g., erosion control), this guidance may be somewhat useful. However, the standards in 10 CFR Part 40, Appendix A, applicable to uranium mills are more prescriptive than the performance-based dose criteria of Part 20, Subpart E. Licensees using the uranium mill guidance should also consider how the guidance can be adapted for applicability to compliance with Part 20, Subpart E. As discussed below, in some cases traditional designs have not performed particularly well with respect to infiltration.

A variety of programs have evaluated and continue to evaluate engineered barrier technology for waste containment. A comprehensive summary of engineered barrier research is not attempted in this document. However, some good examples of programs to understand, design, and support engineered barrier performance applicable to decommissioning sites include:

- The NRC has conducted research at the U.S. Department of Agriculture (USDA) facilities in Beltsville, Maryland, on engineered covers for low-level waste facilities (O'Donnell et al., 1994).

- The Environmental Protection Agency, through the Alternative Cover Assessment Program (ACAP), has supported the field-scale evaluation of engineered covers.

- The Department of Energy (DOE), through the Alternative Landfill Cover Demonstration (ALCD) completed a large-scale field demonstration comparing six landfill cover designs.

- DOE has instrumented engineered covers at some Uranium Mill Tailings Remedial Action (UMTRA) sites to understand and evaluate their performance.

EPA's ACAP evaluated 27 test covers at 12 sites in eight states to characterize the field hydrology of water balance and conventional covers (Albright et al., 2004). The evaluation included 12 conventional covers (7 composite and 5 clay) and 15 water balance covers (9 monolithic and 6 capillary barriers). Nine of the sites had side-by-side comparisons of conventional and alternative covers. Large-scale lysimeters (approximately 10 m by 20 m (30 ft by 70 ft) areal extent) were installed and instrumented to collect detailed water balance information. The main lessons learned for water balance covers were: (1) Percolation rates in semi-arid and sub-humid climates can be very low (<1 mm/yr), provided that there is adequate storage capacity and that the vegetation effectively removes stored water each year, (2) There is a need to better understand the phenology of plants and the response to meteorological and geotechnical conditions, and (3) Low percolation rates may not be achieved at sites with water balance covers, in particular at humid sites, but the water balance covers may still provide some performance benefit. The main lessons learned for composite covers were: (1) Composite covers may be effective at limiting percolation to less than 1 mm/yr while the geomembranes or geosynthetics are intact, and (2) Clay covers are prone to damage over very short periods of time and can transmit percolation at much higher rates than anticipated. The ACAP program provides an excellent example of developing the technical basis for engineered barrier performance.

DOE's ALCD evaluated six cover designs at Sandia National Laboratory in Albuquerque, NM, to obtain large-scale water balance field data subjected to identical field and climatic conditions (Dywer, 2003). The covers evaluated included a RCRA Subtitle D cover, a Geosynthetic Clay Layer cover, a RCRA Subtitle C cover, an Anisotropic Barrier cover, a Capillary Barrier cover, and an Evapotranspiration cover. The RCRA Subtitle D cover had the highest percolation rate (above the 1 mm/yr goal), and the geosynthetic clay layer cover had the second highest average percolation rate. Various damage processes, such as desiccation cracking, led to preferential flow through the RCRA Subtitle D cover. The field data from this project was interpreted to suggest the geosynthetic clay layer cover performance was suspect in this application, potentially due to dessication and ion exchange. The other four cover types all had average annual fluxes less than 0.2 mm/yr (0.01 in/yr).

UMTRA disposal cell covers have been developed over the past 20 years at a variety of sites with different climates. From an erosion control and stability perspective, these covers have required little to no maintenance to prevent erosional release of radioactive materials. At some covers, DOE has removed vegetation, however these actions have primarily been undertaken because of concerns with the impact of the vegetation on water management (e.g., infiltration) and not due to concerns with cover stability. For example, at the Burrell, Pennsylvania, disposal cell, herbicide spraying was used to control plant encroachment (DOE, 1999). However, after additional data collection and analysis it was determined by DOE that this aspect of the long term surveillance and monitoring program was unnecessary and could be discontinued without negative impact on cover performance. DOE has instrumented some of the covers to understand and evaluate their performance. There have been lessons learned from the information that has been collected. In particular, some covers based on resistive type designs (e.g., impermeable layers) similar to unlined RCRA subtitle D have been found to have not achieved the design values for hydraulic conductivity measured in the laboratory and therefore appear to have much higher infiltration rates than anticipated (Waugh, 2004). In fact, at a number of sites the in-situ hydraulic conductivity was measured to be more than two orders of magnitude higher than the design target. However, monitoring data from evapotranspiration type covers suggests that design infiltration rates have been achieved. Lessons learned included that seemingly subtle differences in soil types, sources, and compaction can result in significant differences in performance.

The programs cited above and the documents listed below are not intended to be comprehensive, rather they are intended to provide appropriate examples of studies undertaken to understand and support engineered barrier performance. The examples provided above focused on engineered covers, which are only one type of barrier addressed in this guidance. The documents listed below provide information on a variety of different barriers.

Table 3.2   Summary of Existing Key Documents Related to Engineered Barriers

Document	Brief Summary
NUREG-1573, "A Performance Assessment Methodology for Low-Level Radioactive Waste Disposal Facilities, Recommendations of NRC's Performance Assessment Working Group," U.S. Nuclear Regulatory Commission, Washington, DC, October 2000.	Provides general information pertinent to modeling and assessment of engineered barriers. Provides a bibliography of reports related to engineered barriers.
NUREG/CR-5432, "Recommendations to the NRC for Soil Cover Systems Over Uranium Mill Tailings and Low-Level Radioactive Wastes — Identification and Ranking of Soils for Disposal Facility Covers," U.S. Nuclear Regulatory Commission, Washington, DC, February 1991.	Discusses (1) selecting soil materials, (2) laboratory and field tests for covers, and (3) construction methods.
NUREG/CR-5542, "Models for Estimation of Service Life of Concrete Barriers in Low-Level Radioactive Waste Disposal," U.S. Nuclear Regulatory Commission, Washington, DC, September 1990.	Provides primarily empirically based models for typical concrete formulations to estimate degradation rates.
NUREG-1623, "Design of Erosion Protection for Long-Term Stabilization," U.S. Nuclear Regulatory Commission, Washington, DC, September 2002.	Provides guidance on methods to achieve erosion controls for long-term stabilization. Provides a list of key references including the technical work supporting the guidance.
NUREG -1620, Rev. 1, "Standard Review Plan for the Review of a Reclamation Plan for Mill Tailings Sites Under Title II of the Uranium Mill Tailings Radiation Control Act," U.S. Nuclear Regulatory Commission, Washington, DC, June 2003.	Provides information regarding NRC staff areas of review and the bases for acceptability of a uranium mill reclamation design.
NUREG-1532, "Final Technical Evaluation Report for the Proposed Revised Reclamation Plan for the Atlas Corporation Moab Mill," U.S. Nuclear Regulatory Commission, Washington, DC, March 1997.	Section 4 provides an example of the staff review of a reclamation design and discusses staff bases for acceptability of rock riprap erosion protection and input parameters used for those designs.

Table 3.2   Summary of Existing Key Documents Related to Engineered Barriers (continued)

Document	Brief Summary
NISTIR 89-4086, NUREG/CR-5466, "Service Life of Concrete," National Institute of Standards and Technology (NIST) Gaithersburg, MD, 1995.	Examines degradation processes in cement-based materials and discusses considerations of their occurrence, extent of potential damage, and mechanisms.
NISTIR 7026, "Condition Assessment of Concrete Nuclear Structures Considered for Entombment," National Institute of Standards and Technology (NIST), Gaithersburg, MD, 2003.	Provides assessment of cement-based engineered barrier structures based on characterization of intact concrete and crack properties. Material property uncertainties are incorporated into a Monte Carlo simulation.
NISTIR 6747, "Validation and Modification of the 4SIGHT Computer Program" National Institute of Standards and Technology (NIST) Gaithersburg, MD, 2001.	Discusses the validation and verification of the fluid transport mechanisms incorporated in the concrete degradation code 4SIGHT using reference and laboratory data.
NISTIR 6519, "Effect of Drying Shrinkage Cracks and Flexural Cracks on Concrete Bulk Permeability," National Institute of Standards and Technology (NIST) Gaithersburg, MD, 2000.	Discusses a model for predicting both the width and spacing of flexural and drying-shrinkage cracks to estimate composite (intact and cracked) concrete structure permeability.
NISTIR 5612, "4SIGHT, Manual: A Computer Program for Modeling Degradation of Underground LLW Concrete Vaults, " National Institute of Standards and Technology (NIST) Gaithersburg, MD, 1995.	User Manual for numerical computer modeling of concrete degradation, 4SIGHT, to facilitate assessment of concrete vaults for isolating radioactive waste in Low Level Waste (LLW) disposal applications.
"Barrier Containment Technologies for Environmental Remediation Applications," edited by Ralph R. Rumer and Michael E. Ryan, John Wiley and Sons, 1995.	Review and evaluation of knowledge and practices of containment technologies suitable for remediation. Identifies areas where practical improvements could be developed. NOTE: It is expected that this document will be superseded by a more recent document on waste containment practices to be published in Fall 2005.

Table 3.2   Summary of Existing Key Documents Related to Engineered Barriers (continued)

Document	Brief Summary
National Research Council, National Academy of Sciences, "Barrier Technologies for Environmental Management," Summary of a Workshop, 1997.	Papers presented in the Workshop on the use of Engineered Barriers to prevent the spread of contaminants and its migration.
"Field Water Balance of Landfill Final Covers," Albright, W, Benson, C., Gee, G., Roesler, A., Abichou, T., Apiwantragon, P., Lyles, B., and Rock, S., Journal of Environmental Quality, 33(6), 2317-2332, 2004.	Results of large-scale field research study to assess the ability of landfill final covers to control infiltration into underlying waste. A comprehensive current publication summarizing ACAP experience.
"Assessment and Recommendations for Improving the Performance of Waste Containment Systems," U.S. EPA, EPA/600/R-02/099, 2002.	Discusses issues related to the design, construction and performance of waste containment systems used in landfills, surface impoundments and waste piles and in the remediation of contaminated sites.
National Research Council, National Academy of Sciences, "Research Needs in Subsurface Science," 2000.	Examines gaps in the understanding of the performance of subsurface facilities and recommends research needs in the area.
Waugh, W.J., "Design, Performance, and Sustainability of Engineered Covers for Uranium Mill Tailings," Proceedings of Long-term Performance Monitoring of Metals and Radionuclides in the Subsurface: Strategies, Tools, and Case Studies. U.S. Environmental Protection Agency, U.S. Department of Energy, U.S. Geological Survey, Nuclear Regulatory Commission, April 21-22, 2004, Reston, VA, 2004.	Provides information on experiences with cover designs for DOE's UMTRA Project sites of conventional and alternative covers.
Dwyer, Stephen F., "Water Balance Measurements and Computer Simulations of Landfill Covers," PhD Dissertation, University of New Mexico, 2003.	Provides a comprehensive summary of data collection, analysis, and computer simulations associated with DOE's ALCD program. Also includes a summary of measurements of infiltration at various sites with engineered covers.

Table 3.2   Summary of Existing Key Documents Related to Engineered Barriers (continued)

Document	Brief Summary
O'Donnell, E., R. Ridky, and R. Schulz. "Control of water infiltration into near-surface, low-level waste-disposal units in humid regions," In-situ Remediation: Scientific Basis for Current and Future Technologies, G. Gee and N.R. Wing eds., Battelle Press, Columbus, OH, 295-324, 1994.	Summary of NRC sponsored research at USDA, Beltsville, MD, on engineered covers for low-level waste facilities.
Interstate Technology & Regulatory Council, "Technical and Regulatory Guidance for Design, Installation, and Monitoring of Alternative Final Landfill Covers," Washington, DC, 2003.	Guidance document primarily written for decision makers associated with the plan development, review, and implementation of alternative covers. Focuses on the decisions and facilitating the decision processes related to the design, evaluation, construction, and post-closure care associated with alternative covers.
Interstate Technology & Regulatory Council, "Permeable Reactive Barriers: Lessons Learned/New Directions," Washington, DC, 2005.	Summary of current understanding and experience with permeable reactive barriers, including numerous case studies.
National Research Council, National Academy of Sciences, "Long-Term Institutional Management of U.S. DOE Legacy Waste Sites," 2000.	Discusses long-term management of DOE's waste sites and identifies characteristics and design criteria for effective long-term institutional management.

Appendix P provides an example of a graded approach to the development and implementation of erosion protection covers. The graded approach has been developed to ensure stability of waste for extended periods of time. In other words, by ensuring stability of the waste and designing for the largest event expected over the analysis period, risk from erosional releases can be mitigated. The graded approach is used to manage the uncertainty associated with the timing and magnitude of future events that may cause erosional releases. A risk-informed approach for the development of erosion protection covers would be based on the risk associated with erosional releases and not a stability criteria. In the risk-informed approach, the erosion protection cover may not be designed to achieve stability of the system for the largest event. Rather, the erosion protection cover may be designed to achieve radiological dose criteria, given the timing and magnitude of future events. The risk-informed approach would integrate the impact of all event sizes on the erosion control design and evaluate the resultant radiological doses.

4 FACILITY RADIATION SURVEYS

RADIATION SURVEY AND SITE INVESTIGATION PROCESS

As a framework for collecting the information required for demonstrating compliance identified using the data quality objectives (DQO) process (see Section 3.2 of this Volume), NRC staff recommends using a series of surveys. The radiation site survey and investigation (RSSI) process is an example of a series of surveys designed to demonstrate compliance with the decommissioning regulations of 10 CFR Part 20, Subpart E. Table 4.1 identifies the steps in the RSSI process and indicates where specific guidance on each step can be found.

Table 4.1   Cross-References for Principal Steps in the Radiation Survey and Site Investigation Process

Principal Step	Applicable Guidance
Site Identification	Chapter 16, Volume 1 of this NUREG report Section 2.4 of MARSSIM
Historical Site Assessment	Section 4.0 of this volume Sections 2.4 and Chapter 3 of MARSSIM
Characterization Survey	Sections 2.4 and 5.3 of MARSSIM Section 4.2 of this volume
Remedial Action Support Survey	Sections 2.4 and 5.4 of MARSSIM Section 4.3 of this volume
Final Status Survey	Sections 2.4 and 5.5 of MARSSIM Section 4.4 of this volume

HISTORICAL SITE ASSESSMENT

The RSSI process uses a graded approach that starts with the Historical Site Assessment (HSA) and is later followed by other surveys that lead to the final status survey (FSS). The HSA is an investigation to collect existing information describing a site's complete history from the start of site activities to the present time. The necessity for detailed information and amount of effort to conduct an HSA depends on the type of site, associated historical events, regulatory framework, and availability of documented information. The main purpose of the HSA is to determine the current status of the site or facility, but the data collected may also be used to differentiate sites that need further action from those that pose little or no threat to human health and the environment. This screening process can serve to provide a site disposition recommendation or to recommend additional surveys. Because much of the data collected during HSA activities is qualitative or is analytical data of unknown quality, many decisions regarding a site are the result of professional judgment.

The primary objectives of the HSA include the following:

- identify potential sources of residual radioactivity,

- determine whether or not sites pose a threat to human health and the environment,

- differentiate impacted from non-impacted areas,

- provide input to scoping and characterization survey designs,

- provide an assessment of the likelihood of residual radioactivity migration, and

- identify additional potential radiation sites related to the site being investigated.

The HSA typically consists of three phases: (1) identification of a candidate site, (2) preliminary investigation of the facility or site, and (3) site visits or inspections. The HSA is followed by an evaluation of the site based on information collected during the HSA. Additionally, the HSA should identify special survey situations that may need to be addressed such as subsurface radioactivity; sewer systems, waste plumbing, and floor drains; ventilation ducts; and embedded piping containing residual radioactivity. Refer to Appendix G of this volume for information on special survey situations. Additional guidance on the HSA can be found in Section 2.4.2 and Chapter 3 of the Multi-Agency Radiation Survey and Site Investigation Manual (MARSSIM).

SUMMARY OF SURVEY TYPES

NRC's regulations require a licensee to make or cause to be made surveys that may be necessary for the licensee to comply with the radiological criteria for license termination, Subpart E of 10 CFR Part 20. The licensee would demonstrate compliance with this requirement by performing an FSS. The FSS will demonstrate that the licensee's site or facility, or both meet(s) the radiological criteria for license termination.

Other surveys (e.g., scoping surveys, characterization surveys, and remedial action support surveys) are used for the purpose of locating residual radioactivity, but are not used to demonstrate compliance with the radiological criteria for license termination.

NRC endorses the final status survey methodology described in MARSSIM. The guidance in this chapter does not replace MARSSIM and users of this chapter should be familiar with and use MARSSIM. Thus, it is intended that licensees will use this chapter and MARSSIM as guidance for acceptable approaches or methodologies to conduct remediation surveys and FSSes in particular. The following sections provide references to specific sections of MARSSIM.

The measurement methods applied in assessing radiation and radioactivity levels can vary according to the objectives of the particular survey. It is expected that different types of surveys would be conducted during the course of decommissioning work, with each having different emphasis while at the same time sharing common elements. A brief summary of six survey types is provided below:

Background Survey

Although, not specifically identified as a step in the RSSI process, this survey constitutes measurements of sites in areas surrounding the facility in order to establish the baseline, that is, the normal background levels of radiation and radioactivity. In some situations, historical measurements may be available from surveys performed before the construction and operation of a facility. The background survey takes on added importance if one may ultimately be comparing onsite cleanup units to offsite reference areas. Appendix A of this volume provides guidance on background surveys.

Scoping Survey

This survey, performed to augment the HSA, provides sufficient information for (a) determination if residual radioactivity is present that warrants further evaluation and (b) initial estimates of the level of effort required for remediation and to prepare a plan for a more detailed survey, such as a characterization survey. The scoping survey does not require that all radiological parameters be assessed. Additional guidance on the scoping survey can be found in Sections 2.4 and 5.2 of MARSSIM, and Section 4.2 of this volume.

Characterization Survey

This survey determines the type and extent of residual radioactivity on or in structures, residues, and environmental media. The survey should be sufficiently detailed to provide data for planning decommissioning actions, including remediation techniques, projected schedules, costs, waste volumes, and health and safety considerations during remediation. Additional guidance on characterization surveys can be found in Section 4.2 of this volume.

Remedial Action Support Survey

This monitoring program is conducted in what is effectively a real time mode to guide cleanup efforts and ensure the health and safety of workers and the public. The effectiveness of the remediation efforts as they progress can be assessed. The precision and accuracy of measurements associated with this type of survey are generally not sufficient to determine the final radiological status of the site. Additional guidance on remedial action support surveys can be found in Section 4.3 of this volume.

Final Status Survey

This survey demonstrates that residual radiological conditions satisfy the predetermined criteria for release for unrestricted use or, where appropriate, for use with designated restrictions. It is this survey that provides data to demonstrate that all radiological parameters (e.g., total surface activity, removable surface activity, exposure rate, and radionuclide concentrations in soil and other materials) satisfy the established guidelines and conditions. Additional guidance on final status surveys (FFSes) can be found in Section 4.4 of this volume.

Confirmatory Survey

This survey is performed by the regulator to provide data to substantiate the results of the licensee's FSS. The objective of this type of survey is to verify that characterization, remediation, and final status actions and documentation, conducted as part of the RSSI process, are adequate to demonstrate that the site is radiologically acceptable, relative to applicable criteria. Section 15.4.5 of Volume 1 of this NUREG report provides additional information on confirmatory surveys.

These types of surveys are performed at various stages of the decommissioning process. Early on, and where known residual radioactivity exists, the simplest of measurement approaches can be used to document the need for a specific building surface or parcel of land to be cleaned up. In practice, the simpler methods would generally be applicable to the scoping and remediation control surveys. The more complex methods which produce data with higher precision and accuracy will be required for background, characterization, final status, and confirmatory surveys. In general, wherever measurements are to be performed at or close to background levels, greater sensitivity in the measurement is required.

The conduct of these surveys and the methods applied have some interchangeable elements. It is possible that measurements collected in one survey can be used for another. For instance, if measurements sufficient in spatial coverage and with adequate detection limits were taken, the results of the scoping survey in an unaffected area could be used to support the FSS. The emphasis of the guidance in this volume is on the methodology that can be applied to meet the requirements of the FSS, although they can be applied to other survey work as well.

In late 2004, NRC released the final version of the Multi-Agency Radiological Laboratory Analytical Protocols (MARLAP). MARLAP and MARSSIM are complementary guidance documents in support of cleanup and decommissioning activities. MARSSIM provides guidance on how to plan and carry out a study to demonstrate that a site meets appropriate release criteria. It describes a methodology for planning, conducting, evaluating, and documenting environmental radiation surveys conducted to demonstrate compliance with cleanup criteria. MARLAP provides guidance and a framework for both project planners and laboratory personnel to ensure that radioanalytical data will meet the needs and requirements of cleanup and decommissioning activities.

MARLAP recommends the use of a directed or systematic planning process. A directed planning process is an approach for setting well-defined, achievable objectives and developing a cost effective, technically sound sampling and analysis design that balances the data user's tolerance for uncertainty in the decision process with the resources available for obtaining data to support a decision. For example, NRC and licensees have determined that side-by-side surveys (with subsequent partial site releases) are more efficient then waiting for a final site-wide confirmatory survey (See Appendix O, lesson 4 on inspections). NRC and licensees should plan ahead and coordinate their schedules in order to implement efficient side-by-side confirmatory surveys. See Appendix D for more details on MARLAP and how it can enhance radiation monitoring.

Refer to Appendix D of this volume for information on survey data quality and reporting, Chapter 5 of MARSSIM for survey checklists, Appendix E for information on survey measurements, and Appendix G for information on special survey issues. Also refer to Appendix O for related information on lessons learned from recently submitted DPs and questions and answers to clarify existing license termination guidance.

AREAS OF REVIEW

NRC staff should review the radiological characterization survey results to determine whether the characterization survey provides sufficient information to permit planning for site remediation that will be effective and will not endanger the remediation workers, to demonstrate that it is unlikely that significant quantities of residual radioactivity have gone undetected, and to provide information that will be used to design the FSS.

NRC staff should review the FSS design to determine whether the survey design is adequate for demonstrating compliance with the radiological criteria for license termination.

NRC staff should review the results of the FSS to determine whether the survey demonstrates that the site, area, or building meets the radiological criteria for license termination.

NRC staff should note that NRC regulations require that DPs include a description of the planned final radiological survey. Recognizing the flexible approach discussed in Section 2.2 of this volume and that the MARSSIM approach allows certain information needed to develop the final radiological survey to be obtained as part of the remedial activities at the site, a licensee or responsible party may submit information on facility radiation surveys in one of two ways, as summarized below. Section 2.2 of this volume provides additional relevant guidance.

- Method 1:
  The licensee or responsible party may submit the information contained in Sections 4.1–4.3 of this volume of this NUREG as part of the DP, along with a commitment to use the MARSSIM approach in developing the final radiological survey. The information discussed in Section 4.4 would then be submitted by the licensee or responsible party at the completion of remediation or when the licensee or responsible party has completed developing the design of the final radiological survey for the site. The final status survey report (FSSR) (Section 4.5) will be submitted after the licensee or responsible party has performed the final radiological survey.

- Method 2:
  The licensee or responsible party may submit the information contained in Sections 4.1–4.4 of this volume along with a commitment to calculate the number of sampling points that will be used in the final radiological survey in accordance with the procedure described in MARSSIM. The FSSR (Section 4.5) would then be submitted after the licensee or responsible party has performed the final radiological survey. If this method is used, the licensee or responsible party should include in the FSSR the information contained in the last three bullets under "Information to be Submitted," in Section 4.4 of this chapter.

Acceptance Review

NRC staff should ensure that the licensee's submittal contains the information summarized under the above "Areas of Review," as appropriate for the particular submittal. NRC staff should review the information submitted to ensure that the level of detail appears to be adequate for the staff to perform a detailed technical review, but NRC staff should not review the technical adequacy of the information. The adequacy of this information should be assessed during the detailed review.

Safety Evaluation

The material to be reviewed is both informational in nature and requires specific detailed technical analysis. NRC staff should verify that the survey designs and results are adequate for demonstrating compliance with the radiological criteria for license termination.

4.1   RELEASE CRITERIA

NRC staff review of the release criteria is to verify that the licensee has provided appropriate release criteria, referred to as the derived concentration guideline levels. Generally the licensee should provide the $DCGL_W$, for the survey unit average concentrations, and the applicable $DCGL_{EMC}$, for small areas of elevated concentrations, for all impacted media.

ACCEPTANCE CRITERIA

Regulatory Requirements

10 CFR 20.1402, 20.1403, and 20.1404

Regulatory Guidance

NUREG–1575, "Multi-Agency Radiological Survey and Site Investigation Manual" (MARSSIM)

Information to be Submitted

The licensee should list the DCGL(s) that will be used to design the surveys and to demonstrate compliance with the radiological criteria for release, including:

- a summary table or list of the $DCGL_W$ for each radionuclide and impacted medium of concern;

- a summary table or list of area factors that will be used for determining a $DCGL_{EMC}$ for each radionuclide and media of concern if Class 1 (refer to Appendix A.1 of this volume for classification of site areas) survey units are present;

- the $DCGL_{EMC}$ for each radionuclide and medium of concern if Class 1 survey units are present; and

- the appropriate $DCGL_W$ for the survey method to be used if multiple radionuclides are present.

This information to be submitted is also included as part of the master DP Checklist provided in this NUREG report (see Section XIV.a from Appendix D of Volume 1).

EVALUATION CRITERIA

NRC staff should verify that, for each radionuclide and impacted media of concern, the licensee has provided a $DCGL_W$ and, if Class 1 survey units are present, a table of area factors. NRC staff should verify that the values presented are consistent with the values developed pursuant to the dose modeling, as discussed in Chapter 5 of this volume. If multiple radionuclides are present, MARSSIM Sections 4.3.2, 4.3.3, and 4.3.4 describe acceptable methods to determine DCGLs appropriate for the survey technique.

4.2 SCOPING AND CHARACTERIZATION SURVEYS

SCOPING SURVEYS

Early in the decommissioning process, it is necessary to identify the potential residual radioactivity present at the site, the relative ratios of these nuclides, and the general extent of residual radioactivity—if any—both in activity levels and affected area or volume. Although the license and operational history documentation will assist to varying degrees in providing this information, it will often be necessary to supplement that information with actual survey data. A scoping survey therefore is performed. The scoping survey typically consists of limited direct measurements (exposure rates and surface activity levels) and samples (smears, soil, water, and material with induced activity) obtained (a) from site locations considered to be the most likely to contain residual activity and (b) from other site locations, including immediately adjacent to the radioactive materials use areas. This survey provides a preliminary assessment of site conditions, relative to guideline values. The scoping survey provides the basis for initial estimates of the level of effort required for decommissioning and for planning the characterization survey.

Measurements and sampling in known areas of residual radioactivity need not be as comprehensive or be performed to the same sensitivity level as will be required for the characterization or FSSes. However, when planning and conducting this scoping survey, the licensee should remember that some of the data, particularly that from locations not affected by site operations, may be used as final status results or to supplement the characterization or final survey results, or both. Similar measuring and sampling techniques as used for those categories of surveys therefore may be warranted.

Scoping surveys provide site-specific information based on limited measurements. The following are the primary objectives of a scoping survey:

- perform a preliminary hazard assessment,

- support classification of all or part of the site as a Class 3 area,

- evaluate whether the survey plan can be optimized for use in either the characterization or final stage,

- perform status surveys,

- provide data to address the requirements of other applicable regulations, and

- provide input to the characterization survey design if necessary.

Scoping surveys are conducted after the HSA is completed and consist of judgment measurements based on the HSA data. If the results of the HSA indicate that an area is Class 3 and no residual radioactivity is found, the area may be classified as Class 3, and a Class 3 final status survey is performed. If the scoping survey locates residual radioactivity, the area may be considered as a Class 1 (or Class 2) area for the FSS and a characterization survey is typically performed. Sufficient information should be collected to identify situations that require immediate radiological attention. Licensees should be aware that potential requirements of other applicable regulations (e.g., nonradiological constituents) may differ from NRC requirements. A comparison of MARSSIM guidance to some other requirements is provided in Appendix F of MARSSIM.

CHARACTERIZATION SURVEYS

After locations that are impacted have been identified, a characterization survey is performed to more precisely define the extent and magnitude of residual radioactivity. The characterization survey should be in sufficient detail to provide data for planning the remediation effort, including the remediation techniques, schedules, costs, and waste volumes and necessary health and safety considerations during remediation. The type of information obtained from a characterization survey is often limited to that which is necessary to differentiate a surface or area as containing or not containing residual radioactivity. A high degree of accuracy may not be required for such a decision when the data indicate levels well above the guidelines. On the other hand, when data are near the guideline values, a higher degree of accuracy is usually necessary to assure the appropriate decision regarding the true radiological conditions. In addition, one category of radiological data, to include soil radionuclide concentration or total surface activity, may be sufficient to determine the status as containing residual radioactivity, and other measurements (e.g., exposure rates or removable residual radioactivity levels) may therefore not be performed during characterization. As the scoping survey example demonstrates, the choice of survey technique should be commensurate with the intended use of the data, including considerations for possible future use of the results to supplement the FSS data.

Licensees typically submit site characterization information as part of their DP. However, submission of incomplete site characterization information may result in NRC declining to accept and review the DP until appropriate site characterization information is obtained. The licensee may be requested to submit Site Characterization Plans (SCPs) or other site characterization information prior to submitting the DP or NRC may elect to meet with the licensee prior to, or during, site characterization work. However, it is important to note that, unless required by a license condition, licensees are not required under NRC regulations to submit a separate SCP or Site Characterization Report (SCR), only that site characterization information is required as a component of the DP. So, NRC staff will only request this information when necessary to ensure safety and compliance with NRC regulations.

The characterization survey is generally the most comprehensive of all the survey types and generates the most data. This includes preparing a reference grid, systematic as well as judgment measurements, and surveys of different media to include surface soils, interior and exterior surfaces of buildings. Additionally, the characterization survey should identify all activated materials (typically Decommissioning Groups 4–7) and hard-to-detect radionuclides throughout the site. The decision as to which media will be surveyed is a site-specific decision addressed throughout the RSSI process (see MARSSIM).

Characterization surveys may be performed to satisfy a number of specific objectives. Examples of characterization survey objectives include the following:

- determining the nature and extent of residual radioactivity;

- evaluating remediation alternatives (e.g., unrestricted use, restricted use, onsite disposal, offsite disposal);

- developing input to pathway analysis/dose or risk assessment models for determining site-specific DCGLs (Bq/kg (pCi/g), Bq/m^2 ($dpm/100cm^2$));

- estimating the occupational and public health and safety impacts during decommissioning;

- evaluating remediation technologies;

- developing input to the FSS design; and

- complying with requirements of other applicable regulations.

The scope of this volume precludes detailed discussions of characterization survey design for each of these objectives, and therefore, the user should consult other references for specific characterization survey objectives not covered. For example, the Decommissioning Handbook (DOE 1994) is a good reference for characterization objectives that are concerned with evaluating remediation technologies or unrestricted/restricted use alternatives. Other references (EPA 1988b, 1988c, 1994a; NUREG–1501) should be consulted for planning decommissioning actions, including remediation techniques, projected schedules, costs, and waste volumes, and health and safety considerations during remediation. Also, the types of characterization data needed to support risk or dose modeling should be determined from the specific modeling code documentation.

AREAS OF REVIEW

The purpose of NRC staff review is to verify that the licensee determined the radiological condition of the property well enough to permit planning for a remediation that will be effective and will not endanger the remediation workers, to demonstrate that it is unlikely that significant quantities of residual radioactivity have gone undetected, and to provide sufficient information for designing the FSS. Note that some licensees have used, or may request authorization to use, information developed during the characterization survey to support the final radiological survey.

Licensees may use characterization survey data to support the final radiological survey, as long as they can demonstrate that non-impacted areas at the site have not been adversely impacted by decommissioning operations, and the characterization survey data are of sufficient scope and detail to meet the "Information to be Submitted" guidance for a final survey.

ACCEPTANCE CRITERIA

Regulatory Requirements

10 CFR 30.36(g)(4)(i), 40.42(g)(4)(i), 70.38(g)(4)(i), and 72.54(g)(1)

Regulatory Guidance

NUREG–1575, "Multi-Agency Radiological Survey and Site Investigation Manual" (MARSSIM)

Information to be Submitted

The information supplied by the licensee should be sufficient to allow NRC staff to determine that the characterization survey design is adequate to determine the radiological status of the facility. The licensee should describe the radiation characterization survey design and the results of the survey including:

- a description and justification of the survey measurements for impacted media (for example, building surfaces, building volumetric, surface soils, subsurface soils, surface water, ground water, sediments, etc., as appropriate);

- a description of the field instruments and methods that were used for measuring concentrations and the sensitivities of those instruments and methods;

- a description of the laboratory instruments and methods that were used for measuring concentrations and the sensitivities of those instruments and methods;

- the survey results including tables or charts of the concentrations of residual radioactivity measured;

- maps or drawings of the site, area, or building showing areas classified as non-impacted or impacted and visually summarizing residual radioactivity concentrations in impacted areas;

- the justification for considering areas to be non-impacted;

- a discussion of why the licensee considers the characterization survey to be adequate to demonstrate that it is unlikely that significant quantities of residual radioactivity have gone undetected;

- a discussion of how they were surveyed or why they did not need to be surveyed for areas and surfaces that were considered to be inaccessible or not readily accessible; and

- for sites, areas, or buildings with multiple radionuclides, a discussion justifying the ratios of radionuclides that will be assumed in the FSS or an indication that no fixed ratio exists and each radionuclide will be measured separately (note that this information may be developed and refined during decommissioning and licensees may elect to include a plan to develop and justify final radionuclide ratios in the DP).

This information to be submitted is also included as part of the DP Checklist provided in this NUREG report (see Section XIV.b from Appendix D of Volume 1). For additional information about the characterization of material and components that will be removed prior to license termination and about characterization for initial classification of areas as Class 1, refer to Questions 6 and 7 from Section O.1 of Appendix O of this volume.

Licensees should note that if they elect to dispose of buildings and structures rather then leave the buildings and structures in place (for unrestricted release), the LTR does not apply to the material moved offsite from those buildings and structures. Rather, building and structure deconstruction and dismantlement materials can be released from the site in accordance with existing license conditions. The data from the characterization survey may be sufficient to demonstrate compliance with the conditions of the existing license for releasing material from the site. However, a characterization survey may not be required to demonstrate compliance with the license condition for releasing material from the site. For additional guidance on offsite disposition of materials, refer to Section G.1.1 of Appendix G of this volume.

EVALUATION CRITERIA

NRC staff should verify that the licensee has adequately characterized the site, area, or building relative to the location and extent of residual radioactivity. An adequate characterization is one which permits planning for a remediation that will be effective and will not endanger the remediation workers, demonstrates that it is unlikely that significant quantities of residual radioactivity have gone undetected, and provides information that will be used to design the FSS. The extent of detail in the information provided by the licensee should be appropriate for the specific site, area, or building.

NRC staff should verify that the characterization survey design and results demonstrate that the licensee or responsible party has adequately characterized the site. The characterization survey is adequate if it meets the criteria in the following guidance:

- Section 5.3 of MARSSIM for characterization survey (NRC staff may use the "Example Characterization Survey Checklist" in Section 5.3 of MARSSIM for evaluating the licensee's submittal);

- MARSSIM Chapter 6 and Appendix E for instrument capabilities and sensitivities; and

- MARSSIM Section 4.8.4 for the preparation of areas for surveys.

4.3   REMEDIAL ACTION SUPPORT SURVEYS

The effectiveness of remediation efforts in reducing residual radioactivity to acceptable levels is monitored by a remedial action support survey as the remediation effort is in progress. This type of survey activity guides the cleanup in a real-time mode; it also assures that the remediation workers, the public, and the environment are adequately protected against exposures to radiation and radioactive materials arising from the remediation activities.

The remedial action support survey typically provides a simple radiological parameter such as direct radiation near the surface being remediated. The level of radiation, below which there is reasonable assurance that the guideline values have been attained, is determined and used for immediate, infield decisions. Such a survey is intended for expediency and does not provide thorough or accurate data describing the final radiological status of the site.

The remedial action support survey is applicable to monitoring of surfaces and soils or other bulk materials only if the radionuclides of concern are detectable by field survey techniques. For radionuclides and media which cannot be evaluated at guideline values by field procedures, samples are to be collected and analyzed to evaluate effectiveness of remediation efforts. For large projects, use of mobile field laboratories can provide more timely decisions regarding the effectiveness of remedial actions. Examples of situations for which remedial action support surveys would not be practicable are (a) when soil contains pure alpha or beta emitting radionuclides and (b) when very low energy beta emitters such as H-3 are present on surfaces.

Remedial action support surveys are conducted to:

- support remediation activities,

- determine when a site or survey unit is ready for the FSS, and

- provide updated estimates of site-specific parameters used for planning the FSS.

The determination that a survey unit is ready for an FSS following remediation is an important step in the RSSI Process. Remedial activities may result in changes to the distribution of residual radioactivity within the survey unit. Thus, for many survey units, the site-specific parameters used during FSS planning (e.g., variability in the radionuclide concentration, probability of small areas of elevated activity) may need to be confirmed or re-established following remediation. Obtaining updated values for these critical parameters should be considered when planning a remedial action support survey. In some cases, concentrations of some radionuclides after remediation may be very low. In such cases, it may be useful for licensees to show that certain radionuclides can be considered insignificant; in which case, further detailed evaluation as part of the FSS may not be necessary (see Section 3.3 of this volume). However, the dose from the insignificant radionuclides must be accounted for in demonstrating compliance with the applicable dose criteria.

Note that this survey does not provide information that can be used to demonstrate compliance with the DCGLs and is an interim step in the compliance demonstration process. Areas that are likely to satisfy the DCGLs on the basis of the remedial action support survey will then be surveyed in detail by the FSS. Alternatively, the remedial action support survey can be designed to meet the objectives of an FSS. DCGLs may be recalculated based on the results of the remediation process as the regulatory program allows or permits.

AREAS OF REVIEW

The purpose of the review of the description of the remedial action support surveys is to verify that the licensee has designed these surveys appropriately and to assist the licensee in determining when remedial actions have been successful and that the FSS may commence. In addition, information from these surveys may be used to provide the principal estimate of residual radioactivity variability that will be used to calculate the FSS sample size in a remediated survey unit.

ACCEPTANCE CRITERIA

Regulatory Requirements

10 CFR 30.36(g)(4)(ii), 40.42(g)(4)(ii), and 70.38(g)(4)(ii),

Regulatory Guidance

NUREG–1575, "Multi-Agency Radiological Survey and Site Investigation Manual" (MARSSIM)

Information to be Submitted

NRC staff should verify that included in the licensee's or responsible party's description of the support survey is the following information:

- a description of field screening methods and instrumentation, and

- a demonstration that field screening should be capable of detecting residual radioactivity at the $DCGL_W$.

This information to be submitted is also included as part of the DP Checklist provided in this NUREG report (see Section XIV.c from Appendix D of Volume 1).

EVALUATION CRITERIA

NRC staff should verify that the description of the remedial action support surveys meets (a) the criteria in MARSSIM Section 5.4 for performing remedial action support surveys and (b) the criteria in the applicable MARSSIM chapters listed in this volume for the evaluation of technical issues such as appropriate surveys instruments, and survey instrument sensitivity.

4.4   FINAL STATUS SURVEY DESIGN

Professional judgment and biased sampling are important for locating residual radioactivity and characterizing the extent of residual radioactivity at a site. However, the MARSSIM focus is on planning the FSS which utilizes a systematic approach to sampling. Systematic sampling is based on rules that endeavor to achieve the representativeness in sampling consistent with the application of statistical tests.

The FSS is used to demonstrate compliance with regulations. The primary objectives of the FSS are to perform the following:

- verify survey unit classification,

- demonstrate that the potential dose from residual radioactivity is below the release criterion for each survey unit, and

- demonstrate that the potential dose from small areas of elevated activity is below the release criterion for each survey unit.

Data provided by the FSS can demonstrate that all radiological parameters satisfy the established guideline values and conditions.

AREAS OF REVIEW

The purpose of NRC staff's review is to verify that the design of the FSS is adequate to demonstrate compliance with the radiological criteria for license termination.

ACCEPTANCE CRITERIA

Regulatory Requirements

10 CFR 20.1501(a), 30.36(g)(4)(iv), 40.42(g)(4)(iv), 70.38(g)(4)(iv), and 72.54(g)(4)

Regulatory Guidance

- Draft NUREG–1505, "A Nonparametric Statistical Methodology for the Design and Analysis of Final Status Decommissioning Surveys"

- NUREG–1575, "Multi-Agency Radiological Survey and Site Investigation Manual" (MARSSIM)

- NUREG–1507, "Minimum Detectable Concentrations with Typical Survey Instruments for Various Contaminants and Field Conditions"

Information to be Submitted

The information supplied by the licensee should be sufficient to allow NRC staff to determine that the FSS design is adequate to demonstrate compliance with the radiological criteria for license termination. The information should include all of the following:

- a brief overview describing the FSS design;

- a description and map or drawing of impacted areas of the site, area, or building classified by residual radioactivity levels (Class 1, 2, or 3) and divided into survey units, with an explanation of the basis for division into survey units (maps should have compass headings indicated);

- a description of the background reference areas and materials, if they will be used, and a justification for their selection;

- a summary of the statistical tests that will be used to evaluate the survey results, including the elevated measurement comparison, if Class 1 survey units are present; a justification for any test methods not included in MARSSIM; and the values for the decision errors (α and β) with a justification for α values greater than 0.05;

- a description of scanning instruments, methods, calibration, operational checks, coverage, and sensitivity for each media and radionuclide;

- a description of the instruments, calibration, operational checks, sensitivity, and sampling methods for *in situ* sample measurements, with a demonstration that the instruments and methods have adequate sensitivity;

- a description of the analytical instruments for measuring samples in the laboratory, including the calibration, sensitivity, and methodology for evaluation, with a demonstration that the instruments and methods have adequate sensitivity;

- a description of how the samples to be analyzed in the laboratory will be collected, controlled, and handled; and

- a description of the FSS investigation levels and how they were determined.

This information to be submitted is also included as part of the DP Checklist provided in this NUREG report (see Section XIV.d from Appendix D of Volume 1). For additional information about demonstrating appropriate selection of survey instrumentation, refer to Question 5 from Section O.1 of Appendix O.

EVALUATION CRITERIA

NRC staff review should verify that the FSS design is adequate to demonstrate compliance with the radiological criteria for license termination. The FSS design is adequate if it meets the criteria in the following guidance:

- Appendix A of this volume, for general guidance on implementing the MARSSIM approach for conducting FSSes;

- Appendix B of this volume, for guidance on alternative methods of FSS for simple situations;

- MARSSIM Sections 4.4 and 4.6 for classifying areas by residual radioactivity levels and dividing areas into survey units of acceptable size;

- MARSSIM Section 4.5 for methods to select background reference areas and materials;

- NUREG–1505, Chapter 13, for a method to account for differences in background concentrations between different reference areas;

- MARSSIM Section 5.5.2 for statistical tests;

- Appendix A of this volume, Section A.7.2 for decision errors;

- MARSSIM Sections 6.5.3 and 6.5.4 for selection of acceptable survey instruments, calibration, and operational checkout methods;

- MARSSIM Section 6.7 for methods to determine measurement sensitivity;

- NUREG–1507 for instrument sensitivity information;

- MARSSIM Sections 5.5.2.4, 5.5.2.5, 5.5.3, 7.5, and 7.6 for scanning and sampling;

- MARSSIM Section 7.7 for sample analytical methods (Table 7.2 of Section 7.7 provides acceptable analytical procedural references);

- MARSSIM Sections 7.5 and 7.6 for methods for sample collection;

- MARSSIM Section 5.5.2.6 for survey investigation levels; and

- Appendix G of this volume for surveys for special structural or land situations.

4.5   FINAL STATUS SURVEY REPORT

To the extent possible, the FSSR should stand on its own with minimal information incorporated by reference. Although the FSS is discussed as if it were an activity performed at a single stage of the site investigation process, this does not have to be the case. Data from other surveys conducted during the RSSI Process—such as scoping, characterization, and remedial action support surveys—can provide valuable information for an FSS, provided the data are of sufficient quality.

4.5.1   AREAS OF REVIEW

The purpose of NRC staff review is to verify that the results of the FSS demonstrate that the site, area, or building meets the radiological criteria for license termination. For licensees who have submitted a DP, the FSSR need only include the information described in Section 4.5.2. A licensee who has not submitted a DP should consult with NRC staff to assure its FSSR includes not only the information below, but also any other relevant information the staff needs to carry out its review.

4.5.2   ACCEPTANCE CRITERIA

Regulatory Requirements

10 CFR 20.1402, 20.1403, 20.1501, 30.36(j)(2), 40.42(j)(2), 70.38(j)(2), and 72.54(i)(2)

Regulatory Guidance

NUREG–1575, "Multi-Agency Radiological Survey and Site Investigation Manual" (MARSSIM)

Information to be Submitted

The information submitted by the licensee should be sufficient to allow the staff to determine that the site, area, or building meets the radiological criteria for license termination. The information should include:

- An overview of the results of the FSS.

- A summary of the DCGLs for the facility (if DCGLs are used).

- A discussion of any changes that were made in the FSS from what was proposed in the DP or other prior submittals.

- A description of the method by which the number of samples was determined for each survey unit.

- A summary of the values used to determine the number of samples and a justification for these values.

- The survey results for each survey unit including the following:

  — the number of samples taken for the survey unit;

  — a description of the survey unit, including (a) a map or drawing of the survey unit showing the reference system and random start systematic sample locations for Class 1 and 2 survey units, and random locations shown for Class 3 survey units and reference areas, (b) discussion of remedial actions and unique features, and (c) areas scanned for Class 2 and 3 survey units;

— the measured sample concentrations, in units that are comparable to the DCGLs;

— the statistical evaluation of the measured concentrations;

— judgmental and miscellaneous sample data sets reported separately from those samples collected for performing the statistical evaluation;

— a discussion of anomalous data including any areas of elevated direct radiation detected during scanning that exceeded the investigation level or any measurement locations in excess of $DCGL_W$; and

— a statement that a given survey unit satisfied the $DCGL_W$ and the elevated measurement comparison if any sample points exceeded the $DCGL_W$.

- A description of any changes in initial survey unit assumptions relative to the extent of residual radioactivity (e.g., material not accounted for during site characterization).

- A description of how ALARA practices were employed to achieve final activity levels.

This information to be submitted is also included as part of the DP Checklist provided in this NUREG report (see Section XIV.e from Appendix D of Volume 1).

4.5.3   REVIEW PROCEDURES

After review of the FSSR evaluation, the NRC reviewer should have reasonable assurance that the FSSRs demonstrate that residual radioactivity at the facility complies with the criteria of 10 CFR Part 20, Subpart E. The following guidance discusses the minimal review that should be performed and how the reviewer should select survey units for more detailed reviews.

The minimum information to be submitted in each FSSR is described in Section 4.5.2. Additional information about the recommended level of documentation is in Appendix D of this volume. At individual facilities, there may be site-specific issues and complex technical topics for which additional information from the licensee may be needed by the NRC reviewer to evaluate the FSSR. In addition, the NRC reviewer may need to obtain previous NRC–generated reports regarding the FSS, including but not necessarily limited to inspections, confirmatory surveys, and SERs for the FSS plan.

4.5.3.1   Minimal Technical Review

The NRC reviewer should review all of the following:

- the results of previously conducted inprocess inspections and confirmatory surveys to confirm that the licensee has properly implemented the final status survey plan (FSSP) and associated procedures;

- the licensee's QA/QC program, if it has not been previously reviewed;

- changes made to the DP or LTP, if not previously reviewed, to confirm that the changes are not significant and are technically correct;

- specific parts of the FSS and supporting data that affect the FSS, that were not available when the DP or LTP was approved (such data may include supplemental characterization results, basis for final surrogate ratios for multiple radionuclides, or other data collected to specifically support the FSS);

- issues (a) identified by intervenors and stakeholders and (b) raised in allegations, to assure such issues have been satisfactorily resolved;

- descriptions of the survey units, to determine if any special survey situations (see Appendix G of this volume for examples) are present;

- results of elevated measurement comparisons, to confirm that small areas of residual radioactivity do not exceed the appropriate limits (e.g., $DCGL_{EMC}$); and

- the results of the appropriate statistical tests (e.g., Wilcoxon Rank Sum (WRS) and sign tests), to confirm that results indicate compliance.

The purpose of the NRC staff review of in-process inspections, confirmatory surveys, and licensee procedures is to ensure all of the following:

- The FSSes were implemented in accordance with the approved FSSP.

- Judgmental survey results are not used in the statistical tests and are evaluated separately against the release criteria, and survey results obtained via random start and systematic sampling are statistically treated separately for the purpose of demonstrating compliance.

- The QA/QC program was adequate and implemented for the FSS.

- Inadequacies in the FSS design or implementation were corrected. For example, the licensee improved the overall FSS design and implementation, using information from survey units for which the release criteria were not initially met and resurvey or further remediation was needed.

- Results of confirmatory surveys, including split samples or independent measurements, are consistent with results of licensee surveys.

- Appropriate instrumentation, with sufficient sensitivities, proper calibrations, and adequately trained users, was used for surveys, scans, and measurements, as described in the FSSP.

4.5.3.2   Detailed Technical Review

Along with the minimal review described, the NRC reviewer may perform detailed reviews for a number of survey units. The number of survey units initially chosen for detailed review should use a risk-informed approach and the results of the minimal review. The reviewer should consider past inspection history, results of confirmatory surveys, the relative difference between residual radioactivity concentration and the associated DCGLs, the complexity of the FSSP, and the radionuclide mix. The detailed review could include confirming the selection process and location of measurements using survey unit maps or floor plans, checking measurement results using parameters that are specific to the survey methodology, and re-creating the appropriate MARSSIM statistical test results.

SELECTING SURVEY UNITS FOR DETAILED REVIEWS

Discriminating factors that may be used to select specific survey units for detailed review are listed below. A survey unit that is characterized by one or more of these factors should be considered for potential detailed review. However, the NRC reviewer should focus on survey units for which there are risk-significant issues, issues that are prevalent across a large number of survey units instead of isolated cases, and issues involving an inadequate basis for conclusions.

These factors include any of the following:

- Inconsistencies in defining survey units, including the following:

  — size different from recommended size;

  — multiple areas now combined as one larger Class 1 survey unit;

  — Class 3 survey units that are bordered by Class 1 units;

  — survey units bordered by Partial Site Release (PSR) areas; or

  — gerrymandered survey unit boundaries.

- Application of nonstandard statistical tests (e.g., other than WRS test or sign test).

- Significant inconsistencies between the DP/LTP and implemented FSS including the following examples:

  — use of surface and detector efficiencies that do not match survey methods, surface features, and instrumentation used;

  — type of survey instrumentation;

  — sample collection method;

  — laboratory analytical methods;

  — any survey unit where the scan coverage is less than 100 % for Class 1 areas, or less than minimum commitment for Class 2 or 3 areas; or

  — number of samples per survey unit.

- Survey units that were remediated.

- Survey units for which confirmatory surveys had results inconsistent with the licensee's FSS results.

- Any Class 2 survey unit with final measurement results near the $DCGL_W$ (e.g., >75 %) or any Class 3 survey unit with significant residual radioactivity (e.g., concentrations >10–25 % of the $DCGL_W$).

- Any survey unit which was downgraded in classification (i.e., from Class 1 to 2, Class 2 to 3, or Class 1 to 3, or from impacted to non-impacted).

- Units surveyed prior to resolution of QA/QC concerns.

- Significance of the variability in concentrations (i.e., heterogeneity) across survey units.

- Inconsistent approach or inadequate basis for determining surrogate radionuclide ratios.

- Significant changes to DP or LTP that affect the FSS; that were not previously reviewed.

- Reclassification schemes not approved by NRC staff.

- Use of MARSSIM survey methods and statistical tests when hot particles are present.

- Presence of systems and components, buried and embedded piping, or building foundations slated to remain on the site after license termination.

- Survey units that combine, for demonstrating compliance, the results of random start or systematic sampling patterns with biased or judgmental survey results.

- The survey unit involves surveying or sampling of media other than building surfaces and surface soils (e.g., ground water, surface water, sediments, or deep soils).

- Survey units with areas that are hard to access or have abnormal geometries.

- Any survey unit that combines survey results with a dose assessment or area factors to demonstrate compliance with an EMC test.

DETAILED REVIEW TOPICS

The detailed review could include confirming the selection process and location of measurements using survey unit maps or floor plans, checking measurement results using parameters that are specific to the survey methodology, and re-creating the appropriate MARSSIM statistical test results. In performing detailed reviews, reviewers should consider, but not necessarily be limited to the following questions:

- Are the issues previously discussed under the selection criteria for detailed reviews, immediately above, adequately addressed in the FSSR? For example, if a survey technique was changed from the approved technique, was adequate justification of the new technique provided?

- Are the probabilities of Type I and Type II errors acceptable?

- Does the licensee's analysis rely on a large number of results expressed at MDA or MDC values?

- Are all of the static measurement or sampling locations for a survey unit taken from a single random-start sampling set, without substitution (e.g., in cases where additional remediation was performed)?

- Is there a discernible trend in results within and among survey units (e.g., when comparing survey methods, locations, or media matrices)?

- If there are discernible trends in the results, are the statistical tests appropriate?

- Are there any outliers in the data? How were they detected, and was the disposition of outliers appropriate?

- Are there any assumptions about the variability (variance) of the population?

- What analytical tools (statistical software packages) were used to analyze the data?

- What is the format of the presentation of results? Is it consistent for the survey units reported? For example, are the measurement units consistent with the survey data, media measured, and the DCGLs?

The detailed review of the initially selected survey units may indicate issues that are prevalent across a large number of survey units instead of isolated cases. In this case, the reviewer may decide to evaluate additional survey units in detail.

4.5.4   EVALUATION CRITERIA

The review should verify that the FSSR is adequate to demonstrate compliance with the radiological criteria for license termination. The NRC reviewer should verify that the licensee's FSS results support the conclusion that each survey unit meets the radiological criteria for license termination. The FSS is adequate if it meets the criteria in the following:

- MARSSIM Section 5.5.2 for the acceptable number of samples;

- Appendix D of this volume for information on survey data quality and reporting;

- Section A.9 from Appendix A of this volume for information on determining compliance; and

- MARSSIM Sections 8.3, 8.4, and 8.5 for interpretations of sample results.

4.6   ISSUES NOT COVERED IN MARSSIM

MARSSIM's main focus is on providing guidance for the design of the FSSes for residual radioactivity in surface soils and on building surfaces and evaluating the collected data. However, several issues related to releasing sites are beyond the scope of MARSSIM. MARSSIM does not provide guidance for translating the release criterion into DCGLs. MARSSIM can be applied to surveys performed at vicinity properties—those not under licensee control—but the decision to apply the MARSSIM at vicinity properties is outside the scope of MARSSIM. Other media (e.g., sub-surface soil, volumetrically-contaminated building materials, ground water, surface water, sediments) containing residual radioactivity are not addressed by MARSSIM. In addition, MARSSIM does not address the disposition of components and equipment that are not part of the survey unit. Some of the reasons for limiting the scope of the guidance to surface soils and building surfaces include the following: (a) residual radioactivity is limited to these media for many sites following remediation, (b) since many sites have surface soil and building surfaces as the leading sources of residual radioactivity, existing computer models used for calculating the concentrations based on dose or risk generally consider only surface soils or building surfaces as a source term, and (c) MARSSIM was written in support of cleanup rulemaking efforts for which supporting data are mostly limited to residual radioactivity in surface soil and on building surfaces. Table 4.2 summarizes the scope of MARSSIM. Although this table was taken from MARSSIM, it has been modified to be specific to the needs of NRC licensees.

For some topics beyond the scope of MARSSIM, guidance is provided in this volume. Guidance specific to the characterization of ground water, surface water, and sediments can be found in Appendix F. Other guidance pertaining to dose modeling can be found in Chapter 5 and Appendices H, I, J, K, L, and M. Guidance can be found in Appendix G for special characterization and survey issues such as subsurface residual radioactivity, embedded piping, sewer systems, and paved areas.

Table 4.2   Scope of MARSSIM

Within Scope of MARSSIM	Beyond Scope of MARSSIM
Guidance MARSSIM provides technical guidance on conducting radiation surveys and site investigations.	*Regulation* MARSSIM does not establish new regulations or address non-technical issues (e.g., legal or policy) for site cleanup. Release criterion will be provided rather than calculated using MARSSIM.
Tool Box MARSSIM can be thought of as an extensive tool box with many components—some within the text of MARSSIM, others by reference.	*Tool Box* Many topics are beyond the scope of MARSSIM, including public participation programs, packaging and transportation of wastes for disposal, remediation and stabilization techniques, and training.
Measurement The guidance given in MARSSIM is performance-based and directed toward acquiring site-specific data.	*Procedure* The approaches suggested in MARSSIM vary depending on the various site data needs—there are no set procedures for sample collection, measurement techniques, storage, and disposal established in MARSSIM.
Modeling The interface between environmental pathway modeling and MARSSIM is an important survey design consideration addressed in MARSSIM.	*Modeling* Environmental pathway modeling and ecological endpoints in modeling are beyond the scope of MARSSIM.
Soil and Buildings The two main media of interest in MARSSIM are surface soil and building surfaces with residual radioactivity.	*Other Media* MARSSIM does not cover other media, including subsurface soil, surface or subsurface water, biota, air, sewers, sediments or volumetric building residual radioactivity. *Materials or Equipment* MARSSIM does not cover disposition of materials (including construction materials) or equipment (see Appendix G, Section G.1.1 of this volume).
Final Status Survey (FSS) The focus of MARSSIM is on the FSS as this is the deciding factor in judging if the site meets the release criterion.	*Other Survey Types* Though not the focus, MARSSIM provides less detailed information on scoping, characterization, and remedial action support surveys.
Radiation MARSSIM only considers radiation-derived hazards.	*Chemicals* MARSSIM does not deal with any hazards posed by chemical contamination.
Remediation Method MARSSIM assists in determining when sites are ready for an FSS and provides guidance on how to determine if remediation was successful.	*Remediation Method* MARSSIM does not discuss selection and evaluation of remedial alternatives, public involvement, legal considerations, policy decisions related to planning.
DQO Process MARSSIM presents a systemized approach for designing surveys to collect data needed for making decisions such as whether to release a site.	*DQO Process* MARSSIM does not provide prescriptive or default values of DQOs.
DQA MARSSIM provides a set of statistical tests for evaluating data and lists alternate tests that may be applicable at specific sites.	*DQA* MARSSIM does not prescribe a statistical test for use at all sites.

5 DOSE MODELING EVALUATIONS

INTRODUCTION

Nearly every licensee that submits a DP should provide NRC with estimates of the potential future dose that could be caused by the residual radioactivity remaining on the site after decommissioning activities are completed. Calculating potential doses allows both the licensee and regulator to take site-specific information into account in determining acceptable concentrations of residual radioactivity at the site using dose models and exposure scenarios that are as realistic as necessary for the given facility. This section has been written to maintain this flexibility. It includes the evaluation findings and supporting detailed technical guidance necessary to review the licensee's dose and ALARA analyses. Guidance on information to be submitted is provided by decommissioning group in Volume 1 of this NUREG series.

Dose modeling information is typically submitted as part of a DP or LTP, though in some cases it may be submitted separately or as part of an FSSR or other document. This chapter usually refers to DPs, though other types of reports are implied, if appropriate. NRC staff should review all of the dose modeling information submitted by the licensee. For certain cases, such as screening analyses using default values or a look-up table, most of the review has already been completed in developing these tools and, therefore, the licensee need only submit minimal site information and justification in using these models, parameters, and exposure scenarios. In addition, NRC staff should review the ALARA analyses, which is based, in part, on the dose modeling. Two general approaches exist to provide reasonable assurance that the final concentrations should meet the requirements of Subpart E:

1. The licensee can commit to the scenario(s), model(s), and parameters to be used to evaluate compliance with the dose criterion using the final concentrations. The licensee should project expected final concentrations in the DP to show that there is reasonable assurance that the dose criterion will be met at the time of license termination.

2. The licensee can derive and commit to meeting nuclide-specific concentration limits equivalent to the dose limit.

The "Decommissioning and License Termination Framework" (Figure 1.2), which generalizes the entire decommissioning process (e.g., Step 7 includes FSS and other requirements related to license termination), provides licensees with guidance on how to perform iterative dose analyses. NRC staff review of dose modeling consists of evaluations of four general areas:

1. the source term assumptions,

2. an exposure scenario considering the site environment,

3. the mathematical model/computational method used, and

4. the parameter values and a measure of their uncertainty.

The actions taken as part of the loop suggested by Steps 8 through 12 of Figure 1.2 can result in the licensee modifying one or more of the above four parts. Licensees, generally, should not, and do not need to, provide information on dose modeling iterations that are not the final dose analyses.

Other licensees may wish to include the iterative process as part of the DP. This is, generally, because site characterization is not initially complete enough to provide reasonable justification for assumptions used in modeling the site. Usually, such incorporation would be in the form of license conditions that need to be satisfied before license termination can occur.

For example, a site may have initial data on ground water contamination but, currently, does not have enough data on hydrological conditions to determine which survey units will be affected by the plume. Based on the limited data available, the licensee designates an area around the plume, and all survey units that involve that area will include the dose from the ground water as part of the overall dose analyses. For the purposes of this example, NRC could require the licensee, through a license condition (or other mechanism), to continue to characterize its ground water. If the information validates that the area affected by the ground water contamination is the same or smaller than the assumed area, the licensee can go forward with the decommissioning process. If the licensee wishes to take advantage of the smaller area, or the data points to a larger affected area, the licensee may need to submit a license amendment request to modify the FSS plan, the dose modeling, and any other area of the DP affected by the new assumed ground water contamination-affected area (e.g., adding or subtracting survey units from the list that would consider ground water contributions in complying with Subpart E).

As described by Figure 1.2 and the preceding example, the areas of dose modeling, site characterization, and FSS are interdependent. This is an advantage, as judicious use of dose modeling can help guide site characterization. In addition, both site characterization and FSS can guide development of reasonable scenarios or modeling approaches. For example, the appropriate survey techniques may require more advanced modeling in some areas to make them cost effective to implement. Refer to Question 4 from Section O.1 of Appendix O of this volume, for information to clarify the development and use of input modeling values and to Section O.2.2 for lessons learned regarding dose modeling related to recently submitted decommissioning plans.

This chapter and the associated appendices use a number of different terms describing scenarios. Table 5.1 includes a description and comparison of these scenario terms.

Table 5.1   Comparison and Description of Scenario Terms Used in this Guidance

Types of Scenarios		Evaluation Purpose	Description/Example
Plausible	Screening	All can be compliance scenarios, used to demonstrate compliance with the radiological criteria of the LTR.	A predetermined scenario that can be used with very high confidence, for most facilities, to meet the radiological decommissioning criteria without further analysis. It includes conservative assumptions about land uses or behaviors. The screening scenario for residual radioactivity on building surfaces is building occupancy, and residential farmer for residual radioactivity in surface soils. These scenarios are not site specific.
	Bounding		A scenario with a calculated dose that bounds the doses from other likely scenarios. The screening scenarios represent two bounding scenarios for site specific analyses.
	Reasonably Foreseeable		Land use scenarios that are likely within the next 100 years, considering trends and area land use plans. These scenarios are usually site specific.
	Less Likely but Plausible	Not analyzed for compliance, but is used to risk-inform the decision	Land use scenarios that are plausible, based on historical uses, but are **not** likely within the next 100 years, considering trends and area land use plans (e.g., rural use of property currently in an urban setting). These scenarios are usually site specific.
Implausible		No analysis required	Land uses that because of physical limitations could not occur (e.g., residential land use for an underwater plot of land).

GENERAL APPROACH FOR DOSE MODELING

The following section discusses the basic components that are involved in a dose modeling assessment. It is meant to provide an overview of how the pieces fit together. This discussion should provide both licensees and reviewers with an understanding of the "big picture," while the review components in the following sections focus more on NRC staff review of each part of the dose assessment.

Chapter 4 of this volume addresses characterization of the residual radioactivity currently present at the site and radiological surveys. The information is based on measurements and knowledge of the site history. To perform dose modeling, the licensee should use the site information on residual radioactivity expected to be present at the completion of decommissioning, to develop a generalized view of the site's source term. In developing the source term model, the licensee should consider the site measurements, the intended remedial actions, and the needs of both the conceptual model and the FSS.

For example, a site may have a large number of both historical and current measurements characterizing the residual radioactivity over a 10-hectare (25-acre) site. If the site information shows that residual radioactivity levels do not vary significantly, the licensee may assume that the source term is a uniform layer of residual radioactivity over the site. If the site information shows that most of the residual radioactivity is concentrated in a small area of the site, which may be due to uneven contamination resulting from either a single source or multiple sources, then the licensee may visualize the site as two sources of residual radioactivity. For the purposes of dose modeling, the following are two sources of residual radioactivity:

1. a uniform concentrated source over the smaller area where the assumed concentration is based on that area's measurements; and

2. a second source that uniformly covers the rest of the affected area at some lower concentration.

After a source term model has been developed, the question becomes: "How could humans be exposed either directly or indirectly to residual radioactivity?" or "What is the appropriate exposure scenario?" Each exposure scenario should address the following scenario questions:

• How does the residual radioactivity move through the environment?

• Where can humans be exposed to the environmental concentrations?

• What is the likely land use(s) in the future for these areas?

• What are the exposure group's habits that will determine exposure? (e.g., What do they eat and where does it come from? How much? Where do they get water and how much? How much time do they spend on various activities?)

In most situations, there are numerous possible scenarios of how future human exposure groups could interact with residual radioactivity. The compliance criteria in 10 CFR Part 20 for decommissioning does not require an investigation of all (or many) possible scenarios; its focus is on the dose to members of the critical group for the compliance scenario. The critical group is defined (at 10 CFR 20.1003) as "the group of individuals reasonably expected to receive the greatest exposure to residual radioactivity for any applicable set of circumstances." The compliance scenario is the scenario that leads to the largest peak dose to the average member of the critical group from the mixture of radionuclides. The compliance scenario may be based on a bounding scenario, such as, a screening scenario or another scenario using conservative assumptions about land uses or behaviors, or be based on the reasonably foreseeable land uses for the area.

If the licensee bases its compliance scenario on reasonably foreseeable land use, the licensee also should identify what land uses are less likely but plausible and should evaluate scenarios consistent with these less likely but plausible land uses. The evaluation of less likely but plausible scenarios ensures that, if land uses other than the reasonably foreseeable land use were to occur in the future, significant exposures would not result.

By combining knowledge about the sources of residual radioactivity and the scenario questions, the analyst can develop exposure pathways. Exposure pathways are the routes that residual radioactivity uses to travel from its source, through the environment, until it interacts with a human. They can be fairly simple (e.g., surface soil residual radioactivity emits gamma radiation which results in direct exposure to the individual standing on the soil), or they can be fairly involved (e.g., the residual radioactivity in the surface soil leaches through the unsaturated soil layers into underlying aquifer, and the water from the aquifer is pumped out by the exposed individual for use as drinking water, which results in the exposed individual ingesting the environmental concentrations). Exposure pathways typically fall into three principal categories identified by the manner in which the exposed individual interacts with the environmental concentrations resulting from the residual radioactivity; the three principal categories are ingestion, inhalation, or external (i.e., direct) exposure pathways.

The exposure pathways for many of the exposure groups can be bounded by a smaller number of possible exposure groups. For example, at a rural site with surface soil residual radioactivity, two possible exposure groups are (1) a gardener who grows a small fraction of his or her fruits and vegetables in the soil and (2) a resident farmer who grows a larger fraction of his or her own food (i.e., the site supplies not only vegetables, but also meat and milk). In this case, the resident farmer bounds the gardener exposure group (because it both incorporates the gardener's pathways, but also includes other routes of exposure) and, therefore, the gardener exposure group does not need to be analyzed and the compliance calculation's scenario would involve the resident farmer.

As required by 10 CFR 20.1402, expected doses are evaluated for the average member of the critical group, which is not necessarily the same as the maximally exposed individual. This is not a reduction in the level of protection provided to the public but is an attempt to emphasize the uncertainty and assumptions needed in calculating potential future doses while limiting boundless speculation on possible future exposure scenarios. While it is possible to actually identify, with confidence, the most exposed member of the public in some operational situations (through monitoring, time-studies, distance from the facility, etc.), identification of the specific individual who should receive the highest dose some time (up to 1000 years) in the future is impractical, if not impossible. Speculation on his or her habits, characteristics, age, or metabolism could be endless. The use of the "average member of the critical group" acknowledges that any hypothetical "individual" used in the performance assessment is based, in some manner, on the statistical results from data sets (i.e., the breathing rate is based on the range of possible breathing rates) gathered from groups of individuals. While bounding assumptions could be used to select values for each of the parameters (e.g., the maximum amount of meat, milk, vegetables, possible exposure time), the result could be an extremely conservative calculation of an unrealistic scenario and may lead to excessively low allowable residual radioactivity levels.

Calculating the dose to the critical group is intended to bound the individual dose to other possible exposure groups because the critical group is a relatively small group of individuals, due to their habits, actions, and characteristics, who could receive among the highest potential doses at some time in the future. By using the hypothetical critical group as the dose receptor, coupled with prudently conservative models, it is highly unlikely that any individual would actually receive doses in excess of that calculated for the average member of the critical group. The description of a critical group's habits, actions, and characteristics should be based on credible assumptions and the information or data ranges used to support the assumptions should be limited in scope to reduce the possibility of adding members of less exposed groups to the critical group. An analysis of the average member of the critical group's potential exposure should also include, in most cases, some evaluation of the uncertainty in the parameter values used to represent physical properties of the environment.

Use the definitions in Part 20 when calculating for compliance with the requirements of Subpart E. Use the Federal Guidance Report No. 11 (EPA 1988) when calculating internal exposures by using the intake-to-dose conversion factors, which are based primarily on adults. As stated in EPA's Draft Guidance for Exposure of the General Public (EPA 1994) which proposes a public dose limit of 1.0 mSv (100 mrem) per year from all sources:

> "These dose conversion factors are appropriate for application to any population adequately characterized by the set of values for physiological parameters developed by the [International Committee on Radiological Protection] and collectively known as "Reference Man." The actual dose to a particular individual from a given intake is dependent upon age and sex, as well as other characteristics. As noted earlier, implementing limits for the general public expressed as age and sex dependent would be difficult.... More importantly, the variability in dose due to these factors is comparable in magnitude to the uncertainty in our estimates of the risks which provide the basis for our choice of the [public dose limit]. For this reason EPA believes that, for the purpose of providing radiation protection under the conditions addressed by these recommendations, the assumptions exemplified by Reference Man adequately characterize the general public, and a detailed consideration of age and sex is not generally necessary." [sic]

Since age-based dose conversion factors are not being used, the same dose conversion factors are applied to all individuals. Only in rare scenarios will a non-adult individual receive a higher dose (i.e., take in more radioactive material) than an adult individual in a similar exposure scenario. One example is the milk pathway: children generally drink more milk annually than adults. If milk was the only pathway that would expose the individual to a dose, then the child would have a slightly higher dose than the adult. But in most situations, especially ones involving multiple pathways, the total intake of the adult is greater than that of a child. Therefore, for most multiple pathway scenarios, such as screening analyses, the average member of the critical group should usually be assumed to be an adult with the proper habits and characteristics of an adult. As the licensee eliminates pathways or modifies the scenario, the behavior and dietary habits of children may become important. In such cases, the licensees should consult with NRC staff for guidance.

By integrating the exposure scenario, source term, and knowledge about the applicable environmental transport routes involved in the exposure pathways, a conceptual model of the features and processes at the site can be created. The conceptual model is a qualitative description of the important environmental transport and exposure pathways and their interrelationships. Using this description, a mathematical model quantifying it, or using an off-the-shelf computer code that implements the same (or similar) conceptual model, needs to be identified. Generally, a single mathematical model can be used for several different conceptual models by varying either the boundary conditions or the various parameters.

Going from a conceptual model to a mathematical model involves a number of assumptions and simplifications. For example, one part of a conceptual model of surface soil residual radioactivity involves the leaching of radionuclides through the soil and into the aquifer. In reality, the soil between the surface and the aquifer is usually formed by numerous layers of different types of soils with varying thickness across a site. For the purposes of dose modeling, the conceptual model is more focused on knowing how much activity is entering (and leaving) each major environmental compartment (such as the aquifer) than to precisely predict the level of activity in the intervening material (e.g., any single soil layer between the surface and the aquifer). Therefore, the mathematical model may view the intervening soil layers as one layer or just a few layers, depending on the difficulty of justifying effective parameters that will mimic the real behavior. Users of off-the-shelf codes should be aware of and consider the appropriateness of the assumptions made in the computer model they are using.

Selection of parameter values (or ranges) for features, events, and processes depends not only on the site conditions and the exposure scenario, but also on the computer code (or mathematical model) being used. Nearly any data set will need to be transformed into one appropriate to the situation. This can be as straightforward as generating a site-wide effective soil density value or as complex as converting resuspension factor data into resuspension rates. NRC has already factored these issues in the data used in the screening analyses, but licensees using site-specific information should justify their values.

In the past, the most common computer codes were deterministic and did not explicitly calculate parameter uncertainty. Although it is not always necessary for a licensee to use a probabilistic code to evaluate parameter uncertainty for site-specific analyses, licensees should provide some discussion of the level of uncertainty in the results. It should be noted that the type of uncertainty of prime interest to NRC staff is uncertainty in the physical parameters.

> Licensees using probabilistic dose modeling should use the "peak of the mean" dose distribution for demonstrating compliance with the 10 CFR Part 20, Subpart E. Similar to all regulatory guidance, this NUREG report contains one approach for determining compliance with the regulations using probabilistic analyses. Other probabilistic approaches, such as "mean of the peaks" or other methods, if justified, may also be acceptable for demonstrating compliance. If the licensee intends to use any probabilistic approach to calculate DCGLs, the licensee should discuss their planned approach with NRC staff.

SCOPING REVIEW

As part of the DP review, NRC staff should evaluate the basis for each of the calculated doses used by the licensee in the various decommissioning options. NRC staff should organize this review by first looking at the overall scope of the dose modeling provided (possibly for several decommissioning options and/or critical groups). This scoping review should help NRC staff identify which specific dose modeling sections are applicable for a given DP. After the scoping review, NRC staff should review each of the scenarios that the licensee or responsible party is using to show compliance with the regulations.

An acceptable way to organize the scoping review is (a) to identify and confirm the principal sources (before and after remediation) of residual radioactivity and (b) to identify the decommissioning goal of the DP. Coupling the two sets of information, NRC staff should have a good indication of the appropriate sections. For decommissioning goals involving unrestricted release, NRC staff should quickly evaluate to what decommissioning group the licensee belongs. Section 5.1 is used for evaluating screening dose assessments for Decommissioning Groups 1–3. Section 5.2 is used for evaluating site-specific dose assessments for unrestricted release for Decommissioning Groups 4 and 5.

Next, NRC staff should verify that conditions at the site are consistent with the approach chosen by the licensee and the decommissioning group's requirements (i.e., whether the licensee may use a screening analysis approach or whether site-specific dose modeling should be performed). Licensees may not be able to use a screening analysis approach at sites exhibiting any of the following conditions (excluding those caused by sources of background radiation):

- soil residual radioactivity greater than 30 cm (12 in) below the ground surface,

- radionuclide residual radioactivity present in an aquifer,

- buildings with volumetrically contaminated material,

- radionuclide concentrations in surface water sediments, and

- sites that have an infiltration rate that is greater than the vertical saturated hydraulic conductivity (i.e., resulting in the water running off the surface rather than only seeping into the ground).

These are limitations caused by the conceptual models used in developing the screening analysis. In other words, the conceptual model, parameters, and scenarios in the DandD computer code are generally incompatible with such conditions. Situations do exist where you can still utilize the analyses using scenario assumptions to modify the source term. For example, by assuming buried radioactive material is excavated and spread across the surface, the screening criteria may be applicable for use at the site.

When evaluating any decommissioning option that has a goal of terminating the license under the unrestricted release requirements of 10 CFR 20.1402, the primary scenarios generally involve individuals exposed on the site. A licensee needs to evaluate scenarios consistent with the reasonably foreseeable land use over the next decades, or use a bounding scenario such as a resident farmer. In rare instances, a scenario involving offsite use of residual radioactivity may be the critical scenario. A bounding exposure scenario for residual radioactivity in the environment (versus building surfaces) is usually a residential farmer, because this group usually includes a nearly comprehensive number of exposure pathways. In addition to pathways that may be limited by land use assumptions, site conditions, such as soil type, or ground water quality, may remove potential exposure pathways from consideration with the appropriate level of justification by the licensee.

A decommissioning option that results in the license being terminated under the restricted use provisions of 10 CFR 20.1403 will require, at a minimum, two different exposure scenarios. One scenario should evaluate the performance of the proposed restrictions by assuming these restrictions never fail. Depending on where the residual radioactivity is and what the proposed restrictions are, the exposure location(s) for the critical group could be either onsite or offsite. The second scenario should be performed similarly to the analyses for unrestricted release, in which it assumes that the restrictions put in place by the licensee have failed to work properly (or effectively), and the site will be used without knowledge of the presence of residual radioactivity.

5.1 UNRESTRICTED RELEASE USING SCREENING CRITERIA (DECOMMISSIONING GROUPS 1–3)

NRC staff should review the information provided in the DP, FSSR, or other document pertaining to the licensee's assessment of the potential doses resulting from the residual radioactivity remaining at the end of the decommissioning process. The findings and conclusions of the review under this chapter should be used to evaluate the compliance with the dose limit specified in Subpart E. This chapter addresses decommissioning options involving unrestricted release using the default screening models or derived tables. These will be licensees from Decommissioning Groups 1–3. Decommissioning Groups 4–7 may utilize the screening criteria described here as part of their dose modeling.

The evaluation criteria in this section on screening analyses have been divided into two categories based on the location of the residual radioactivity:

- Building Surface Evaluation Criteria, and

- Surface Soil Residual Radioactivity.

CALCULATION OF RADIOLOGICAL IMPACTS ON INDIVIDUALS

The overall objective of NRC staff's review is to determine if the screening criteria were used correctly by the licensee and whether the calculations provide reasonable assurance that potential doses would not exceed the dose limits. Specific impacts to be calculated include those associated with exposures using the default building scenario and model.

ACCEPTANCE CRITERIA

Regulatory Requirements

10 CFR 20.1402

Regulatory Guidance

- Appendix H of this NUREG report

- NUREG–1757, Vol. 1, "Consolidated NMSS Decommissioning Guidance: Decommissioning Process for Materials Licensees"

Information to be Submitted

NRC staff should organize this review by first looking at the overall scope of the dose modeling contained in the DP, FSSR, or other document (possibly for several decommissioning options and/or critical groups). This scoping review, discussed in Chapter 5, should help the reviewers identify which review sections are applicable for a given DP. After the scoping review, NRC staff should review each of the scenarios that the licensee is using to show compliance with the regulations using the appropriate review section.

The licensee's dose modeling for building surfaces or surface soil using the default screening criteria should include both of the following:

- the general conceptual model (for both the source term and the building or outside environment) of the site, and

- a summary of the screening method (i.e., running DandD or using the look-up tables) used.

This information to be submitted is also included as part of the DP Checklist provided in this NUREG report (see Checklist Section V.a from Appendix D of Volume 1).

EVALUATION CRITERIA

When licensees use the default screening methods and parameters inherent in the DandD code by either running the computer code or using look-up tables, the review and acceptance of nearly all areas of the analysis have already been done by NRC staff in developing the screening tool and reviewers should only need to review the source term model and the overall applicability of using the screening method with the associated residual radioactivity.

NRC staff will determine the acceptability of the licensee's projections of (a) radiological impacts on future individuals from residual radioactivity and (b) compliance with regulatory criteria. The information may be considered acceptable if it is sufficient to ensure a reasonable assessment of the possible future impacts from the residual radioactivity on building surfaces or surface soil. The information should allow an independent staff evaluation of the justifications and assumptions used.

NRC staff should review the information identified in Sections 5.1.1 and 5.1.2, as necessary, for each dose assessment of residual radioactivity on building surfaces or surface soil that the licensee has submitted in the various decommissioning options. If the licensee did not directly calculate the dose from residual radioactivity, but instead derived, or proposed to use, unit concentration values, NRC staff may need only to review the information on the configuration of the residual radioactivity and the appropriate screening criteria section, below. Review of the spatial variability should be performed as part of the final survey. Detailed guidance is in Appendix H.

5.1.1   BUILDING SURFACE EVALUATION CRITERIA

- Source Term Configuration

  NRC staff should confirm that the actual measurements, facility history, and planned remedial action(s) support the source term configuration used in the modeling by reviewing the information in the facility history, radiological status, and planned remedial action(s) portions of the DP. The NRC reviewer should review both the areal extent of residual radioactivity and the depth of penetration of the residual radioactivity into the building surfaces. The NRC reviewer should determine if the physical configuration of the residual radioactivity can adequately be assumed to be a thin layer of residual radioactivity on the building surfaces. If the residual radioactivity is not limited to the building surfaces, then use of the default screening criteria are not warranted without additional justification. The NRC reviewer should reclassify the licensee as a Group 4 licensee and evaluate the modeling using Section 5.2.

- Residual Radioactivity Spatial Variability

  NRC staff should review the information provided by the licensee for conditions both before and those projected after the decommissioning alternative. Based on this information, NRC staff should determine whether it is appropriate to make an assumption of homogeneity (a) for the whole facility or (b) for subsections of the facility. NRC staff should then review the adequacy of the licensee's determination of a representative value (or range of values) for the residual radioactivity concentration in the source term modeled. To evaluate the final survey, as a general guideline, NRC staff could use the concepts related to area factors included in MARSSIM and in Section I.3.3.3.5 of Appendix I.

- Execution of the DandD Computer Code Dose Calculations

  If the licensee has used the DandD computer code to calculate the dose based on either current concentrations or projected final concentrations, NRC staff should verify that:

  1. The residual radioactivity is limited to building surfaces.

  2. If the appropriate annual peak dose is greater than 0.025 mSv (2.5 mrem), the removable fraction of the residual radioactivity is 10 % or less at the time of license termination, or the removable fraction has been adjusted as explained in footnote a in Table H.1.[4]

  3. The output reports verify that no parameters (other than source term concentrations) were modified.

  4. The licensee has used the 90th percentile of the dose distribution to compare with the dose limit.

- DCGLs from the DandD Code or Look-up Tables

  The licensee may use either the DandD computer code or the published look-up table for beta and gamma emitters (see Appendix H) to establish radionuclide-specific DCGLs equivalent to 0.25 mSv/y (25 mrem/y).

  If the licensee is proposing to use radionuclide-specific DCGLs, NRC staff should verify that the following three conditions are true:

  1. The residual radioactivity is limited to building surfaces.

  2. If the residual radioactivity is greater than 10 % of the respective screening DCGLs (Table H.1 from Appendix H of this volume), the removable fraction is 10 % or less at license termination, or the removable fraction has been adjusted as explained in footnote a in Table H.1.

---

[4] The DandD default scenario assumes that only 10 % of the surface residual radioactivity is removable and available for resuspension. Only at 10 % of the dose limit does the assumption begin to become important because in the extreme case of 100 % removable, for radionuclides that produce the majority of dose from the inhalation pathway, the code result may be underestimating the result by a factor as great as 10.

3. If more than one radionuclide is involved, there is reasonable assurance that the sum of fractions (concentrations divided by DCGLs) (see Section 2.7) is no greater than 1.

If the licensee has used the DandD Version 2 computer code to calculate the radionuclide-specific DCGLs, NRC staff should also verify that the following two conditions are true:

1. The output reports verify that no parameters (other than entering unit concentrations) were modified.

2. The licensee has used the 90th percentile of the dose distribution to derive the concentrations.

- Compliance with Regulatory Criteria

The licensee's projections of compliance with regulatory criteria, if that decommissioning option is pursued, are acceptable provided that NRC staff has reasonable assurance that at least one of the following is true:

1. The only residual radioactivity is on building surfaces, and the level of removable residual radioactivity does not violate the assumptions in the model.

2. The final concentrations result in a peak annual dose of less than 0.25 mSv (25 mrem) and the licensee has committed to calculating the annual dose using a screening analysis at license termination.

3. The planned DCGLs are equal to or less than those provided by the screening criteria, and the licensee has committed to maintaining the sum of fractions, if applicable.

5.1.2   SURFACE SOIL EVALUATION CRITERIA

- Source Term Configuration

NRC staff should confirm that the actual measurements, facility history, and planned remedial action(s) support the source term configuration used in the modeling by reviewing the information in the facility history, radiological status, and planned remedial action(s) portions of the DP. The NRC reviewer should review both the areal extent of residual radioactivity and the depth of penetration of the residual radioactivity into the soil. The NRC reviewer should determine if the physical configuration of the residual radioactivity can adequately be assumed to be a layer of surface soil containing residual radioactivity without overlying surface layers. If the residual radioactivity is not limited to the surface soil, then use of the default screening criteria are not warranted without additional justification. The NRC reviewer should reclassify the licensee as a Group 4 licensee and evaluate the modeling using Section 5.2.

- Residual Radioactivity Spatial Variability

    NRC staff should review the information provided by the licensee for conditions both before and those projected after the decommissioning alternative is complete. Based on this information, NRC staff should determine whether it is appropriate to make an assumption of homogeneity (a) for the entire affected area or (b) for major subsections of the site. NRC staff should then review the adequacy of the licensee's determination of a representative value (or range of values) for the residual radioactivity concentration in the source term model. At the time of the final survey, as a general guideline, NRC staff can use the concepts related to area factors included in the MARSSIM and in Section I.3.3.3.5 of Appendix I.

- Conceptual Models

    Detailed NRC staff review of the information is not necessary as these topics were previously addressed by NRC staff establishing the default screening methods. NRC staff should verify that the site and DandD's conceptual models are compatible. Situations that would not allow the use of the DandD code as a screening tool for environmental concentrations of radionuclides would include those where the source is not predominantly present in the surface soil, residual radioactivity in the aquifer, or sites with infiltration rates higher than the vertical saturated hydraulic conductivity (i.e., resulting in surface runoff or a bathtub effect) without additional justification showing that the results would still calculate a conservative dose estimate. A complete list of screening values can be found in Appendix H.

- Execution of DandD Computer Code for Dose Calculations

    If the licensee has used the DandD computer code, NRC staff should verify that all of the following is true:

    1. The residual radioactivity is limited to surface soil.

    2. The total dose calculated includes all sources of residual radioactivity.

    3. The output reports verify that no parameters (other than source term concentrations) were modified.

    4. The licensee has used the 90th percentile of the dose distribution to compare with the dose limit.

- DCGLs from the DandD Code or Look-up Tables

    The licensee may use either the DandD computer code or the published look-up table (see Appendix H) to establish nuclide-specific DCGLs equivalent to 0.25 mSv/y (25 mrem/y). If the licensee is proposing to use radionuclide-specific DCGLs, NRC staff should verify that both of the following conditions are true:

1. The residual radioactivity (for the action under review) is limited to surface soil.

2. If more than one radionuclide is involved, there is reasonable assurance that the sum of fractions (see Section 2.7) will be maintained.

If the licensee has used the DandD Version 2 computer code to calculate the radionuclide-specific DCGLs, NRC staff should also verify that both of the following conditions are true:

1. The output reports verify that no parameters (other than entering unit concentrations) were modified.

2. The licensee has used the 90th percentile of the dose distribution to derive the concentrations.

- Compliance with Regulatory Criteria

  The licensee's projections of compliance with regulatory criteria (if the decommissioning option is pursued) are acceptable, if NRC staff has reasonable assurance of all the following:

1. The licensee has applied an appropriate source term.

2. The only residual radioactivity is surface soil.

The final concentrations result in a peak annual dose of less than 0.25 mSv (25 mrem) and the licensee has committed to calculating the annual dose using a screening analysis at license termination, or the planned DCGLs are equal to or less than those provided by the screening criteria, and the licensee has committed to maintaining the sum of fractions, if applicable.

5.2 UNRESTRICTED RELEASE USING SITE-SPECIFIC INFORMATION (DECOMMISSIONING GROUPS 4–5)

The following guidance is for reviewing DPs submitted by licensees from Decommissioning Groups 4 and 5.

AREAS OF REVIEW

NRC staff should review the information provided in the DP pertaining to the licensee's assessment of the potential doses resulting from exposure to residual radioactivity remaining at the end of the decommissioning process. The findings and conclusions of the review under this section should be used to evaluate the DP's compliance with 10 CFR 20.1402. NRC staff should ensure that, at a minimum, information on the source term, exposure scenario(s), conceptual model(s), numerical analyses (e.g., hand calculations or computer models), and uncertainty have been included. NRC staff should review the abstraction and assumptions regarding the source term, the conceptual model of the site or building as appropriate, the exposure scenario(s), the mathematical method employed, and the parameters used in the analysis and their uncertainty.

The amount of information provided by the licensee and the depth of the reviewer's investigation of that information will depend on the complexity of the case and the amount of site-specific information (versus default assumptions) being used by the licensee. This section has been written for review of the most complex analyses; most analyses should not need in-depth review of all parts of the evaluation criteria.

REVIEW PROCEDURES

Acceptance Review

NRC staff should review the DP to ensure that, at a minimum, the DP contains the information summarized under "Areas of Review," above. NRC staff should review the dose modeling portion of the DP without assessing the technical accuracy or completeness of the information contained therein. The adequacy of the information should be assessed during the detailed technical review. NRC staff should review the DP table of contents and the individual descriptions under "Areas of Review" as shown above (a) to ensure that the licensee or responsible party has included this information in the DP and (b) to determine if the level of detail of the information appears to be adequate for NRC staff to perform a detailed technical review.

Safety Evaluation

The material to be reviewed is technical in nature, and NRC staff should review the information provided by the licensee to ensure that the licensee used defensible assumptions and models to calculate the potential dose to the average member of the critical group. NRC staff should also verify that the licensee provided (a) enough information to allow an independent evaluation of the potential dose resulting from the residual radioactivity after license termination and (b) reasonable assurance that the decommissioning option will comply with regulations.

ACCEPTANCE CRITERIA

Regulatory Requirements

10 CFR 20.1402

Regulatory Guidance

- Appendix I of this NUREG report
- NUREG–1549, "Decision Methods for Dose Assessment to Comply with Radiological Criteria for License Termination"

- NUREG/CR–5512, Volume 1, "Residual Radioactive Contamination from Decommissioning: Technical Basis for Translating Contamination Levels to Annual Total Effective Dose Equivalent"

- Draft NUREG/CR–5512, Volume 2, "Residual Radioactive Contamination from Decommissioning: User's Manual"

- Draft NUREG/CR–5512, Volume 3, "Residual Radioactive Contamination from Decommissioning: Parameter Analysis"

- Federal Guidance Report Number 11, "Limiting Values of Radionuclide Intake and Air Concentration and Dose Conversion Factors for Inhalation, Submersion, and Ingestion" (EPA 1988)

- Federal Guidance Report Number 12, "External Exposure to Radionuclides in Air, Water, and Soil" (EPA 1993)

Information to be Submitted

NRC staff should organize this review by first looking at the overall scope of the dose modeling contained in the DP (possibly for several decommissioning options and/or critical groups). This scoping review, discussed in Chapter 5, should help the reviewers identify which section is applicable for a given dose assessment. After the scoping review, NRC staff should review each of the scenarios that the licensee is using to show compliance with the regulations.

In describing the licensee's dose modeling analysis methods, "site-specific" is used in a very general sense to describe all dose analyses except those based only on the default screening tools. This may be as simple as a few parameter changes, in the DandD computer code from their default ranges, to licensees using scenarios, models, and parameter ranges that are only applicable at the licensee's site. The information submitted should include the following:

- the source term information including nuclides of interest, configuration of the source, areal variability of the source, and so forth;

- a description of the compliance scenario including a description of the critical group;

- a description of any other reasonably foreseeable or less likely but plausible scenarios considered;

- a description of the conceptual model of the site including the source term, physical features important to modeling the transport pathways, and the critical group;

- the identification, description and justification of the mathematical model used (e.g., hand calculations, DandD v2.1, RESRAD v6.1);

- a description of the parameters used in the analysis;

- a discussion about the effect of uncertainty on the results; and

- input and output files or printouts, if a computer program was used.

This information to be submitted is also included as part of the DP Checklist provided in this NUREG report (see Checklist Section V.b from Appendix D of Volume 1).

EVALUATION CRITERIA

NRC staff should determine the acceptability of the licensee's projections of radiological impacts on the average member of the critical group during the compliance period from residual radioactivity. The information in the DP is acceptable if it is sufficient to ensure a defensible assessment of the possible future impacts from the residual radioactivity. The licensee's assessment can be either realistic or prudently conservative. The information should allow an independent NRC staff evaluation of the assumptions used (e.g., source term configuration, applicable transport pathways) and possible doses to the average member of the critical group.

NRC staff should review the following information, as necessary, for each dose assessment of residual radioactivity that the licensee has submitted in the various decommissioning options.

- Source Term

  NRC staff should review the assumptions used by the licensee to characterize the facility's source term of residual radioactivity for dose modeling purposes. NRC staff should compare the assumptions with the current site information and the decommissioning alternative's goal. The model should be an appropriate generalization of this information. Three key areas of review for the source term assumptions are the (1) configuration, (2) residual radioactivity spatial variability, and (3) chemical form(s). For additional guidance, refer to Section I.2 from Appendix I of this volume.

  1. Configuration

     NRC staff should confirm that the actual measurements, facility history, and planned remedial action(s) support the source term configuration used in the modeling by reviewing the information in the facility history, radiological status, and planned remedial action(s) portions of the DP. NRC staff should review the provided information for both the areal extent of residual radioactivity and the depth (for soil or buildings) or volume (for ground water or buried material) of the residual radioactivity. The NRC reviewer should determine if the information provided supports the configuration assumptions used in the exposure scenario and mathematical model (e.g., a thin layer of residual radioactivity on the building surfaces).

  2. Residual Radioactivity Spatial Variability

     NRC staff should review residual radioactivity concentration values provided by the licensee for conditions both before, and projected after, the decommissioning alternative is complete. For this subsection, NRC staff should review the spatial extent and the degree of heterogeneity in the values. Based on this information, NRC staff should determine whether it is reasonable to make an assumption of homogeneity for each source for either (a) the whole site or (b) subsections of the site. NRC staff should then review the adequacy of the licensee's determination of a representative value (or range of values) for the residual radioactivity concentration in the source term model. At the time of final survey, as a general guideline, NRC staff could use the concepts related to area factors included in the MARSSIM and in Section I.3.3.3.5 from Appendix I of this volume.

If the licensee used dose modeling to develop DCGLs, instead of estimating final concentrations and then entering them into the code, the licensee need not specifically address the spatial variability acceptance criteria at this time. The licensee should provide this information in the FSS for NRC staff review. NRC staff should verify that the spatial variability is compatible with the assumptions made for dose modeling.

3. Chemical Form

The licensee's assumptions regarding the chemical form of the residual radioactivity should be reviewed for its adequacy by NRC staff. NRC staff should determine whether the licensee has considered possible chemical changes that may occur during the time period of interest. Without any justification of possible chemical forms, the analysis should use the bounding chemical form(s) (e.g., the chemical form(s) that give the individual the highest dose per unit intake as described in Federal Guidance Report Number 11 (EPA 1988)). Acceptable rationale for other assumptions should be provided by the licensee. Some acceptable rationales for using other chemical forms are (a) chemical forms that would degrade quickly in the environment (e.g., UF_6) or (b) the unavailability of an element or conditions to realistically form that molecule (e.g., $SrTiO_4$ or high-fired UO_2).

- Critical Groups, Scenarios, Pathways, Identification and Selection

In its review, NRC staff should confirm that the licensee has identified and quantified the most significant scenarios based on available site- or facility-specific information. NRC staff should review the basis and justification for the licensee's selected critical group. For scenarios in which possible environmental pathways have been modified or eliminated, NRC staff should review the justifications provided by the licensee. For additional guidance on these subjects, refer to Section I.3 from Appendix I of this volume.

1. Scenario Identification

The compliance exposure scenario is based on the location and type of source (e.g., contaminated walls), the reasonably foreseeable land use, the general characteristics and habits of the critical group (e.g., an adult light industry worker) and the possible pathways which describe how the residual radioactivity would incur dose in humans. The licensee should provide justification on the scenario(s) evaluated.

The licensee should justify the possible land use(s) the site might experience in the future and create exposure scenarios consistent with these uses. The licensee should provide justification for selecting the compliance scenario from the possible exposure scenarios derived from the land use. The compliance scenario should result in the greatest exposure to the average member of the critical group for all scenarios given the mixture of radionuclides. A licensee may choose to make a bounding assumption for land use to derive the scenario (e.g., assuming a rural land use for an urban location) or base the scenario on the reasonably foreseeable land use that results in the highest dose.

If the compliance scenario is based on the reasonably foreseeable land use, the licensee should provide justification for the scenario, based on discussions with land planners, meetings with local stakeholders, trending analysis of land use for the region, or comparisons with land use in similar alternate locations. The time period of interest for possible land use is changes within 100 years, depending on the rate of change in the region, and the peak exposure time. Note that the 100-year timeframe described here is only for estimating future land uses; the licensee must evaluate doses that could occur over the 1000-year time period specified in the LTR. The licensee should identify what land uses are less likely but plausible, and evaluate scenarios consistent with these less likely but plausible land uses. In some cases, the use of reasonably foreseeable land use may require the licensee to evaluate offsite uses of materials containing residual radioactivity as alternate scenarios in defining the compliance scenario.

The licensee needs to provide a quantitative analysis of or a qualitative argument discounting the need to analyze all scenarios generated from the reasonably foreseeable land uses. The level of detail can vary between scenario and it is expected for the licensee to use simple analyses to limit the number of detailed scenarios. The licensee may use screening or generic analyses to assist in determining the critical scenario for compliance. With a mixture of radionuclides, more than one compliance scenario may need to be used. The peak dose from the compliance scenario(s) should exceed the exposures resulting from other scenarios.

Similarly, the licensee needs to provide either a quantitative analysis of or a qualitative argument discounting the need to analyze all scenarios generated from the less likely but plausible land uses. The results of these analyses will be used by the staff to evaluate the degree of sensitivity of dose to overall scenario assumptions (and the associated parameter assumptions). The reviewer will consider both the magnitude and time of the peak dose from these scenarios. If peak doses from the less likely but plausible land use scenarios are significant, the licensee would need to provide greater assurance that the scenario is less likely to occur, especially during the period of peak dose.

The screening scenarios for building surface residual radioactivity and soil residual radioactivity are described in NUREG–1549 and NUREG/CR–5512, Volumes 1, 2 and 3. Dose evaluations that use these scenarios (i.e., the licensee changes parameter values or mathematical method but does not change the general scenario) are acceptable, if the scenario is appropriate for the situation. In DPs where the licensee eliminates certain pathways, with justification, but still maintains the same general scenario category, NRC staff should find the scenario identification to be acceptable. For example, a licensee may eliminate the use of ground water because the near surface aquifer has total dissolved solids of 30,000 mg/L. The licensee still evaluates the impacts from crops grown in the residual radioactivity but irrigation is provided by a noncontaminated source and therefore, the screening scenario, a residential farmer, is maintained.

2. Critical Group Determination

In general, critical groups that are exposed to multiple exposure pathways result in higher doses than groups with more limited interaction with the residual radioactivity. NUREG–1549 and the NUREG/CR–5512 series, details the critical group assumptions for the screening scenarios. In DPs where the licensee has used the screening scenarios, the reviewer should verify that the critical group is the same as listed in NUREG–1549 and the NUREG/CR–5512 series.

The licensee should provide either qualitative or quantitative justification that the critical group is the highest exposed group for the assumed land use(s). The selection of the critical group may be dependent on the assumption of the relative mixture of radionuclides and sources of residual radioactivity present at the site. The licensee should justify its compliance approach in these cases. A similar justification should be provided by the licensee for the critical scenario for less likely but plausible scenarios.

3. Exposure Pathways

The DP should describe the exposure pathways to which the critical group is exposed, except for cases where the licensee is using the screening scenarios and critical groups without modification. If the licensee has chosen to modify the screening scenario, the changes should be justified. In general, the justification should be based on physical limitations or situations that would not allow individuals to be exposed as described in the scenario. For other scenarios, the exposure pathways should be consistent with the land use assumptions, exposure group behavior, and physical site conditions.

For example, acceptable justifications for removing the ground water pathway include (a) the near surface ground water is neither potable nor allowed to be used for irrigation; (b) aquifer volume is insufficient to provide the necessary yields; and (c) there are current (and informed consideration of future) land use patterns that would preclude ground water use. Justification of water quality and quantity of the saturated zone should be based on the classification systems used by EPA or the State, as appropriate. In cases where the aquifer is classified as not being a source of drinking water but is considered adequate for stock watering and irrigation, the licensee can eliminate (i.e., does not need to consider) the drinking water pathway (and the fish pathway—depending on the scenario), but the licensee should still maintain the irrigation and meat/milk pathways, if appropriate for the land use assumptions.

Another example would be a rural site with a relatively small discrete outdoor area of residual radioactivity (compared to the area assumed in the default scenarios). In this situation, it may be appropriate, based on the area of residual radioactivity, that gardening of some vegetables and fruits would still be an assumption, but the area is not large enough to allow one to grow grain or raise animals for meat or milk.

- Conceptual Models

  NRC staff should review the adequacy of the conceptual model used by the licensee. For additional guidance on these subjects, refer to Section I.4 from Appendix I of this volume.

  The conceptual model should qualitatively describe the following:

  1. the relative location and activities of the critical group;

  2. both the hydrologic and environmental transport processes important at the site;

  3. the dimensions, location and spatial variability of the source term used in the model; and

  4. the major assumptions made by the licensee in developing the conceptual model (e.g., recharge of the aquifer is limited to the infiltration through the site's footprint).

  The NRC reviewer should verify that the site conditions are adequately addressed in the conceptual model and simplifying assumptions.

- Calculations and Input Parameters

  In its review, NRC staff will confirm that the licensee has used a mathematical model that is an adequate representation of the conceptual model and the exposure scenario. For additional guidance on these subjects, refer to Sections I.5 and I.6 from Appendix I of this volume.

  1. Execution of DandD Computer Code

     If the licensee has used the DandD computer code in its analysis, NRC staff should verify the following points:

     a. The residual radioactivity is limited to the surface (either building or near surface soil, as appropriate).

     b. The site conceptual model is adequately represented by DandD's inherent conceptual model.

c.   For building surfaces, if the total dose is greater than 10 % of the dose limit, the licensee has modified the resuspension factor to account for the removable fraction to be present at the time of decommissioning.

d.   For sites eliminating pathways, the licensee has used the appropriate parameters in the DandD code as "switches" to turn off the pathways without unintentionally removing others. For example, to remove the ground water pathways, the licensee should set the drinking water rate, irrigation rate, and pond volume to zero.

e.   For each parameter modified, the licensee has adequately justified the new parameter value or range and has evaluated the effect on other parameters.

f.   For modifications of behavioral parameters, the changes should be based on acceptable changes in the critical group, and the mean values of the behavioral parameters should be used, although use of the ranges is also acceptable.

g.   If the licensee has randomly sampled the parameter ranges in DandD, the licensee has used the "peak of the mean" dose distribution to either calculate the dose or derive the DCGLs.

2.   Other Mathematical Methods

The NRC reviewer should verify the following:

a.   The mathematical method's conceptual model is compatible with the site's conceptual model (e.g., RESRAD v. 6.0 would not be an acceptable mathematical method for sites with building surface residual radioactivity).

b.   For each parameter or parameter set, the licensee has adequately justified the parameter value or range. For modifications of behavioral parameters, the changes should be based on acceptable changes in the critical group, and the mean value (or full range) of the behavior should be used.

c.   For residual radioactivity resulting in alpha decay (e.g., uranium or thorium) and present on building surfaces, NRC staff should review the resuspension factor/rate and the assumptions regarding the degree of removable residual radioactivity. For example, if the licensee has assumed that 10 % of the residual radioactivity will be removable at the time of unrestricted release, the model's parameters should either implicitly or explicitly include this assumption (see NUREG/CR–5512, Volume 3, on how it has been done for the DandD code).

d.   If the licensee has randomly sampled the parameter ranges, the licensee has used the "peak of the mean" dose distribution to either calculate the dose or derive the DCGLs.

- Uncertainty Analysis

  NRC staff should review the licensee's discussion of the uncertainty resulting from the physical parameter values used in the analysis. The review should focus on the uncertainty analysis for the critical pathways or parameters. NRC reviewers should expect that the degree of uncertainty analysis should depend on the level of complexity of the modeling (e.g., generally qualitative discussions for simple modeling to quantitative analyses for more complex sites). The overall acceptability of the uncertainty analysis should be evaluated on a case-by-case basis. For additional guidance on these subjects, refer to Section I.7 from Appendix I of this volume.

  If the licensee evaluated scenarios based on reasonably foreseeable land uses, the licensee needs to provide either a quantitative analysis of or a qualitative argument discounting the need to analyze all scenarios generated from the less likely but plausible land uses. The results of these analyses will be used by the staff to evaluate the degree of sensitivity of dose to overall scenario assumptions (and the associated parameter assumptions). The reviewer will consider both the magnitude and time of the peak dose from these scenarios. If peak doses from the less likely but plausible land use scenarios are significant, the licensee would need to provide greater assurance that the scenario is unlikely to occur, especially during the period of peak dose.

- Compliance with Regulatory Criteria

  The licensee's projections of compliance with regulatory criteria are acceptable provided that NRC staff has reasonable assurance of the following:

  1. The licensee has adequately characterized and applied its source term.

  2. The licensee has analyzed the appropriate scenario(s) and that the exposure group(s) adequately represents a critical group.

  3. The mathematical method and parameters used are appropriate for the scenario and parameter uncertainty has been adequately addressed.

  4. For deterministic analyses, the peak annual dose to the average member of the critical group for the appropriate exposure scenario(s) for the option is less than (or equal to) 0.25 mSv (25 mrem), or was used to calculate $DCGL_W$.

  5. For probabilistic analyses, the "peak of the mean" dose distribution to the average member of the critical group for the appropriate exposure scenario(s) for the option is less than (or equal to) 0.25 mSv (25 mrem), or was used to calculate $DCGL_W$.

6. Either one of the following:

    a.   The licensee has committed to using a specific scenario, model and set of parameters with the final survey results to show final compliance with the dose limit.

    b.   The licensee has committed to using radionuclide-specific DCGLs and will ensure that the total dose from all radionuclides will meet the requirements of Subpart E by using the sum of fractions.

5.3   RESTRICTED RELEASE (DECOMMISSIONING GROUP 6)

The following guidance is for reviewing DPs submitted by licensees from Decommissioning Group 6.

AREAS OF REVIEW

NRC staff should review the information provided in the DP pertaining to the licensee's assessment of the potential doses resulting from exposure to residual radioactivity remaining at the end of the decommissioning process. The findings and conclusions of the review under this section should be used to evaluate the DP's compliance with 10 CFR 20.1403. NRC staff should ensure that, at a minimum, information on the source term, exposure scenario(s), conceptual model(s), numerical analyses (e.g., hand calculations or computer models), and uncertainty have been included. NRC staff should review the abstraction and assumptions regarding the source term, the conceptual model of the site or building as appropriate, the exposure scenario(s), the mathematical method employed, and the parameters used in the analysis and their uncertainty.

The amount of information provided by the licensee and the depth of the reviewer's investigation of that information should depend on the complexity of the case and the amount of site-specific information being used by the licensee. This section has been written for review of the most complex analyses; most analyses should not need in-depth review of all parts of the evaluation criteria.

REVIEW PROCEDURES

Acceptance Review

NRC staff should review the DP to ensure that, at a minimum, the DP contains the information summarized under "Areas of Review," above. NRC staff should review the dose modeling portion of the DP without assessing the technical accuracy or completeness of the information contained therein. The adequacy of the information should be assessed during the detailed technical review. NRC staff should review the DP table of contents and the individual descriptions under "Areas of Review," above, to ensure that the licensee or responsible party has included this information in the DP and to determine if the level of detail of the information appears to be adequate for NRC staff to perform a detailed technical review.

Safety Evaluation

The material to be reviewed is technical in nature, and NRC staff should review the information provided by the licensee to ensure that the licensee used defensible assumptions and models to calculate the potential dose to the average member of the critical group. NRC staff should also verify that the licensee provided enough information to allow an independent evaluation of the potential dose resulting from the residual radioactivity after license termination and provide reasonable assurance that the decommissioning option will comply with regulations.

ACCEPTANCE CRITERIA

Regulatory Requirements

10 CFR 20.1403

Regulatory Guidance

- Appendix I of this NUREG Report

- NUREG–1200, "SRP for the review of a license application for a Low-Level Radioactive Waste Disposal Facility" [sic], Chapter 6

- NUREG–1549, "Decision Methods for Dose Assessment to Comply with Radiological Criteria for License Termination"

- NUREG–1573, "A Performance Assessment Method for Low-Level Waste Disposal Facilities: Recommendations of NRC's Performance Assessment Working Group"

- NUREG/CR–5512, Volume 1, "Residual Radioactive Contamination from Decommissioning: Technical Basis for Translating Contamination Levels to Annual Total Effective Dose Equivalent"

- Draft NUREG/CR–5512, Volume 2, "Residual Radioactive Contamination from Decommissioning: User's Manual"

- Draft NUREG/CR–5512, Volume 3, "Residual Radioactive Contamination from Decommissioning: Parameter Analysis"

- Federal Guidance Report Number 11, "Limiting Values of Radionuclide Intake and Air Concentration and Dose Conversion Factors for Inhalation, Submersion, and Ingestion" (EPA 1988)

- Federal Guidance Report Number 12, "External Exposure to Radionuclides in Air, Water, and Soil" (EPA 1993)

Information to be Submitted

NRC staff should organize this review by first looking at the overall scope of the dose modeling contained in the DP (possibly for several decommissioning options and/or critical groups). This scoping review, discussed in Chapter 5, should help the reviewers identify which section is applicable for a given dose assessment. After the scoping review, NRC staff should review each of the scenarios that the licensee is using to show compliance with the regulations.

In describing the licensee's dose modeling analysis methods, "site-specific" is used in a very general sense to describe all dose analyses except those based only on the default screening tools. This may be as simple as a few parameter changes, in the DandD computer code from their default ranges, to licensees using scenarios, models, and parameter ranges that are only applicable at the licensee's site. The information submitted should include the following:

- the source term information including nuclides of interest, the configuration of the source, the areal variability of the source, and so forth;

- a description of the compliance scenarios (for institutional controls both in place and not in place) including a description of the critical group;

- a description of any other reasonably foreseeable or less likely but plausible scenarios considered;

- a description of the conceptual model of the site including the source term, physical features important to modeling the transport pathways, and the critical group;

- the identification, description and justification of the mathematical model used (e.g., hand calculations, DandD v2.1, RESRAD v6.1);

- a description of the parameters used in the analysis;

- a discussion about the effect of uncertainty on the results; and

- input and output files or printouts, if a computer program was used.

This information to be submitted is also included as part of the master DP Checklist provided in this NUREG report (see Section V.c from Appendix D of Volume 1).

EVALUATION CRITERIA

NRC staff should determine the acceptability of the licensee's projections of radiological impacts on the average member of the critical group during the compliance period from residual radioactivity. The information in the DP is acceptable if it is sufficient to ensure a defensible assessment of the possible future impacts from the residual radioactivity. The licensee's assessment can be either realistic or prudently conservative. The information should allow an independent NRC staff evaluation of the assumptions used (e.g., source term configuration, applicable transport pathways) and possible doses to the average member of the critical group.

NRC staff should review the following information, as necessary, for each dose assessment of residual radioactivity that the licensee has submitted in the various decommissioning options.

- Source Term

  NRC staff should review the assumptions used by the licensee to characterize the facility's source term of residual radioactivity for dose modeling purposes. NRC staff should compare the assumptions with the current site information and the decommissioning alternative's goal. The model should be an appropriate generalization of this information. Three key areas of review for the source term assumptions are the (1) configuration, (2) the residual radioactivity spatial variability, and (3) the chemical form(s). For additional guidance, refer to Section I.2 from Appendix I of this volume.

  1. Configuration

     NRC staff should confirm that the actual measurements, facility history, and planned remedial action(s) support the source term configuration used in the modeling by reviewing the information in the facility history, radiological status, and planned remedial action(s) portions of the DP. The NRC reviewer should review the provided information for both the areal extent of residual radioactivity and the depth (for soil or buildings) or volume (for ground water or buried material) of the residual radioactivity. The NRC reviewer should determine if the information provided supports the configuration assumptions used in the exposure scenario and mathematical model (e.g., a thin layer of residual radioactivity on the building surfaces).

  2. Residual Radioactivity Spatial Variability

     NRC staff should review residual radioactivity concentration values provided by the licensee for conditions both before, and projected after, the decommissioning alternative is complete. For this subsection, NRC staff should review the spatial extent and the degree of heterogeneity in the values. Based on this information, NRC staff should determine whether it is reasonable to make an assumption of homogeneity for each source for either (a) the whole site or (b) the specific subsections of the site. NRC staff should then review the adequacy of the licensee's determination of a representative value (or range of values) for the residual radioactivity concentration in the source term model. At the time of the FSS, NRC staff could use, as a general guideline, the concepts related to area factors included in the MARSSIM and in Section I.3.3.3.5 of Appendix I.

     If the licensee develops DCGLs as a result of dose modeling, instead of estimating final concentrations and then, entering them into the code, the licensee need not specifically address the spatial variability acceptance criteria at this time. The licensee should provide information for NRC staff review of the FSS. NRC staff should verify that the spatial variability is compatible with the assumptions made for dose modeling.

3. Chemical Form

The licensee's assumptions regarding the chemical form of the residual radioactivity should be reviewed for its adequacy by NRC staff. NRC staff should determine whether the licensee has considered possible chemical changes that may occur during the time period of interest. Without any justification of possible chemical forms, the analysis should use the bounding chemical form(s) (i.e., the chemical form(s) that give(s) the individual the highest dose per unit intake as described in Federal Guidance Report Number 11). Acceptable rationale for other assumptions should be provided by the licensee. Some acceptable rationales for using other chemical forms are (a) chemical forms that would degrade quickly in the environment (e.g., UF_6) or (b) elements or conditions that are unavailable to realistically form that molecule (e.g., $SrTiO_4$ or high-fired UO_2).

• Critical Groups, Scenarios, and Pathways Identification and Selection

In its review, NRC staff should confirm that the licensee has identified and quantified the most significant scenarios based on available site- or facility-specific information including proposed site restrictions. A minimum of two scenarios will be necessary to evaluate both dose limits. One addresses the situation when the restrictions are in place and working properly. The other addresses the possible doses that may occur if restrictions were to fail. NRC staff should review the basis and justification for the licensee's selected critical group for each scenario. For scenarios in which possible environmental pathways have been modified or eliminated, NRC staff should review the justifications provided by the licensee for those modifications or eliminations. For additional guidance on these subjects, NRC staff is directed to Section I.3 of Appendix I and Appendix M of this volume.

1. Scenario Identification

The compliance exposure scenarios are based on the location and type of source (e.g., contaminated walls), the reasonably foreseeable land use, the general characteristics and habits of the critical group (e.g., an adult light industry worker), the possible pathways which describe how the residual radioactivity would incur dose in humans, and the potential limitations on use because of institutional controls. The licensee should provide justification for the scenario(s) evaluated.

The licensee should justify the possible land use(s) the site might experience in the future and create exposure scenarios consistent with these uses. One compliance scenario will consider the effect of potential institutional controls limiting the potential uses. The other compliance scenario will analyze the impact of the site with no institutional controls. The licensee should provide justification for selecting each of the compliance scenarios from the possible exposure scenarios derived from the land uses. The compliance scenario should result in the greatest exposure for all scenarios for the mixture of radionuclides. A licensee may chose to make a bounding assumption for land use to derive the scenarios (e.g., assuming a rural land use for an urban location) or base the scenario on the reasonably foreseeable land use that results in the highest dose.

If the compliance scenarios are based on the reasonably foreseeable land use, the licensee should provide justification for the scenarios, based on discussions with land planners, meetings with local stakeholders, trending analysis of land use for the region, or comparisons with land use in similar alternate locations. The time period of interest for possible land use is for land use changes within 100 years, depending on the rate of change in the region, and the peak exposure time. The licensee should identify what land uses are less likely but plausible, and evaluate scenarios consistent with these less likely but plausible land uses. In some cases, the use of reasonably foreseeable land use, especially those limited by institutional controls, may require the licensee to evaluate off-site uses of materials containing residual radioactivity as alternate scenarios in defining the compliance scenario.

The licensee needs to provide either a quantitative analysis of or a qualitative argument discounting the need to analyze all scenarios generated from the reasonably foreseeable land uses. The level of detail can vary between scenarios, and it is expected for the licensee to use simple analyses to limit the number of detailed scenarios. The licensee may use screening or generic analyses to assist in determining the critical scenario for compliance. With a mixture of radionuclides, more than one compliance scenario may need to be used. The compliance scenario(s) should exceed the exposures from other scenarios.

Similarly, the licensee needs to provide either a quantitative analysis of or a qualitative argument discounting the need to analyze all the scenarios generated from the less likely but plausible land uses. The results of these analyses will be used by the staff to evaluate the degree of sensitivity of dose to overall scenario assumptions (and the associated parameter assumptions). The reviewer will consider both the magnitude and time of the peak dose from these scenarios. One goal of the staff's review is to ensure that, if land uses other than the reasonably foreseeable land use were to occur in the future, significant exposures would not result. If peak doses from the less likely but plausible land use scenarios are significant, the licensee would need to provide greater assurance that the scenario is unlikely to occur, especially during the period of peak dose.

The screening scenarios for building surface residual radioactivity and soil residual radioactivity are described in NUREG–1549 and NUREG/CR–5512, Volumes 1, 2 and 3. Dose evaluations that use these scenarios (i.e., the licensee changes parameter values or mathematical method but does not change the general scenario) are acceptable, if the scenario is appropriate for the situation. In DPs where the licensee eliminates certain pathways, with justification, but still maintains the same general scenario category, NRC staff should find the scenario identification to be acceptable. For example, a licensee may eliminate the use of ground water because the near surface aquifer has total dissolved solids of 30,000 mg/L. The licensee still evaluates the impacts from crops grown in the residual radioactivity but irrigation is provided by a noncontaminated source and therefore, the screening scenario, a residential farmer, is maintained.

The restrictions at a site may result in the evaluation of an offsite exposure scenario as the compliance scenario. NUREG–1573 and Chapter 6 of NUREG–1200 provide sources to use for additional guidance focused on assessing offsite exposure.

2. Critical Group Determination

The critical group represents a group that could receive the highest dose from the residual radioactivity. In general, critical groups that are exposed to multiple exposure pathways result in higher doses than groups with more limited interaction with the residual radioactivity. NUREG–1549 and the NUREG/CR–5512 series detail the critical group assumptions for the screening scenarios. In instances where the licensee has used the screening scenarios, NRC staff should verify that the critical group is the same as that listed in NUREG–1549 and the NUREG/CR–5512 series. For example, it may be acceptable to use the screening critical group for contaminated surface soil in offsite exposure scenarios (e.g., a resident farmer using contaminated ground water flowing from the site).

The licensee should provide either a qualitative or quantitative justification that the critical group is the highest exposed group for the assumed land use(s). Separate critical groups are necessary for the two primary analysis situations: restrictions working and restrictions not in place. The selection of the critical group may be dependent on the assumption of the relative mixture of radionuclides and sources of residual radioactivity present at the site. The licensee should justify its compliance approach in these cases. A similar justification should be provided by the licensee for the critical scenario for less likely but plausible scenarios.

3. Exposure Pathways

The DP should describe the exposure pathways to which the critical group is exposed, except for cases where the licensee is using the screening scenarios and critical groups without modification. If the licensee has chosen to modify the screening scenario, the changes should be justified. In general, the justification should be based on physical limitations or situations that would not allow individuals to be exposed as described in the scenario. For other scenarios, the exposure pathways should be consistent with the land use assumptions, exposure group behavior, and physical site conditions. The licensee may also use proposed restrictions to eliminate or change exposure pathways.

For example, acceptable justifications for removing the ground water pathway based on physical limitations include any of the following: (a) the near surface ground water is neither potable nor allowed to be used for irrigation, (b) aquifer volume is insufficient to provide the necessary yields, (c) there is current (and informed consideration of future) land use patterns that would preclude ground water use, or (d) site restrictions would preclude ground water use. Justification of water quality and quantity of the saturated zone should be based on the classification systems used by EPA or the State, as appropriate.

For cases where the aquifer is classified as not being a source of drinking water, but is adequate for stock watering and irrigation, the licensee can eliminate the drinking water pathway and generally, the fish pathway, depending on the scenario. The licensee, however, should still maintain the irrigation and meat/milk pathways, if consistent with the land use assumptions.

Another example would be a rural site with a relatively small, discrete, outdoor area of residual radioactivity (compared with the area assumed in the screening scenarios). In this situation, it may be appropriate, based on the area of residual radioactivity, that gardening of some vegetables and fruits would still be an assumption, but the area is not large enough to allow one to grow grain, or raise animals for meat or milk.

- Conceptual Models

NRC staff should review the adequacy of the conceptual model(s) used by the licensee for each exposure scenario, as appropriate. For additional guidance on these subjects, refer to Appendix I, Section I.4, of this volume.

The conceptual model should qualitatively describe the following:

1. the relative location and activities of the critical group;

2. both the hydrologic and environmental transport processes important at the site;

3. the dimensions, location and spatial variability of the source term used in the model;

4. major assumptions made by the licensee in developing the conceptual model (e.g., recharge of the aquifer is limited to the infiltration through the site's footprint); and

5. the effects of the site restrictions on transport or exposure pathways.

The NRC license reviewer should verify that the site conditions and effects of site restrictions are adequately addressed in the conceptual model and simplifying assumptions.

- Calculations and Input Parameters

In its review, NRC staff should confirm that the licensee has used a mathematical model that is an adequate representation of the conceptual model and the exposure scenario. For additional guidance on these subjects, refer to the Sections I.5 and I.6 from Appendix I of this volume.

1. Execution of DandD Computer Code

    If the licensee has used the DandD computer code in its analysis, NRC staff should verify the following points:

a. The residual radioactivity is limited to the surface (building or near surface soil, as appropriate).

b. The site conceptual model is adequately represented by DandD's inherent conceptual model.

c. For building surfaces, if the total dose is greater than 10 % of the dose limit, the licensee has modified the resuspension factor to account for the removable fraction to be present at the time of decommissioning.

d. For sites eliminating pathways, the licensee has used the appropriate parameters in the DandD code as "switches" to turn off the pathways without unintentionally removing others. For example, to remove the ground water pathways, the licensee should set the drinking water rate, irrigation rate, and pond volume to zero.

e. For each parameter modified, the licensee has adequately justified the new parameter value or range and has evaluated the effect on other parameters.

f. For modifications of behavioral parameters, the changes should be based on acceptable changes in the critical group, and the mean value of the behavior should be used, although use of the range is also acceptable.

g. If the licensee has randomly sampled the parameter ranges in DandD, the licensee has used the "peak of the mean" dose distribution to either calculate the dose or derive the DCGLs.

2. Other Mathematical Methods

The NRC license reviewer should verify the following:

a. The mathematical method's conceptual model is compatible with the site's conceptual model (e.g., RESRAD Ver.6.0 would not be an acceptable mathematical method for sites with building surface residual radioactivity).

b. For each parameter or parameter set, the licensee has adequately justified the parameter value or range. For modifications of behavioral parameters, the changes should be based on acceptable changes in the critical group, and the mean value (or full range) of the behavior should be used.

c. For residual radioactivity resulting in alpha decay (e.g., uranium or thorium) and present on building surfaces, NRC staff should review the resuspension factor/rate and the assumptions regarding the degree of removable residual radioactivity. For example, if the licensee has assumed that 10 % of the residual radioactivity will be removable at the time of unrestricted release, the model's parameters should either implicitly or explicitly include this assumption (see NUREG/CR–5512, Volume 3, on how it has been done for the DandD code).

     d.   If the licensee has randomly sampled the parameter ranges, the licensee has used the "peak of the mean" dose distribution to either calculate the dose or derive the DCGLs.

- Uncertainty Analysis

  NRC staff should review the licensee's discussion of the uncertainty resulting from the physical parameter values used in the analysis. The review should focus on the uncertainty analysis for the critical pathways or parameters. NRC license reviewers should expect that the degree of uncertainty analysis will depend on the level of complexity of the modeling (i.e., generally, qualitative discussions should be for simple modeling, and quantitative discussions should be for more complex sites). The overall acceptability of the uncertainty analysis should be evaluated on a case-by-case basis. For additional guidance on these subjects, refer to Appendix I, Section I.7.

  Similarly, the licensee needs to provide either a quantitative analysis of or a qualitative argument discounting the need to analyze all scenarios generated from the less likely but plausible land uses. The results of these analyses will be used by the staff to evaluate the degree of sensitivity of dose to overall scenario assumptions (and the associated parameter assumptions). The reviewer will consider both the magnitude and time of the peak dose from these scenarios. If peak doses from the less likely but plausible land use scenarios are significant, the licensee would need to provide greater assurance that the scenario is unlikely to occur, especially during the period of peak dose.

- Compliance with Regulatory Criteria

  The licensee's projections of compliance with regulatory criteria are acceptable provided that NRC staff has reasonable assurance of all the following:

  1.   The licensee has adequately characterized and applied its source term.

  2.   The licensee has analyzed the appropriate scenario(s) and that the exposure group(s) adequately represents a critical group.

  3.   The mathematical method and parameters used are appropriate for the scenario and parameter uncertainty has been adequately addressed.

  4.   For deterministic analyses, the peak annual dose to the average member of the critical group is in compliance with the 10 CFR 20.1403(b) or 20.1403(e) dose criteria, as appropriate.

  5.   For probabilistic analyses, the "peak of the mean" dose distribution to the average member of the critical group for the appropriate exposure scenario(s) for the option is in compliance with the 10 CFR 20.1403(b) or 20.1403(e) dose criteria, as appropriate.

6. Either one of the following:

    a. The licensee has committed to using a specific scenario, model and set of parameters with the final survey results to show final compliance with the dose limit.

    b. The licensee has committed to using radionuclide-specific DCGLs and should ensure that the total dose from all radionuclides will meet the requirements of Subpart E by using the sum of fractions.

5.4 RELEASE INVOLVING ALTERNATE CRITERIA (DECOMMISSIONING GROUP 7)

The following guidance is for reviewing DPs submitted by licensees from Decommissioning Group 7.

AREAS OF REVIEW

NRC staff should review the information provided in the DP pertaining to the licensee's proposed alternate criteria. The findings and conclusions of the review under this section should be used to evaluate the DP's compliance with 10 CFR 20.1404. NRC staff should ensure that, at a minimum, information on the source term, exposure scenario(s), conceptual model(s), numerical analyses, and uncertainty have been included, if appropriate. NRC staff should review the abstraction and assumptions regarding the source term, the conceptual model of the site or building as appropriate, the exposure scenarios, the mathematical method employed, and the parameters used in the analyses and their uncertainty. NRC staff should also review the health, safety, and protection of the environment basis for the alternate criteria proposed.

The amount of information provided by the licensee and the extent of NRC staff's review of that information should depend on the complexity of the case and the amount of site-specific information being used by the licensee.

REVIEW PROCEDURES

Acceptance Review

NRC staff should review the DP to ensure that, at a minimum, the DP contains the information summarized in the above "Areas of Review." NRC staff should review the dose modeling portion of the DP without assessing the technical accuracy or completeness of the information contained therein. The adequacy of the information should be assessed during the detailed technical review. NRC staff should review the DP table of contents and the individual descriptions under the above "Areas of Review" to ensure that the licensee has included this information in the DP and to determine if the level of detail of the information appears to be adequate for NRC staff to perform a detailed technical review. NRC staff should use Section 5.3

of this volume and Chapter 6 of NUREG–1200, "SRP for the review of a license application for a Low-Level Radioactive Waste Disposal Facility" [sic], as guidelines, in developing site-specific acceptance review criteria for the proposed alternate criteria and the licensee's compliance evaluation.

Safety Evaluation

The material to be reviewed is technical in nature, and NRC staff should review the information provided by the licensee to ensure that the licensee used defensible assumptions and models to establish and demonstrate compliance with the proposed alternate criteria. NRC staff should also verify that the licensee provided enough information to allow an independent evaluation of the assessment resulting from the residual radioactivity after license termination and provide reasonable assurance that the decommissioning option should comply with regulations. Each evaluation should be performed on a case-by-case basis. NRC staff should use Section 5.3 of this volume and Chapter 6 of NUREG–1200, as guidelines, in developing site-specific review criteria for the proposed alternate criteria and the licensee's compliance evaluation.

An alternative release proposal in accordance with 10 CFR 20.1404 may allow a dose of up to 1.0 mSv/y (100 mrem/y) with restrictions in place. However, if the restrictions fail, the dose may not exceed the values in 10 CFR 20.1403(e). Furthermore, all of the other provisions of 10 CFR 20.1403 must be met.

6 ALARA ANALYSES

This chapter is applicable to Decommissioning Groups 2–7. Licensees in Decommissioning Groups 2 and 3 may only have to refer to the discussion of good housekeeping practices in the text box in Section 6.3.

6.1 SAFETY EVALUATION REVIEW PROCEDURES

AREAS OF REVIEW

NRC staff should review the information supplied by the licensee or responsible party to determine if the licensee has developed a DP that ensures that doses to the average member of the critical group are as low as is reasonably achievable (ALARA). Information submitted should include (a) a cost-benefit analysis(or qualitative arguments) for the preferred option of removing residual radioactivity to a level that meets or exceeds the applicable limit and (b) a description of the licensee's preferred method for showing compliance with the ALARA requirement at the time of decommissioning.

REVIEW PROCEDURES

Acceptance Review

NRC staff should review the DP to ensure that, at a minimum, the DP contains the information summarized under "Areas of Review," above. NRC staff should review the ALARA portion of the DP without assessing the technical accuracy or completeness of the information contained therein. The adequacy of the information should be assessed during the detailed technical review. NRC staff should review the DP table of contents and the individual descriptions under "Areas of Review," above, to ensure that the licensee or responsible party has included this information in the DP and to determine if the level of detail of the information appears to be adequate for the staff to perform a detailed technical review.

Safety Evaluation

The material supporting the optimized DP to be reviewed is technical in nature, and specific detailed technical analysis may be necessary. NRC staff should evaluate the licensee's dose estimates for various alternatives using the appropriate guidance in Chapter 5 of this volume. NRC staff should evaluate the licensee's cost estimates using the guidance in Section 4.1 from NUREG–1757, Volume 3.

6.2 ACCEPTANCE CRITERIA

REGULATORY REQUIREMENTS

10 CFR 20.1402, 20.1403(a), 20.1403(e), and 20.1404(a)(3)

REGULATORY GUIDANCE

Appendix N of this NUREG report

INFORMATION TO BE SUBMITTED

The information supplied by the licensee should be sufficient to allow NRC staff to fully understand the basis for the licensee's conclusion that projected dose limit/residual radioactivity concentrations (hereafter, the decommissioning goal) are ALARA. The decommissioning goal should be established at the point that the incremental benefits equal the incremental costs. NRC staff review should verify that the following information is included in the description of the development of the decommissioning goal:

- a description of how the licensee will achieve a decommissioning goal below the dose limit,

- a quantitative cost benefit analysis,

- a description of how costs were estimated, and

- a demonstration that the doses to the average member of the critical group are ALARA.

This information to be submitted is also included as part of the master DP Checklist provided in this NUREG report (see Section VII from Appendix D of Volume 1).

6.3   EVALUATION CRITERIA

Good Housekeeping

For ALARA during decommissioning, all licensees should use typical good practice efforts such as floor and wall washing, removal of readily removable radioactivity in buildings or in soil areas, and other good housekeeping practices. In addition, licensees should provide a description in the FSSR of how these practices were employed to achieve the final activity levels.

NRC staff review should verify that the qualitative descriptions provide reasonable assurance that the activities and decommissioning goal should result in doses that are ALARA. In light of the conservatism in the building surface and surface soil generic screening levels developed by NRC staff, the NRC staff presumes, absent information to the contrary, that licensees who remediate building surfaces or soil to the generic screening levels do not need to provide analyses to demonstrate that these screening levels are ALARA. In addition, both the "Statements of Consideration" for Subpart E and the Final Generic Impact Statement (NUREG–1496) provide that an ALARA analysis for unrestricted release of soil need not be done. See Example 3 in Section N.1.4 in Appendix N of this volume.

For those situations in which a licensee prepares cost-benefit analyses, NRC staff should ensure that the analyses are developed using the methodology described in Appendix N and applied as described in the following text.

CALCULATION OF BENEFITS

Appendix N of this volume discusses five different possible benefits: (1) collective dose averted, (2) regulatory costs avoided, (3) changes in land values, (4) esthetics, and (5) reduction in public opposition. Numerical estimates will generally only be available for the first three benefits, if they are appropriate. Qualitative analysis of the benefits can be done especially if the costs are large (e.g., no matter what the change in land value is, the costs will exceed the benefits). In most comparisons between alternatives in the same class (e.g., both alternatives result in unrestricted release), the only important benefit should be the collective dose averted. In comparisons between restricted and unrestricted release, the other benefits can become important.

The collective dose averted is generalized as the incremental dose difference between the licensee's approach (hereafter, preferred option) and the alternative under analysis. Therefore, NRC staff needs to ensure that the licensee has calculated the benefits correctly by using the correct population density, area, and averted dose. This may require technical analysis of the dose modeling, and the NRC license reviewer should use Chapter 5 for these cases. If the licensee has used discounting, NRC staff should ensure that the proper rates were used. The licensee is not required to discount because the discount reduces the benefits of adverting dose in later time periods.

For compliance with 10 CFR 20.1403(a), one acceptable method of compliance is to demonstrate that cleanup to the unrestricted release criteria is beyond ALARA considerations. In this case, a beneficial estimate should include costs that would be avoided if the site were to be released for unrestricted use, including calculation of site control and maintenance costs and should include estimation of the additional regulatory costs associated with termination of a restricted site (e.g., development of an environmental impact statement, public meetings).

NRC staff should ensure that the licensee has properly documented the basis for any estimates of changes in land values. Acceptable sources of such estimates include real estate agents familiar with the local area and the issues involved or governmental assessors (e.g., county, State).

CALCULATION OF COSTS

NRC staff should verify that the licensee has adequately estimated the effective monetary costs of the incremental remediation by using the equations in Appendix N of this volume. To review the calculated monetary costs of the incremental remediation, NRC staff should use Section 4.1 of NUREG–1757, Volume 3, with the following changes (this may require calculating total cost estimates for the preferred option and each alternative):

- The cost estimate should be based on actual costs expected to be incurred by decommissioning the facility and should not assume that the work will be performed by an independent third-party contractor.

- The cost estimate *does* take credit for (a) any salvage value that might be realized from the sale of potential assets during or after decommissioning or (b) any tax reduction that might result from payment of decommissioning costs and/or site control and maintenance costs.

- The decommissioning cost estimates should reflect the actual situation rather than maximized assumptions.

For each of the cost terms (e.g., disposal costs, worker fatalities) the incremental difference between the preferred and the alternate options may be negative (i.e., the alternative "costs" less than the preferred option).

NRC staff should verify that the licensee's proposed demonstration that doses to the average member of the critical group are ALARA. There are two approaches to demonstrate compliance with the ALARA requirement at the end of decommissioning: (1) a predetermined acceptable dose limit or concentration guideline(s) or (2) an acceptable preferred option and decommissioning goal with organizational oversight and review during decommissioning. Both options have their own advantages and disadvantages. Establishment of the compliance method needs to be made by the licensee, with the staff reviewing the applicability, given the site-specific information.

PREDETERMINED COMPLIANCE MEASURE

Under the predetermined compliance measure, the licensee would agree to meet the dose calculated for the preferred option or the radiological concentrations associated with this dose. This could be met by either establishing deterministic concentration limits for the site or agreeing to use a specified dose scenario with associated parameters and assumptions. If the licensee's final survey results meet the self-imposed concentration limits (or dose limit), the licensee has met the ALARA requirement.

PERFORMANCE-BASED COMPLIANCE

Performance-based compliance allows a licensee to adjust its ALARA assessment during decommissioning to deal with actual site conditions experienced and actual costs incurred. The philosophy behind this compliance measure is very similar to how ALARA is handled during routine operations. The licensee's DP needs to meet all of the following criteria to use this approach:

- The preferred option, based on valid assumptions, would result in reducing residual activity to ALARA levels, as described above.

- The licensee has established decommissioning guidelines (either dose or concentrations) based on the DP's analysis.

- The licensee has a documented method to review the effectiveness of the remediation activities. This method should include all of the following:

  — An ALARA committee or RSO, for small licensees, similar to operations requirements.

  — An establishment of appropriate review frequency established.

  — An acceptable set of criteria on the scope of activities/commitments that the ALARA committee can change.

  — A commitment for acceptable documentation of ALARA findings that result in the licensee making changes in its remediation activities or decommissioning guidelines.

  — A commitment to provide annually to NRC, all necessary page changes to the DP because of ALARA findings.

At the end of remediation, a licensee using the performance-based approach should meet the following criteria:

- The final survey results satisfy the appropriate dose limit(s).

- Any substantial weaknesses in the ALARA program that were found during licensee audits or NRC inspections have been resolved.

- Any deviation from the decommissioning goal presented in the DP is properly justified by the ALARA committee findings. For long-term projects, these should be reviewed annually by the NRC license reviewer or inspection staff.

Appendix A

Implementing the MARSSIM Approach
for Conducting
Final Radiological Surveys

This appendix is applicable to Decommissioning Groups 2–7.

NRC regulations in 10 CFR 20.1501(a) require licensees to make or cause to be made surveys that may be necessary for the licensee to comply with the regulations in Part 20.

The final status survey (FSS) is the radiation survey performed after an area has been fully characterized, remediation has been completed, and the licensee believes that the area is ready to be released. The purpose of the FSS is to demonstrate that the area meets the radiological criteria for license termination. The FSS is not conducted for the purpose of locating residual radioactivity; the historical site assessment (HSA) and the characterization survey perform that function.

NRC endorses the FSS method described in MARSSIM. This appendix (a) provides an overview of the MARSSIM approach for conducting a final radiological survey, (b) provides additional specific guidance on acceptable values for use in the MARSSIM method, (c) describes how to use the MARSSIM method in a way that is consistent with the dose modeling, (d) describes how to use the MARSSIM method to meet NRC's regulations, and (e) describes how to extend or supplement the MARSSIM method to certain complex situations that may be encountered, such as how to address subsurface residual radioactivity. Note that the guidance in this appendix does not replace the MARSSIM, and licensees and reviewers should refer to, and use, the MARSSIM for designing final radiological surveys to support decommissioning. This guidance assumes a working knowledge of the MARSSIM approach and terminology and does not attempt to provide a comprehensive overview of the entire MARSSIM. In addition, for Decommissioning Groups 1–3, licensees may also use the alternative, simpler final survey methods described in Appendix B of this volume. Refer to Appendix O of this volume for information on lessons learned from recently submitted DPs and questions and answers to clarify existing license termination guidance related to implementing the MARSSIM approach.

Survey checklists are found in Chapter 5 of MARSSIM. These checklists are useful in implementing the steps of the Radiation Site Survey and Investigation (RSSI) process (Decommissioning Groups 3–7). These checklists present a useful tool for visualizing the sequential steps (i.e., design, performance, and evaluation) of the survey process. Furthermore, the use of these checklists should ensure that the necessary information is collected for each type of survey. Sites not using the Radiation Site Survey and Investigation (RSSI) process, such as Decommissioning Groups 1 and 2, should also find these checklists or parts of these checklists useful.

A.1   Classification of Areas by Residual Radioactivity Levels

The licensee should classify site areas based on levels of residual radioactivity from licensed activities. The area classification method contained in Section 4.4 of MARSSIM is acceptable to NRC staff. Its essential features are described below.

The licensee should first classify site areas as impacted or non-impacted. *Impacted areas* are areas that may have residual radioactivity from the licensed activities. *Non-impacted areas* are areas without residual radioactivity from licensed activities. Impacted areas should be identified by using knowledge of past site operations together with site characterization surveys. In the FSS, radiation surveys do not need to be conducted in non-impacted areas. The licensee should classify impacted areas into one of the three classes, listed below, based on levels of residual radioactivity.

- **Class 1 Areas**: Class 1 areas are impacted areas that are expected to have concentrations of residual radioactivity that exceed the $DCGL_W$. ($DCGL_W$ is defined in the Glossary of this volume.)

- **Class 2 Areas**: Class 2 areas are impacted areas that are not likely to have concentrations of residual radioactivity that exceed the $DCGL_W$.

- **Class 3 Areas**: Class 3 areas are impacted areas that have a low probability of containing residual radioactivity.

Surveys conducted during operations or during characterization at the start of decommissioning are the bases for classifying areas. If the available information is not sufficient to designate an area as a particular class, the area either should be classified as Class 1 or should be further characterized. Areas that are considered to be on the borderline between classes should receive the more restrictive classification.

NRC staff recognizes that there may be a need for a licensee to reclassify Class 1 Areas to Class 2, when insufficient information was available for the initial classification. If more information becomes available to indicate that another classification is more appropriate, the guidance in MARSSIM allows for classifications to be changed at any time before the FSS. For more guidance on criteria for downgrading classifications (e.g., from Class 1 to Class 2), a licensee should refer to MARSSIM, in particular, Sections 2.2, 2.5.2, and 5.5.3. If a licensee plans to make use of reclassification during the RSSI process, the licensee should provide in the DP the criteria and methodology the licensee plans to use for reclassification. In addition, a licensee contemplating use of reclassification is encouraged to consult with NRC staff.

For soils, impacted areas in Classes 1 and 2 should also be classified by whether they have substantial amounts of subsurface residual radioactivity. This classification should be based on the HSA and site characterization. In this context "substantial amounts of subsurface residual radioactivity" would be defined as an amount of radioactivity, or contaminated material (such as soil) that could contribute at least 10 % of the potential dose to the average member of the critical group or soil that exceeded the $DCGL_{EMC}$.

Determining whether there is a substantial amount of subsurface residual radioactivity (deeper than 15 centimeters) should not require a complex set of characterization measurements. In most cases there will be either substantial amounts of residual radioactivity or only traces (such as in occasional small pockets or from leaching from surface layers by rainwater). When there are small amounts of residual radioactivity below 15 centimeters, the MARSSIM survey methods for surface measurements are acceptable. When there are substantial amounts of residual radioactivity below 15 centimeters, the dose modeling and the survey methods should be modified to account for the subsurface residual radioactivity.

The presence of subsurface residual radioactivity is usually determined by the HSA (see Chapter 3 of MARSSIM), applying knowledge of how the residual radioactivity was deposited. Characterization surveys to detect subsurface residual radioactivity in soil are not routinely conducted unless there is reason to expect that subsurface residual radioactivity may be present. The need to survey or sample subsurface soil will depend, in large part, on the quality of the information used to develop the HSA, the environmental conditions at the site, the types and forms (chemical and radiological) of the radioactive material used at the site, the authorized activities and the manner in which licensed material was managed during operations.

NRC staff's experience has shown that submittal of the DP should occur only after sufficient site characterization has occurred. NRC staff suggests that the DP provide sufficient information demonstrating the characterization of the radiological conditions of site structures, facilities, surface and subsurface soils, and ground water. NRC staff has observed that some DPs have been submitted with incomplete or inadequate characterizations of radiological conditions. A review of such DPs has shown that the lack of information makes it difficult to agree with the rationale justifying the proposed classification of survey units. NRC staff suggests that the following issues related to the use of characterization survey results and classification of survey units be considered when developing a DP:

- use of operational, post-shutdown scoping, or turnover surveys as characterization surveys;

- reclassification of survey units; and

- completeness of characterization survey design and results.

Regulatory Issue Summary 2002–02 provides a detailed discussion of this issue.

A.2   Selection and Size of Survey Units

The licensee should divide the impacted area into survey units based on the classification described above. A survey unit is a portion of a building or site that is surveyed, evaluated, and released as a single unit. The entire survey unit should be given the same area classification. Section 4.6 of MARSSIM contains a method acceptable to NRC staff for dividing impacted areas into survey units. The important features of this method are summarized here.

For buildings, it is normally appropriate to designate each separate room as either 1 or 2 survey units (e.g., floors with the lower half of walls and upper half of walls with ceiling) based on the pattern of potential of residual radioactivity. It is generally not appropriate to divide rooms of normal size (100 m² area or less) into more than two survey units because the dose modeling is based on the room being considered as a single unit. However, very large spaces such as warehouses may be divided into multiple survey units.

For soil, survey units should be areas with similar operational history or similar potential for residual radioactivity to the extent practical. Survey units should be formed from areas with the same classification to the extent practical, but if areas with more than one class are combined into one survey unit, the entire survey unit should be given the more restrictive classification. Survey units should have relatively compact shapes and should not have highly irregular (gerrymandered) shapes unless the unusual shape is appropriate for the site operational history or the site topography.

Suggested survey unit areas from MARSSIM are given in Table A.1. These areas are suggested in MARSSIM because they give a reasonable sampling density and they are consistent with most commonly used dose modeling codes. However, the size and shape of a particular survey unit may be adjusted to conform to the existing features of the particular site area.

Table A.1   Suggested Survey Unit Areas (MARSSIM, Roadmap Table 1)

Suggested Survey Unit Area		
Class	Structures	Land
1	up to 100 m²	up to 2000 m²
2	100 to 1000 m²	2000 to 10,000 m²
3	no limit	no limit

A.3   Selection of Background Reference Areas and Background Reference Materials

A.3.1   Need for Background Reference Areas

Background reference areas are not needed when radionuclide-specific measurements will be used to measure concentrations of a radionuclide that is not present in background. Background reference areas are needed for the MARSSIM method if (a) the residual radioactivity contains a radionuclide that occurs in background, or (b) the sample measurements to be made are not radionuclide-specific. However, a licensee may find cost benefits to consider the background for a particular radionuclide as zero or some other appropriately low value approved by the staff, recognizing that this is a risk-informed approach. The survey unit itself may serve as the reference area when a surrogate radionuclide in the survey unit can be used to determine background. For example, it may be possible to use radium-226 as a surrogate for natural uranium. (More information on the use of surrogate radionuclides is provided in Section 4.3.2 of MARSSIM.)

Multiple reference areas may be used if reference areas have significantly different background levels because of the variability in background between areas. (See Section A.3.4 below and Section 13.2 of NUREG–1505.) A derived reference area may be used when it is necessary to extract background information from the survey unit because a suitable reference area is not readily available. For example, it may be possible to derive a background distribution based on areas of the survey unit where residual radioactivity is not present.

A.3.2   Characteristics of Soil Reference Areas

The objective is to select non-impacted background reference areas where the distribution of measurements should be the same as that which would be expected in the survey unit if that survey unit had never been contaminated. An acceptable method for selecting background areas is contained in Section 4.5 of MARSSIM and is briefly described below.

For soils, reference areas should have a soil type as similar to the soil type in the survey unit as possible. If there is a choice of possible reference areas with similar soil types, consideration should be given to selecting reference areas that are most similar in terms of other physical, chemical, geological, and biological characteristics. Each reference area should have an area at least as large as the survey unit, if practical, in order to include the full potential spatial variability in background concentrations. Reference areas may be offsite or onsite, as long as they are non-impacted. NUREG–1506 provides additional information on reference area selection. Licensees should consult with NRC staff when they are unable to find a reference area that satisfies the above criteria.

A.3.3   Different Materials in a Survey Unit

Survey units may contain a variety of materials with markedly different backgrounds. An example might be a room with drywall walls, concrete floor, glass windows, metal doors, wood trim, and plastic fixtures. It is not appropriate to make each material a separate survey unit because the dose modeling is based on the dose from the room as a whole and because a large number of survey units in a room would require an inappropriate number of samples.

When there are different materials with substantially different backgrounds in a survey unit, the licensee may use a reference area that is a non-impacted room with roughly the same mix of materials as the survey unit.

If a survey unit contains several different materials, but one material is predominant or if there is not too great a variation in background among materials, a background from a reference area containing only a single material may still be appropriate. For example, a room may be mostly concrete but with some metal beams, and the residual radioactivity may be mostly on the concrete. In this situation where the concrete predominates, it would be acceptable to use a reference area that contained only concrete. However, the licensee should demonstrate that the selected reference area will not result in underestimating the residual radioactivity on other materials.

The licensee may also use measured backgrounds for the different materials or for groups of similar materials. When the licensee decides to use different measured backgrounds for different materials or for a group of materials with similar backgrounds, it is acceptable to perform a one-sample test on the difference between the paired measurements from the survey unit and from the appropriate reference material. An acceptable method to do this is described in detail in Chapter 2 of NUREG–1505.

For onsite materials, present either in buildings or as nonsoil materials present in outdoor survey units (e.g., concrete, brick, drywall, fly ash, petroleum product wastes), the licensee should attempt to find non-impacted materials that are as similar as possible to the materials on the site. Sometimes such materials will not be available. In those situations, the licensee should make a good faith effort to find the most similar materials readily available or use appropriate published estimates.

A.3.4   Differences in Backgrounds Between Areas

When using a single reference area, any difference in the mean radionuclide concentration between the survey unit and the reference area would be interpreted as caused by residual radioactivity from site operations. This interpretation may not be appropriate when the variability in mean background concentrations among different reference areas is a substantial fraction of the $DCGL_W$. When there may be a significant difference in backgrounds between different areas, a Kruskal–Wallis test, as described in Chapter 13 of NUREG–1505, can be conducted to determine whether there are, in fact, significant differences in mean background concentrations among potential reference areas.

While NUREG–1505 does not recommend specific values for the Kruskal–Wallis test, NRC staff recommends at least 15 samples in each of at least 4 reference areas and a Type I error rate of $\alpha_{KW} = 0.2$ to provide an adequate number of measurements for the determination of whether there is a significant difference in the background concentrations. However, different values may be appropriate on a site–specific basis.

If significant differences in backgrounds among reference areas are found, NRC staff recommends that a value of three times the standard deviation of the mean of the reference area background values should be added to the mean of the reference area background to define a background concentration. A value of three times the standard deviation of the mean is chosen to minimize the likelihood that a survey unit that contains only background would fail the statistical test for release. A two-sample test (see Section A.4, below) should then be used to test whether the survey unit meets the radiological criteria for license termination. This method is described in detail in Chapter 13 of NUREG–1505.

A.3.5 Background Survey Design

This survey constitutes measurements of non-impacted areas on and surrounding the site in order to establish the baseline, that is, the normal background levels of radiation and radioactivity. In some situations, historical measurements may be available from surveys performed before the construction and operation of a facility. Areas such as roads, parking lots, and other large paved surfaces that may have been impacted or disturbed by site-related activities should be avoided. The background survey takes on added importance since the licensee may decide to use a statistical test that compares impacted areas to off or onsite reference areas in order to demonstrate compliance with the release criteria in 10 CFR Part 20, Subpart E. To minimize systematic biases in the comparison, the same sampling procedure, measurement techniques, and type of instrumentation (e.g., detection sensitivity and accuracy) should be used at both the survey unit and the reference area.

NUREG–1505 provides additional guidance on survey design, the methods of accounting for background radiation, and the nonparametric statistical methods for testing compliance with the decommissioning criteria in 10 CFR Part 20, Subpart E. Formulas contained in NUREG–1505 can be used to compute the required number of samples (measurement points) that will be needed in both the background reference and survey areas.

A.4 Methods to Evaluate Survey Results

All survey units should be evaluated to determine whether the average concentration in the survey unit as a whole is below the $DCGL_W$. If the radionuclide is not present in background and the measurement technique is radionuclide-specific so that comparison with a reference area is not necessary, a one-sample test, the Sign test, should be used. This test is described in Section 8.3 of MARSSIM.

When the residual radioactivity contains a radionuclide present in the environment or when the measurements are not radionuclide-specific, the survey unit should be compared to a reference area. When the survey unit will be compared to a reference area, a two-sample test, the Wilcoxon Rank Sum (WRS) test, should be used. This test is described in Section 8.4 of MARSSIM.

A.4.1 A Case for Not Subtracting Background

An exception to using a two-sample test when a radionuclide is present in background is when the licensee plans to assume that all the radionuclide activity in the survey unit is caused by licensed operations and none is from background. This could be the case for cesium-137, for example, because the levels in the environment are often so much less than the $DCGL_W$ that background concentrations may be ignored.

A.4.2   Elevated Measurements Comparison

Class 1 survey units that pass the Sign test or WRS test but have small areas with concentrations exceeding the $DCGL_W$ should also be tested to demonstrate that those small areas meet the dose criteria for license termination. This test is called the elevated measurement comparison (EMC). It is described in Section 8.5.1 of MARSSIM and summarized here.

To perform the EMC, the size of the area in the survey unit with a concentration greater than the $DCGL_W$ is determined, then the area factor for an area of that size is determined. (The area factor is the multiple of the $DCGL_W$ that is permitted in a limited area of a survey unit. See Section A.7.5.) The average concentration in the area is also determined. The EMC is acceptable if the following condition is met as shown in Equation A–1 (adapted from MARSSIM Equation 8-2):

$$\frac{\delta}{DCGL_W} + \frac{average\ concentration\ in\ the\ elevated\ area\ -\ \delta}{area\ factor\ for\ elevated\ area\ \times\ DCGL_W} < 1 \qquad \text{(A–1)}$$

where δ =   the average residual radioactivity concentration for all sample points in the survey unit only.

If there is more than one elevated area, a separate term should be included for each one.

As an alternative to the unity rule expressed in Equation A–1, the dose from the actual distribution of residual radioactivity can be calculated if there is an appropriate exposure pathway model available.

A.5   Instrument Selection and Calibration

To demonstrate that the radiological criteria for license termination have been met, the measurement instruments should have an adequate sensitivity, be calibrated properly, and be checked periodically for proper response.

A.5.1   Calculation of Minimum Detectable Concentrations

The licensee should determine the MDC for the instruments and techniques that will be used. The MDC is the concentration that a specific instrument and technique can be expected to detect 95 % of the time under actual conditions of use.

For scanning building surfaces for beta and gamma emitters, the MDC_{scan} should be determined from the following equation (obtained by combining MARSSIM Equations 6-8, 6-9, and 6-10 and using a value recommended in this appendix for the index of sensitivity d' of 1.38, which is for 95 % detection of a concentration equal to MDC_{scan} with a 60 % false-positive rate).

$$MDC_{scan} \text{ (building surfaces)} = \frac{270{,}000 \times 1.38 \sqrt{B}}{\sqrt{p} \; \epsilon_i \; \epsilon_s \; A \; t} \qquad (A-2)$$

where MDC_{scan} = minimum detectable concentration for scanning building surfaces in pCi/m^2

271,000 ... 270,000 = conversion factor to convert to pCi/m^2

1.38 = index of sensitivity d'

B = number of background counts in time interval t

p = surveyor efficiency

ϵ_i = instrument efficiency for the emitted radiation

ϵ_s = source efficiency in emissions/disintegration

A = probe's sensitive area in cm^2

t = time interval of the observation while the probe passes over the source in seconds

Based on the measurements described in NUREG/CR–6364, a surveyor efficiency p of 0.5 represents a mean value for normal field conditions and its use is generally acceptable. If the licensee wants to determine a value appropriate for particular measurement techniques, the information in NUREG/CR–6364 describes how the value can be determined. NUREG–1507 provides additional information on the interpretation of results reported in NUREG/CR–6364.

For scanning soil with a sodium iodide gamma detector, the MDC_{scan} values given in Table 6.7 of MARSSIM provide an acceptable estimate of MDC_{scan}.

For static measurements of surface concentrations, the MDC_{static} may be calculated using the following equation (from NUREG–1507, Equation 3-10):

$$MDC_{static} = \frac{3 + 4.65 \sqrt{B}}{K \; t} \qquad (A-3)$$

where MDC_{static} = minimum detectable concentration in pCi/m^2 or pCi/g

B = background counts during measurement time interval t

t = counting time in seconds

K = a calibration constant (best estimate) to convert counts/second to pCi/m^2 or pCi/g and is discussed further in NUREG–1507.

An example using this equation is shown in Section 6.7.1 of MARSSIM.

The instruments used for sample measurements at the specific sample locations should have an MDC_{static} less than 50 % of the $DCGL_W$ as recommended in Section 4.7.1 of MARSSIM. There is no specific recommendation for the MDC_{scan}, but the MDC_{scan} will determine the number of samples needed, as discussed in Section A.7.6 of this appendix.

The licensee should record all numerical values measured, even values below the "minimum detectable concentration" or "critical level," including values that are negative (when the measured value is below the average background). Entries for measurement results should not be "nondetect," "below MDC," or similar entries because the statistical tests can only tolerate a maximum of 40% nondetects.

A.5.2 Instrument Calibration and Response Checks

NRC regulations at 10 CFR 20.1501(b) require that the licensee periodically calibrate radiation measurement instruments used in surveys such as the FSS.

For *in situ* gamma measurements, the detector efficiency (count rate per unit fluence rate) should be determined for the gamma energies of interest and the assumed representative depth distribution. The surface and volumetric distributions should be explicitly considered to evaluate potential elevated areas. To calibrate for the representative depth distribution, acceptable methods are to (a) use a test bed with radioactive sources distributed appropriately or (b) use primarily theoretical calculations that are normalized or verified experimentally using a source approximating a point source. The calibration of the source used for the verification source should be traceable to a recognized standards or calibration organization, for example, the National Institute of Standards and Technology.

Some modern instruments are very stable in their response. Thus, as long as instrument response checks are performed periodically to verify that the detector is operating properly, it may be acceptable to calibrate only initially without periodic recalibrations. The initial calibration may be performed by either the instrument supplier or the licensee, but in either case, 10 CFR 20.2103(a) requires that a record describing the calibration be available for inspection by NRC.

A.5.3 Instrument Response Checks

The response of survey instruments should be checked with a check source to confirm constancy in instrument response each day before use. Licensees should establish criteria for acceptable response. If the response is not acceptable, the instrument should be considered as not responding properly and should not be used until the problem has been resolved. Measurements made after the last acceptable response check should be evaluated and discarded, if appropriate.

The check source should emit the same type of radiation (i.e., alpha, beta, gamma) as the radiation being measured and should give a similar instrument response, but the check source does not have to use the same radionuclide as the radionuclide being measured.

A.6 Scanning Coverage Fractions and Investigation Levels

Scanning is performed to locate small areas of elevated concentrations of residual radioactivity to determine whether they meet the radiological criteria for license termination. The licensee should perform scanning in each survey unit to detect areas of elevated concentrations. The licensee should establish investigation levels for investigating significantly elevated concentrations of residual radioactivity. Acceptable scanning coverage fractions and scanning investigation levels for buildings and land areas are shown in Table A.2. This table is based on MARSSIM Roadmap Tables 2 and 5.8.

Systematic scans are those conducted according to a preset pattern. Judgmental scans are those conducted to include areas with a greater potential for residual radioactivity. In Class 2 areas, a 10 % scanning coverage would be appropriate when there is high confidence that all locations would be below the $DCGL_W$. A coverage of 25 % to 50 % would be appropriate when there may be locations with concentrations near the $DCGL_W$. A coverage of 100 % would be appropriate if there is any concern that the area should have had a Class 1 classification rather than a Class 2 classification. In Class 3 areas, scanning coverage is usually less than 10 %. If any location exceeds the scanning investigation level, scanning coverage in the vicinity of that location should be increased to delineate the elevated area.

Table A.2 Scanning Coverage Fractions and Scanning Investigation Levels

Class	Scanning Coverage Fraction	Scanning Investigation Levels
1	100 %	$>DCGL_{EMC}$
2	10 to 100 % for soil and for floors and lower walls of buildings, 10 to 50 % for upper walls and ceilings of buildings, systematic and judgmental	$>DCGL_W$ or $>MDC_{scan}$ if MDC_{scan} is greater than $DCGL_W$
3	Judgmental	$>DCGL_W$ or $>MDC_{scan}$ if MDC_{scan} is greater than $DCGL_W$

Sometimes the sensitivity of static measurements at designated sample points is high enough to detect significantly elevated areas between sample points. If the sensitivity is high enough, only this single set of measurements is necessary. For example, both scanning and sampling for cobalt-60, which emits an easily detectable gamma, can be done with a single set of *in situ* measurements in some cases.

A.7   Determining the Number of Samples Needed

A minimum number of samples are needed to obtain sufficient statistical confidence that the conclusions drawn from the samples are correct. The method described below from Chapter 5 of MARSSIM is acceptable for determining the number of samples needed.

A.7.1   Determination of the Relative Shift

The number of samples needed will depend on a ratio involving the concentration to be measured relative to the variability in the concentration. The ratio to be used is called the relative shift, Δ/σ_s. The relative shift, Δ/σ_s, is defined in Section 5.5.2.2 of MARSSIM as:

$$\Delta/\sigma_s = \frac{DCGL_W - LBGR}{\sigma_s} \qquad \text{(A–4)}$$

where $DCGL_W$ = derived concentration guideline

$LBGR$ = concentration at the lower bound of the gray region. The $LBGR$ is the concentration to which the survey unit must be cleaned in order to have an acceptable probability of passing the test (i.e., $1-\beta$).

σ_s = an estimate of the standard deviation of the concentration of residual radioactivity in the survey unit (which includes real spatial variability in the concentration as well as the precision of the measurement system)

The value of σ_s is determined either from existing measurements or by taking limited preliminary measurements of the concentration of the residual radioactivity in the survey unit at about 5 to 20 locations as recommended in Section 5.5.2.2 of MARSSIM. If a reference area will be used and the estimate of the standard deviation in the reference area, σ_r, is larger than the estimate of the standard deviation in the survey unit, σ_s, then the larger value should be used in the equation.

The $LBGR$ should be set at the mean concentration of residual radioactivity that is estimated to be present in the survey unit. However, if no other information is available regarding the survey unit, the $LBGR$ may be initially set equal to 0.5 $DCGL_W$, as recommended by the MARSSIM. If the relative shift, Δ/σ_s, exceeds 3, the $LBGR$ should be increased until Δ/σ_s is equal to 3. The licensee may refer to Section 5.5.2.2 of MARSSIM for additional details and information.

A.7.2   Determination of Acceptable Decision Errors

A decision error is the probability of making an error in the decision on a survey unit by failing a survey unit that should pass or by passing a survey unit that should fail. When using the statistical tests, larger decision errors may be unavoidable when encountering difficult or adverse measuring conditions. This is particularly true when trying to measure residual radioactivity concentrations close to the variability in the concentration of those materials in natural background.

The α decision error is the probability of passing a survey unit whose actual concentration exceeds the release criterion. A decision error α of 0.05 is acceptable under the more favorable conditions when the relative shift, Δ/σ_s, is large (about 3 or greater). Larger values of α may be considered when the relative shift is small to avoid an unreasonable number of samples. The β decision error is the probability of failing a survey unit whose actual concentration is equal to *LBGR*. Any value of β is acceptable to NRC.

A.7.3 Number of Samples Needed for the Wilcoxon Rank Sum (WRS) Test

The minimum number of samples, *N*, needed in each survey unit for the WRS test may be determined from the following equation (adapted from MARSSIM Equation 5–1 with *N* redefined as the number of samples in the survey unit):

$$N \ = \ \frac{1}{2} \times \frac{(Z_{1-\alpha} + Z_{1-\beta})^2}{3 \, (P_r - 0.5)^2} \tag{A–5}$$

where N =   the number of samples in the survey unit
$Z_{1-\alpha}$ =   the percentile represented by the decision error α
$Z_{1-\beta}$ =   the percentile represented by the decision error β
P_r =   the probability that a random measurement from the survey unit exceeds a random measurement from the background reference area by less than the $DCGL_W$ when the survey unit median is equal to the *LBGR* concentration above background
½ =   a factor added to MARSSIM Equation 5–1 because *N* always is defined in this guide as the number of samples in the survey unit

Values of P_r, $Z_{1-\alpha}$, and $Z_{1-\beta}$, are tabulated in Tables 5.1 and 5.2 of MARSSIM. *N* is the minimum number of samples necessary in each survey unit. An additional *N* samples will also be needed in the reference area. If *N* is not an integer, the number of samples is determined by rounding up. In addition, the licensee should consider taking some additional samples (MARSSIM recommends 20 %) to protect against the possibility of lost or unusable data. Fewer samples increase the probability of an acceptable survey unit failing to demonstrate compliance with the radiological criteria for release.

A.7.4 Number of Samples Needed for Sign Test

The number of samples N needed in a survey unit for the sign test may be determined from the following equation (adapted from MARSSIM Equation 5–2):

$$N = \frac{(Z_{1-\alpha} + Z_{1-\beta})^2}{4 \, (Sign \ p - 0.5)^2} \qquad (A–6)$$

where
N = number of samples needed in a survey unit
$Z_{1-\alpha}$ = percentile represented by the decision error α
$Z_{1-\beta}$ = percentile represented by the decision error β
$Sign \ p$ = estimated probability that a random measurement for the survey unit will be less than the $DCGL_W$ when the survey unit median concentration is actually at the LBGR.

Values of $Z_{1-\alpha}$, $Z_{1-\beta}$, and $Sign \ p$ are tabulated in Tables 5.2 and 5.4 of MARSSIM. In addition, the licensee should consider taking some additional samples (MARSSIM recommends 20 %) to protect against the possibility of lost or unusable data. Fewer samples increase the probability of an acceptable survey unit failing to demonstrate compliance with the radiological criteria for release. If a survey unit fails to demonstrate compliance because there were not enough samples taken, a totally new sampling effort may be needed unless resampling was anticipated.

A.7.5 Use of Two-Stage or Double Sampling

It may be desirable for a licensee to sample a survey unit a second time to determine compliance. "Two-stage sampling" and "double sampling" are two methods by which additional survey unit data can be acquired. Two-stage sampling refers to survey designs specifically intended to be conducted in two stages. Double sampling refers to the case when the survey unit design is a one-stage design, but allowance is made for a second set of samples to be taken if the retrospective power of the test using the first set of samples does not meet the design objectives. Use of either method should be considered as part of the DQO process when developing the design of the FSS. Refer to Appendix C of this volume for information on the use of two-stage or double sampling.

A.7.6 Additional Samples for Elevated Measurement Comparison in Class 1 Areas

Additional samples may be needed when the concentration that can be detected by scanning, MDC_{scan}, is larger than the $DCGL_W$. The licensee should determine whether additional samples are needed in Class 1 survey units for the elevated measurement comparison when the concentration that can be detected by scanning, MDC_{scan}, is larger than the $DCGL_W$. The method in Section 5.5.2.4 of MARSSIM to determine whether additional samples are needed is acceptable to NRC staff and is described here.

The area factor is the multiple of the $DCGL_W$ that is permitted in a limited portion of the survey unit. In Equation A–7, the ratio of the MDC_{scan} to the $DCGL_W$ establishes the area factor (the multiple of the $DCGL_W$) that can be detected by scanning (adapted from MARSSIM Equation 5–4):

$$area\ factor = \frac{MDC_{scan}}{DCGL_W} \tag{A–7}$$

Using the methods in NUREG–1549, the size of the area corresponding to the area factor, A_{EC}, can be determined. The number of sample points that may be needed to detect this area of elevated measurement concentration, N_{EMC}, in a survey unit is:

$$N_{EMC} = \frac{A}{A_{EC}} \tag{A–8}$$

where A = the area of the survey unit
A_{EC} = the area of concentration greater than $DCGL_W$

If N_{EMC} is larger than N, additional samples may be needed to demonstrate that areas of elevated concentrations meet the radiological criteria for license termination. However, the number of samples needed is not necessarily N_{EMC}. To determine how many additional samples may be needed, the HSA and site characterization should be considered. Based on what is known about the site, it may be possible to estimate a concentration that is unlikely to be exceeded. If there is a maximum concentration, the size of the area corresponding to this area factor for this concentration may be used for A_{EC} in Equation A–8. Similarly, based on knowledge of how the radioactive material was handled or dispersed on the site, it may be possible to estimate the smallest area likely to have elevated concentrations. If this is so, that area can be used in Equation A–8. Likewise, knowledge of how the residual radioactivity would be likely to spread or diffuse after deposition could be used to determine an area A_{EC} for Equation A–8.

It has been shown in Figure D–7 of Appendix D to MARSSIM and in Section 3.7.2 of NUREG–1505 that a triangular grid is slightly more effective in locating areas of elevated concentrations. Therefore, a triangular grid generally should be used if N_{EMC} is significantly larger than N and if areas similar in size or smaller than the grid spacing are expected to have concentrations at or above the area factor.

A.8   Determining Sample Locations

The licensee should establish a reference coordinate system for the impacted areas. A reference coordinate system is a set of intersecting lines referenced to a fixed site location or benchmark. Reference coordinate systems are established so that the locations of any point in the survey unit can be identified by coordinate numbers. A reference coordinate system does not establish the number of sample points or determine where samples are taken. A single reference coordinate system may be used for a site, or different coordinate systems may be used for each survey unit or for a group of survey units. Section 4.8.5 of MARSSIM describes an acceptable method to establish a reference coordinate system.

In Class 1 and Class 2 areas, the sampling locations are established in a regular pattern, either square or triangular. The method described below is from in Section 5.5.2.5 of MARSSIM.

After the number of samples needed in the survey unit has been determined and the licensee has decided whether to use a square or triangular grid, sample spacings, L, are determined from Equations A–9 and A–10 (adapted from MARSSIM Equations 5–5, 5–6, 5–7, and 5–8).

$$L = \sqrt{\frac{A}{0.866\ N}} \quad \textit{for a triangular grid} \qquad \text{(A–9)}$$

$$L = \sqrt{\frac{A}{N}} \quad \textit{for a square grid} \qquad \text{(A–10)}$$

where A =   the survey unit area
N =   the number of samples needed (in Class 1 areas, the larger of the number for the statistical test or the EMC).

The calculated value of L is then often rounded downward to a shorter distance that is easily measured in the field.

A random starting point should be identified for the survey pattern. The coordinate location of the random starting point should be determined by a set of two random numbers with one representing the x axis and the other, the y axis. The random numbers can be generated by calculator or computer or can be obtained from a table of random numbers. Each random number should be multiplied by the appropriate survey unit dimension to provide a coordinate relative to the origin of the survey unit reference coordinate system.

Beginning at the random starting point, a row of points should be identified parallel to the x axis at intervals of L. For a square grid, the additional rows should be parallel to the first row at a distance of L from the first row. For a triangular grid, the distance between rows should be 0.866 L, and the sample locations in the adjacent rows should be midway on the x axis between the sample locations in the first row. Sample locations selected in this manner that either do not fall within the survey unit area or cannot be surveyed because of site conditions should be replaced with other sample locations determined using the same random selection process that was used to select the starting point. An example illustrating the triangular grid pattern is shown in MARSSIM in Figure 5.5.

In Class 3 survey units and in reference areas, all samples should be taken at random locations. Each sample location should be determined by a set of two random numbers, one representing the x axis and the other the y axis. Each set of random numbers should be multiplied by the appropriate survey unit dimension to provide coordinates relative to the origin of the survey unit reference coordinate system. Coordinates identified in this manner that do not fall within the survey unit area or that cannot be surveyed because of site conditions should be replaced with other sample locations determined in the same manner. MARSSIM Figure 5.4 illustrates a random sample location pattern.

A.9   Determination of Compliance

The licensee should first review the measurement data to confirm that the survey units were properly classified. MARSSIM Section 8.2.2, contains methods for this review that are acceptable to NRC staff. If the FSS shows that an area was misclassified with a less restrictive classification, the area should receive the correct classification and the FSS for the area should be repeated. A pattern of misclassifications that are not restrictive enough indicates that the characterization was not adequate. In this case, the site or portions of the site in question should be characterized again, reclassified, and resurveyed for the new classification.

The licensee should then determine whether the measurement results demonstrate that the survey unit meets the radiological criteria for license termination. Tables A.3 and A.4, below, summarize an acceptable way to interpret the sample measurements. The WRS test is described in Section 8.4 of MARSSIM. The Sign test is described in Section 8.3 of MARSSIM. The EMC is described in Section 8.5 of the MARSSIM. The elevated measurement is applied to all sample measurements and all scanning results that exceed the $DCGL_W$.

In some cases, licensees may choose to use scanning or fixed measurement techniques that assess 100 % of the population of potential direct measurements or samples within the survey unit. For these cases, it may be reasonable to demonstrate compliance by directly comparing the average radionuclide concentrations determined from the survey with the appropriate $DCGL_W$, without the need for performing statistical tests. Guidance has not yet been developed for using such techniques without performing statistical tests; therefore, licensees should discuss such techniques with NRC staff on a case-by-case basis.

Table A.3 Interpretation of Sample Measurements when a Reference Area is Used

Measurement Results	Conclusion
Difference between maximum survey unit concentration and minimum reference area concentration is less than $DCGL_W$.	Survey unit meets release criterion.
Difference between survey unit average concentration and reference area average concentration is greater than $DCGL_W$.	Survey unit fails.
Difference between any survey unit concentration and any reference area concentration is greater than $DCGL_W$ and the difference of survey unit average concentration and reference area average concentration is less than $DCGL_W$.	Conduct WRS test and EMC.

Table A.4 Interpretation of Sample Measurements when No Reference Area is Used

Measurement Results	Conclusion
All concentrations are less than $DCGL_W$.	Survey unit meets release criterion.
Average concentration is greater than $DCGL_W$.	Survey unit fails.
Any concentration is greater than $DCGL_W$ and average concentration less than $DCGL_W$.	Conduct Sign test and EMC.

Some facilities may have residual radioactivity composed of more than one radionuclide. When there are multiple radionuclides rather than a single radionuclide, the dose contribution from each radionuclide needs to be considered. Refer to Section 2.7 of this volume for information about using the sum of fractions approach for compliance when multiple radionuclides are present.

When there is a fixed ratio among the concentrations of the nuclides, a $DCGL_W$ for each nuclide can be calculated. Compliance with the radiological criteria for license termination may be demonstrated by comparing the concentration of the single surrogate radionuclide that is easiest to measure with its $DCGL_W$ (which has been modified to account for the other radionuclides present). For example, if Cs-137 and Sr-90 are present, using measured concentrations of Cs-137 as a surrogate for the mix of Cs-137 and Sr-90 may be simpler than separately measuring Cs-137 and Sr-90, and may thus save labor and analytical expenses. When using a surrogate radionuclide to represent the presence of other radionuclides, a sufficient number of measurements, spatially distributed throughout the survey unit, should be used to establish a consistent ratio between the surrogate and the other radionuclides. Section 4.3.2 of MARSSIM provides additional information on the use of surrogate radionuclides for surveys.

When there is no fixed ratio among the concentrations of the nuclides, it is necessary to evaluate the concentration of each nuclide. Compliance with the radiological criteria for license termination is then demonstrated by considering the sum of the concentration of each nuclide relative to its $DCGL_W$, calculated as if it were the only nuclide present. An acceptable method for performing the evaluation is described in Chapter 11 of NUREG–1505.

In some cases in which multiple nuclides are present with no fixed ratio in their concentrations, the dose contribution from one or more of the nuclides in the mixture will dominate the total dose, and the dose from other radionuclides will be insignificant. For example, at a nuclear power plant, many different radionuclides could be present with no fixed ratio in their concentrations, but almost all of the dose would come from just one or two of the nuclides. For guidance on elimination of radionuclides or pathways from consideration, refer to Section 3.3 of this volume.

If a survey unit fails, the licensee should evaluate the measurement results and determine why the survey unit failed. MARSSIM, in Sections 8.2.2 and 8.5.3 and in Appendix D, provides acceptable methods for reviewing measurement results. If it appears that the failure was caused by the presence of residual radioactivity in excess of that permitted by the radiological release criteria, the survey unit should be re-remediated and resurveyed. However, some failures may not be caused by the presence of residual radioactivity. If it can be determined that this is the case, the survey unit may be released.

A.10   References

- Nuclear Regulatory Commission (U.S.) (NRC). NUREG–1505, Rev. 1, "A Proposed Nonparametric Statistical Methodology for the Design and Analysis of Final Status Decommissioning Surveys–Interim Draft Report for Comment and Use." NRC: Washington, DC. June 1998.

- —————. NUREG–1506, "Measurement Methods for Radiological Surveys in Support of New Decommissioning Criteria, Draft Report for Comment." NRC: Washington, DC. August 1995.

- —————. NUREG–1507, "Minimum Detectable Concentrations with Typical Radiation Survey Instruments for Various Contaminants and Field Conditions." NRC: Washington, DC. June 1998.

- —————. NUREG–1549, "Decision Methods for Dose Assessment to Comply with Radiological Criteria for License Termination, Draft Report for Comment." NRC: Washington, DC. July 1998.

- —————. NUREG–1575, Rev. 1, "Multi-Agency Radiation Survey and Site Investigation Manual (MARSSIM)." EPA 402–R–97–016, Rev. 1, DOE/EH–0624, Rev. 1. U.S. Department of Defense, U.S. Department of Energy, U.S. Environmental Protection Agency, and NRC: Washington, DC. August 2000. Corrected pages for MARSSIM, Revision 1 (August 2000) with the June 2001 updates, are available at the EPA Web site: http://www.epa.gov/radiation/marssim.

- —————. NUREG/CR–6364, "Human Performance in Radiological Survey Scanning." NRC: Washington, DC. March 1998.

- —————. Regulatory Issue Summary 2002–02, "Lessons Learned Related to Recently Submitted Decommissioning Plans and License Termination Plans." NRC: Washington, DC. January 2002.

Appendix B

Simple Approaches for
Conducting Final Radiological Surveys

A large number of licensees may use a simplified method to demonstrate regulatory compliance for decommissioning, avoiding complex final status surveys (FSSes). For Decommissioning Groups 1–3, licensees may use the simplified FSS method described in Appendix B of MARSSIM or the alternative protocol described in this volume below.

B.1   MARSSIM Simplified Method

The simplified method in Appendix B of MARSSIM may be used by Decommissioning Group 1 and some of Decommissioning Group 2 licensees. These are sites where radioactive materials have been used or stored only in the form of (a) non-leaking, sealed sources; (b) short half-life radioactive materials (e.g., $T_{1/2} \leq 120$ days) that have since decayed to insignificant quantities; (c) small quantities exempted or not requiring a specific license; or (d) combination of the above. Refer to Appendix B of MARSSIM for the details of this simplified method.

B.2   Alternative Simplified Method

This alternative method may be used by Decommissioning Groups 1–3 and is applicable only for surfaces of building structures and for surface soils. The following conditions are prerequisite to the use of this method:

- Use of screening DCGLs (including DandD code using default distributions).

- No complex or special surveys are included (e.g., volumetric building structure residual radioactivity, duct work, embedded piping, ground water residual radioactivity, subsurface soil residual radioactivity, buried conduit, sewer pipes, or prior onsite disposals).

- Not to be applied to land areas where soil has been previously remediated.

- Removable residual radioactivity for building surfaces must comply with the screening $DCGL_W$ basis of 10 % removable or adjusted per Screening Table (see Appendix H of this volume).

- MDC between 10 to 50 % of the $DCGL_W$ for scans, static or direct measurements, and sampling and analysis (using NUREG–1507 guidance).

If the above conditions are met, then the following simplified method may be used to design and conduct the FSS for each survey unit.

- Size is limited to 2000 m^2 for land areas and 100 m^2 for structures.

- Scanning and sampling to be performed:

    — 100 % scan and

    — 30 samples.

- Hot spot criteria is three times the $DCGL_W$, applied to any sampling location.

- A quality control program to ensure results are accurate and sources of uncertainty are identified and controlled.

- The average concentration for the survey unit is compared to the $DCGL_W$.

- Statistical tests may be the Student's t test, Sign test, or Wilcoxon Rank Sum test, with $\alpha = 0.05$ (no statistics are needed if all measurements are less than the $DCGL_W$).

The final status survey report (FSSR) should provide a complete and unambiguous record of the radiological status of the site and should stand on its own with minimal information incorporated by reference (see Appendix D of this volume for additional information on reporting survey results).

B.3 References

Nuclear Regulatory Commission (U.S.) (NRC). NUREG–1507, "Minimum Detectable Concentrations with Typical Radiation Survey Instruments for Various Contaminants and Field Conditions." NRC: Washington, DC. June 1998.

—————. NUREG–1575, "Multi-Agency Radiation Survey and Site Investigation Manual." NRC: Washington, DC. August 2000.

Appendix C

Use of
Two-Stage or Double Sampling
for Final Status Surveys

When might it be desirable to allow a licensee to sample a survey unit a second time to determine compliance? In the statistical literature this is called either two-stage sampling or double sampling. Resampling is something else altogether. The terms "double sampling" and "two-stage sampling" seem to appear interchangeably in the literature. More recently, the latter seems to have gained favor, so this appendix will use two-stage sampling when referring to survey designs specifically intended to be conducted in two stages. The term double sampling will be used to refer to the case when the survey design is a one-stage design, but allowance is made for a second set of samples to be taken if the retrospective power of the test using the first set of samples does not meet the design objectives. Such allowance, if given, should be specifically mentioned in preparing the Data Quality Objectives (DQOs) and in advance of any sampling and analysis. During the DQO process, double sampling could be considered as an option in setting the Type I error rates. The reasoning behind this is discussed in the next section.

C.1   Double Sampling

Suppose it is thought that a survey unit might have passed the final status statistical test had the initial sampling design been powerful enough. That is, a retrospective examination of the power of the statistical tests used reveals that the probability of detecting that the survey unit actually meets the release criterion was lower than that planned for during the DQO process. This could occur if the spatial variability in residual radioactivity concentrations was larger than anticipated. The power of the test specified during the DQO process depends on an estimate of the uncertainty. The power of the statistical test will be less than planned if the standard deviation is higher than expected. If samples were lost, did not pass analytical QA/QC, or are otherwise unavailable for inclusion in the analysis, the power will also be lower than was planned. Might additional samples be taken in the survey unit to improve the power of the test?

The approach of draft NUREG/CR–5849 allowed the licensee to take additional samples in a survey unit if, after the first sampling, the mean was less than the $DCGL_W$, and the desired upper confidence level on the mean was greater than the $DCGL_W$. Because a 95 % confidence interval is constructed using Student's t statistic rather than using a hypothesis test, Type II errors are not considered in the survey design. The second set of samples was taken so that a t test on the combined set of samples would have 90 % power at the mean of the first set of samples, given the estimated standard deviation from the first set of samples. Such double sampling was to be allowed only once.

Increasing the probability that a clean survey unit passes (power) by the use of double sampling will also tend to increase the probability that a survey unit that is not clean will pass (Type I error). In addition, the two tests are not independent because the data from the first set of samples is used in both. The increase in the Type I error rate is probably less than a factor of two. The fact that this is possible when double sampling is allowed should be clearly understood at the beginning. Thus, the issue of whether or not to allow double sampling is properly a part of the DQO process used to set the acceptable error rates.

Two-stage or double sampling is not usually expected (nor is it encouraged) when the DQO process is used, as in the MARSSIM. This is because the Type II error and the power desired are explicitly considered in the survey design process. If higher power in the test is desired, it should be specified as such. Sufficient samples should be taken to achieve the specified power. The value of this approach lies in the greater objectivity and defensibility of the decision made using the data. Nonetheless, it is recognized that there may be instances when some sort of double sampling is considered desirable. An example is when it is difficult to estimate the standard deviation of the concentrations in a survey unit. A first set of data may be taken with an estimated standard deviation that is too low, and thus, the power specified in the DQO process may not be achieved. Similarly, some scoping data may be taken to estimate the standard deviation in a survey unit. Under what circumstances may this data also be used in the test of the final status?

In such cases, it will be useful for planning if there is an estimate of how much the Type II error rate might increase as a result of double sampling.

Consider the Sign test, as indicated in NUREG–1505. Suppose N_1 samples are taken. Recall that for the Sign test in Scenario A, the test statistic, S_1, was equal to the number of survey unit measurements below the $DCGL_W$. If S_1 exceeds the critical value k_1, then the null hypothesis that the median concentration in the survey unit exceeds the $DCGL_W$ is rejected, i.e., the survey unit passes this test. The probability that any single survey unit measurement falls below the $DCGL_W$ is found from

$$p(C) = \int_{-\infty}^{DCGL_W} f(x)dx = \frac{1}{\sqrt{2\pi}\sigma} \int_{-\infty}^{DCGL_W} e^{-(x-C)^2/2\sigma^2} dx = \Phi\left(\frac{DCGL_W - C}{\sigma}\right) \tag{C--1}$$

C is the true, but unknown, mean concentration in the survey unit. When $C = DCGL_W$, $p = 0.5$.

The probability that more than k_1 of the N_1 survey unit measurements fall below the $DCGL_W$ is simply the following binomial probability:

$$\sum_{t=k_1+1}^{N_1} \binom{N_1}{t} p^t (1-p)^{N_1-t} = 1 - \sum_{t=0}^{k_1} \binom{N_1}{t} p^t (1-p)^{N_1-t} \tag{C--2}$$

This is the probability that the null hypothesis will be rejected, and it will be concluded that the survey unit meets the release criterion. When the mean concentration in the survey unit is at the $DCGL_W$, this is just the Type I error rate, α. When $C = DCGL_W$, $p = (1-p) = 0.5$, so

$$\alpha = \sum_{t=k_1+1}^{N_1} \binom{N_1}{t} (0.5)^t (0.5)^{N_1-t} = (0.5)^{N_1} \sum_{t=k_1+1}^{N_1} \binom{N_1}{t} \tag{C--3}$$

Now, suppose it is decided to allow the licensee to take a second set of samples of size N_2. The test statistic, S, is equal to the number of the total of $N = N_1 + N_2$ survey unit measurements below the $DCGL_W$. If S exceeds the critical value k, then the null hypothesis that the median concentration in the survey unit exceeds the $DCGL_W$ is rejected, i.e., the survey unit passes this test. Now the overall probability that the null hypothesis is rejected (i.e., the survey unit passes) is equal to the sum of the probabilities of two events that are mutually exclusive:

a. The probability that more than k_1 of the N_1 survey unit measurements fall below the $DCGL_W$, and

b. The probability that fewer than k_1 of the first N_1 survey unit measurements fall below the $DCGL_W$ but that more than k of the N total survey unit measurements fall below the $DCGL_W$.

Now $S = S_1 + S_2$, where S_2 is the number of the second set of N_2 survey unit measurements that fall below the $DCGL_W$. S_1 and S_2 are independent, but S_1 and $S = S_1 + S_2$ are not.

The covariance of S_1 and S, using $E(\cdot)$ to denote expected value, is

$$
\begin{aligned}
Cov(S_1, S) &= E(S_1, S) - E(S_1)E(S) \\
&= E(S_1(S_1 + S_2)) - E(S_1)E(S) \\
&= E(S_1^2) + E(S_1 S_2) - E(S_1)E(S) \\
&= (N_1^2 p(1-p) + N_1^2 p^2) + N_1 N_2 p^2 - N_1 p(N_1 + N_2)p \\
&= N_1 p(1-p)
\end{aligned}
\qquad \textbf{(C–4)}
$$

Therefore the correlation coefficient between S_1 and S is

$$
\begin{aligned}
\rho(S_1, S) \quad &= \frac{N_1 p(1-p)}{\sqrt{N_1 p(1-p)(N_1 + N_2)p(1-p)}} \\
&= \frac{N_1}{\sqrt{N_1(N_1 + N_2)}} \\
&= \sqrt{N_1/(N_1 + N_2)} = \sqrt{N_1/N}
\end{aligned}
\qquad \textbf{(C–5)}
$$

To calculate the overall probability that the survey unit passes, one requires the joint probability of S_1 and S,

$$
\begin{aligned}
\Pr(S_1 = s_1, S = s) &= \Pr(S_1 = s_1)\Pr(S_2 = s - s_1) \\
&= \binom{N_1}{s_1} p^{s_1}(1-p)^{N_1-s_1} \binom{N_2}{s-s_1} p^{s-s_1}(1-p)^{N_2-(s-s_1)} \\
&= \binom{N_1}{s_1}\binom{N_2}{s-s_1} p^{s}(1-p)^{N-s}
\end{aligned}
\tag{C-6}
$$

Therefore, the overall probability that the survey unit passes is

$$
\begin{aligned}
\Pr(S_1 > k_1 \text{ or } S > k) &= \Pr(S_1 > k_1) + \Pr(S_1 \le k_1 \text{ and } S > k) \\
&= \sum_{s_1=k_1+1}^{N_1} \binom{N_1}{s_1} p^{s_1}(1-p)^{N_1-s_1} \\
&\quad + \sum_{s_1 \le k_1} \sum_{s_2 > k-s_1} \binom{N_1}{s_1}\binom{N_2}{s_2} p^{s_1+s_2}(1-p)^{(N_1+N_2)-(s_1+s_2)}
\end{aligned}
\tag{C-7}
$$

The first term is equal to (or slightly less than) the Type I error rate α specified during the DQO process. The second term is the additional probability of a Type I error introduced by allowing double sampling.

Note that

$$
\begin{aligned}
\Pr(S_1 \le k_1 \text{ and } S > k) &= \sum_{s>k}^{N} p^{s}(1-p)^{N-s} \sum_{s_1=0}^{k_1} \binom{N_1}{s_1}\binom{N_2}{k-s_1} \\
&\le \sum_{s>k}^{N} p^{s}(1-p)^{N-s} \sum_{s_1=0}^{k} \binom{N_1}{s_1}\binom{N_2}{k-s_1} \\
&= \sum_{s>k}^{N} p^{s}(1-p)^{N-s} \binom{N}{s} = \Pr(S > k) \le \alpha
\end{aligned}
\tag{C-8}
$$

Thus, the Type I error rate would be at most doubled when double sampling is allowed.

For example, if a survey is designed so that $N_1 = 30$, and $\alpha = 0.05$, then the critical value for the Sign test is $k_1 = 19$. Suppose the first survey results in 19 or fewer measurements that are less than the $DCGL_W$. In addition, suppose the survey unit is sampled again, taking an additional $N_2 = 30$ samples. Then the total number of samples is $N = N_1 + N_2 = 60$. The critical value for the Sign test with $\alpha = 0.05$ and $N = 60$ is $k = 36$. When the survey unit concentration is equal to the $DCGL_W$, $p = 0.5$, one has

$$
\begin{aligned}
\Pr(S_1 > 19 \text{ or } S > 36) &= \Pr(S_1 > 19) + \Pr(S_1 \leq 19 \text{ and } S > 36) \\
&= \sum_{s_1=20}^{30} \binom{30}{s_1} (0.5)^{s_1} (1-0.5)^{30-s_1} \\
&\quad + \sum_{s_1=0}^{19} \binom{30}{s_1} \sum_{s_2=(37-s_1)}^{30} \binom{30}{s_2} (0.5)^{s_1+s_2} (1-0.5)^{(30+30)-(s_1+s_2)} \\
&= 0.049 + 0.027 = 0.076
\end{aligned}
\tag{C–9}
$$

Thus, the total Type I error rate is about 50 % greater than originally specified.

In conclusion, double sampling should not be used as a substitute for adequate planning. If it is to be allowed, this should be agreed upon with NRC staff as part of the DQO process. The procedure for double sampling, i.e., the size of the second set of samples, N_2, should be specified, recognizing that the Type I error rate could be up to twice that specified for the Sign test when only one set of samples is taken.

Similar considerations apply for the WRS test; however, the calculation of the exact effect on the Type I error rate is considerably more complex.

Finally, double sampling should never be necessary for Class 2 or Class 3 surveys, which are not expected to have concentrations above the $DCGL_W$. These classes of survey unit should always pass after the first set of samples because every measurement should be below the $DCGL_W$. The very need for a second set of samples (i.e., failure to reject the null hypothesis) in Class 2 or Class 3 survey units would raise an issue of survey unit misclassification. In addition, double sampling is generally not appropriate for Class 1 survey units where elevated areas have been found.

A better solution to the issue of double sampling is to plan for data collection in two stages and design the final status survey accordingly, as is discussed in the remainder of this appendix.

C.2   Two-Stage Sequential Sampling

Suppose there are a large number of survey units of a similar type to be tested. In this case a two-stage sampling procedure may result in substantial savings of time and money by reducing the average number of samples required to achieve a given level of statistical power.

To plan a two-stage Sign test, let N_1 be the size of the first set of samples taken, and let S_1 be the number of these less than the $DCGL_W$. Similarly, let N_2 be the size of the second set of samples taken, and let S_2 be the number of these less than the $DCGL_W$. Let $N = N_1 + N_2$, and let $S = S_1 + S_2$. The procedure is as follows:

- if $S_1 > u_1$ then the survey unit passes (reject H_0),

- if $S_1 < l_1$ then the survey unit fails,

- if $l_1 \leq S_1 \leq u_1$ then the second set of samples is taken.

- If $S = S_1 + S_2 > u_2$ after the second set of samples is analyzed, then the survey unit passes.

What is the advantage of two-stage testing? For given error rates α and β, the number of samples, N_1, taken in the survey unit during the first stage of sampling will be less than the number, N_0, required in the MARSSIM tables. Unless the result is "too close to call," this will be the only sampling needed. When the result is "too close to call," $l_1 \leq S_1 \leq u_1$, a second sample of size N_2 is taken, and the test statistic S_2 is computed using the combined data set, $N_1 + N_2$. While the size of the combined set, $N = N_1 + N_2$, will generally be larger than the number, N_0, from in the MARSSIM tables, the expected sample size over many survey units will still be lower. Thus, two-stage sampling scheme will be especially useful when there are many similar survey units for which the final status survey design is essentially the same. Two-stage sampling may be used whether or not a reference area is needed. It may be used with either the Sign or the WRS test.

The remaining major issue is how to choose the critical values l_1, u_1, and u_2. Hewett and Spurrier (1983) suggest three criteria:

1. Match the power curve of the two-stage test to that of the one-stage test. The curves are matched at three points. The points with power equal to α, $1-\beta$, and 0.5 are generally well enough separated to assure a good match over the entire range of potential survey unit concentrations.

2. Maximize the power at the LBGR for given values of α and average sample size.

3. Minimize the sample size for given values of α, and $1-\beta$.

While any one of these criteria could be used, the first has received more attention in the literature. Thus, it may be more readily applied to the case of FSS design. The other criteria would require further development.

Spurrier and Hewett (1975) initially developed a two-stage sampling methodology using criteria 1 assuming the data are normally distributed. They matched power at α, 0.5, and 0.9. Table C.1 shows the values of l_1, u_1, and u_2 they obtained for six different sets of sample sizes, N_1/N_0, N_2/N_0, expressed as fractions of the sample size, N_0, that would be required for the one-stage test with equivalent power. The term $E(N)/N_0$, is the maximum expected combined sample size for the two-stage test relative to the sample size, N_0, that would be required for the one-stage test with equivalent power. This number is almost always less than one, but it depends on how close the actual concentration in the survey unit is to the $DCGL_W$. Clearly, if the concentration is over the $DCGL_W$, the survey unit is likely to fail on the first set of samples. If the concentration is much lower than the $DCGL_W$, the survey unit is likely to pass on the first set of samples. It is only when the true concentration in the survey unit falls within the gray region that there will be much need for the second set of samples. The fact that the maximum $E(N)/N_0$ is almost always less than one indicates that the overall number of samples required for a two-stage final status survey will almost never exceed the number required for a one-stage test, even if the true concentration of the survey unit falls in the gray region between the $LBGR$ and the $DCGL_W$.

Recall that the power to distinguish clean from dirty survey units is relatively low when the true concentration is in the gray region. It falls from $1-\beta$ at the $LBGR$ to α at the $DCGL_W$. Thus, when the true concentration is in the gray region, there will be a larger amount of cases when the second set of samples is needed. The gray region is exactly where the results are "too close to call." However, if the true concentration of the survey unit is below the $LBGR$ or above the $DCGL_W$, the actual average number of samples will be closer to N_1, because the second set of samples will seldom be needed.

In 1976, Spurrier and Hewett dropped the assumption of normality and extended their methodology to two-stage Wilcoxon Signed Rank (WSR) and Wilcoxon Rank Sum (WRS) tests. The procedure depends on an extension of the Central Limit Theorem to the joint distribution of the test statistics S_1 and $S = S_1 + S_2$. Spurrier and Hewett suggest that the approximation works reasonably well for sample sizes as small as nine.

In this appendix, their method is also applied to the Sign test.

For the Sign test, one computes

$$S_1 = \frac{S_1^+ - N_1/2}{\sqrt{N_1/4}} \tag{C–10}$$

where S_1^+ is the usual Sign test statistic, i.e., the number of measurements less than the $DCGL_W$.

Using Table C.1,

- if $S_1 > u_1$ then reject the null hypothesis (the survey unit passes)
- if $S_1 < l_1$ then do not reject the null hypothesis (the survey unit fails)
- if $l_1 \leq S_1 \leq u_1$ then take the second set of samples.

If a second set of samples is taken, then compute

$$S = \frac{(S_1^+ + S_2^+) - (N_1 + N_2)/2}{\sqrt{(N_1 + N_2)/4}} = \frac{S^+ - N/2}{\sqrt{N/4}} \tag{C–11}$$

Using Table C.1,

- if $S > u_2$ then reject the null hypothesis (the survey unit passes)

- if $S \leq u_2$ then do not reject the null hypothesis (the survey unit fails).

This test relies on "a large sample approximation." That is, one is assuming that the sample size is large enough that the joint distribution of S_1 and S is bivariate standard normal with correlation coefficient $\rho(S_1, S) = \sqrt{N_1 / N}$. Some simulation studies may be done to determine quantitative bounds on the accuracy of this approximation.

The choice of which set of sample sizes should be used is dependent on how confident one is of passing.

For Class 2 and Class 3 survey units (discussed in Appendix A of this volume), case 3 with $N_1/N_0 = 0.2$ and $N_2/N_0 = 1.0$ might be reasonable. In these classes of survey units no individual sample concentrations in excess of the $DCGL_W$ are expected. The probability of passing on the first set of samples should be close to one. Therefore, it makes sense to choose a design with the minimum number of samples required in the first set.

For Class 1 survey units (discussed in Appendix A of this volume), case 2 with $N_1/N_0 = 0.4$ and $N_2/N_0 = 0.8$ might be more appropriate. There is some chance that the survey unit will not pass on the first set of samples, so it may be desirable to reduce Max $E(N)/N_0$ from 0.999 to 0.907 by taking more samples in the first set.

If the gray region has been expanded in order to increase Δ/σ, case 1 or 4 would be a more conservative choice. In this situation, statistical power has been compromised somewhat, so it may be important to reduce the risk of having a larger average total number of samples (as indicated by the potential Max $E(N)/N_0$ even further.

Scan sensitivity will also impact the ability to use two-stage designs in Class 1 survey units. It would have to be determined if the $DCGL_{EMC}$ can be detected when only N_1 samples are taken. If not, the sample size would have to be increased until the MDC_{scan} is lower than the $DCGL_{EMC}$. In this situation, the choice of N_1, and the average savings possible with two-stage sampling may be severely limited.

Table C.1   Critical Points for Two-Stage Test of Normal Mean for a One-Sided Alternative

	N_1/N_0	N_2/N_0	a = 0.05				a = 0.01			
			u_1	l_1	u_2	Max E(N)/N_0	u_1	l_1	u_2	Max E(N)/N_0
1	0.6	0.6	1.886	0.71	1.783	0.866	2.499	1.259	2.493	0.879
2	0.4	0.8	1.984	0.179	1.782	0.907	2.558	0.635	2.496	0.931
3	0.2	1	2.073	−0.482	1.784	0.999	2.6	−0.146	2.502	1.03
4	0.55	0.55	2.05	0.438	1.716	0.869	2.635	0.966	2.411	0.878
5	36193	36193	1.781	0.95	1.868	0.882	2.415	1.52	2.6	0.897
6	0.7	0.7	1.749	1.045	1.909	0.893	2.39	0.628	2.651	0.908

Source: Spurrier and Hewett (1975).

For the WRS test, at each stage one sets the number of measurements required in the survey unit, n_1 and n_2, and in the reference area m_1 and m_2 relative to the number required for the one-stage test $n_0 = m_0 = N_0/2$ specified in Table 5.3 of the MARSSIM. There is an additional requirement that $n_1/n_2 = m_1/m_2$, which should be satisfied with sufficient accuracy for most MARSSIM designs. Minor departures due to small differences in sample size caused by filling out systematic grids or the loss of a few samples should not severely impact the results.

One now computes

$$S_1 = \frac{W_1^R - m_1(n_1+m_1+1)/2}{\sqrt{n_1 m_1(n_1+m_1+1)/12}} \tag{C-12}$$

where W_1^R is the usual WRS test statistic, i.e., the sum of the ranks of the adjusted reference area measurements.

Using Table C.1,

- if $S_1 > u_1$ then reject the null hypothesis (the survey unit passes),

- if $S_1 < l_1$ then do not reject the null hypothesis (the survey unit fails),

- if $l_1 \le S_1 \le u_1$ then take the second set of samples.

If a second set of samples is taken, then compute

$$S = \frac{(W_1^+ + W_2^+) - (m_1+m_2)(m_1+m_2+n_1+n_2+1)/2}{\sqrt{(m_1+m_2)(n_1+n_2)(m_1+m_2+n_1+n_2+1)/12}} = \frac{W^R - m(m+n+1)/2}{\sqrt{mn(m+n+1)/12}} \tag{C-13}$$

Using Table C.1,

- if $S > u_2$ then reject the null hypothesis (the survey unit passes),

- if $S \le u_2$ then do not reject the null hypothesis (the survey unit fails).

This test relies on "a large sample approximation." That is, one is assuming that the sample size is large enough that the joint distribution of S_1 and S is bivariate standard normal with correlation coefficient

$$\rho(S_1,S) = \sqrt{(m_1 + n_1)/(m + n)} \tag{C–14}$$

Some simulation studies would be needed to determine some quantitative bounds on the accuracy of this approximation.

C.3 An Alternative Two-Stage, Two-Sample Median Test

A different approach to this testing problem has been suggested by Wolfe (1977). In his procedure, a specific number of sample measurements are made in a reference area, and the median, M, is calculated, and the $DCGL_W$ added. Survey unit samples are then analyzed until r of them are found to be below M. The test statistic, n_r, is the number of survey unit samples that have been analyzed. Smaller values of n_r indicate that the survey unit meets the release criterion. For Class 2 and Class 3 survey units in particular, one would expect that $n_r = r$. In that case, the number of reference area measurements, m, and the value of r are chosen to meet the DQO for the Type I error rate. In each survey unit, r samples are taken. If all are less than M, one rejects the null hypothesis that the survey unit exceeds the release criterion. If any one of them exceeds M, the null hypothesis will not be rejected. Thus, the total number of samples needed in each survey unit may be relatively small. In addition, as soon as one sample is measured above M, the result of the test is known. Thus, it may not be necessary to analyze every survey unit sample. Of course, the need to identify elevated areas may preclude the use of this method in some circumstances. However, the potential savings when the analytical costs are high may make this procedure attractive. It merits further investigation.

C.4 References

Hewett, J.E. and J.D. Spurrier. "A Survey of Two Stage Tests of Hypothesis: Theory and Application." *Communications on Statistics – Theory and Methods*. Vol. 12, No. 20: pp. 2307–2425. 1983.

Nuclear Regulatory Commission (U.S.) (NRC). NUREG–1505, Rev 1, "A Proposed Nonparametric Statistical Methodology for the Design and Analysis of Final Status Decommissioning Surveys–Interim Draft Report for Comment and Use." NRC: Washington, DC. June 1998.

—————. NUREG/CR–5849, "Manual for Conducting Radiological Surveys in Support of License Termination." Draft Report for Comment. NRC: Washington, DC. June 1992.

Spurrier, J.D., and J.E. Hewitt. "Double Sample Tests for the Mean of a Normal Distribution." *Journal of the American Statistical Association.* Vol. 70, No. 350: pp. 448–450. 1975.

Spurrier, J.D., and J.E. Hewitt. "Two-Stage Wilcoxon Tests of Hypotheses." *Journal of the American Statistical Association.* Vol. 71, No. 356: pp. 982–987. 1976.

Wolfe, D.A. "Two-Stage Two Sample Median Test." *Technometrics.* Vol. 19, No. 4: pp. 459–501. 1977.

Appendix D

Survey Data Quality and Reporting

Introduction

The Multi-Agency Radiological Laboratory Analytical Protocols (MARLAP, 2004) and the Multi-Agency Radiation Survey and Site Investigation Manual (MARSSIM, 2000) are complementary guidance documents in support of cleanup and decommissioning activities. MARSSIM provides guidance on how to plan and carry out a study to demonstrate that a site meets appropriate release criteria. It describes a methodology for planning, conducting, evaluating, and documenting environmental radiation surveys conducted to demonstrate compliance with cleanup criteria. See Chapter 4 and Appendix A for more details on MARSSIM. MARLAP provides guidance and a framework for both project planners and laboratory personnel to ensure that radioanalytical data will meet the needs and requirements of cleanup and decommissioning activities.

Radioanalytical data are commonly generated to support activities such as: characterization and survey of radiologically contaminated sites, effluent and environmental monitoring of nuclear facilities, emergency response to accidents involving radiological materials, cleanup and decommissioning of nuclear facilities, and radioactive waste management. Numerous significant decisions, impacting the health and safety of the public and the environment, are frequently based on the available radioanalytical data. Considering these activities, the decisions associated with the radioanalytical data may involve issues pertaining to the extent and depth of contamination and associated remedial actions, demonstration of compliance with the cleanup criteria, demonstration of compliance with the effluent release criteria, assessment of effluent radiological releases and corrective measures, assessment of actions in response to incidents or accidental releases of radiological materials, and issues involving waste storage, transport, and disposal. In addition, radioanalytical data commonly influence decisions related to the cost of remedial actions as well as decisions involving environmental monitoring strategies and designs.

MARLAP was developed to provide guidance and framework for project planners, managers, technical reviewers, and laboratory personnel to ensure that the radioanalytical data produced by surveys will meet the needs and requirements for cleanup and decommissioning activities. MARLAP addresses the need for a nationally consistent approach to producing radioanalytical laboratory data that meet a project's or program's data requirements. The guidance provided by MARLAP is both scientifically rigorous and flexible enough to be applied to a diversity of projects and programs. The MARLAP manual (NRC document number NUREG-1576 and U.S. Environmental Protection Agency (EPA) document number EPA 402-B-04-001A-C) is issued in three volumes (printed and CD-ROM versions) and is available through the Internet at: http://www.nrc.gov/reading-rm/doc-collections/nuregs/staff/sr1576.

The NRC staff encourages licensees to follow the recommendations provided in the MARLAP.

D.1 An Overview of Marlap

MARLAP is divided into two main parts, Part I and Part II. Part I provides guidance on using a performance-based approach for the three phases of radioanalytical projects, including: (1) the planning phase; (2) the implementation phase; and (3) the assessment phase. These three main phases and associated processes should result in analytical data of known quality appropriate for the intended use. Table 1 provides an overview of the three main phases, the processes associated with each phase, and the anticipated outputs for each process. Figure 1 illustrates an overview of MARLAP terms and processes and interactions of the radioanalytical project manager with the laboratory performing the analysis. MARLAP processes and terms described in Table 1 and Figure 1 are consistent with standard practices of the American Society for Testing of materials (ASTM) for generation of environmental data. Chapters 3 through 9 of the MARLAP manual provide a detailed description of MARLAP phases and specific processes. It should be noted that it is not a regulatory requirement to follow or use MARLAP processes as described in Figure 1; however, these processes are believed to be flexible and scientifically rigorous to be applied for generation of radioanalytical data of the desired quality for the intended use.

Part II of MARLAP provides technical information on the laboratory analysis of radionuclides. Specifically, Part II highlights common radioanalytical problems and how to correct them. It also provides options for analytical protocols and discusses the pros and cons of these options. It should be noted that Part II does not provide a step-by-step instructions on how to perform certain laboratory procedures or tasks. However, Part II provides guidance to assist laboratory personnel in selection of the best approach for a particular laboratory task. For example, Chapter 13 does not contain a step-by-step instruction on how to dissolve a soil sample; however, it does provide information on acid digestion, fusion techniques, and microwave digestion, to help the analyst select the most appropriate technique or approach for a particular sample characteristics and project needs. Part II presents detailed technical information in the following areas: (1) field sampling that affect laboratory measurements; (2) sample receipt, inspection, and tracking; (3) laboratory sample preparation; (4) sample dissolution; (5) separation techniques; (6) quantification of radionuclides; (7) data acquisition, reduction, and reporting for nuclear counting instrumentation; (8) waste management in a radioanalytical laboratory; (9) laboratory quality control; (10) measurement uncertainty; and (11) detection and quantification capabilities. MARLAP adopted the International Organization for Standardization (ISO) processes, terms, and expressions for analytical measurements, quantifications, and estimation of uncertainty.

MARLAP also presents additional technical details on specific topics outlined in Parts I and II. Appendices A through E support Part I for the following specific topics: Appendix A, Directed Planning Approaches; Appendix B, The Data Quality Objective Process; Appendix C, Measurement Quality Objectives for Method Uncertainty and Detection; Appendix D, Content of Project Plan Documents; and Appendix E, Contracting Laboratory Services. Appendix F supports Part II for the specific topic on laboratory sub-sampling, whereas Appendix G provides a compilation of statistical tables.

D.2   Use of Marlap in Decommissioning and Cleanup Projects

MARLAP presents a useful approach and methodology applicable to radianalytical projects for cleanup and decommissioning activities. The major processes of the data life cycle are described briefly below for application in cleanup and decommissioning activities:

D.2.1   The Planning Phase

The directed planning process for cleanup and decommissioning typically involves the following radioanalytical aspects:

- **Stating the cleanup problem**: Identify the analytes of concern, matrix of concern, regulatory requirements, sampling constraints, primary decisions maker, available resources, existing data and its reliability.

- **Identifying the cleanup decision**: Assess different analytical protocols, identify items of the analytical protocols specifications (APS), and determine how sample collection will affect the measurement quality objectives (MQOs).

- **Identifying the inputs to the cleanup decisions:** Define characteristics of the analytes and matrix, assess the concentration range for the analyte of interest, and define action levels.

- **Defining the decision boundaries:** Identify background, temporal and spatial trends of data and determine limitations of current analytical protocols.

- **Developing a decision rule and tolerable decision error rates**: For example the decision rule may be defined as: "If the mean concentration of analyte x in the upper 15 cm of the soil is greater than z Bq/g, then an action would be taken to remove the soil from the site." Estimates of uncertainties in the data considering action levels and/or derived concentration guidelines should be made.

- **Specifying limits on decision error rates:** Evaluate range of possible parameter values and allowable difference between the action level and the actual value.

- **Optimizing the strategy for obtaining data**: This process may involve optimization of the design for data collection through coordination with the different team members. The process also involves development of analytical protocols specifications and establishing performance measures of the MQOs.

D.2.2   The Implementation Phase

The radioanalytical process is a compilation of activities starting from the time a sample is collected and ending with the data reduction and reporting. Figure 2 illustrates the typical components of an analytical process used for radiological characterization and survey of contaminated sites. Certain cleanup or decommissioning projects may not include all of the components listed in Figure 2. The analytical protocols usually comprise a compilation of specific procedures or methods and are performed in succession depending on the particular analytical process. Using a performance based approach, there will be a number of alternative protocols that might be appropriate for a particular analytical process. A major component of the analytical protocol is the analytical method. The radioanalytical process should also include the analytical uncertainty, the analytical error, the precision, the bias, and the accuracy of the method used.

D.2.3   The Assessment Phase

The assessment phase focuses on three major steps including:

- **Data verification:** This step assures that the laboratory conditions and operations are in compliance with the statement of work (SOW) and the project quality assurance project plan (QAPP). The verification process would examine the laboratory standard operating procedures. It would also check for consistency and comparability of the data, correctness of the data calculations, and completeness of the results and data documentation.

- **Data validation:** This step addresses the reliability aspects of the radioanalytical data. It addresses the analyte and matrix types as well as the uncertainty of the measurement to support the intended use. Validation flags (qualifiers) are typically applied to data that do not meet the acceptance criteria established to meet the project data quality objectives (DQOs) and MQOs.

- **Data quality assessment (DQA):** This step represents the scientific and statistical data evaluation aspects to determine if data are of the right type, quality, and quantity to support the intended use. The DQA is more global in its purview such that it considers the combined impacts of all project activities on data quality and its usability.

D.3   Benefits of Using Marlap in Decommissioning and Cleanup Projects

MARLAP is an extensive document which presents a comprehensive guidance and information on the three phases of the radioanalytical data life cycle. MARLAP emphasizes the importance of establishing the proper linkages among these phases. Use of MARLAP in decommissioning and cleanup projects can benefit the user in the following aspects:

- MARLAP ensures generation of radioanalytical data of acceptable quality for the intended use.

- MARLAP minimizes time and effort expended in generation of unacceptable data.

- MARLAP enhances public trust in radioanalytical data generated by licensees and regulators.

- MARLAP minimizes efforts applied to justifying data and may limit any litigation costs.

- Because MARLAP uses an early coordinated approach to develop the radioanalytical data DQOs and MQOs, this approach would require early coordination and inputs from the decision makers, the project manager, the shareholders, the concerned team members, and the analyst (see Figure 1). Therefore, this approach should resolve issues or difficulties related to sampling, sample tracking, sample preservation, analysis, data quality, time, and costs early in the process.

- MARLAP provides flexibility in selection of the appropriate analytical method using a performance based approach considering the DQOs, the MQOs, and the available resources.

- MARLAP enhances regulatory reviews of radioanalytical data and saves time and effort for site characterization, environmental monitoring, decommissioning, and remediation.

Table D.1   The Radioanalytical Data Life Cycle

PHASE	PROCESS	PROCESS OUTPUTS
PLANNING	Directed Planning Process	Development of DQOs and MQOs including Optimized Sampling and Analytical Designs
	Plan Documents	Project Plan Documents Including QAPP Work Plan, or Sampling and Analysis Plan (SAP), Data Validation Plan; Data Quality Assessment Plan
	Contracting Services	SOW and Other Contractual Documents
IMPLEMENTATION	Sampling	Laboratory Samples
	Analysis	Laboratory Analysis including QC Samples and Complete Data Package
ASSESSMENT	Verification	Verified Data and Data Verification Report
	Validation	Validated Data and Data Validation Report
	Data Quality Assessment	Assessment Report
Data of Known Quality Appropriate for the Intended Use		

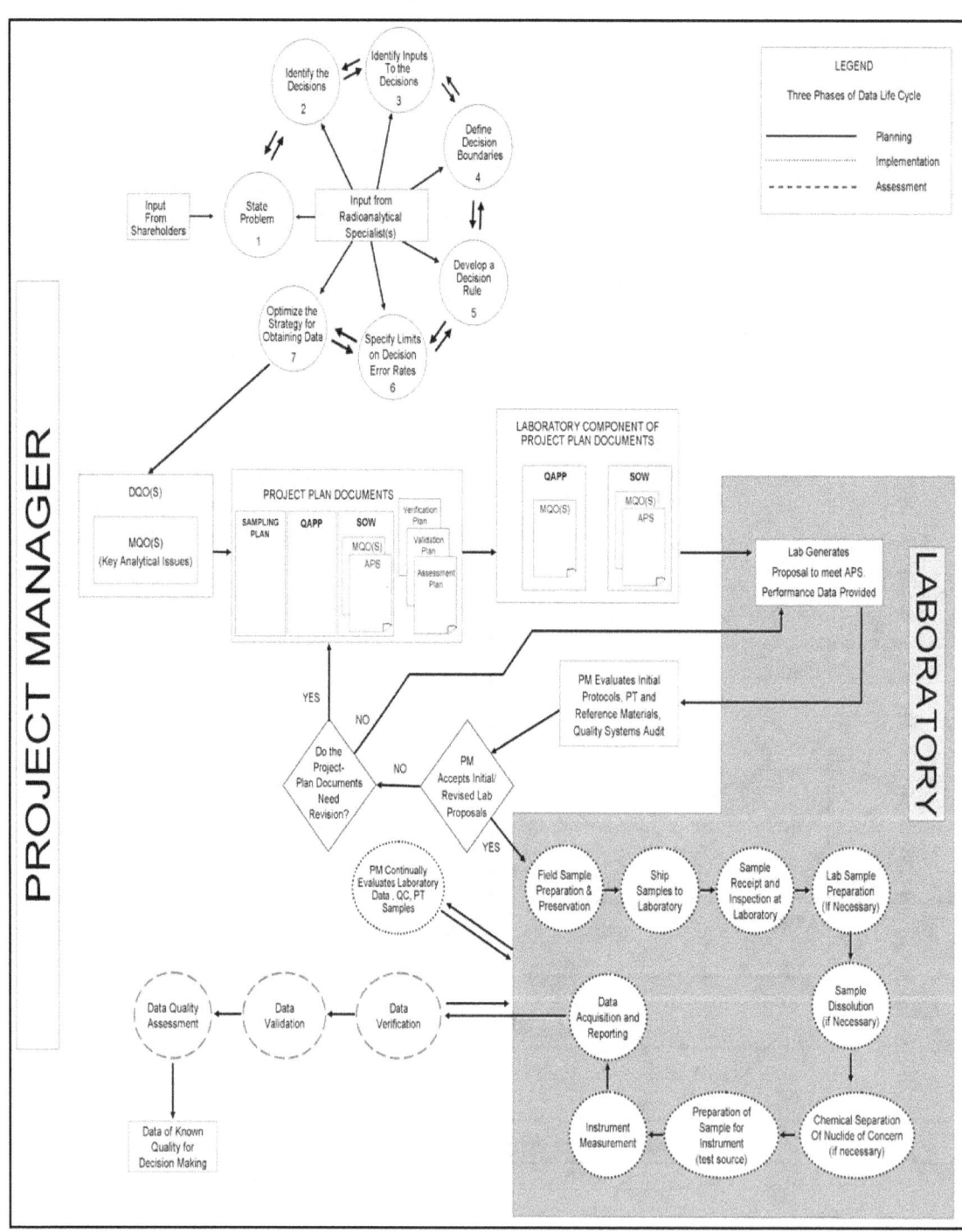

Figure D.1 MARLAP Road Map – Key Terms and Processes.

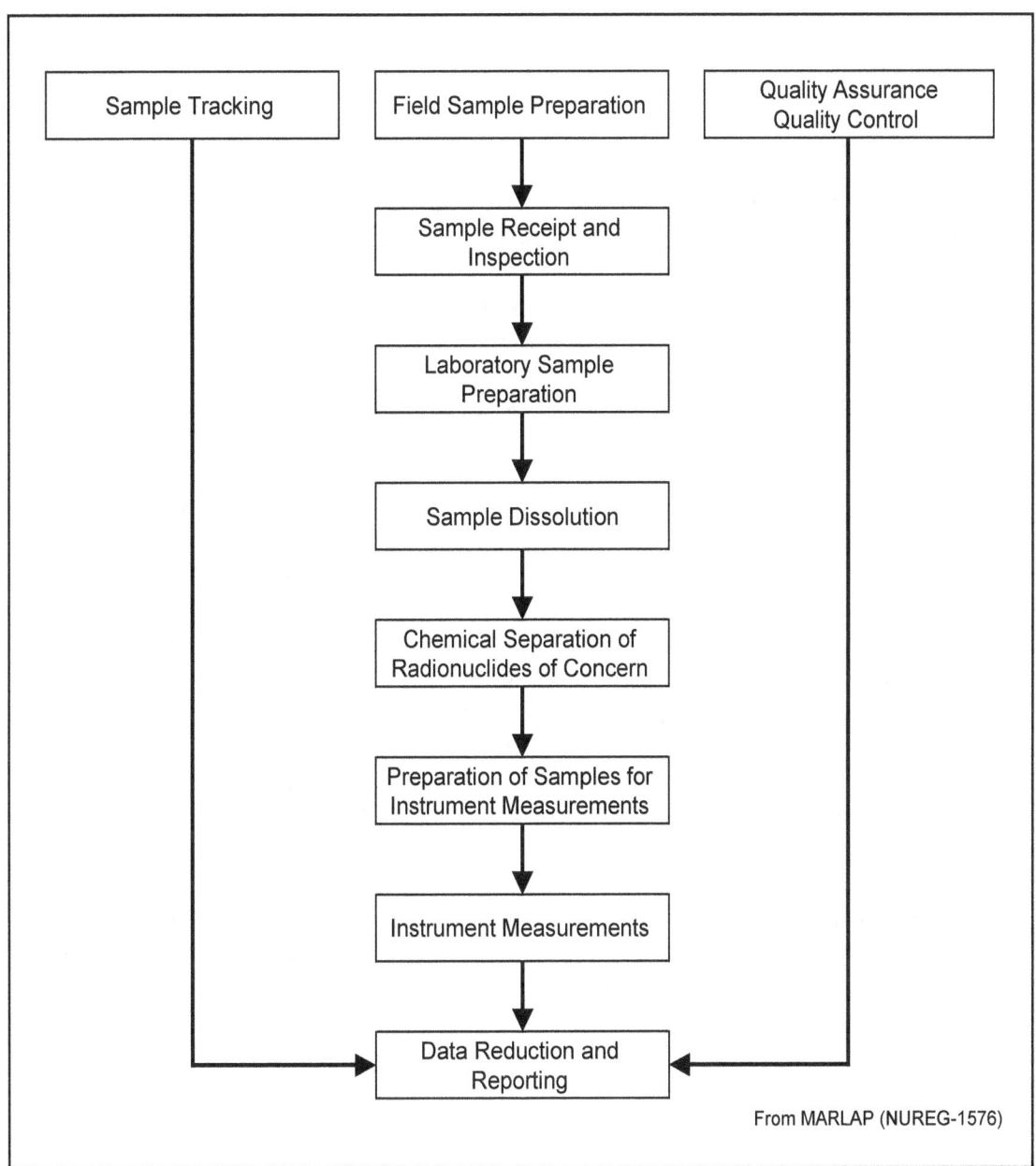

Figure D.2 Typical Components of the Radioanalytical Process.

D.4   References

American Society for Testing and Materials (ASTM) D5792, "Standard Practice for Generation of Environmental Data Related to Waste Management Activities: Development of Data Quality Objectives," 1995.

American Society for Testing and Materials (ASTM) D6233, "Standard Guide for Data Assessment for Environmental Waste Management Activities," 1998.

International Organization for Standardization (ISO), "International Vocabulary of Basic and General Terms in Metrology," Geneva, Switzerland, 1993.

International Organization for Standardization (ISO), "Guide to the Expression of Uncertainty in Measurement," Geneva, Switzerland, 1995.

MARSSIM, "Multi-Agency Radiation Survey and Site Investigation Manual, Revision," NUREG-1575, Rev 1, EPA 402-R-97-016, Rev 1, DOE/EH-0624, Rev 1. August. Available from www.epa.gov/radiation/marssim/.

U.S. Environmental Protection Agency (EPA), "Guidance for the Data Quality Objective Process (EPA QA/G-4)," EPA/600/R-96/055, Washington, DC, 2000. Also available at www.epa.gov/quality/qa_docs.html.

U.S. Environmental Protection Agency (EPA). 2000b. Guidance for Data Quality Assessment: Practical Methods for Data Analysis (EPA QA/G-9). EPA/600/R-96/084, Washington, DC. Available from www.epa.gov/quality/qa_docs.html.

U.S. Environmental Protection Agency, U.S. Department of Defense, U.S. Department of Energy, U.S. Department of Homeland Security, U.S. Nuclear Regulatory Commission, U.S. Food and Drug Administration, U.S. Geological Survey, and National Institute of Standards and Technology, "Multi-Agency Radiological Laboratory Analytical Protocols (MARLAP)," NUREG-1576, EPA402-B-04-001A, NTIS PB2004-105421, Vol. I, II, and III and Supp. 1. Washington, DC, 2004.

Appendix E

Measurements for
Facility Radiation Surveys

E.1 Introduction

This appendix is applicable to all decommissioning groups. All surveys, whether simple or complex final status surveys (FSSes), require information on the basis for instrument selection, the nature of the radionuclides, measurement techniques and procedures, MDCs of the instruments (measurement systems), and instrument calibration. Therefore, the information presented in this appendix would apply to a simple survey used to demonstrate compliance with regulatory decommissioning criteria as well as a complex FSS.

This appendix contains limited, general information on survey techniques and survey measurements. The information presented here is related to the process of implementing a survey plan and refers to the appropriate sections of MARSSIM, MARLAP, and various NUREGs for more detailed information. These are important areas for the conduct of surveys in the RSSI process and include the basic modes for determining levels of radiation and radioactivity at a site, instrument and scanning detection limits, instrument calibration, and laboratory measurements for samples. The data from the FSS is the deciding factor in judging if the site meets the release criteria.

Radiological conditions that should be determined for license termination purposes include any combination of total surface activities, removable surface activities, exposure rates, radionuclide concentrations in soil, or induced activity levels. To determine these conditions, field measurements and laboratory analyses may be necessary. For certain radionuclides or radionuclide mixtures, both alpha and beta radiations may have to be measured. In addition to assessing the average radiological conditions, small areas with elevated levels of residual radioactivity should be identified and their extents and activities determined. There are three basic modes in which one can operate in determining the levels of radiation and radioactivity at a site. They are scanning with hand-held survey instruments, direct measurements with these same or larger instruments, and sample collection at the site followed by analysis in the laboratory. In many cases, some combination of these modes would be used to obtain data, although the exact mix would be expected to vary according to the application.

In practice, the DQO process is used to obtain a proper balance among the uses of various measurement techniques. In general, there is an inverse correlation between the cost of a specific measurement technique and the detection levels being sought. Depending on the survey objectives, important considerations include survey costs and choosing the optimum instrumentation and measurement mix.

The decision to use a measurement method as part of the survey design is determined by the survey objectives and the survey unit classification. Scanning is performed to identify areas of elevated activity that may not be detected by other measurement methods. Direct measurements are analogous to collecting and analyzing samples to determine the average activity in a survey unit. Refer to Appendix O of this volume for information on lessons learned from recently submitted DPs and questions and answers to clarify existing license termination guidance related to measurements for facility radiation surveys.

E.2   Direct Measurements (Fixed Measurements)

To conduct direct measurements of alpha, beta, and photon surface activity, instruments and techniques providing the required detection sensitivity are selected. The type of instrument and method of performing the direct measurement are selected as dictated by the type of residual radioactivity present, the measurement sensitivity requirements, and the objectives of the radiological survey.

Direct measurements may be collected at random locations in the survey unit. Alternatively, direct measurements may be collected at systematic locations and supplement scanning surveys for the identification of small areas of elevated activity. Direct measurements may also be collected at locations identified by scanning surveys as part of an investigation to determine the source of the elevated instrument response. Professional judgment may also be used to identify locations for direct measurements to further define the areal extent of residual radioactivity and to determine maximum radiation levels within an area, although these types of direct measurements are usually associated with preliminary surveys (i.e., scoping, characterization, remedial action support). All direct measurement locations and results shall be documented.

If the equipment and methodology used for scanning is capable of providing data of the same quality required for direct measurement (e.g., detection limit, location of measurements, ability to record and document results), then scanning may be used in place of direct measurements. Results should be documented for at least the number of locations required for the statistical tests. In addition, some direct measurement systems may be able to provide scanning data, provided they meet the objectives of the scanning survey.

Refer to Chapter 6 of MARSSIM for information on radiation measurements. Specifically, Section 6.4.1 of MARSSIM contains information on direct measurements for alpha, beta, and gamma emitting radionuclides.

E.3   Scanning Measurements

Scanning is the process by which the operator uses portable radiation detection instruments to detect the presence of radionuclides on a specific surface (i.e., ground, wall, floor, equipment). The term scanning survey is used to describe the process of moving portable radiation detectors across a suspect surface with the intent of locating residual radioactivity. Investigation levels for scanning surveys are determined during survey planning to identify areas of elevated activity. Scanning surveys are performed to locate radiation anomalies indicating residual gross activity that may require further investigation or action.

Areas of elevated activity typically represent a small portion of the site or survey unit. Thus, random or systematic direct measurements or sampling on the commonly used grid spacing may have a low probability of identifying these areas. Scanning surveys are often relatively quick and inexpensive to perform. For these reasons, scanning surveys are typically performed before direct measurements or sampling. In this way, time is not spent fully evaluating an area that may quickly prove to contain residual radioactivity above the investigation level during the scanning process. Based on the historical site assessment (HSA), surfaces to be surveyed, and survey design objectives, scans are conducted which would be indicative of all radionuclides potentially present. Surrogate measurements may be utilized where appropriate. Documenting scanning results and observations from the field is very important. For example, a scan that identified relatively sharp increases in instrument response or identified the boundary of an area of increased instrument response should be documented. This information is useful when interpreting survey results.

Refer to Chapter 6 of MARSSIM for information on radiation measurements. Specifically, Section 6.4.2 of MARSSIM contains information on scanning measurements for alpha, beta, and gamma emitting radionuclides.

E.4   Sampling

For certain radionuclides that cannot be effectively measured directly in the field, samples of the medium under investigation (e.g., soil) should be collected and then analyzed with a laboratory-based procedure. On the simplest level, this would include the analysis of a smear sample using a gross alpha-beta counter. More involved analyses would include gamma spectrometry, beta analysis using liquid scintillation counting, or alpha spectrometry following separation chemistry.

Samples from a variety of locations may be required, depending upon the specific facility conditions and the results of scans and direct measurements. Inaccessible surfaces cannot be adequately evaluated by direct measurements on external surfaces alone; therefore, those locations which could contain residual radioactive material should be accessed for surveying. Residue can be collected from drains using a piece of wire or plumber's "snake" with a strip of cloth attached to the end; deposits on the pipe interior can be loosened by scraping with a hard-tipped tool that can be inserted into the drain opening. Particular attention should be given to "low-points" or "traps" where activity would likely accumulate. The need for further internal monitoring and sampling is determined on the basis of residue samples and direct measurements at the inlet, outlet, cleanouts, and other access points to the pipe interior.

Residual activity will often accumulate in cracks and joints in the floor. These are sampled by scraping the crack or joint with a pointed tool such as a screwdriver or chisel. Samples of the residue can then be analyzed; positive results of such an analysis may indicate possible subfloor residual radioactivity. Checking for activity below the floor will require accessing a crawl space (if one is present) or removal of a section of the flooring. Coring, using a commercially available unit, is a common approach to accessing the subfloor soil. After removing the core (whose diameter may range from a few centimeters to up to 20 centimeters), direct monitoring of the underlying surface can be performed and samples of soil collected.

Coring is also useful for collecting samples of construction material which may contain activity that has penetrated below the surface, or activity induced by neutron activation. This type of sampling is also applicable to roofing material which may contain embedded or entrapped contaminants. The profile of the distribution and the total radionuclide content can be determined by analyzing horizontal sections of the core.

If residual activity has been coated by paint or some other treatment, the underlying surface and the coating itself may contain residual radioactivity. If the activity is a pure alpha or low-energy beta emitter, measurements at the surface will probably not be representative of the actual residual activity level. In this case, the surface layer is removed from a known area, usually 100 cm^2, using a commercial stripping agent or by physically abrading the surface. The removed coating material is analyzed for activity content and the level converted to units of dpm/100 cm^2 for comparison with guidelines for surface activity. Direct measurements are performed on the underlying surface, after removal of the coating.

MARSSIM and MARLAP contain information on sampling and laboratory analysis for decommissioning. Refer to Chapter 10 of MARLAP for field and sampling issues that affect laboratory measurements.

E.5   Minimum Detectable Concentrations

Detection limits for field survey instrumentation are an important criteria in the selection of appropriate instrumentation and measurement procedures. For the most part, detection limits need to be determined in order to evaluate whether a particular instrument and measurement procedure is capable of detecting residual activity at the regulatory release criteria (DCGLs). One may demonstrate compliance with decommissioning criteria by performing surface activity measurements and directly comparing the results to the surface activity DCGLs. However, before any measurements are performed, the survey instrument and measurement procedures to be used must be shown to possess sufficient detection capabilities relative to the surface activity DCGLs.

The measurement of residual radioactivity during surveys in support of decommissioning often involves measurement of residual radioactivity at near-background levels. Thus, the minimum amount of radioactivity that may be detected by a given survey instrument and measurement procedure must be determined. In general, the minimum detectable concentration (MDC) is the minimum activity concentration on a surface or within a material volume, that an instrument is expected to detect (i.e., activity expected to be detected with 95 % confidence). It is important to note that this activity concentration, the MDC, is determined *a priori* (i.e., before survey measurements are conducted).

As generally defined, the detection limit, which may be a count or count rate, is independent of field conditions such as scabbled, wet, or dusty surfaces. That is, the detection limit is based on the number of counts and does not necessarily equate to measured activity under field conditions. These field conditions do, however, affect the instrument's "detection sensitivity" or MDC. Therefore, the terms MDC and detection limit should not be used interchangeably.

In MARSSIM, MARLAP, and other NRC NUREGs, the MDC corresponds to the smallest activity concentration measurement that is practically achievable with a given instrument and type of measurement procedure. That is, the MDC depends not only on the particular instrument characteristics (instrument efficiency, background, integration time, etc.) but also on the factors involved in the survey measurement process (EPA 1980), which include surface type, source-to-detector geometry, and source efficiency (e.g., backscatter and self-absorption).

See MARLAP Section 3.3.7, "Method Performance Characteristics and Measurement Quality Objectives" and Chapter 20, "Detection and Quantification Capabilities" for a discussion of MDCs.

E.6   Survey MDCs

During radiological surveys in support of decommissioning, scanning is performed to identify the presence of any locations of elevated direct radiation. The probability of detecting residual radioactivity in the field is affected not only by the sensitivity of the survey instrumentation when used in the scanning mode of operation, but also by the surveyor's ability. The surveyor must decide whether the signals represent only the background activity, or whether they represent residual radioactivity in excess of background.

The minimum detectable concentration of a scan survey, referred to as scan MDC or MDC_{scan}, depends on the intrinsic characteristics of the detector (efficiency, window area, etc.), the nature (e.g., type and energy of emissions) and relative distribution of the residual radioactivity (e.g., point versus distributed source and depth of residual radioactivity), scan rate, and other characteristics of the surveyor. Some factors that may affect the surveyor's performance include the costs associated with various outcomes—e.g., cost of missed residual radioactivity versus cost of incorrectly identifying areas as containing residual radioactivity—and the surveyor's *a priori* expectation of the likelihood of residual radioactivity present. For example, if the surveyor believes that the potential for residual radioactivity is very low, as in an unaffected area, then a relatively large signal may be required for the surveyor to conclude that residual radioactivity is present. NUREG/CR–6364, "Human Performance in Radiological Survey Scanning," provides a complete discussion of the human factors as they relate to the performance of scan surveys.

Signal detection theory provides a framework for the task of deciding whether the audible output of the survey meter during scanning was due to background or signal plus background levels. An index of sensitivity (d') that represents the distance between the means of the background and background plus signal, in units of their common standard deviation, can be calculated for various decision errors—Type I error (α) and Type II error (β). As an example, for a correct detection or true positive rate of 95 % ($1-\beta$) and a false positive rate (α) of 5 %, d' is 3.29 (similar to the static MDC for the same decision error rates). The index of sensitivity is independent of human factors, and therefore, the ability of an ideal observer (i.e., theoretical construct) may be used to determine the minimum d' that can be achieved for particular decision errors. The ideal observer makes optimal use of the available information to maximize the percent correct responses and thus provides an effective upper bound against which to compare actual surveyors. Computer simulations and field experimentation can then be performed to evaluate the surveyor efficiency (p) relative to the ideal observer. The resulting expression for the ideal observer's minimum detectable count rate ($MDCR$), in cpm, can be written:

$$MDCR = d' \times \sqrt{b_i} \times (60/i) = s_i \times (60/i) \qquad (E{-}1)$$

where $MDCR$ =   minimum detectable (net) count rate in cpm,
    b_i =   background counts in the observation interval,
    s_i =   minimum detectable number of net source counts in the observation interval, and
    i =   observational interval (in seconds), based on the scan speed and areal extent of the residual radioactivity.

Scan MDCs are determined from the *MDCR* by applying conversion factors to obtain results in terms of measurable surface activities and soil concentrations. As an example, the *scan MDC* for a structure surface can be expressed as:

$$Scan\ MDC = \frac{MDCR}{\sqrt{p}\ \epsilon_i\ \epsilon_s\ \dfrac{probe\ area}{100\ cm^2}} \qquad (E{-}2)$$

Chapter 6 of NUREG-1507 contains an excellent discussion of survey MDCs. Included in this discussion are *scan MDC* equations for both building/structure surface scans and land area scans.

E.7   Survey Instrument Calibration

Before the MDC for a particular instrument and survey procedure can be determined, it is necessary to introduce the expression for total alpha or beta surface activity per unit area. In the International Organization for Standardization's (ISO) Guide 7503–1, "Evaluation of Surface Contamination," the ISO recommends that the total surface activity, A_S, be calculated similarly to the following expression:

$$A_s = \frac{R_{S+B} - R_B}{(\epsilon_i)(W)(\epsilon_s)} \qquad (E{-}3)$$

where R_{S+B} = the gross count rate of the measurement in cpm,
R_B = the background count rate in cpm,
ϵ_I = the instrument or detector efficiency (unitless),
ϵ_s = the efficiency of the residual radioactivity source (unitless), and
W = the area of the detector window (cm^2).

(For instances in which W does not equal 100 cm^2, probe area corrections are necessary to convert the detector response to units of dpm per 100 cm^2.)

This expression clearly distinguishes between instrument (detector) efficiency and source efficiency. The product of the instrument and source efficiency yields the total efficiency, ϵ_{tot}. Currently, surface residual radioactivity is assessed by converting the instrument response to surface activity using one overall total efficiency. This is not a problem provided that the calibration source exhibits characteristics similar to the surface residual radioactivity—including radiation energy, backscatter effects, source geometry, self-absorption, etc. In practice this is hardly the case; more likely, total efficiencies are determined with a clean, stainless steel source, and then those efficiencies are used to measure residual radioactivity on a dust-covered concrete surface. By separating the efficiency into two components, the surveyor has a greater ability to consider the actual characteristics of the surface residual radioactivity.

The instrument efficiency is defined as the ratio between the net count rate of the instrument and the surface emission rate of a source for a specified geometry. The surface emission rate, $q_{2\pi}$, is defined as the "number of particles of a given type above a given energy emerging from the front face of the source per unit time" (ISO 7503–1). The surface emission rate is the 2π particle fluence that embodies both the absorption and scattering processes that affect the radiation emitted from the source. Thus, the instrument efficiency is determined by

$$\epsilon_i = \frac{R_{S+B} - R_B}{q_{2\pi}} \qquad (E\text{–}4)$$

The instrument efficiency is determined during calibration by obtaining a static count with the detector over a calibration source that has a traceable activity or surface emission rate or both. In many cases, it is the source surface emission rate that is measured by the manufacturer and certified as National Institute of Standards and Technology (NIST) traceable. The source activity is then calculated from the surface emission rate based on assumed backscatter and self-absorption properties of the source. The theoretical maximum value of instrument efficiency is one.

The source efficiency, ϵ_s, is defined as the ratio between the number of particles of a given type emerging from the front face of a source and the number of particles of the same type created or released within the source per unit time (ISO 7503–1). The source (or surface) efficiency takes into account the increased particle emission due to backscatter effects, as well as the decreased particle emission due to self-absorption losses. For an ideal source (no backscatter or self-absorption), the value of ϵ_s is 0.5. Many real sources will exhibit values of ϵ_s less than 0.5, although values greater than 0.5 are possible, depending on the relative importance of the absorption and backscatter processes. Source efficiencies may either be determined experimentally or simply selected from the guidance contained in ISO 7503–1.

Some of the factors that affect the instrument efficiency, ϵ_I, include detector size (probe surface area), window density thickness, geotropism, instrument response time, and ambient conditions such as temperature, pressure, and humidity. The instrument efficiency also depends on the radionuclide source used for calibration and the solid angle effects, which include source-to-detector distance and source geometry.

Some of the factors that affect the source efficiency, ϵ_s, include the type of radiation and its energy, source uniformity, surface roughness and coverings, and surface composition (e.g., wood, metal, concrete).

Surface activity levels are assessed by converting detector response, through the use of a calibration factor, to radioactivity. Once the detector has been calibrated and an instrument efficiency (ϵ_I) established, several factors must still be carefully considered when using that instrument in the field. These factors involve the background count rate for the particular surface and the surface efficiency (ϵ_s), which addresses the physical composition of the surface and any surface coatings. Ideally, the surveyor should use experimentally determined surface efficiencies for the anticipated field conditions. The surveyor needs to know how and to what degree these different field conditions can affect the sensitivity of the instrument. A particular field condition may significantly affect the usefulness of a particular instrument (e.g., wet surfaces for alpha measurements or scabbled surfaces for low-energy beta measurements).

One of the more significant implicit assumptions commonly made during instrument calibration and subsequent use of the instrument in the field is that the composition and geometry of residual radioactivity in the field is the same as that of the calibration source. This may not be the case, considering that many calibration sources are fabricated from materials different from those that comprise the surfaces of interest in the field [e.g., activity plated on a metallic disc (Walker 1994)]. This difference usually manifests itself in the varying backscatter characteristics of the calibration and field surface materials.

Generally, it will not be necessary to recalculate the instrument MDC to adjust for the field conditions. The instrument detection limit (in net counts or net count rate) remains the same, but the surface activity MDC may be different (due to the varying ϵ_s).

Refer to Chapter 4 of NUREG–1507 for a discussion of survey instrument calibration and the effects of efficiency changes on MDC. Chapter 5 of NUREG–1507 discusses variables affecting efficiencies in the field. Chapter 20 of MARLAP discusses instrument efficiency and the minimum detectable net instrument signal.

E.8   Laboratory Measurements

Frequently during surveys in support of decommissioning it is not feasible, or even possible, to detect the residual radioactivity with portable field instrumentation; thus arises the need for laboratory analysis of media samples. This is especially the case for such media samples as soil, that result in significant self-absorption of the radiation from the residual radioactivity. Another common situation that necessitates the use of laboratory analyses occurs when the residual radioactivity is difficult to detect even under ideal conditions. This includes residual radioactivity that emits only low-energy beta radiation (e.g., H-3 and Ni-63) or X–ray radiation (e.g., Fe-55). Laboratory analyses for radionuclide identification, using spectrometric techniques, are often performed during scoping or characterization surveys. Here the principal objective is to simply determine the specific radionuclides present in the residual radioactivity, without necessarily having to assess the quantity of residual radioactivity. Once the residual radioactivity has been identified, sufficiently sensitive field survey instrumentation and techniques are selected to demonstrate compliance with the DCGLs.

Samples collected during surveys for decommissioning purposes should be analyzed by trained individuals using the appropriate equipment and procedures at a well-established laboratory, which uses either inhouse or contractor laboratory services. There should be written procedures that document both (a) the laboratory's analytical capabilities for the radionuclides of interest and (b) the QA/QC program which assures the validity of the analytical results. Many of the general types of radiation detection measuring equipment used for survey field applications are also used for laboratory analyses, usually under more controlled conditions which provide for lower detection limits and greater delineation between radionuclides. Laboratory methods often also involve a combination of both chemical and instrumental technique to quantify the low levels expected to be present in samples from decommissioning facilities.

To reemphasize, a thorough knowledge of the radionuclides present, along with their chemical and physical forms and their relative abundance, is a prerequisite to selecting laboratory methods. With this information, it may be possible to substitute certain gross (i.e., nonradionuclide- specific) measurement techniques for the more costly and time-consuming wet chemistry separation procedures and relate the gross data back to the relative quantities of specific contaminants. The individual responsible for the survey should be aware that radiochemical analyses require lead times which will vary, according to the nature and complexity of the request. For example, a lab may provide fairly quick turnaround on gamma spectrometry analysis because computer-based systems are available for interpretation of gamma spectra. On the other hand, soil samples, which must be dried and homogenized, will require much longer lead time. Some factors influencing the analysis time include (a) the nuclides of concern, (b) the type of samples to be analyzed, (c) the QA/QC considerations required, (d) the availability of adequate equipment and personnel, and (e) the required detection limits.

For relatively simple analyses, such as gross alpha and gross beta counting of smears and water samples, liquid scintillation spectrometry for low-energy beta emitters in smear and water samples, and gamma-spectrometry of soil, it is usually practical to establish in-house laboratory capabilities. The more complicated and labor-intensive procedures, such as alpha spectrometry, Sr-90 and low-energy beta emitters (H-3, Ni-63, etc.) in soil samples, should be considered candidates for contract laboratory analyses.

Analytical methods should be capable of measuring levels below the established release guidelines, detection sensitivities of 10 to 25 % of the guideline should be the target. Although laboratories will state detection limits, these limits are usually based on ideal situations and may not be achievable under actual measurement conditions. Also, remember that detection limits are subject to variation from sample to sample, instrument to instrument, and procedure to procedure depending upon sample size, geometry, background, instrument efficiency, chemical recovery, abundance of the radiations being measured, counting time, self-absorption in the prepared sample, and interference from other radionuclides present.

MARSSIM and MARLAP contain information on sampling and laboratory analysis for decommissioning. MARLAP Sections 12, 13, 14, and 15 discuss laboratory sample preparation, sample dissolution, separation techniques, and quantification of radionuclides.

E.9   References

Nuclear Regulatory Commission (U.S.) (NRC). NUREG–1575, "Multi-Agency Radiation Survey and Site Investigation Manual." NRC: Washington, DC. August 2000.

— — — — —. NUREG–1506, "Measurement Methods for Radiological Surveys in Support of New Decommissioning Criteria–Draft." NRC: Washington, DC. August 1995.

— — — — —. NUREG–1507, "Minimum Detectable Concentrations with Typical Radiation Survey Instruments for Various Contaminants and Field Conditions." NRC: Washington, DC. June 1998.

— — — — —. NUREG/CR–6364, "Human Performance in Radiological Survey Scanning." NRC: Washington, DC. December 1997.

Department of Energy (U.S.) (DOE). HASL–300, "EML Procedures Manual, 28th ed." DOE: Washington, DC. 1997.

Environmental Protection Agency (U.S.) (EPA). EPA 520/1–80–012, "Upgrading Environmental Radiation Data." EPA: Washington, DC. August 1980.

International Organization for Standardization. "Evaluation of Surface Contamination – Part 1: Beta Emitters and Alpha Emitters (first edition)." ISO 7503–1. ISO: Geneva, Switzerland. 1988.

Walker, E. "Proper Selection and Application of Portable Survey Instruments for Unrestricted Release Surveys." Bechtel Environmental, Inc. Presented at 1994 International Symposium on D&D. April 24–29, 1994.

Appendix F

Ground and Surface Water Characterization

The majority of the information in this appendix is taken directly from the *Draft Branch Technical Position on Site Characterization for Decommissioning* (NRC 1994). The checklist in Section F.3 of this appendix regarding potential indicators for ground water contamination is taken directly from NUREG–1496. This chapter is applicable, either in total or in part, to Decommissioning Group 4 for surface water and Decommissioning Groups 5–7.

Characterization of surface and ground water is an essential component of the dose modeling used in the estimation of doses to demonstrate compliance with the license termination requirements in 10 CFR Part 20, Subpart E. If contaminated surface or ground water is identified, the screening DCGLs for soil are inappropriate since they are usually based on initially uncontaminated surface and ground water. Appendix I of this volume discusses the aspects of dose modeling that are specific to site hydrology.

F.1   Planning for Surface Water and Ground Water Characterization

Surface and ground water characterization should be planned in a manner that maximizes the utility of the information to be collected and optimizes its adequacy and quality during the characterization process. For example, a licensee may show for a particular site that the surface water pathway is not likely to be significant in terms of existing and potential future exposure to the public. In such a case, the need for detailed characterization of the surface water system is decreased. As an example of effective interactions during site characterization, identification of ground water contamination during preliminary scoping survey may warrant installation and sampling of additional monitoring wells to define the extent and migration status of the contamination.

In some instances, ground water may be unsuitable for specific uses, such as human and livestock consumption, but may be acceptable for crop irrigation. In addition, some aquifers may not have the yield to support crop irrigation but may produce enough water for human consumption. In some instances, EPA or a State agency may have declared that the aquifer in question is unfit for human or livestock use. Accordingly, this type of information needs to be addressed since it will be used to support site scenario development and dose modeling. Refer to Section I.3.3.3.2 from Appendix I of this volume for guidance on modification of waterborne exposure pathways.

NRC staff experience has shown that some decommissioning plans (DPs) have not adequately provided ground water characterization data. Additional environmental monitoring data may be needed because there may not be enough operational monitoring of ground water for adequate site characterization and dose assessment. Regulatory Issue Summary 2002–02 provides a detailed discussion of lessons learned regarding ground water characterization. The information contained in this RIS has been included in Section O.2.2 from Appendix O of this volume.

F.2   Ground Water Characterization

The need for surveys to characterize ground water should be determined from the historical site assessment (HSA). If the HSA indicates that residual radioactivity may have reached potable water, surveys of ground water would be appropriate. The nature of appropriate ground water surveys should be determined on a site-specific basis. In addition to that which is discussed below, MARSSIM (Sections 3.6.3.4 and 5.3.3.3) provides some guidance on evaluating the likelihood for release of radionuclides into ground water and on evaluating related concerns regarding characterization and sampling.

Characterization of ground water contamination, including all significant radiological constituents, along with inorganic and organic constituents and related parameters, should be adequate to determine the following:

- extent and concentration distribution of contaminants;

- source (known or postulated) of radioactive contaminants to ground water;

- background ground water quality;

- rate(s) and direction(s) of contaminated ground water migration;

- location of ground water plume and concentration profiles (i.e., maximum concentration in the vertical and lateral extent);

- assessment of present and potential future effects of ground water withdrawal on the migration of ground water contaminants;

- potential safety and environmental issues associated with remediating the surface and ground water;

- effect of the nonradiological constituents on the mobility of the radionuclides;

- whether the remediation activities and radiation control measures proposed by the licensee are appropriate for the type and amount of radioactive material present in the surface and ground water;

- whether the licensee's waste management practices are appropriate; and

- whether the licensee's cost estimates are plausible.

Besides licensee process discharges, there are other mechanisms that may affect ground water. For example, sumps that are used to capture infiltrating ground water may affect the local ground water elevation during pumping. In some situations, sumps are used to collect ground water at the lowest elevation of a building, with pumping going on continuously. Such pumping has been shown to affect the local ground water elevation (i.e., cone of depression).

Characterization of the nonradiological constituents and related parameters may also be required by other regulatory Agencies that have jurisdiction over the decommissioning effort. Therefore, licensees should contact Federal, State, or local government bodies responsible for regulating water. Typical analytical parameters include gross alpha particle activity, gross beta particle activity, specific radionuclide concentrations, gamma spectrum analysis for all gamma-emitting radionuclides suspected to be present, sulfate, chloride, carbonate, alkalinity, nitrate, TDS, Total Organic Carbon (TOC), Eh, pH, calcium, sodium, potassium, iron, and dissolved oxygen. Additional analytical parameters may be necessary to characterize any suspected contamination.

The extent of contamination and background ground water quality should be determined based on ground water monitoring data from a suitable monitoring well network. Guidance documents on acceptable ground water monitoring techniques are listed under References [Korte and Ealey (1983), Korte and Kearl (1984), NUREG–1383 and NUREG–1388, USGS (1977 and 1996), and EPA (1977, 1980, 1985, and 1986)]. The actual number, location, and design of monitoring wells depend on the size of the contaminated area, the type and extent of contaminants, the background ground water quality, the hydrogeologic system, and the objectives of the monitoring program. For example, if the objective of monitoring is only to indicate the presence of ground water contamination, relatively few downgradient and upgradient monitoring wells are needed. In contrast, if the objective is to develop a detailed characterization of the distribution of constituents within a complex aquifer as the design basis for a corrective action program, a large number of suitably designed and installed monitoring wells and well points may be necessary. Planned site characterization activities should be flexible enough to allow for the installation of additional monitoring wells during the characterization effort if either:
(a) preliminary characterization indicates contamination where previously unanticipated; or
(b) there is a need to delineate the vertical or lateral extent of contaminant plumes. Monitoring well locations, contaminant concentrations, and contaminant sources should be plotted on a map (or a series of maps for multiple contaminants) to show the relationship among contamination, sources, hydrogeologic features and boundary conditions, and property boundaries. At sites with significant vertical migration of contaminants, the DP should also provide hydrogeologic cross-sections that depict the vertical distribution of contaminants in ground water. The vertical exaggeration of the sections should not exceed 10 times.

The DP should also describe the ground water characterization program used to characterize the extent and distribution of contaminants in the ground water. Depending on the complexity of the site, the DP can include the detailed information as described below or can summarize this information and then reference the documents where the supporting details are contained. The description should provide monitoring well completion diagrams explaining elevation, internal and external dimensions, types of casings, type of backfill and seal, type of the screen and its location and size, borehole diameter and elevation and depth of hole, and type and dimension of riser pipe and other necessary information on the wells. An acceptable generic completion design is illustrated in Figure F.1.

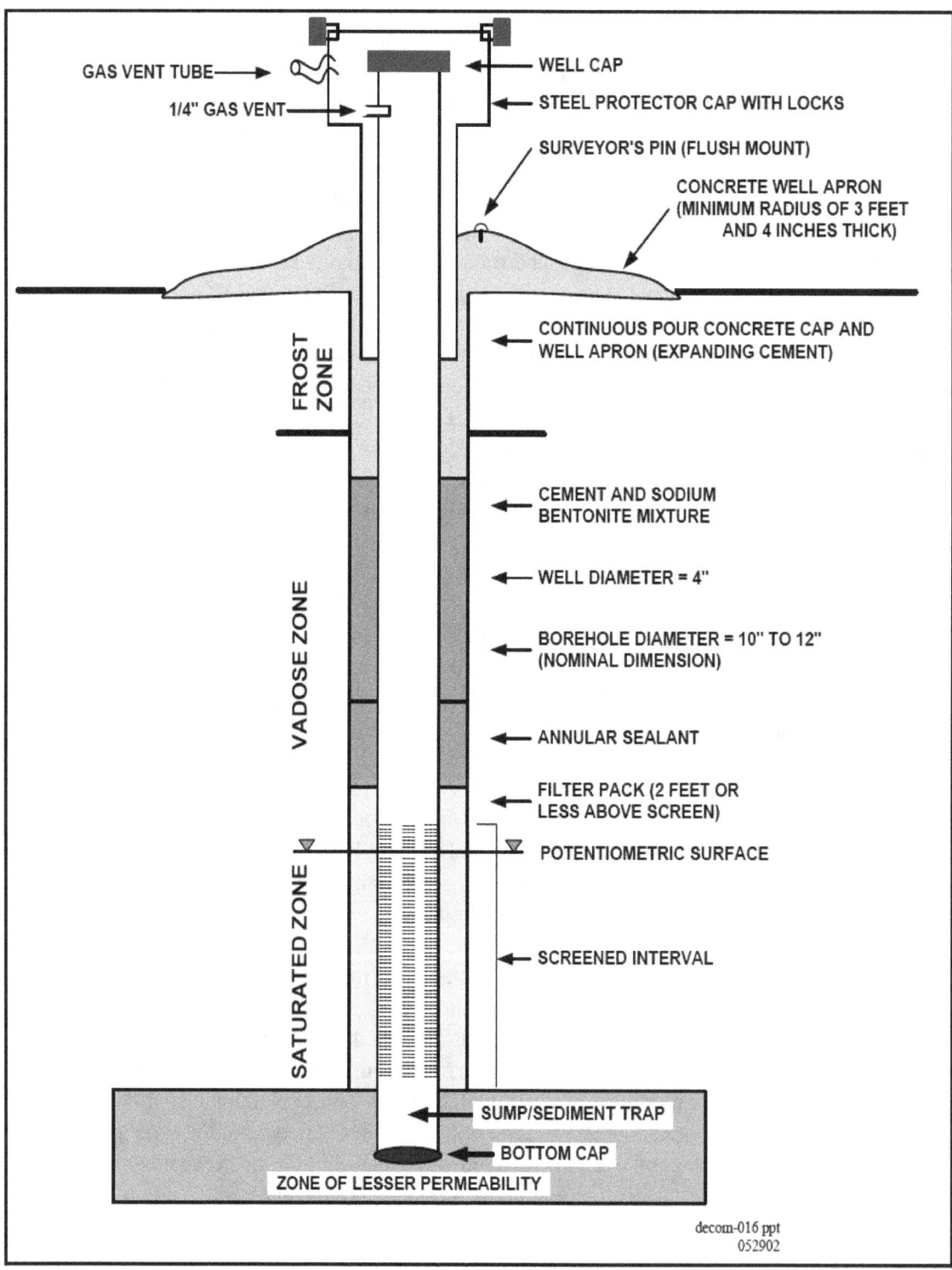

GAS VENT TUBE →

1/4" GAS VENT →

← WELL CAP

← STEEL PROTECTOR CAP WITH LOCKS

SURVEYOR'S PIN (FLUSH MOUNT)

CONCRETE WELL APRON
(MINIMUM RADIUS OF 3 FEET
AND 4 INCHES THICK)

FROST ZONE

← CONTINUOUS POUR CONCRETE CAP AND
WELL APRON (EXPANDING CEMENT)

VADOSE ZONE

← CEMENT AND SODIUM
BENTONITE MIXTURE

← WELL DIAMETER = 4"

← BOREHOLE DIAMETER = 10" TO 12"
(NOMINAL DIMENSION)

← ANNULAR SEALANT

← FILTER PACK (2 FEET OR
LESS ABOVE SCREEN)

POTENTIOMETRIC SURFACE

SATURATED ZONE

← SCREENED INTERVAL

SUMP/SEDIMENT TRAP

BOTTOM CAP

ZONE OF LESSER PERMEABILITY

decom-016.ppt
052902

Figure F.1 General Monitoring Well Cross-Section.

Sampling techniques, methodology, and procedures should be documented or referenced in the DP. Site characterization procedures and methods should generally adhere to acceptable national practices and standards [e.g., American Society for Testing and Materials (ASTM), U.S. Geological Survey (USGS), U.S. Environmental Protection Agency (EPA), U.S. Department of Energy Environmental Monitoring Laboratory (DOE/EML), and National Institute of Standards and Technology (NIST)]. The DP should identify specific analytical methods that conform to generally accepted protocols and methods, such as those endorsed by NIST, DOE/EML, or other methods established through comprehensive peer review and recommendation process (e.g., ANSI/ASME 1986). Korte and Kearl (1984) provides forms for documenting well summary information, samples, chain of custody, quality assurance information for field chemical analyses, and sample location and identifier.

The site characterization program should include sufficient sampling and analysis of ground water samples collected upgradient from the site to develop a representative characterization of background ground water quality. Background ground water quality should not exhibit any influence from contaminants released by the site and should be representative of the quality of ground water that would exist if the site had not been contaminated. The site characterization should also assess any temporal or spatial variations in background ground water quality. If sources of contamination other than the site are present, the potential impact of such sources should be evaluated to determine the degree of ground water contamination caused by these sources.

F.3 Indicators for Potential Ground Water Contamination

When evaluating ground water contamination, it is especially necessary to consider the site-specific factors that permit radionuclides to migrate through the ground water pathway, and thus contribute to the dose to an individual from residual radioactivity.

As described in Table 1.1 of Volume 1 of this NUREG report, Decommissioning Groups 5–7 are sites that have the potential for residual radioactivity in ground water. Based on the experience gained from operational and decommissioning NRC–licensed sites, the following is a list of potential indicators for ground water contamination at decommissioning sites. The following are intended to be illustrative only and not intended to constitute a complete list:

- High Potential: if a site has a history of, or currently has:
  - unlined lagoons, pits, canals, or surface-drainage ways that received radioactively contaminated liquid effluent;
  - lined lagoons, pits, canals, or surface drainage ways that received radioactively contaminated liquid effluent, where the lining has leaked or ruptured, or where overflow has occurred;
  - septic systems, dry wells, or injection wells that received radioactively contaminated liquid effluent;

— storage tanks, waste tanks, and/or piping (above or below ground) that held or transported radioactively contaminated fluids and are known to have leaked;

— liquid or wet radioactive waste buried onsite (i.e., burial under 10 CFR 20.302 or 20.304 (or the current 10 CFR 20.2002));

— an accident or spill onsite where radioactive material was released exterior to a building;

— wet bulk waste (e.g., sludge or tailings) stored exterior to buildings or used as backfill; and

— containerized-liquid waste, stored exterior to buildings, that has leaked.

- Medium Potential: if a site has a history of or currently has:

— surface water or atmospheric discharge of radioactive effluents;

— radioactive contamination detected on the roof of a building;

— radioactive contamination detected in the floor cracks or sump of a building;

— an accident or spill onsite, where liquid radioactive material was released to the interior of a building;

— the presence of greater than 10-year-old underground storage tank or underground piping that held radioactively contaminated fluids, not known to have leaked, but never tested;

— a history of incineration of radioactive waste exterior to onsite buildings;

— dry bulk waste (e.g., sludge or tailings) stored exterior to buildings or used as backfill;

— solid containerized waste, stored exterior to buildings, that has leaked.

- Low Potential: if a site has a history of or currently has:

— less than 10-year-old underground storage tanks or underground piping that has held radioactively contaminated fluids and is known not to have leaked;

— dry bulk waste stored inside the buildings;

— a sealed-source-only license.

The potential for ground water contamination at any of these sites is conditioned by certain site characteristics such as depth of ground water, amount of yearly precipitation and hydraulic conductivity, and by certain source characteristics such as half-life, solubility, and distribution coefficient.

F.4   Monitoring Practices and Procedures

Depending on the complexity of the site, the DP can include the detailed information as described below, or the DP can summarize this information and then reference the documents where the supporting details are contained.

The site characterization should include a description of all surface and ground water characterization activities, methods, and monitoring installations sufficient to demonstrate that the methods and devices provided data that are representative of site conditions. Also included should be a description of the monitoring practices, procedures, and quality assurance programs used to collect water quality data. Monitoring well descriptions, for example, should include location, elevation, screened interval(s), depth, construction and completion details, and the hydrologic units monitored. Aquifer test descriptions should include testing configuration, test results, and a discussion of the assumptions, analytical techniques, test procedures, pretesting baseline conditions, limitations, errors in measurements, and final results. The description of the water quality sampling and analysis program should include or reference the procedures for sampling, preserving, storing, and analyzing the samples, including QA/QC protocols implemented. All methods used should be consistent with current standard methods and practices (e.g., ASTM, USGS, EPA, NIST, and ANSI/ASME). Some additional guidance on acceptable methods for sampling and analyzing water quality samples can be found in Korte and Ealey (1983); Korte and Kearl (1984); DOE (1988 and 1993); ANSI/ASME (1986); EPA (1977, 1985, 1986, 1987a, 1991); and NUREG–1293, NUREG–1383, and Regulatory Guide 4.15. Any deviations from standard methods should be appropriately justified.

F.5   Sampling Frequencies

Surface and ground water quality and water levels should be determined on a set frequency established based on site-specific considerations. For sites with extensive ground water contamination, a network of monitoring wells should be designed and installed to provide a high probability of detecting and characterizing existing contamination and determining background ground water quality. Ground water levels should be measured in piezometers and monitoring wells that provide a sufficiently accurate indication of hydraulic head to characterize the hydraulic gradient within the uppermost aquifer and adjacent units. Water levels should be measured on a quarterly basis for a minimum one year to determine temporal variations in the hydraulic gradient. After this period, the frequency of water level measurements should be adjusted to reflect anticipated temporal variation in hydraulic heads (e.g., tides, river bank storage, water year variations). Acceptable methods for ground water sampling and for measuring water levels are described in EPA and USGS documents (EPA 1977, 1985, 1986, and 1987a; USGS 1977) and in Korte and Kearl (1984).

The sampling frequency for determining variations in ground water quality should be determined based on the temporal variation in hydraulic gradients, as well as temporal variations in hydrochemistry and migration of radiological and associated nonradiological constituents. After an initial sampling round in which each monitoring well is sampled, representative samples should be collected and analyzed on a quarterly basis from key monitoring wells to estimate the temporal variation of water quality in the uppermost aquifer and adjacent units. After this initial period, sampling frequency should be adjusted to reflect variations in the hydraulic gradient and hydrochemistry. Concentrations of principal radiological constituents should not change by more than about 10–20 % between sampling events. If the concentrations change by more than 10–20 %, the frequency of sampling should be increased in an attempt to characterize the temporal variability of ground water quality. For most sites, sampling on a quarterly basis (i.e., one sample per well per calendar quarter) should be sufficiently frequent to characterize temporal changes in water quality. More frequent sampling may be necessary, however, especially at sites involving offsite or potential offsite contamination of ground water resources. Acceptable methods for ground water sampling are described in Korte and Kearl (1984), USGS (1977) and the EPA references mentioned above.

Quarterly sampling of surface water and sediments should be sufficient at most sites. This sampling should be supplemented by additional sampling to characterize the surface system at representative low or high stage flow conditions (e.g., minimum annual, 7-day average low flow or maximum annual, 7-day average high flow). This information should be used to bound the existing and projected impacts of the release of contamination on adjacent surface water bodies.

F.6   Surface Water and Sediments

Surface water can include ponds, creeks, streams, rivers, lakes, coastal tidal waters, oceans, and other bodies of water. Note that certain ditches and intermittently flowing streams qualify as surface water. The need for surface water samples should be evaluated on a case-by-case basis. Surveys for water should be based on appropriate environmental standards for water sampling. If the body of water is included in a larger survey unit, then sediment samples should be taken at sample locations selected by the normal method without taking the body of water into consideration. In addition to that which is discussed below, MARSSIM (Sections 3.6.3.3 and 5.3.3.3) provides some guidance on evaluating the likelihood for release of radionuclides into surface water and sediments and on related concerns regarding characterization and sampling.

For sites that are located near surface water streams and could reasonably affect surface water pathways, the site characterization program should establish background surface water quality by sampling upstream of the site being studied or areas unaffected by any known activity at the site. Water should be collected as grab samples from the stream bank in a well-mixed zone. Depending on the significance and the potential for surface water contamination, it may be necessary for certain sites to collect stratified samples from the surface water to determine the distribution of contaminants within the water column. Surface water quality sampling should be accompanied by at least one round of stream sediment quality sampling to assess the relationship between the composition of the dissolved solids, the suspended sediment, and the bedload sediment fractions. Water levels and discharge rates of the stream should be determined at the time samples are collected. Licensees should also consider the effects of variability of the surface water flow rate. Based on the results of the HSA and/or preliminary investigation surveys, surface scans for gamma activity should be conducted in areas likely to contain residual activity (e.g., along the banks). Acceptable methods for surface water and sediment sampling are described in Korte and Kearl (1984) and USGS (1977). In addition, Fleishhauer and Engelder (1984) present suggested procedures for stream sediment sampling. The EPA guidance documents mentioned above are also applicable. In some cases, the Radiological Environmental Monitoring Program (REMP) data from a facilities operating period may provide useful information to support the characterization program. Appendix O of this volume discusses the limitations of the use of REMP data to support site characterization.

Surface water sampling should be conducted in areas of runoff from active operations. In case of direct discharge into a stream, the outfall and the stream should be monitored and sampled upstream and downstream from the outfall. Preliminary characterization of the contamination levels should be conducted by measuring gross alpha and total beta particle activity (total and dissolved) and by obtaining a gamma spectrum for surface water samples. It should be noted that determination of gross alpha activity (and low energy beta emitters as well) may be of limited value for samples containing elevated total or dissolved solids concentrations because of sample attenuation. In such instances, gamma spectroscopy might be the only recourse. Specific radionuclide analysis may be needed depending on level of activities and type of radionuclides. Nonradiological parameters, such as specific conductance, pH, and total organic carbon, may be used as surrogate indicators of potential contamination, provided a clear relationship is established between radionuclide concentration and the level of the surrogate. Additional analysis for other parameters like volatile and semi-volatile compounds, chelating agents, pesticides, and polychlorinated biphenyls (PCBs) may also be necessary if they affect the mobility of radiological constituents and to evaluate potential environmental effects of the decommissioning.

Each of the surface water and sediment sampling locations should be carefully recorded on the appropriate survey form. Additionally, surface water flow models can be used to assist in estimating contaminant concentrations or migration rates.

F.7   Geochemical Conditions

Geochemical conditions at the site and their association with ground water and contaminants should also be described. Specifically, geochemical conditions that enhance or retard contaminant transport should be given special consideration. Geochemical data should include information on solid composition, buffering capacity, redox potential, sorption (represented as a range of distribution coefficients (K_d) for each radiological constituent), and other relevant geochemical data. Piper and Stiff diagrams may be useful for visualizing the geochemistry of the water. In general, licensees or responsible parties may estimate the values of K_d through laboratory column or batch sorption measurements [e.g., ASTM methods D4319 (ASTM 2002), D4646 (ASTM 2001), and D4874 (ASTM 2001)] or by using a conservative value to represent the values of K_d from available literature references [e.g., Sheppard and Thibault (1990) and NUREG/CR–5512, Volume 3]. If necessary, licensees (or responsible parties) may use appropriate geochemical codes to understand and quantify geochemical mechanisms that significantly affect transport of radiological and nonradiological contaminants and their potential fate (e.g., MINTEQ (EPA 1984); EQ3/6 (Daveler and Woolery 1992)). Additional information on ground water parameters necessary for dose modeling is discussed in Appendix I of this volume.

F.8 Surface Water and Ground Water Models

As a joint effort, EPA, NRC, and DOE have developed specific guidance on selecting and applying surface and ground water models (EPA 1994a, b, c). Supporting details may be found elsewhere (NUREG–3332 and NUREG/CR–5454, Volumes 1–5; NCRP 1985, 1996; DOE 1995; EPA 1987a, b, 1994a, b, c; NAS 1999).

F.9 References

American National Standards Institute/American Society of Mechanical Engineers (ANSI/ASME) NQA–1, "Quality Assurance Program Requirements for Nuclear Facilities." ANSI/ASME: New York. 1986.

American Society for Testing and Materials (ASTM). D4646–87, "Standard Test Method for 24-h Batch-Type Measurement of Contaminated Sorption by Soils and Sediments." ASTM: West Conshohocken, Pennsylvania. 2001.

—————. ASTM D4874–95, "Standard Test Method for Leaching Solid Material in a Column Apparatus." ASTM: West Conshohocken, Pennsylvania. 2001.

—————. ASTM D4319–93,"Standard Test Method for Distribution Ratios by the Short-Term Batch Method." ASTM: West Conshohocken, Pennsylvania. 2002.

Daveler, S.A. and T.J. Wolery. UCLRL–MA–110662 Part II, "EQPT, A Data File Preprocessor for the EQ3/6 Software Package: Users Guide and Related Documentation (Version 7.0)." Lawrence Livermore National Laboratory: Livermore, California, 1992.

Department of Energy (U.S.) (DOE). DOE/LLW–67T (DE91 000751), "Site Characterization Handbook." DOE: Washington, DC. June 1988.

—————. DOE/EH–94007658 (DE94007658), "Remedial Investigation/Feasibility Study (RI/FS) Process, Elements, and Techniques Guidance." DOE: Washington, DC. December 1993.

—————. DOE, PNL–10395, "Department of Energy, Multimedia Environmental Pollutant Assessment System (MEPAS) Application Guidance." Washington, DC. February 1995.

Environmental Protection Agency (U.S.) (EPA). SW–611, "Procedures Manual for Ground Water Monitoring at Solid Waste Disposal Facilities." EPA: Washington, DC. 1977.

—————. EPA 520/1–80–012, "Upgrading Environmental Radiation Data." EPA: Washington, DC. August 1980.

—————. EPA–600/3–84–032, "MINTEQ: A Computer Program for Calculating Aqueous Geochemical Equilibria." EPA: Washington, DC. 1984.

—————. EPA/600/2–85/104, "Practical Guide for Groundwater Sampling." EPA: Washington, DC. September 1985.

—————. EPA, OSWER–9950.1, "RCRA Ground-Water Monitoring Technical Enforcement Guidance Document." EPA: Washington, DC. 1986.

—————. EPA 540/G–87/004, "Data Quality Objectives for Remedial Response Activities: Example Scenario RI/FS Activities at a Site with Contaminated Soils and Groundwater." EPA: Washington, DC. 1987(a).

—————. EPA/540/G–87/003, OSWER Directive 9355.07B, "Data Quality Objectives for Remedial Response Activities – Development Process." EPA: Washington, DC. 1987(b).

—————. EPA/600/D–91/141, "Importance of Quality for Collection of Environmental Samples: Planning, Implementation, and Assessing Field Sampling Quality at CERCLA Sites." EPA: Washington, DC. June 1991.

—————. EPA 402–R–94–012, "A Technical Guide to Ground-Water Model Selection at Sites Contaminated with Radioactive Substances." EPA: Washington, DC. June 1994(a).

—————. EPA 500–B–003, "Environmental Protection Agency, Assessment Framework for Ground-Water Model Applications." EPA: Washington, DC. 1994(b).

—————. EPA 500–B–004, "Environmental Protection Agency, Ground-Water Modeling Compendium: Fact Sheets, Descriptions, Applications, and Costs Guidelines." EPA: Washington, DC. 1994(c).

Fleishauer, C. and T. Engelder, "Procedures for Reconnaissance Stream-Sediment Sampling." Bendix Field Engineering Corporation. GJ/TMC–14. 1984.

Korte, N. and D. Ealey. "Procedures for Field Chemical Analyses of Water Samples." Bendix Field Engineering Corporation. GJ/TMC–07. 1983.

Korte, N. and P. Kearl. "Procedures for the Collection and Preservation of Groundwater and Surface Water Samples and For the Installation of Monitoring Wells." Bendix Field Engineering Corporation. GJ/TMC–08. 1984.

National Academy of Sciences (NAS). "Risk Assessment of Radon in Drinking Water." National Academy Press: Washington, DC. 1999.

National Council on Radiation Protection and Measurements (NCRP). NCRP Report No. 76, "Radiological Assessment: Predicting the Transport, Bioaccumulation and Uptake by Man of Radionuclides Released to the Environment." NCRP Publications: Bethesda, Maryland. 1985.

— — — — —. NCRP Report No. 123, Vol 2, "Screening Models for Releases of Radionuclides to Atmosphere, Surfaces, Water, and Ground." NCRP Publications: Bethesda, Maryland. 1996.

Nuclear Regulatory Commission (U.S.) (NRC). "Draft Branch Technical Position on Site Characterization for Decommissioning." NRC: Washington, DC. November 1994.

— — — — —. NUREG–1293, "Quality Assurance Guidance for Low-Level Radioactive Waste Disposal Facility." NRC: Washington, DC. January 1989.

— — — — —. NUREG–1383, "Guidance on the Application of Quality Assurance for Characterizating a Low-Level Radioactive Disposal Site." NRC: Washington, DC. November 1989.

— — — — —. NUREG–1388, "Environmental Monitoring of Low-Level Radioactive Waste Disposal Facility." NRC: Washington, DC. December 1989.

— — — — —. NUREG–1496, "Generic Environmental Impact Statement in Support of Rulemaking on Radiological Criteria for License Termination of NRC–Licensed Nuclear Facilities." NRC: Washington, DC. July 1997.

— — — — —. NUREG/CR–3332, "Radiological Assessment—A Textbook on Environmental Dose Analysis." NRC: Washington, DC. 1983.

— — — — —. NUREG/CR–5454, Vol. 1–5, "Nuclear Regulatory Commission, Background Information for the Development of a Low-Level Waste Performance Assessment Methodology." NRC: Washington, DC. 1989.

— — — — —. NUREG/CR–5512, Vol. 3, "Residual Radioactive Contamination from Decommissioning, Parameter Analysis, Draft Report for Comment." NRC: Washington, DC. October 1999.

— — — — —. Regulatory Guide 4.15, Rev. 1, "Quality Assurance for Radiological Monitoring Programs (Normal Operations)—Effluent Streams and Environment." NRC: Washington, DC. February 1979.

— — — — —. Regulatory Issue Summary 2002–02, "Lessons Learned: Related to Recently Submitted Decommissioning Plans and License Termination Plans." NRC: Washington, DC. January 2002.

Sheppard, M.I. and D.H. Thibault. "Default Soil Solid/Liquid Partition Coefficients, K_d's for Four Major Soil Types–A Compendium." *Health Physics*. Vol. 59, No. 4: pp. 471–478. 1990.

U.S. Geological Survey (USGS), "National Handbook of Recommended Methods for Water-Data Acquisition." USGS: Washington, DC. 1977 (with updates).

— — — — —. Water-Resources Investigation Report 96–4233, "Guidelines and Standard Procedures For Studies of Ground-Water Quality Selection and Installation of Wells, and Supporting Documentation." USGS: Washington, DC. 1996.

Appendix G

Special Characterization and Survey Issues

There are several special situations during the decommissioning process that are not, or are only minimally, addressed in NRC regulatory guidance and MARSSIM (NUREG–1575), for which licensees may need to perform characterization and an FSS in order to demonstrate compliance with the license termination criteria in 10 CFR Part 20, Subpart E. As part of NRC staff's review and approval of DPs, these special situations are evaluated on a case-by-case basis at this time. Additional guidance that covers these special situations may be developed in the future and these will be included in revisions to the consolidated guidance. Refer to Appendix O of this volume for information on lessons learned from recently submitted DPs and questions and answers to clarify existing license termination guidance related to special characterization and survey issues.

This Appendix G is applicable, either in total or in part, to Decommissioning Groups 4–7.

G.1   Surveys for Special Situations in Buildings

The survey method described in this volume thus far (e.g., Chapter 4 and Appendix A) can be applied to simple ideal geometries in a straightforward manner; however, there are likely to be some additional special situations at actual sites that will need further consideration. For each situation discussed below, it is assumed that the HSA and minimal site characterization have located and given a rough estimate of the concentration of residual radioactivity present.

G.1.1   Structures Versus Equipment

Background

The NRC staff acknowledges that the relationship between the License Termination Rule (LTR) for unrestricted use of a site [dose criteria of 0.25 mSv/y (25 mrem/y) and ALARA], and existing guidance for unrestricted releases of solid materials from a site on a case-by-case basis may have been unclear. In particular, the criteria for the LTR and for releases of solid materials prior to license termination are different. Consistent with the LTR, once a site meets the radiological criteria for unrestricted use [0.25 mSv/y (25 mrem/y) plus ALARA] and the NRC terminates the license, solid material may be removed from a site. However, before license termination, material cannot be removed from the site for unrestricted use unless it meets either (a) criteria already approved for the licensed facility (e.g., in a license condition), for surficially contaminated materials; or (b) the few mrem/y criterion for the case-by-case approach for volumetrically contaminated materials. One rationale for the difference in criteria is that the technical basis for the LTR assumed that individuals are generally exposed to residual radioactivity at a single location (the site), while for releases of solid material, an individual may be exposed to materials through several scenarios at offsite locations (NRC 2003). For more information about the relationship between the LTR and the case-by-case approaches to release of solid materials from a site, see the LTR Analysis Commission Paper SECY-03-0069 (NRC 2003) and the associated Regulatory Issue Summary (NRC 2004).

This section focuses on compliance with the LTR, in particular what building structure-related materials may be left onsite at license termination, and what criteria should apply. For more information about current approaches to releases of solid material before license termination, see Section 15.11 of Volume 1 of this NUREG report.

Implementation

The LTR applies to building structures that remain in place after decommissioning and does not apply to releases of equipment from the facility before license termination. If licensees elect to dismantle building structures and dispose of the associated materials offsite (in accordance with applicable regulatory requirements), rather then leave the building structures in place (for unrestricted use), the LTR does not apply to the associated materials moved offsite prior to license termination. Materials licensees may release equipment and building structure deconstruction and dismantlement materials in accordance with existing license conditions. Reactor licensees (10 CFR Part 50 licensees) may release equipment and building structure deconstruction and dismantlement materials in accordance with the guidance in Inspection and Enforcement Circular 81–07, Information Notice 85–92, and Information Notice 88–22. Licensees should refer to Section 15.11 of Volume 1 of this NUREG report and should consult with NRC staff for further guidance on equipment and solid material releases.

When the LTR was developed, it was assumed that decommissioning generally would include the removal of systems and components from onsite buildings prior to license termination. However, with experience, it has become clear that each licensee uses a different approach for decommissioning, and these approaches are not necessarily consistent with the original assumptions of the LTR. Differences are the result of factors such as: (1) the potential for re-use of systems and components; (2) cost of recycling/price of scrap metal and concrete; and (3) cost and availability of disposal options.

It is clear from the LTR Technical Basis, provided in the Generic Environmental Impact Statement (NRC 1997), and NRC draft Regulatory Guide DG-4006, "Demonstrating Compliance with the Radiological Criteria for License Termination," dated July 1998, that the LTR was not intended to apply to releases of "equipment" from the facility. "Equipment" includes anything *not attached to*, or not an integral part of, the building structure. On the other hand, the previous guidance (the previous version of Section G.1.1 of this Volume) was not prescriptive enough to provide a definitive answer about whether systems and components must be considered "building structures" or "equipment." The previous guidance considered "doors, windows, sinks, lighting fixtures, utility lines, built-in laboratory hoods and benches and other types of built-in furniture" to be part of the structure. Under that guidance, those items could be included in the Final Status Survey (FSS) and left in place at license termination. It could be argued that, based on the examples provided, many plant systems and components also could be considered "building structures," and, therefore, left in place at license termination. This previous guidance may have been inconsistent with the discussion in the LTR Analysis Commission Paper SECY-03-0069 (NRC 2003), which described an expectation that removable materials and equipment would generally not be present at the time of license termination.

In order to clarify for licensees and the NRC staff what building structure-related materials may be left onsite at the time of license termination and what criteria should be applied to those materials, the staff has identified a number of acceptable approaches.

For this discussion only, NRC staff uses the following descriptions of building structures, systems and components, and equipment:

- "Building structures" include floors, walls, and roofs; components embedded in floors, walls, and roofs (e.g., embedded piping); and items that are attached to and are an integral part of the buildings (e.g., doors and windows).

- "Systems and components" include items attached to a building structure that are not an integral part of the building, but provide important functions to the building (e.g., utility lines, sinks, lighting fixtures, built-in laboratory hoods and benches, polar cranes (in power reactors), and major process equipment).

- "Equipment" includes items not attached to the building structure, that are generally readily removable from the building. Examples of equipment include furniture or appliances that are not built into or attached to the structure; stocks of chemicals, reagents, metals, and other supplies; motor vehicles; and any other items that normally would not be conveyed with a building when it is sold.

Building Structures, and Systems and Components that May Be Left in Place at License Termination

The NRC staff finds the following approaches acceptable to determine what materials may be left in buildings at license termination.

1. **Materials Left Onsite Meet Previously Approved Release Criteria**—Building structures and systems and components may be left in place if residual radioactivity in all materials is within the licensee's previously approved criteria for releases of solid materials for unrestricted use. Such criteria may have been approved in license conditions, technical specifications, or generic NRC guidance. The criteria could include use of the "no-detect" policy for reactor licensees, or policy FC-83-23 (or Regulatory Guide 1.86) for materials licensees (see also Section 15.11 of Volume 1 of this NUREG report for more information about the current approaches to releases of solid materials).

2. **Materials Left Onsite Meet "Few Millirem per Year"**—Building structures and systems and components may be left in place if residual radioactivity in all materials is volumetrically distributed (not surficial) and if the potential dose from offsite use scenarios is no greater than a few hundredths mSv per year (few mrem per year).

3. **Materials Left Onsite Meet 0.25 mSv/y (25 mrem/y)**—Building structures may be left in place if the potential dose from the residual radioactivity in or on the structures is within the

applicable dose criteria of the LTR (for unrestricted use, no greater than 0.25 mSv/y (25 mrem/y) and ALARA).

4. **Alternative Approaches** — Licensees also may propose alternative approaches, which the staff will review on a case-by-case basis. Before submitting such alternative approaches, licensees should consult with the NRC staff.

For all approaches, the residual radioactivity in building structures, systems and components, and all other media at the site (e.g., soils or ground water) must be in compliance with the applicable criteria of the LTR (e.g., for unrestricted use, doses must not exceed 0.25 mSv/y (25 mrem/y) and must be ALARA).

Licensees will perform dose assessments (or use NRC-approved screening dose assessments) to demonstrate compliance with the dose criteria. Typically, licensees may not need to evaluate potential offsite future use scenarios, such as removal of soil for fill material or road base or reuse of concrete as road bed material, because such offsite use scenarios are usually bounded by onsite use scenarios. However, for some of the dose assessments needed for the above approaches, when less conservative and more realistic exposure scenarios are selected, the onsite scenarios may no longer bound potential offsite use scenarios. Thus, in these cases, the licensee should evaluate offsite use scenarios. For additional guidance, see Section I.3.3.3.6 of Appendix I of this Volume.

Equipment Not Covered by the LTR

The LTR does not apply to equipment, so equipment should not be left on the site at license termination. Equipment should be released under the current approaches for releases of solid materials, as discussed in Section 15.11 of Volume 1 of this NUREG report, or could be disposed as radioactive waste.

G.1.2   Residual Radioactivity Beneath the Surface

The historical site assessment (HSA) and characterization surveys may indicate that residual radioactivity is present beneath the surface. In the dose modeling, the parameters for resuspension and ingestion are normally derived for residual radioactivity on the surface. However, if the residual radioactivity is beneath rather than on the surface, that may be considered in the dose modeling, and the survey results may be interpreted in a manner consistent with the dose modeling.

For the FSS, cracks and crevices are surveyed in the same manner as other building surfaces, except that these areas should receive judgmental scans when scanning coverage is less than 100 %.

For painted-over residual radioactivity, the HSA and characterization surveys should be used to determine whether residual radioactivity was fixed in place by being painted over. If so, the process for its removal may be considered in developing the parameters for the dose modeling, and the survey results may be interpreted in a manner consistent with the dose modeling.

G.1.3   Sewer Systems, Waste Plumbing Systems and Floor Drains

The HSA and characterization surveys are used to determine whether there are unusual or unexpected levels of residual radioactivity in sewer systems and floor drains. Residual radioactivity in sewer systems and floor drains generally does not contribute to the dose pathways in the building occupancy scenario or the residential scenario; thus, the dose from residual radioactivity in sewer pipes should be calculated using a site-specific scenario. The FSS should then be conducted in a manner consistent with the site-specific dose scenario. If the sewer water is sent to an onsite drainage field or cesspool, any residual radioactivity should be evaluated and surveyed as subsurface residual radioactivity. If unusual or unexpected results are found during the characterization survey, the situation should be dealt with on a case-by-case basis.

If sewage is sent to an onsite drainage field, any residual radioactivity is subsurface and the survey methods discussed in Section G.2.1 are appropriate.

G.1.4   Ventilation Ducts

The HSA and characterization surveys should be used to indicate whether residual radioactivity may be present. External duct surfaces of ventilation ducts are surveyed as if they are a part of the building surface. For internal duct surfaces, surveys should be performed in a manner consistent with the dose modeling assumptions.

G.1.5   Piping and Embedded Pipping

Embedded piping is piping embedded in a durable material, typically concrete, that cannot be easily removed without significant effort and tools. The HSA and characterization surveys should be used to indicate whether residual radioactivity is present in piping. The normal room surveys will adequately account for direct (external gamma) radiation from the pipes when the pipes are in place and undisturbed. The direct (external gamma) dose from the pipes will be in addition to the dose from the residual radioactivity on surfaces in the room. It may also be necessary to take into consideration building renovation that would disturb the piping as described in "Residual Radioactive Contamination from Decommissioning" NUREG/CR–5512, Volume 1. If this is done, the survey should be consistent with the dose modeling assumptions.

NRC staff experience has shown that some DPs have not adequately described the methods the licensee plans to use when surveying the embedded piping planned to be left behind. Often, licensees have not provided a discussion on the methodology for conducting surveys of embedded pipe planned to be left behind, nor have they provided sufficient justification for the assumptions considered in the dose modeling analysis. Regulatory Issue Summary 2002–02 provides a detailed discussion of this issue. The information contained in this Regulatory Issues Summary has been included in Section O.2.2 from Appendix O of this volume.

G.2   Surveys for Special Situations on Land

G.2.1   Subsurface Residual Radioactivity

The MARSSIM final status survey method was designed specifically for residual radioactivity in the top 15 centimeters of soil. If significant amounts of residual radioactivity are deeper than 15 centimeters, this should be taken into consideration in performing the FSS.

The licensee should first determine whether there is a need for surveys of subsurface residual radioactivity. The HSA will usually be sufficient to indicate whether there is likely to be subsurface residual radioactivity. If the HSA indicates that there is no likelihood of substantial subsurface residual radioactivity, subsurface surveys are not necessary.

If the HSA indicates that there is substantial subsurface residual radioactivity and the licensee plans to terminate the license with some subsurface residual radioactivity in place, the FSS should consider the subsurface residual radioactivity in order to demonstrate compliance with the radiological criteria for license termination. To prepare for the FSS, the characterization survey determines the depth of the residual radioactivity. In addition to conventional drilling, the licensee may consider the use of exploratory trenches and pits, where the patterns, locations, and depths are determined using prior survey results or HSA data. The DCGL may be based on the assumption that the residual radioactivity may be excavated some day and that mixing of the residual radioactivity will occur during excavation. When the subsurface residual radioactivity is mixed and brought to the surface, most of the dose pathways will depend only on the average concentration. Only the ground water pathways are affected by the total inventory of residual radioactivity, including that deeper than 15 centimeters. The direct, inhalation, ingestion, and crop pathways are determined by concentration only, not total inventory.

When the appropriate DCGLs and mixing volumes based on an acceptable site-specific dose assessment are established, the FSS is performed by taking core samples to the measured depth of the residual radioactivity. The number of cores to be taken is initially the number (N) required for the WRS or Sign test, as appropriate. The adjustment to the grid spacing for an elevated measurement comparison (EMC) is more complicated than for surface soils, because scanning is not applicable. The core samples should be homogenized over a soil thickness that is consistent with assumptions made in the dose assessment, typically not exceeding 1 meter in depth. It is not acceptable to average radionuclide concentrations over an arbitrary soil thickness. The appropriate test (WRS or Sign) then is applied to the sample results. Triangular grids are recommended, because they are slightly more effective in locating areas of elevated concentrations. Site-specific EMCs may also need to be developed to demonstrate regulatory compliance. Generic guidance has not yet been developed for performing an EMC for subsurface samples; therefore, licensees should discuss this matter with NRC staff on a case-by-case basis.

The sampling approach described above may not be necessary if sufficient data to characterize the subsurface residual radioactivity are available from other sources. For example, for some burials conducted under prior NRC regulations, the records on the material buried may be sufficient to demonstrate compliance with the radiological criteria for license termination.

G.2.2 Rubble, Debris, and Rocks

Rubble, debris, and rocks can include naturally occurring rocks (either in place or in piles), pieces of concrete or rubble from buildings that have been razed, sheet metal disposed of as trash, asphalt, fly ash, and similar material. The HSA and characterization surveys should be used to determine the volumetric extent and residual radioactivity concentration. If the materials are contaminated, they would be disposed of as radioactive waste. If the radioactivity is not substantially elevated, the rubble, debris, and rocks may be evaluated as part of a larger survey unit. When these materials will be evaluated as part of a larger survey unit and when they are found on a relatively small fraction of the area of a survey unit, the volumetric soil DCGL should be used uniformly throughout the survey unit. However, the reasonableness of modeling rocks and rubble as soil should be justified by the licensee.

G.2.3 Paved Parking Lots, Roads, and Other Paved Areas

The HSA and characterization surveys should be used to determine whether the residual radioactivity is on or near the surface of the paving and whether there are significant concentrations of residual radioactivity beneath the paving. If the residual radioactivity is primarily on top of the paving, then the measurements should be taken as if the area were normal soil. Depending on how large the paved area is, the paved area may be included as part of a larger survey unit or may be its own survey unit. If the residual radioactivity is primarily beneath the paving, it should be surveyed as subsurface residual radioactivity, as discussed above.

G.3   References

Nuclear Regulatory Commission (U.S.) (NRC). Inspection and Enforcement Circular No. 81–07, "Control of Radioactively Contaminated Material." NRC: Washington, DC. May 14, 1981.

—————. Information Notice No. 85–92, "Surveys of Wastes Before Disposal from Nuclear Reactor Facilities." NRC: Washington, DC. December 2, 1985.

—————. Information Notice No. 88–22, "Disposal of Sludge from Onsite Sewage Treatment Facilities at Nuclear Power Stations." NRC: Washington, DC. May 12, 1988.

—————. NUREG-1496, "Generic Environmental Impact Statement in Support of Rulemaking on Radiological Criteria for License Termination of NRC-Licensed Nuclear Facilities." NRC: Washington, DC. November 1996.

—————. NUREG–1575, Rev. 1, "Multi-Agency Radiation Survey and Site Investigation Manual (MARSSIM)." EPA 402–R–97–016, Rev. 1, DOE/EH–0624, Rev. 1. U.S. Department of Defense, U.S. Department of Energy, U.S. Environmental Protection Agency, and NRC: Washington, DC. August 2000.

—————. NUREG/CR–5512, Vol. 1, "Residual Radioactive Contamination From Decommissioning: Technical Basis for Translating Contamination Levels to Annual Total Effective Dose Equivalent." NRC: Washington, DC. October 1992.

—————. Regulatory Issue Summary 2002–02, "Lessons Learned Related to Recently Submitted Decommissioning Plans and License Termination Plans." NRC: Washington, DC. January 2002.

—————. Regulatory Issue Summary 2004–08, "Results of the License Termination Rule Analysis." NRC: Washington, DC. May 2004.

—————. SECY-03-0069, "Results of the License Termination Rule Analysis." Policy Issue Memorandum from W.D. Travers to the Commissioners. NRC: Washington, DC. May 2, 2003.

Appendix H

Criteria for Conducting Screening Dose Modeling Evaluations

H.1   Introduction

This appendix consists of the technical guidance for the use of the screening criteria, applicable to Decommissioning Groups 1–3. References cited are detailed in Appendix I.8.

This section pertains to NRC staff's review of a licensee's demonstration of compliance with the dose criteria in Part 20, Subpart E, using a screening approach dose analysis. NRC staff review of the screening analysis should be performed using one or more of the currently available screening tools:

1. a look-up table for common beta- and gamma-emitting radionuclides for building surface residual radioactivity (63 FR 64132, November 18, 1998);

2. a look-up table for common radionuclides for soil surface residual radioactivity (64 FR 68395, December 7, 1999); and

3. screening levels derived using DandD, Version 2.1, or the most current version for the specific radionuclide(s) that use the code's default parameters.

Other tools for performing a screening analysis might become available in the future, depending on further NRC staff efforts to develop additional look-up tables. Based on the merits and level of conservatism of the alternate screening approaches or procedures, other alternate screening approaches or procedures might be appropriate.

A screening analysis is usually conducted for simple sites with building surface (i.e., non-volumetric) and/or with surficial soil [approximately 15 cm (6 in)] residual radioactivity. Simple and conservative models/codes and parameters, under generic scenarios and default site conditions, are usually employed to define the screening DCGLs equivalent to the dose criteria. Because of the conservative nature of the screening analysis approach, the screening DCGLs are expected to be more restrictive than the site-specific DCGLs. Screening analysis may save licensees time and effort by reducing the amount of site characterization, modeling analysis, and reviews needed, versus those needed when using a site-specific analysis approach.

To conduct a screening analysis review, NRC staff first needs to make a generic assessment and evaluation of a licensee's justification that the site is qualified for screening. In addition, NRC staff should be familiar with the tools (e.g., models, codes, and calculations) and embedded assumptions used in derivation of the screening DCGLs. This section addresses the major issues that NRC staff may encounter in the generic screening analysis reviews and includes recommendations of approaches for addressing and resolving these issues. Refer to Section O.2.2 from Appendix O of this volume for information on lessons learned from recently submitted decommissioning plans (DPs) related to dose modeling evaluations.

H.2 Issues in Performing Screening Analysis

The major issues associated with the screening analysis that NRC staff may encounter include the following: (a) the definition of screening and the transition from a screening to a site-specific analysis; (b) qualification of the site for screening, in terms of site physical conditions and compatibility with the modeling code's assumptions and default parameters; and (c) the acceptable screening tools (e.g., code, look-up tables), approaches, and parameters that NRC staff can use to translate the dose into equivalent screening concentration levels. Each one of these issues is discussed in the following subsections:

H.2.1 Screening Definition and Approaches for the Transition from Screening to Site-Specific Analysis;

H.2.2 Qualification of the Site for Screening; and

H.2.3 Acceptable Screening Tools.

H.2.1 Screening Definition And Approaches for the Transition from Screening to Site-Specific Analysis

NRC staff may encounter some inconsistencies regarding the definition of the term "screening" in dose analysis which may cause confusion regarding the transition from a screening to a site-specific analysis. These inconsistencies become more apparent when dividing screening approaches into multiple levels (NCRP 1996b, 1999). In some cases screening and site-specific terms are mixed, and the term "site-specific screening" is used (NRC 1992). In certain cases screening is categorized on the type of models used (e.g., simple and conservative models vs. more advanced and complex models) and the extent of data and information needed to support the dose analysis.

Within the context of NUREG–1757, NRC staff should consider the definition of screening as the process of developing DCGLs at a site using either (a) NRC's look-up tables in 63 FR 64132 and 64 FR 68395, or (b) the latest version (e.g., Version 2.1) of the DandD code developed by NRC to perform the generic screening analysis.

It should be noted that, in the future, NRC staff may modify current look-up tables or develop additional look-up tables for the common alpha-emitters for building surfaces (based on the DandD code and modification of sensitive parameters). In addition, NRC staff may also consider the use of other screening tools (e.g., other look-up tables or other conservative codes/models) after evaluating and comparing these screening tools with the current screening codes.

When licensees either (a) select other approaches or models for the dose analysis or (b) modify the DandD code default parameters, scenarios, or pathways, NRC staff considers licensees to be performing site-specific analyses. With regard to footnote a of Table H.1, use of values of the fraction of removable surface contamination other than 0.1 or 1.0 (as described in the footnote) in the DandD code is considered a site-specific analysis. For a site-specific analysis, staff should use Section 5.2 to review the analysis.

While there is no requirement that the licensee consider the use of screening criteria, licensees should recognize the advantages and disadvantages of selecting a screening approach for demonstrating compliance with the dose criteria. The merits of using screening versus using site-specific analysis are discussed in Section 2.6 of this volume.

H.2.2 Qualification of the Site for Screening

NRC staff should be aware that a screening analysis, for demonstrating compliance with the dose criteria in Part 20, Subpart E, may not be applicable for certain sites because of the status of contaminants (e.g., location and distribution of radionuclides), or because of site-specific physical conditions. Therefore, NRC staff should assess the site source-term (e.g., radionuclide distribution) characteristics to ensure consistency with the source-term assumptions in DandD. Further, NRC staff may determine that there could be conditions, at the specific site, that cannot be handled by the simple screening model, because of the complex nature of the site, or because of the simple conceptual model in the DandD screening code.

When using the screening approach for demonstrating compliance with the dose criteria in Part 20, Subpart E, licensees need to demonstrate that the particular site conditions (e.g., physical and source-term conditions) are compatible and consistent with the DandD model assumptions (NRC 1992). In addition, the default parameters and default scenarios and pathways must also be used in the screening dose analysis. Therefore, reviewers should examine the site conceptual model, the generic source-term characteristics, and other attributes of the site to ensure that the site is qualified for screening.

NRC staff should verify that the following site conditions exist for each of the residual radioactivity conditions:

- Building Surface Residual Radioactivity:

  1. The residual radioactivity on building surfaces (e.g., walls, floors, ceilings) should be surficial and non-volumetric [e.g., ≤10 mm (0.39 in) of penetration].

  2. Residual radioactivity on surfaces is mostly fixed (not loose), with the fraction of loose (removable) residual radioactivity no greater than 10 % of the total surface activity. Note that for cases when the fraction of removable contamination is undetermined or higher than 0.1, licensees may assume for screening purposes that 100 % of surface contamination is removable, and therefore the screening values should be decreased by a factor of 10 (see footnote a to Table H.1).

  3. The screening criteria are not being applied to surfaces such as buried structures (e.g., drainage or sewer pipes) or equipment within the building without adequate justification; such structures, buried surfaces, and clearance of equipment should be treated on a case-by-case basis.

- Surface Soil Residual Radioactivity:

    1. The initial residual radioactivity (after decommissioning) is contained in the top layer of the surface soil [e.g., approximately 15 cm (5.9 in)].

    2. The unsaturated zone and the ground water are initially free of residual radioactivity.

    3. The vertical saturated hydraulic conductivity at the specific site is greater than the infiltration rate (e.g., there is no ponding or surface run-off).

Questions have also been raised about the appropriateness of using a screening analysis at sites with contaminated areas larger than the current default cultivated area [e.g., 2400 m^2 (25,800 ft^2)]. Initially, NRC staff evaluated the effect of a large contaminated area on the derived screening dose and determined that this effect is trivial for sites with the dominant dose arising from direct exposure or inhalation. As modeled by DandD with its default parameter set, this effect could be appreciable for sites with a significant dose contribution associated with the ingestion pathway (specifically ingestion associated with the drinking water and irrigation pathways). NRC staff determined that for sites with contaminated areas of 6000–7200 m^2 (64,600–77,500 ft^2) the dose may be underestimated under worst-case conditions by a factor of 2 to 3. However, further analysis by NRC staff showed that, because of the conservative assumptions of the DandD code, it is more likely that the derived dose (based on the use of other codes or the use of a site-specific analysis) would be far less than the derived dose using these default conditions. Therefore, for sites with areas larger than 7200 m^2 (77,500 ft^2), the change in actual risk due to this effect is not appreciable. In summary, assuming that the site is qualified for screening based on the above listed criteria, the screening approach would be accepted for sites with areas larger than the default cultivated area [i.e., 2400 m^2 (25,800 ft^2)].

It should be noted that NRC staff should also evaluate complex site conditions that may disqualify the site for screening. Examples of such complex site conditions may include highly fractured formation, karst conditions, extensive surface water contamination, and highly non-homogeneous distribution of residual radioactivity. Therefore, reviewers should ensure that the site meets the definition of a "simple site" to qualify for screening (see Section 1.2 for details).

H.2.3   Acceptable Screening Tools

In the past, it may not have been clear what screening tools NRC has determined to be acceptable. Some may believe that using simple, common codes (other than DandD), with their deterministic default parameters may be acceptable to derive the desired screening DCGLs. Others may believe that use of any look-up tables published by certain scientific committees or authorities may be used to convert concentration levels directly into doses for purposes of complying with Subpart E. Questions regarding use of the DandD code for screening, particularly whether modification of input default parameters is acceptable for screening, have also been raised.

NRC staff should accept for a screening analyses the following currently available screening tools:

- A look-up table (Table H.1) for common beta- and gamma-emitting radionuclides for building-surface residual radioactivity (63 FR 64132, November 18, 1998).

- A look-up table (Table H.2) for common radionuclides for soil surface residual radioactivity (64 FR 68395, December 7, 1999).

The screening values in Tables H.1 and H.2 are intended for single radionuclides. For radionuclides in mixtures, the "sum of fractions" rule should be used (see Section 2.7 of this volume). These values were derived using DandD screening methodology based on selection of the 90th percentile of the output dose distribution for each specific radionuclide or radionuclide with the specific decay chain. Behavior parameters were set at the mean of the distribution of the assumed critical group. The metabolic parameters were set either at the Reference Man or at the mean of the distribution for an average human.

NOTE

For a radionuclide with its progeny present at equilibrium, the "+C" values of Table H.2 should be interpreted carefully. As described in footnote c to Table H.2, these "+C" values are concentrations of the parent radionuclide only, but account for dose contributions from the complete chain of progeny in equilibrium with the parent radionuclide. For example, Uranium-238+C lists the soil screening value as 18.5 Bq/kg (0.5 pCi/g). This means that it is also assumed that there is 18.5 Bq/kg (0.5 pCi/g) of U-234, 18.5 Bq/kg (0.5 pCi/g) Th-230, and so forth, present.

- Screening levels derived using the latest version of DandD Version 2 for the specific radionuclide and using code default parameters and parameter ranges.

  In August 1998, NRC staff issued DandD, Version 1.0 for screening and simple site-specific analysis. NRC staff and users (through public workshops) identified several areas where DandD, Version 1, may be overly conservative. One such conservatism was the methodology used for establishing a single default parameter set for all radionuclides listed in the DandD code. That is, if the default parameter set were tailored for each specific radionuclide, the dose calculated using the DandD model would, in most cases, be lower. DandD, Version 2, was developed to address these issues. A detailed discussion of the way the default parameters were selected is included in NUREG/CR–5512, Volume 3; the conservatism of DandD code, Version 1, is discussed in Appendix I of this volume.

- Potential use of other tools or approaches for screening.

  The current NRC staff position is to limit screening to the look-up tables developed by NRC staff and the execution of the latest version of DandD code with the default parameter ranges. As indicated above, NRC staff may develop additional look-up tables or modify the screening tables based on refining certain sensitive parameters. NRC staff may evaluate the possibility of using other simple codes/models for screening, such as the probabilistic RESRAD and RESRAD–BUILD codes currently under development. Furthermore, NRC staff may evaluate requests by licensees to use other look-up tables developed by specific consensus professional or technical groups or authorities. Usually, NRC staff should treat these approaches as "site-specific" since the review is very similar. NRC staff will examine the screening approaches, methodologies, scenarios, and assumptions in these other approaches to ensure compatibility with the current screening methodology using DandD. NRC staff will also assess the site conditions to ensure that the screening analysis is appropriate for the site. In certain cases, NRC staff may need to examine and compare the default screening parameters with the site-specific conditions.

Table H.1   Acceptable License Termination Screening Values of Common Radionuclides for Building-Surface Contamination

Radionuclide	Symbol	Acceptable Screening Levels[a] for Unrestricted Release (dpm/100 cm^2)[b]
Hydrogen-3 (Tritium)	^{3}H	120000000
Carbon-14	^{14}C	3700000
Sodium-22	^{22}Na	9500
Sulfur-35	^{35}S	13000000
Chlorine-36	^{36}Cl	500000
Manganese-54	^{54}Mn	32000
Iron-55	^{55}Fe	4500000
Cobalt-60	^{60}Co	7100
Nickel-63	^{63}Ni	1800000
Strontium-90	^{90}Sr	8700
Technetium-99	^{99}Tc	1300000
Iodine-129	^{129}I	35000
Cesium-137	^{137}Cs	28000
Iridium-192	^{192}Ir	74000

Notes:

a   Screening levels are based on the assumption that the fraction of removable surface contamination is equal to 0.1. For cases when the fraction of removable contamination is undetermined or higher than 0.1, users may assume for screening purposes that 100 % of surface contamination is removable, and therefore the screening levels should be decreased by a factor of 10. Users may calculate site-specific levels using available data on the fraction of removable contamination and DandD Version 2.

b   Units are disintegrations per minute (dpm) per 100 square centimeters (dpm/100 cm^2). One dpm is equivalent to 0.0167 becquerel (Bq). Therefore, to convert to units of Bq/m^2, multiply each value by 1.67. The screening values represent surface concentrations of individual radionuclides that would be deemed in compliance with the 0.25 mSv/y (25 mrem/y) unrestricted release dose limit in 10 CFR 20.1402. For radionuclides in a mixture, the "sum of fractions" rule applies (see Part 20, Appendix B, Note 4).

Table H.2   Screening Values[a] (pCi/g) of Common Radionuclides for Soil Surface Contamination Levels

Radionuclide	Symbol	Surface Soil Screening Values[b]
Hydrogen-3	^{3}H	110
Carbon-14	^{14}C	12
Sodium-22	^{22}Na	4.3
Sulfur-35	^{35}S	270
Chlorine-36	^{36}Cl	0.36
Calcium-45	^{45}Ca	57
Scandium-46	^{46}Sc	15
Manganese-54	^{54}Mn	15
Iron-55	^{55}Fe	10000
Cobalt-57	^{57}Co	150
Cobalt-60	^{60}Co	3.8
Nickel-59	^{59}Ni	5500
Nickel-63	^{63}Ni	2100
Strontium-90	^{90}Sr	1.7
Niobium-94	^{94}Nb	5.8
Technetium-99	^{99}Tc	19
Iodine-129	^{129}I	0.5
Cesium-134	^{134}Cs	5.7
Cesium-137	^{137}Cs	11
Europium-152	^{152}Eu	8.7
Europium-154	^{154}Eu	8
Iridium-192	^{192}Ir	41
Lead-210	^{210}Pb	0.9
Radium-226	^{226}Ra	0.7
Radium-226+C[c]	^{226}Ra+C	0.6
Actinium-227	^{227}Ac	0.5
Actinium-227+C	^{227}Ac+C	0.5
Thorium-228	^{228}Th	4.7

Table H.2   Screening Values[a] (pCi/g) of Common Radionuclides for Soil Screening Surface Contamination Levels (continued)

Radionuclide	Symbol	Surface Soil Screening Values[b]
Thorium-228+C[c]	^{228}Th+C	4.7
Thorium-230	^{230}Th	1.8
Thorium-230+C	^{230}Th+C	0.6
Thorium-232	^{232}Th	1.1
Thorium-232+C	^{232}Th+C	1.1
Protactinium-231	^{231}Pa	0.3
Protactinium-231+C	^{231}Pa+C	0.3
Uranium-234	^{234}U	13
Uranium-235	^{235}U	8
Uranium-235+C	^{235}U+C	0.29
Uranium-238	^{238}U	14
Uranium-238+C	^{238}U+C	0.5
Plutonium-238	^{238}Pu	2.5
Plutonium-239	^{239}Pu	2.3
Plutonium-241	^{241}Pu	72
Americium-241	^{241}Am	2.1
Curium-242	^{242}Cm	160
Curium-243	^{243}Cm	3.2

Notes:

a   These values represent surficial surface soil concentrations of individual radionuclides that would be deemed in compliance with the 25 mrem/y (0.25 mSv/y) unrestricted release dose limit in 10 CFR 20.1402. For radionuclides in a mixture, the "sum of fractions" rule applies; see Section 2.7 of this volume.

b   Screening values are in units of (pCi/g) equivalent to 25 mrem/y (0.25 mSv/y). To convert from pCi/g to units of becquerel per kilogram (Bq/kg), divide each value by 0.027. These values were derived using DandD screening methodology (NUREG/CR–5512, Volume 3 (NRC 1999)). They were derived based on selection of the 90th percentile of the output dose distribution *for each specific radionuclide* (or radionuclide with the specific decay chain). Behavioral parameters were set at the mean of the distribution of the assumed critical group. The metabolic parameters were set at "Reference Man" or at the mean of the distribution for an average human.

c   "Plus Chain (+C)" indicates a value for a radionuclide with its decay progeny present in equilibrium. The values are concentrations of the parent radionuclide but account for contributions from the complete chain of progeny in equilibrium with the parent radionuclide (NUREG/CR–5512, Volumes 1, 2, and 3).

Appendix I

Technical Basis for Site-Specific Dose Modeling Evaluations

I.1   Introduction

This appendix consists of the technical guidance for the use of the site-specific dose modeling, applicable to Decommissioning Groups 4–7.

For guidance on lessons learned regarding use of site-specific dose modeling evaluations, refer to Question 4 from Section O.1 and Lesson 6 from Section O.2.2 from Appendix O of this volume.

Appendix I Table of Contents

APPENDIX I

I.1.1 Background

On July 21, 1997, the U.S. Nuclear Regulatory Commission (NRC) published a final rule on "Radiological Criteria for License Termination," in the *Federal Register* (62 FR 39058), which was incorporated as Subpart E to 10 CFR Part 20. In 1998 NRC staff developed a draft regulatory guide, Demonstrating Compliance with the Radiological Criteria for License Termination (DG–4006) (NRC 1998), and a draft document Decision Methods for Dose Assessment to Comply With Radiological Criteria for License Termination (NUREG–1549) (NRC 1998a) in support of the final rule. In addition, staff developed a screening code "DandD" for demonstrating compliance with the dose criteria in Part 20, Subpart E.

On July 8, 1998, the Commission approved publication of the draft guidance, DG–4006, the draft NUREG–1549, and the DandD screening code for interim use for a 2-year period (i.e., from July 8, 1998, through July 7, 2000) (NRC 1998b). In addition, the Commission directed NRC staff to (a) develop a standard review plan (SRP) for decommissioning and provide the Commission with a timeline for developing the SRP; (b) maintain a dialogue with the public during the interim period; (c) address areas of excessive conservatism, particularly in the DandD screening code; (d) develop a more user-friendly format for the guidance; and (e) use a probabilistic approach to calculate the total effective dose equivalent (TEDE) to the average member of the critical group (NRC 1998b).

NRC staff completed development of the SRP, and it was published in 2000 of September as NUREG–1727. Chapter 5 of the SRP (which is incorporated into Chapter 5 of this volume) addresses NRC staff review of licensee's dose modeling to demonstrate compliance with the criteria in 10 CFR Part 20, Subpart E. Appendix C of the SRP (Appendix I of this volume) was developed by NRC staff as a technical information support document for performing NRC staff evaluations of the licensee's dose modeling. It presents detailed technical approaches, methodologies, criteria, and guidance to staff reviewing dose modeling for compliance demonstration with the dose criteria in 10 CFR Part 20, Subpart E. Appendix C of the SRP was developed through an iterative process with the public including, licensees, Federal agencies, States, and other interested individuals. To support this process, NRC staff conducted seven public workshops and gave several presentations at national and international professional meetings, stakeholder meetings, Interagency Steering Committee on Radiation Standards (ISCORS) meetings, Conference of Radiation Control Program Directors (CRCPD) meetings, as well as presentations to NRC's Advisory Committee on Nuclear Waste (ACNW). In addition, NRC posted the draft Appendix C (formerly the Technical Basis Document) on NRC's Web site and requested interested individuals to provide NRC with comments.

Since the publication of the license termination rule (LTR), NRC staff has tested the DandD code for complex sites and addressed the issue of excessive conservatism in the DandD code. In addition, NRC staff developed a new probabilistic DandD code (i.e., DandD Version 2.1) to reduce the excessively conservative approach in the initial version of the DandD code. Further, NRC staff developed RESRAD and RESRAD–BUILD probabilistic codes for site-specific analysis. Development of the probabilistic DandD and RESRAD/RESRAD–BUILD codes also responds to the Commission's direction to use a probabilistic approach to calculate the TEDE to the average member of the critical group.

> **Licensees using probabilistic dose modeling should use the "peak of the mean" dose distribution (see Section I.7.3.2.2 from Appendix I of this volume) for demonstrating compliance with the 10 CFR Part 20, Subpart E. Similar to all regulatory guidance, this NUREG report contains one approach for determining compliance with the regulations using probabilistic analyses. Other probabilistic approaches, such as, "mean of the peaks" or other methods, if justified, may also be acceptable for demonstrating compliance.**

I.1.2   Brief Description and Scope

This section is divided into the following different topic areas, as summarized below.

- Section I.2 presents NRC approaches for reviewing the conceptual representation of the radioactive source term at the site. This section describes the areas of reviews pertaining to the existing radioactive material contamination and physical and chemical characteristics of the material. In addition, the section presents recommended approaches for source-term abstraction for the purpose of performing the dose analysis.

- Section I.3 focuses on areas of review and criteria for accepting modifications of pathways of the two generic critical group scenarios, the "resident farmer" and the "building occupancy" scenarios. Section I.3, also, along with Appendices L and M, discusses the information that should be provided for a licensee's justification for modifying default screening scenarios and associated pathways. It also presents approaches for establishing site-specific scenarios, critical groups, and/or sets of exposure pathways based on specific land use, site restrictions, and/or site-specific physical conditions.

- Section I.4 provides approaches for developing site-conceptual models for dose analysis. This section presents approaches—via the linkage of the source term with the critical group receptor and the use of applicable pathways and site-characterization data—for the assimilation of data to establish a site conceptual model. It also presents approaches for employing applicable mathematical models to simulate and calculate the release and transport of contaminants from the source to the receptor. This section also presents discussions of the typical conceptual models used in the DandD and RESRAD codes. Additionally, the section provides (a) information on the limitations of the DandD and RESRAD models and (b) review areas to ensure compatibility of the site conceptual model with the conceptual models embedded in the DandD and RESRAD codes.

- Section I.5 presents approaches and criteria for NRC staff acceptance of computer codes/models. This section discusses review aspects pertaining to specifications, testing, verification, documentation, and QA/QC of the licensee's codes/models. This section also addresses reviews applicable to embedded numerical models for the source term, the exposure pathway models, the transport models, and the intakes or dose conversion models. In addition, the section provides a discussion of the development of and a description of the DandD code, particularly the excessive conservatism of the Version 1 of the DandD code. Section I.5 also presents a generic description of the RESRAD/RESRAD–BUILD codes.

- Section I.6 describes approaches for the selection and modification of input parameters for dose modeling analysis and includes the use of default parameters from the DandD code in other models.

- Section I.7 addresses the acceptable criteria for treating uncertainties in the dose modeling analysis. Issues pertaining to uncertainty and sensitivity are described, and NRC staff recommended approaches for the resolution of these issues are addressed. Policy positions are presented regarding approaches both to uncertainty/sensitivity treatments and to specific percentile dose-distribution selection for the screening and site-specific analysis. NRC staff review of input parameter distributions for Monte Carlo analysis and generic description of sensitivity analysis, including statistical techniques, are also described.

- Section I.8 compiles the references used throughout the appendix.

- Appendix J integrates the guidance in Appendix I and discusses methods that licensees may use in analyzing former burials with a very simple approach.

I.2   Source-Term Abstraction

I.2.1   Introduction

Source-term abstraction is the process of developing a conceptual representation of the radioactive source at a site. Typically, the radiological conditions at a site proposed for decommissioning are relatively complex. Source-term abstraction is necessary to allow the detailed radiological characterization of the site to be incorporated into the mathematical and computer models that are used to estimate radiological impacts (e.g., dose). The abstraction process involves generalizing the radiological characteristics across the site to produce a simplified representation, which should facilitate the modeling of radiological impacts. The conceptual representation of the source developed in the abstraction process, however, should not be simplified to the extent that radiological impacts are significantly underestimated or unrealistically overestimated.

As discussed in Chapter 5 of this volume, source-term abstraction serves as the starting point for the dose modeling process. The conceptual abstraction of the source term is combined with the physical characteristics of the site and characteristics of the critical group receptor to develop the conceptual model for the site. This conceptual model provides the basis for identifying applicable exposure scenarios, pathways, and selection of computer models. These other elements of dose modeling are discussed in subsequent sections of this document.

Volume 1 of NUREG–1757 and Chapter 4 of this volume discuss the information the licensee is expected to provide regarding the existing radiological characterization of the site. The licensee should provide a description of the types, levels, and extent of radioactive material contaminated at the site. This should include residual radioactivity in all media (including buildings, systems and equipment that will remain after license termination, surface and subsurface soil, and surface and subsurface ground water). The source-term abstraction should be based on the characterization of the radiological status (e.g., process historical development, records of leakage or disposal). The licensee should explicitly relate the information provided in the discussion of radiological status of the site with the discussion of source-term abstraction. The reviewer should be able to clearly interpret the relationship.

Generally, in the source-term abstraction process, the licensee may focus on several specific elements of the source term, which include the following:

1. The licensee should identify the radionuclides of concern. This should be taken directly from the description of the site's radiological status. The radionuclides should be identified based on pre-remediation radiological status. All radionuclides potentially present at the site should be included, so that their presence or absence may be verified during the FSS, except as noted in Chapter 4 and Section 3.3 of this volume.

2. The licensee should describe the physical/chemical form(s) of the contaminated media *anticipated at the time of FSS and site release.* The licensee should indicate whether the residual radioactivity will be limited to building surfaces and/or surface soil, or whether the residual radioactivity will involve other media such as subsurface soil, debris or waste materials (e.g., sludge, slag, tailings), or ground and surface water.

3. The licensee should delineate the spatial extent of the residual radioactivity *anticipated at the time of FSS and site release.* The delineation of the spatial extent should include descriptions of (a) the areal extent of radionuclides throughout the site and (b) the vertical extent of soil residual radioactivity of radionuclides below the ground surface. The delineation of spatial extent and depth should establish the source areas and volumes. Depending on the presence of specific radionuclides, source areas and volumes may be radionuclide-specific.

4. The licensee should define the distribution of each radionuclide throughout the delineated source areas and volumes *anticipated at the time of FSS and site release.* The distribution of a radionuclide through the source should be defined in terms of representative volumetric or areal concentrations. In addition, for volumetrically contaminated soil, the licensee may provide an estimate of total radioactivity of each radionuclide.

5. The licensee should define sources in ground water or surface water, if any, based on environmental monitoring and sampling of aquifers and surface water bodies. A site with ground water or surface water contamination may be categorized as "complex" and may require more advanced dose modeling analysis (see Section 1.3 of this volume).

In the source-term abstraction process, the licensee should address the first two of these five elements. Whether the licensee needs to address the other elements depends on the objective of the licensee's dose modeling. This is discussed later in this section.

I.2.2 Issues Associated with Source-Term Abstraction

The level of effort that a licensee expends to develop a conceptualization of a source term should be commensurate with the licensee's approach to demonstrating compliance with the release criterion. Also, the focus should be on the source-term characteristics anticipated to exist at the site at the time of FSS and release, after any planned remediation.

If a licensee plans to use the screening DCGLs published by NRC in the *Federal Register*, a licensee should only have to identify the radionuclides that may be present at the site, and demonstrate that the conditions at the site meet the prerequisites for using the screening values [i.e., residual radioactivity is limited to building surfaces or the uppermost 15 to 30 cm (6 to 12 in) of surface soil and no contamination of ground water or surface water]. The licensee's source-term abstraction would not have to address issues such as existing radiological conditions, areal and volumetric extent of residual radioactivity, or spatial variability or radiological conditions for such sources. This is discussed further in Section I.2.3 of this appendix.

If a licensee anticipates that residual radioactivity will be limited to building surfaces or surface soils at the time of FSS, but considers the published DCGLs overly restrictive, the licensee may develop site-specific DCGLs. In this case, the licensee would most likely have to delineate the anticipated areal extent of residual radioactivity. However, the licensee would not have to discuss the anticipated spatial variability of radionuclide concentrations within the anticipated area of residual radioactivity.

A licensee should provide a site-specific dose assessment if the residual radioactivity is not limited to building surfaces or surface soil. In this case, the licensee would have to delineate the spatial extent (laterally and vertically) of the residual radioactivity, and the licensee would have to provide a discussion of the spatial variability of the physical, chemical, and radiological characteristics of the contaminated media.

Ideally, the source characteristics at a site would be relatively uniform, justifying simplified abstraction. However, this is generally not the case. Issues may arise when the residual radioactivity projected at a site at the time of release falls short of the ideal case. These issues may include the following:

1. Spatial extent

   — limited areal extent of residual radioactivity;

   — irregular areal shape; and

   — varying depth of residual radioactivity in soil.

2. Spatial variability

— nonuniform distribution of radioactivity throughout a site;

— limited areas of relatively elevated radionuclide concentrations;

— multiple noncontiguous areas of residual radioactivity; and

— nonuniform physical and chemical characteristics.

The following approach to source-term abstraction addresses most of these issues. Others (e.g., irregular areal shape) are best addressed by appropriate selection of computer codes.

I.2.3 Approach to Source-Term Abstraction

A licensee's approach to source-term abstraction will depend on the objective of the dose modeling presented in the decommissioning plan (DP). Generally, the licensee's dose modeling should have one of the following objectives:

• Develop DCGLs commensurate with demonstrating compliance with the dose-based release criterion, and then demonstrate through FSS that residual radioactivity concentrations at the site are equal to or below the DCGLs.

• Assess dose associated with actual concentrations of residual radioactivity distributed across the site to determine whether the concentrations will result in a dose that is not equal to or below the regulatory dose criterion.

In the first objective, the licensee intends to demonstrate at the time of FSS before release that residual radionuclide concentrations across the site are below a prespecified concentration limit with some prespecified degree of confidence. The design of the FSS would be based on the proposed DCGLs, in accordance with MARSSIM. The MARSSIM process does not require that the licensee incorporate information regarding the existing (i.e., pre-remediation or pre–FSS) spatial distribution of radioactivity into the source-term abstraction. The identification of DCGLs may involve site-specific model and parameter assumptions, or may use "screening" analyses.

In the second objective, the licensee intends to assess potential radiation doses that may result from specified levels of radioactive material. The contaminated material may not be limited to building surfaces or surface soils, but may include contaminated subsurface soil, debris, and waste. The licensee's dose modeling should demonstrate that the residual radioactivity should not result in radiation doses in excess of applicable regulatory limits. This modeling would likely be site-specific. Most likely, this modeling objective would require that the licensee incorporate information regarding both the spatial extent and spatial variability of radioactivity into the source-term abstraction.

Table I.1 summarizes the approach to source-term abstraction that the licensee should adopt, depending on the licensee's dose modeling objective and whether the licensee is providing screening or site-specific analyses. This table can serve as an index for the reviewer of the licensee's source-term abstraction.

Table I.1 Summary of Source-Term Abstraction Approaches Based on Dose-Modeling Objective

Objective	Screening	Site-Specific
Identify DCGLs.	No source-term abstraction is necessary beyond radionuclide identification. (Assume unit radionuclide concentrations.)	Delineate proposed lateral and vertical extent of residual contamination. (Assume unit radionuclide concentrations.)
Provide Dose Assessment.	Use actual concentrations with DandD v2.1 and assure that spatial variability is minimal.	Site-specific source-term abstraction incorporating spatial extent and variability.

I.2.3.1   Dose Modeling Objective One: Identify DCGLs

The MARSSIM approach, as documented in NUREG–1575 (NRC 1997) and discussed in Chapter 4 of this volume, requires that a licensee establish a set of DCGLs before conducting an FSS. In fact, the design of the FSS should be based on the identified DCGLs. DCGL is defined in MARSSIM as:

> "...*a derived, radionuclide-specific activity concentration within a survey unit corresponding to the release criterion.... DCGLs are derived from activity/dose relationships through various exposure pathway scenarios.*"

The $DCGL_W$ is the concentration of a radionuclide which, if distributed uniformly across a survey unit, would result in an estimated dose equal to the applicable dose limit. The $DCGL_{EMC}$ is the concentration of a radionuclide which, if distributed uniformly across a smaller limited area within a survey unit, would result in an estimated dose equal to the applicable dose limit.

Two approaches are possible for developing DCGLs: screening and site-specific analysis.

SCREENING DCGLs

NRC has published radionuclide-specific screening DCGLs in the *Federal Register* for residual building-surface radioactivity and residual surface-soil radioactivity. The DCGLs in the *Federal Register* are intended to be concentrations which, if distributed uniformly across a building or soil surface, would individually result in a dose equal to the dose criterion. The licensee may adopt these screening DCGLs without additional dose modeling, if the site is suitable for screening analysis. Alternatively, the licensee may use the DandD computer code to develop screening DCGLs. The licensee would use the code to determine the dose attributable to a unit concentration of a radionuclide and scale the result to determine the $DCGL_W$ for the radionuclide. Either of these methods for identifying screening DCGLs requires the licensee (a) to identify the radionuclides of concern for the site and (b) to demonstrate that the source term and model screening assumptions are satisfied. Thus, this approach requires essentially no source-term abstraction. The screening process and the source-term screening assumptions are discussed in detail in Appendix H of this volume.

Before designing an FSS, the licensee may likely need to identify a $DCGL_{EMC}$ for each radionuclide over a range of smaller limited areas. Since the conservative screening models of DandD are not appropriate for modeling small limited areas of residual radioactivity, use of the DandD screening code would likely result in $DCGL_{EMC}$ values that are overly conservative. Therefore, licensees may likely use other codes or approaches to develop $DCGL_{EMC}$ values. These would be considered "site-specific" analyses in that they would not be using the DandD code with the default screening values. See Section I.3.3.3.5 of this appendix for more information.

SITE-SPECIFIC DCGLs

The licensee may choose to identify site-specific DCGLs if (a) the site conditions are not consistent with screening criteria or (b) the licensee believes the screening DCGLs are unnecessarily restrictive. As defined in MARSSIM, the site-specific DCGLs may be derived from activity/dose relationships through various exposure pathway scenarios. "Site-specific" in this context may refer to the selection of conceptual models/computer models, physical (site) input parameter values, or behavioral/metabolic input parameter values. These aspects of site-specific analyses are discussed in other sections of this document. "Site-specific" may also refer to the source-term abstraction.

From the MARSSIM perspective, identifying a site-specific $DCGL_W$ still begins with assuming a uniform radionuclide concentration across some source area (building surface) or volume (surface soil). The site-specific $DCGL_W$ for a particular radionuclide may be identified by evaluating the dose resulting from a unit concentration and then scaling the result. Spatial variability of the radionuclide concentration within the area or volume is not evaluated in identifying the DCGLs, but is taken into account in the statistical analysis of the data collected during the FSS. In identifying the site-specific DCGLs, the licensee may, however, take the spatial extent into account.

If the licensee is certain that the residual radionuclide concentration is limited to a specific lateral extent, the licensee may incorporate the "area of residual radioactivity" into the identification of DCGLs. Computer modeling codes, such as RESRAD or DandD, allow the user to directly specify the area of residual radioactivity. Through the FSS, the licensee would have to demonstrate that the DCGL$_W$ is satisfied within the specified area of residual radioactivity, and would have to demonstrate that residual radioactivity is not present outside the specified area of residual radioactivity. In order to adequately design the FSS, the licensee should develop DCGL$_{EMC}$ values for smaller areas within the area of residual radioactivity.

In addition to specifying a limited area of residual radioactivity in developing the site-specific DCGLs for soil, the licensee should also appropriately represent the vertical extent of residual radioactivity within the area. The screening DCGLs and the DandD code assume that residual radioactivity is contained within the uppermost 15 to 30 cm (6 to 12 in) of soil. If the licensee intends to leave residual radioactivity at depths below 15 to 30 cm (6 to 12 in), this should be reflected in the calculation of the DCGL$_W$. Otherwise, leaving residual radioactivity below 15 to 30 cm (6 to 12 in) may not be acceptable.

For subsurface residual radioactivity [i.e., residual radioactivity at depths greater than 15 to 30 cm (6 to 12 in)], the NRC license reviewer should evaluate whether the licensee has reviewed existing historical site data (including previous processes or practices) and site characterization data to establish an adequate conceptual model of the subsurface source specifically regarding horizontal and vertical extent of residual radioactivity. Lateral and vertical trends of variation in concentration for each specific radionuclide should be evaluated. Since certain radionuclides have higher mobility than others, radionuclide ratios may not be maintained as constant across subsurface soil. In other words, radionuclide concentration within the unsaturated zone may vary depending on the original source location and the time since contamination existed. The NRC license reviewer should evaluate whether the licensee has reviewed the physical and chemical properties of the source and the surface/subsurface formation to assess potential for leaching or retardation within the natural physical system of the concerned site. In this context, the NRC license reviewer should evaluate the selected physical parameters and the physical conceptual model of the site versus actual subsurface geologic units or formation to ensure conservative selection of pertinent sensitive physical parameters. The NRC license reviewer should also consider (a) the physical variability in subsurface soil and the unsaturated zone and (b) the selected depth to the water table considering the lower boundary of the subsurface source term.

If the thickness of residual radioactivity that the licensee intends to leave at the site is generally uniform across the site, the licensee may choose to use an upper bounding value for modeling the thickness. Alternatively, the licensee may choose to adopt an area-weighted approach to calculate an representative thickness. The representative thickness may be the area-weighted average value, or may reflect a conservative upper-percentile value. The NRC license reviewer should ensure that the representative thickness value proposed by the licensee does not significantly underestimate localized thicknesses at sites where the thickness of the proposed residually contaminated soil varies greatly across the site.

If appropriate, the licensee should provide maps and cross-sections detailing the proposed lateral and vertical extent of residual radioactivity left on the site.

I.2.3.2   Dose Modeling Objective Two: Assess Dose

An alternative objective that a licensee may have for performing and submitting dose modeling may be to assess doses attributable to specific quantities of radioactive material. Although the development of DCGLs focuses on the determination of radionuclide concentrations corresponding to a specified dose, the dose assessment objective focuses on the determination of doses corresponding to specified radionuclide concentrations.

In this situation, the licensee should give much more attention to the source-term abstraction. The licensee should address all elements of the source-term abstraction:

- identify the radionuclides of concern;

- delineate the spatial extent of residual radioactivity;

- represent the spatial variability of residual radioactivity; and

- incorporate spatial variability of physical and chemical characteristics of the contaminated media.

The licensee should focus on the distribution of radioactive material expected to be present at the time of FSS and subsequent site release. The licensee may assess doses attributable to existing radiological conditions at the site if the licensee can demonstrate that the existing radiological conditions reasonably bound conditions expected at FSS, from a dose perspective.

The first two elements of source-term abstraction—radionuclides of concern and spatial extent — were considered in the discussion of source-term abstraction for development of DCGLs. Spatial variability was not considered since it is statistically evaluated after FSS. In dose assessment, however, spatial variability should be factored into the source-term abstraction before dose modeling.

Assuming that the licensee has identified the radionuclides of concern and delineated the spatial extent of residual radioactivity, the licensee should provide a projection of residual radionuclide concentration distribution and total residual radionuclide inventory across the site. This projection should be directly tied to the characterization of existing radiological conditions at the site. The site may then be divided into relatively large areas that are radiologically distinct, based on radionuclide concentration or depth of residual radioactivity. The licensee should statistically demonstrate that the radionuclide concentrations or depth within an area may be relatively uniform, taking into account the spatial distribution of the data. Similarly, within the larger areas, the licensee should statistically delineate relatively small areas of projected elevated radionuclide concentrations or increased depth. (The licensee should discuss the reason for leaving the elevated concentrations in place as residual radioactivity.)

When complete, the licensee's source-term abstraction should define a site divided into relatively large areas of statistically uniform radionuclide concentrations and residual radioactivity depth. Within these areas may be relatively small areas of elevated concentration or increased depth. Assuming that the physical and chemical conditions across the site are relatively uniform, the licensee may use this source-term abstraction for modeling and proceed with the dose assessment. The following is a suggested approach:

- Consider each relatively large area independently, and initially ignore the relatively small elevated areas within each large area.

- Assess dose based on the properties of a large area, taking the areal extent into account.

- Repeat the dose assessment, but assume essentially infinite areal extent. The specific approach will depend on the computer modeling code used. This should quantify the impact of dividing the site into artificial modeling areas.

- Assess dose attributable to each limited area of elevated concentration, assuming no residual radioactivity exists outside the limited area. This may then be combined with the dose attributable to the surrounding larger area, to assess the impact of leaving the elevated concentrations.

In some cases, it may not be practical to separate a site into areas with relatively uniform radionuclide concentrations; sometimes areas to be evaluated will have non-uniform distributions of concentrations. In such cases, for performing the second step above, there may be a question about what statistical value best represents the radionuclide concentration for the large area. Log-normal distributions occur frequently in nature and are not unexpected when surveying contaminated sites. For log-normal distributions, the geometric mean is often used as a descriptor of the distribution. However, the geometric mean concentration should not be used as the average value for the source term for dose calculations. Arithmetic means reflect that (1) the dose rate is proportional to radionuclide concentration; (2) the dose receptor generally spends an equal amount of time in each area of the site; and (3) each characterization data point represents an equal area. If samples are not taken randomly or systematically (and thus data points represent unequal areas), weighted means may be appropriate, with application of weighting factors consistent with the assumptions of receptor exposures. Therefore, the arithmetic mean or weighted mean is the appropriate statistic to use for calculating source term average concentrations for the large areas (second step above) for dose modeling.

The above discussion does not specifically address the determination of relatively significant large or small areas. This designation will depend on the areal assumptions underlying the computer modeling code used. For example, the DandD code considers the area of cultivation to be uniformly contaminated and irrigated. The area of cultivation depends on the cultivation requirements defined by the specific exposure scenario. Conversely, the RESRAD code considers a range of exposure-pathway specific areas [e.g., 400 m^2 (4300 ft^2) for soil ingestion; 1000 m^2 (11,000 ft^2) for plant ingestion; and 20,000 m^2 (5 ac) for milk and meat ingestion]. Therefore, the licensee should discuss and justify the designation of relatively large and relatively small areas, based on the computer code used. However, by providing the additional assessments identified above, where alternative areas are evaluated, the sensitivity of the dose modeling results to the area designation can be determined.

The licensee may also have to consider the impact of multiple areas of elevated concentration within a single larger area. In general, modeling two small areas independently and combining the results of the two dose assessments should result in a higher dose than if the two areas were combined and modeled as a single area. The higher dose is unrealistic in that it assumes that the receptor location relative to each contaminated area is such that the dose is maximized from each contaminated area independently. For a more reasonable estimate of potential dose, these smaller areas may be combined into a single larger area if the concentrations within the smaller areas are comparable. If this is not the case, then the licensee may model each smaller area individually and modify the scenario and critical group assumptions for each area (e.g., time spent on each area) and combine the results.

I.3 Criteria for Selecting and Modifying Scenarios, Pathways, and Critical Groups

I.3.1 Introduction

After the source term has been evaluated, the question becomes: "How could humans be exposed either directly or indirectly to residual radioactivity?" or "What is the appropriate exposure scenario?" Each exposure scenario should address the following questions:

1. How does the residual radioactivity move through the environment?

2. Where can humans be exposed to the environmental concentrations?

3. What is the likely land use(s) in the future for these areas?

4. What are the exposure group's habits that will determine exposure? (e.g., what do they eat and where does it come from? How much? Where do they get water and how much? How much time do they spend on various activities? etc.)

The ultimate goal of dose modeling is to estimate the dose to a specific receptor. Broad generalizations of the direct or indirect interaction of the affected receptors with the residual radioactivity can be identified for ease of discussion between the licensee, regulator, public, and other interested parties. Scenarios are defined as reasonable sets of activities related to the future use of the site. Therefore, scenarios provide a description of future land uses, human activities, and behavior of the natural system.

In most situations, there are numerous possible scenarios of how future human exposure groups could interact with residual radioactivity. The compliance criteria in Part 20 for decommissioning does not require an investigation of all (or many) possible scenarios; its focus is on the dose to members of the critical group. The critical group is defined (at 10 CFR 20.1003) as "…the group of individuals reasonably expected to receive the greatest exposure to residual radioactivity for any applicable set of circumstances."

By combining knowledge about the answers to Questions 1 and 2, the licensee can develop exposure pathways. Exposure pathways are the routes that residual radioactivity travels, through the environment, from its source, until it interacts with a human. They can be fairly simple (e.g., surface-soil residual radioactivity emits gamma radiation, which results in direct exposure to the individual standing on the soil) or they can be fairly involved (e.g., the residual radioactivity in the surface soil leaches through the unsaturated soil layers into the underlying aquifer and the water from the aquifer is pumped out by the exposed individual for use as drinking water, which results in the exposed individual ingesting the environmental concentrations). Exposure pathways typically fall into three principal categories, identified by the manner in which the exposed individual interacts with the environmental concentrations resulting from the residual radioactivity: ingestion, inhalation, or external (i.e., direct) exposure pathways.

As required under Subpart E, the dose from residual radioactivity is evaluated for the average member of the critical group, which is not necessarily the same as the maximally exposed individual. This is not a reduction in the level of protection provided to the public, but an attempt to emphasize the uncertainty and assumptions needed in calculating potential future doses, while limiting boundless speculation on possible future exposure scenarios. Although it is possible to actually identify with confidence the most exposed member of the public in some operational situations (through monitoring, time studies, distance from the facility, etc.), identification of the specific individual who may receive the highest dose some time (up to 1000 years) in the future is impractical, if not impossible. Speculation on his or her habits, characteristics, age, or metabolism could be endless. The use of the "average member of the critical group" acknowledges that any hypothetical "individual" used in the performance assessment is based, in some manner, on the statistical results from data sets (e.g., the breathing rate is based on the range of possible breathing rates) gathered from groups of individuals. Although bounding assumptions could be used to select values for each of the parameters (i.e., the maximum amount of meat, milk, vegetables, possible exposure time, etc.), the result could be an extremely conservative calculation of an unrealistic scenario and may lead to excessively low allowable residual radioactivity levels, compared to the actual risk.

Calculating the dose to the critical group is intended to bound the individual dose to other possible exposure groups because the critical group is a relatively small group of individuals, because of their habits, actions, and characteristics, who could receive among the highest potential dose at some time in the future. By using the hypothetical critical group as the dose receptor, coupled with prudently conservative models, it is highly unlikely that any individual would actually receive doses in excess of that calculated for the average member of the critical group. The description of a critical group's habits, actions, and characteristics should be based on credible assumptions and the information or data ranges used to support the assumptions should be limited in scope to reduce the possibility of adding members of less exposed groups to the critical group.

ALARA analyses should use the dose based on the reasonably foreseeable land use for any cost-benefit calculations performed.

I.3.2 Issues in Selecting and Modifying Scenarios, Pathways, and Critical Groups

The definition of scenarios, identification of a critical group with its associated exposure pathways, and the dose assessment based on that definition can be generic or site specific. Licensees might:

- Use screening scenarios, screening groups, and pathway parameters as described in NUREG–1549 (NRC 1998a) and the NUREG/CR–5512 series. This can be used for either screening or site-specific analyses.

- Use the default screening scenarios as a starting point to develop more site-specific pathway analyses or critical group habits.

- Develop site-specific scenarios, critical groups, and identify associated exposure pathways from scratch.

To establish either site-specific scenarios, critical groups, and/or sets of exposure pathways, the licensee may need to provide justifications defending its selections. For some licensees, this may require minimum amounts of site-specific data to support the assumptions inherent in the existing default screening scenarios or for removing specific exposure pathways. For others, the licensee may need to thoroughly investigate and justify the appropriateness of the selected scenarios and/or critical groups, which may include evaluation of alternate scenarios and/or critical groups. If a licensee creates the exposure scenario and associated critical group based on site-specific conditions (e.g., at a site that is grossly different than the assumptions inherent in the default scenarios), the licensee should provide documentation that provides a transparent and traceable audit trail for each of the assumptions used in developing the exposure scenario and critical group [e.g., justify the inclusion (or exclusion) of a particular exposure pathway].

I.3.3 Recommended Approaches

I.3.3.1 Screening Analyses

In the case of screening, the decisions involved in identifying the appropriate scenario and critical group, with their corresponding exposure pathways, have already been made. Scenario descriptions acceptable to NRC staff for use in generic screening are developed and contained in NUREG/CR–5512, Volume 1. NUREG/CR–5512, Volume 3, and NUREG–1549, provide the rationale for applicability of the generic scenarios, critical groups, and pathways at a site; the rationale and assumptions for scenarios and pathways included (and excluded); and the associated parameter values or ranges (only from NUREG/CR–5512, Volume 3). A summary of the scenarios is in Table I.2. The latest version of the DandD computer code should contain the latest default data values for the critical group's habits and characteristics.

Table I.2 Pathways for Generic Scenarios

Building Occupancy Scenario

This scenario accounts for exposure to fixed and removable residual radioactivity on the walls, floor, and ceiling of a decommissioned facility. It assumes that the building may be used for commercial or light industrial activities (e.g., an office building or warehouse).

Pathways include:

- external exposure from building surfaces;

- inhalation of (re)suspended removable residual radioactivity; and

- inadvertent ingestion of removable residual radioactivity.

Resident Farmer Scenario

This scenario accounts for exposure involving residual radioactivity that is initially in the surficial soil. A farmer moves onto the site and grows some of his or her diet and uses water tapped from the aquifer under the site.

Pathways include:

- external exposure from soil;

- inhalation to (re)suspended soil;

- ingestion of soil;

- ingestion of drinking water from aquifer;

- ingestion of plant products grown in contaminated soil and using aquifer to supply irrigation needs;

- ingestion of animal products grown onsite (using feed and water derived from potentially contaminated sources); and

- ingestion of fish from a pond filled with water from the aquifer.

I.3.3.2   Site-Specific Analyses

Site-specific analyses give licensees greater flexibility in developing the compliance scenario. The licensee should justify its selection of the compliance scenario based on reasonably foreseeable land use at the site. The compliance scenario should result in an exposure to the public, such that no other scenario, using reasonably foreseeable land use assumptions, will result in higher doses to its exposure group(s). The level of justification and analysis provided by the licensee will be depend on the how close the analysis is to the "real" dose. The more realistic the analysis, greater degrees of justification and, potentially ancillary analyses, will be required. For example, a site is currently zoned industrial and the local area is a mix of suburban, commercial, and industrial. Rural uses of the property are less likely but plausible for the foreseeable future. If the licensee chose to use the generic screening scenario, the licensee would need to provide limited justification for the bounding scenario. If the licensee proposed to use a maintenance worker scenario assuming industrial land use as the compliance scenario, the licensee would need to provide quantitative analyses of or a qualitative argument discounting the need to analyze other competing scenarios (based on industrial land use and on suburban or commercial land use) to justify the selection of the compliance scenario. In addition, the licensee would need to provide analyses of the rural use of the land to show what impacts would occur from this less likely but plausible situation.

Site-specific analyses can use the generic screening scenario(s) with a little justification. The licensee may need to justify that the site contains no physical features nor locations of residual radioactivity, other than those assumed in the screening analyses, that would invalidate the assumptions made in developing the scenarios. The NRC license reviewer should evaluate the justification to provide reasonable assurance that the generic scenario would still be appropriate for the site. A site can fail to meet the requirements of the conceptual model (see Section I.4 of this appendix) without invalidating the generic scenario, and situations can arise where the default scenario is no longer the limiting case. For example, the site may have pre-existing ground water contamination, which is counter to the assumptions in the conceptual model inherent in the screening models, but this may not require any change in the exposure scenario because the residential farmer scenario may still be an appropriate scenario, as it contains all of the appropriate exposure pathways, including ground water use for drinking, irrigation, and for animals. Alternately, if the residual radioactivity were a volumetric source in the walls of a building, rather than on the building surfaces, the generic exposure scenario of an office worker may not be the scenario leading to the critical group. For certain sets of radionuclides, a building renovation scenario may be more limiting because of the exposure to airborne concentration of material as the walls are modified.

Site-specific scenarios, critical groups, and pathways can be developed, for any situation, and would occur in cases where, for example:

1. Major pathways (e.g., the ground water pathway, or agricultural pathways) associated with the default screening scenarios could be eliminated, either because of physical reasons or site-use reasons.

2. The location of the residual radioactivity and the physical features of the site are outside the major assumptions used in defining the default critical group and/or scenarios.

3. Restricted use was proposed for a site.

The second situation listed above can be ambiguous, as a number of assumptions key to the development of the DandD screening tool do not affect the scenario description, and may require an NRC reviewer to evaluate whether the initial generic scenario would still be appropriate for the site.

Modifying scenarios or developing a site-specific critical group requires information regarding plausible uses of the site and demographic information. Such information might include considerations of the prevailing (and future) uses of the land, and physical characteristics of the site that may constrain site use. Potential land uses should be categorized as reasonably foreseeable, less likely but plausible, or implausible. Any land uses that similar property in the region currently has, or may have in the near future (e.g., approximately 100 years), should be characterized as reasonably foreseeable. Consideration should be given to trends and area land use plans in determining the likelihood of potential land use. Land uses that are plausible, generally because similar land historically was used for the purpose, but are counter to the current trends or regional experience could be characterized as less likely but plausible (e.g., rural use of property currently in an urban setting). Implausible land uses are those that because of physical limitations could not occur (e.g., residential land use for an underwater plot of land). It may be necessary to evaluate several potential critical groups, based on different combinations of site-specific scenarios developed from expected land use, pathways and demographics, to determine the group receiving the highest exposure.

Depending on the resulting exposure scenarios, considerations of offsite exposure by either transport (e.g., through ground water) or material transfer (e.g., soil being taken from the site and used elsewhere) may be necessary to identify the critical group. Thus, the licensee should consider if offsite uses are reasonably foreseeable. If they are, such offsite uses should also be analyzed to determine if the critical group might be an offsite user instead of an onsite user.

Similar considerations apply for restricted release. Thus, when analyzing the dose under restricted conditions, the nature of the critical group is likely to change because of these restrictions and controls. Site restrictions and institutional controls can restrict certain kinds of activities and land or water uses associated with the physical features of the site. The detailed definition of the scenarios considered for restricted release need to include the impact of the control provisions on the location and behavior of the average member of the appropriate critical group.

For restricted use, licensees must also evaluate doses assuming the loss of institutional controls. This evaluation should address: (a) the associated degradation of engineered barriers without active maintenance; and (b) inadvertent intruder scenarios. See Section 3.5.2 of this volume for additional information.

The NRC license reviewer should evaluate the justifications provided by the licensee on its scenarios using the following appropriate guidance. The guidance is characterized by the general approach used in development of the scenarios: (a) modifying existing generic exposure scenarios or (b) developing site-specific scenarios from "scratch."

I.3.3.2.1   Modification of Generic Scenarios

First, the NRC license reviewer should evaluate whether the generic scenario was applicable to the site before modification. If the scenario was applicable before the licensee started modifying the scenario based on physical features or restrictions, go to the next step and evaluate the justifications for the various modifications performed by the licensee. If the scenario was not initially applicable, that does not mean that a final modified scenario is inappropriate for the site conditions. It just means that the review may be more complex than a simple modification of a scenario and that the NRC license reviewer should evaluate whether it may be more appropriate to evaluate the scenario using the guidance below.

The NRC license reviewer should identify the modifications done by the licensee to the scenario and evaluate the licensee's justification for those changes. Table I.3 lists some common exposure scenarios, but is by no means comprehensive. The Sandia Letter Report, "Process for Developing Alternate Scenarios at NRC Sites Involved in D&D and License Termination" (Thomas, et al., 2000), which is included in this volume as Appendix M, provides a series of flow charts and sources of information to assist a licensee or reviewer in modifying the default scenarios using site-specific information. See below for specific guidance on acceptable justifications using different types of site-specific information, which was adapted from the letter report. Additionally, if the licensee's intent is restricted release, the final scenario should be reviewed looking at the effect of site restrictions. The licensee's justifications should support, based on either site restrictions or site-specific data, the elimination of scenarios and pathways from the analysis. The NRC license reviewer should focus the review on the pathways, and models associated with those pathways, that have the highest likelihood of significant exposures to the critical group.

Table I.3 Potential Scenarios for Use in Dose Assessments

<table>
<tr><td>

General Scenario Classification

- Building occupancy (Generic screening – NUREG/CR–5512-based).

- Residential farmer (Generic screening – NUREG/CR–5512-based).

- Urban construction (contaminated soil, no suburban or agricultural uses). This scenario is meant for small urban sites cleared of all original buildings; only contaminated land and/or buried waste remains.

- Residential (a more restricted subset of the residential farmer scenario, for those urban or suburban sites where farming is not a realistic projected future use of the site).

- Recreational User (where the site is preserved for recreational uses only).

- Maintenance Worker (tied to the Recreational User scenario but involves the grounds keepers maintaining or building on the site).

- Hybrid industrial building occupancy (adds contaminated soil, building may or may not be contaminated).

- Drinking water (e.g., no onsite use of ground water; offsite impacts from the contaminated plume).

</td></tr>
</table>

The licensee may need to evaluate whether the final modified scenario is still the limiting reasonable representation of the critical group at the site. This may involve investigation of exposure pathways not covered in the default scenarios.

I.3.3.2.2   Development of Alternate Scenarios

In some decommissioning cases, either the location of the residual radioactivity, the physical characteristics of the site, and/or planned institutional restrictions may make the default scenarios inappropriate. In other cases, the licensee may wish to provide a transparent and traceable development of the compliance and other exposure scenarios, starting with the potential land use and the site conditions. Development (and review) of alternate scenarios may involve iterative steps involving the development of the conceptual model of the site. For example, the licensee may (a) develop a generic list of exposure pathways, (b) develop the site conceptual model to screen the generic list, (c) aggregate or reduce the remaining exposure pathways to the major exposure pathways, and (d) re-evaluate the conceptual model to verify that all the necessary processes are included.

A brief summary of the NRC–recommended pathway analysis process follows. An example development of exposure scenarios, while developed for partial site release, is listed in Appendix K.

- The licensee compiles a list of exposure pathways applicable to any contaminated site. There are a number of existing sources of information that can be used. One source is NUREG/CR–5512, Volume 1 (NRC 1992), and the list is summarized in Appendix C.1 of NUREG–1549 (NRC 1998a). Another source, although the guidance is more focused on offsite exposures, is NUREG/CR–5453, Volumes 1 and 2, "Background Information for the Development of a Low-Level Waste Performance Assessment Methodology" (Shipers 1989; Shipers and Harlan 1989). Another potential source is the international "Features, Events and Processes," list which is an expansive generic list that does not strictly deal with decommissioning issues (SSI 1996).

- Categorize the general types of residual radioactivity at the site (e.g., sediment or soil, deposits in buildings, surface residual radioactivity, surface water, ground water, industrial products such as slag).

- Screen out pathways, for each contaminant type, that do not apply to the site.

- Identify the physical processes pertinent to the remaining pathways for the site.

- Separate the list of exposure pathways into unique pairs of exposure media (e.g., source to ground water, ground water to surface water, etc.). Determine the physical processes that are relevant for each exposure media pair and combine the processes with the pathway links.

- Reassemble exposure pathways for each source type, using the exposure media pairs as building blocks, thus associating all the physical processes identified with the individual pairs with the complete pathway.

The licensee's documentation of the decisions made regarding inclusion (or exclusion) of the various pathways should be transparent and traceable. An international working group of Biospheric Model Validation Study, Phase II (BIOMOVS II) (SSI 1996), established a methodology for developing models to analyze radionuclide behavior in the biosphere and associated radiological exposure pathways (i.e., the Reference Biospheres Methodology). BIOMOVS II published the methodology in its Technical Report No.6, "Development of a Reference Biospheres Methodology for Radioactive Waste Disposal" (SSI 1996), and it may be useful as a guide for additional information on a logical method to complete the pathway analysis sets above and include proper justification. Generally, the Reference Biospheres Methodology is more useful for complex sites that may have numerous physical processes that interact in such as a way that a number of different exposure groups may need to be investigated to identify the critical group. Additional work has been done on implementing the Reference Biospheres Methodology by a working group of the International Atomic Energy Agency's Biosphere Modeling and Assessment (BIOMASS) program (IAEA 1999a). Specifically, IAEA Working Document BIOMASS/T1/WD03, "Guidance on the Definition of Critical and Other Hypothetical Exposed Groups for Solid Radioactive Waste Disposal," may provide additional information on developing a site-specific critical group for situations where the generic critical group is inappropriate.

I.3.3.3   Guidance on Specific Issues

I.3.3.3.1   Land Use

A licensee's assumptions for land use should focus on current practice in the region. The region of concern can be as large as an 80-kilometer (50-mile) radius. To narrow the focus of current land practices, the licensees can use information on how land use has been changing in the region, and more weight should be given to land-use practices either close to the site or in similar physical settings. This can be very important for semi-rural sites that are being encroached by suburban residential development. Reviewers may wish to involve State and local land-use planning agencies in discussions, if the licensee has not already requested their involvement.

Potential land uses should be categorized as reasonably foreseeable, less likely but plausible, or implausible. Any land uses that similar properties in the region currently have, or may have in the near future (e.g., approximately 100 years), should be characterized as reasonably foreseeable. Consideration should be given to trends and area land use plans in determining the likelihood of potential land use. The time frame of interest for scenario development could be less than 100 years in certain cases and would depend on such factors as the rate of change in land use patterns in the area, radionuclides of interest and the time of peak dose. For example, a site with residual Cobalt-60, which has approximately a 5 year half-life, would not likely need to explore possible land uses that may exist at the site beyond a few decades, because of the natural decay of the residual material. Note that the 100-year timeframe described here is only for estimating future land uses; the licensee must evaluate doses that could occur over the 1000-year time period specified in the LTR.

Land use that are plausible, generally because similar land historically was used for the purpose, but are counter to the current trends or regional experience should be characterized as less likely but plausible (e.g., rural use of property currently in an urban setting).

Implausible land uses are those that because of generally physical limitations could not occur (e.g., residential land use for an underwater plot of land).

Land use justifications by licensees often rely on State or local codes, in building or well development to constrain future use. In general, licensees requesting unrestricted release should not rely solely on these factors as reasons to remove pathways or justify the scenario unless (a) the radionuclides have a relatively short-half life (approximately 10 years or less) or (b) the dose from long-lived radionuclides reaches its peak before 100 years. Similarly, licensees requesting unrestricted release should not limit land use scenarios based on commitments, or require the enforcement of limitations by the licensee or another party (e.g., a licensee states that the land will remain industrial because the licensee states that the land will not be sold by the licensee after the license is terminated).

Licensees should base justifications of land use on (1) the nature of the land and reasonable predictions based on its physical and geologic characteristics, and (2) societal uses of the land based on past historical information, current uses of it and similar properties, and what is reasonably foreseeable in the near future. The societal uses of the site in the future should be based on advice from local land planners and other stakeholders on what possible land uses are likely within a time period of around a hundred years. The level of justification for the final land uses is inversely proportional to the level of realism assumed by the licensee. Limited justification may be required for bounding analyses while much more detailed justification including alternate reasonably foreseeable and less likely but plausible scenario analyses may be needed for a situation with a smaller degree of conservatism in the analyses.

Additional guidance is available on potential sources of land use information in Appendix M.5.

I.3.3.3.2   Waterborne Exposure Pathways

Removal of waterborne exposure pathways can range from being global (e.g., all ground water pathways) to being specific (e.g., no drinking water but still have agricultural/fish pond use). Acceptable justifications are generally based on physical conditions at the site rather than local codes. Justification of water quality and quantity of the saturated zone should be based on the classification systems used by the U.S. Environmental Protection Agency (EPA) or the State, as appropriate. Arguments involving depth to water table, or well production capacity, should have supporting documentation from either the U.S. Geological Survey (USGS), an appropriate State agency, or an independent consultant.

NRC license reviewers should evaluate the reasons for the classification. Tables M.5–M.12 in Appendix M provide details regarding water quality standards. For example, where the aquifer is classified as not being a source of drinking water, but is adequate for stock watering and irrigation, the licensee can eliminate the drinking water pathway, but should still maintain the irrigation and meat/milk pathways. Aquifers may exceed certain constituents and still be able to be used for various purposes because those constituents may easily be treatable (e.g., turbidity). In cases where the water may be treatable or because the degree of connection between the aquifer and surface water may make the use of the aquifer questionable, the NRC license reviewer should involve the EPA and/or the State, as appropriate, in discussions on reasonable assumptions for the aquifer use.

I.3.3.3.3   Agricultural Pathways

Agricultural pathways may be removed or modified for various reasons: (a) land use patterns, (b) poor-quality soil, (c) topography, and (d) size of contaminated area. Many justifications may result in modification of the pathways, rather than complete elimination. For example, the soil may of inappropriate quality to support intensive farming activities, but residential gardening may still be reasonable.

Licensees using poor-quality soil as a justification for modifying the agricultural pathways should provide the reviewer with supporting documentation from the Soil Conservation Service, appropriate State or local agency, or an independent consultant. Reviewers should carefully consider whether the state of the soil would reasonably preclude all activities (e.g., because of high salinity of soil) or only certain activities. In most cases, soil quality can reasonably preclude activities such as intensive farming, but could allow grazing or small gardens.

When reviewing justifications involving topography, the NRC license reviewer should limit speculation of future topographical changes from civil engineering projects. The NRC license reviewer should evaluate the reasonableness of the critical group performing its activities on the current topography, for example, a slope. Supporting documentation should be provided by the licensee in the form of pictures, USGS or similar topographic maps, hand-drawn maps, or a detailed description of how the topography would limit farming. NRC reviewers may wish to perform a site visit to evaluate the topography firsthand.

I.3.3.3.4   Age-Dependent Critical Groups

Use the definitions in Part 20 when calculating for compliance with the requirements of Subpart E. Use the Federal Guidance Report No. 11 when calculating internal exposures by using the intake-to-dose conversion factors, which are based primarily on adults. As stated in the Environmental Protection Agency's *Federal Register* notice (59 FR 66414, Dec. 23, 1994) on "Federal Radiation Protection Draft Guidance for Exposure of the General Public," which proposes a public dose limit of 1.0 mSv/y (100 mrem/y) from all sources:

> "These dose conversion factors are appropriate for application to any population adequately characterized by the set of values for physiological parameters developed by the [International Committee on Radiological Protection] and collectively known as "Reference Man." The actual dose to a particular individual from a given intake is dependent upon age and sex, as well as other characteristics. As noted earlier, implementing limits for the general public expressed as age and sex dependent would be difficult.... More importantly, the variability in dose due to these factors is comparable in magnitude to the uncertainty in our estimates of the risks which provide the basis for our choice of the [public dose limit]. For this reason EPA believes that, for the purpose of providing radiation protection under the conditions addressed by these recommendations, the assumptions exemplified by Reference Man adequately characterize the general public, and a detailed consideration of age and sex is not generally necessary." (59 FR 66423, Dec. 23, 1994) [sic]

Since age-based dose conversion factors are not being used, the same dose conversion factors are applied to all individuals. Only in rare scenarios will a non-adult individual receive a higher dose (i.e., intake more radioactive material) than an adult individual in a similar exposure scenario. One example is the milk pathway, children generally drink more milk annually than adults. If milk was the only pathway that would expose the individual to a dose, then the child would have a slightly higher dose than the adult. But in most situations, especially ones involving multiple pathways, the total intake of the adult is greater than that of a child.

Therefore, for most multiple pathway scenarios, such as screening analyses, the average member of the critical group should usually be assumed to be an adult, with the proper habits and characteristics of an adult. As the licensee eliminates pathways or modifies the scenario, the behavior and dietary habits of children may become important. In such cases, the licensees should consult with NRC staff for guidance.

I.3.3.3.5   Area Factors

The $DCGL_W$ is the average concentration across the site that is calculated to result in the average member of the critical group receiving a dose at the appropriate dose limit [e.g., 0.25 mSv/y (25 mrem/y) for unrestricted release]. The general assumption is that the concentration of the radionuclides in the source are fairly homogenous. The degree to which any single localized area can be elevated above the average, assuming the average is at the $DCGL_W$, and not invalidate the homogenous assumption is characterized by the $DCGL_{EMC}$ (see Chapter 4 of this volume and MARSSIM). One method for determining values for the $DCGL_{EMC}$ is to modify the $DCGL_W$ using a correction factor that accounts for the difference in area and the resulting change in dose. The area factor is then the magnitude by which the concentration within the small area of elevated activity can exceed $DCGL_W$ while maintaining compliance with the release criterion.

The area factor works by taking into consideration how a smaller area would affect the dose to the average member of the critical group. For example, a smaller area could mean that external dose is more limited because it is not reasonable to expect the individual to be exposed the same amount of time as the individual would be to a larger area.

The default scenario for surface soil assumes large areas of homogeneous surface residual radioactivity. If the area of residual radioactivity is smaller than the defaults [e.g., 2400 m^2 (0.6 ac) for DandD], the licensee may propose modifying the exposure pathways to account for the effect on the critical group's activities. The licensee can follow either of two methods:

- Reduce the calculated dose by modifying the exposure time or usage parameters accordingly.

- Modify the exposure scenario and pathways and/or modify the calculational method to account for the size of the residual radioactivity.

These methods may be built into some dose assessment codes for surficial soil, but the user should verify proper use of the method. When the user changes the size of the contaminated area, the code will modify the appropriate usage factors and remove pathways if they are no longer viable.

When the extent of residual radioactivity becomes smaller, some of the activities are no longer viable as reasonable assumptions for exposure. Generally, the first pathways affected are animal husbandry activities, because of the larger area needs for grazing and growing fodder. As a general rule, as the area gets smaller, the more the scenario transforms into a residential gardener scenario, so long as the initial residual radioactivity begins in the surface soil. For cases where the residual radioactivity is not in the surficial soil, the original area of residual radioactivity may not be as important in scenario development, because some of the primary transport mechanisms result in redistribution of the radionuclides over larger areas (i.e., ground water used as irrigation).

One common mistake in licensee submittals is that area factors are typically not provided for residual radioactivity on building surfaces. The primary reason for this is that such factors could not be calculated by using the DandD, Version 1. Therefore, when the screening $DCGL_W$ values were published in the *Federal Register* (see Appendix H), which were derived from an improved DandD, Version 1, the associated area factors were not published. An alternative approach should be used to calculate area factors for residual radioactivity on building surfaces.

One approach is to use DandD, Version 2.1, to calculate the area factors, although it models area factors conservatively. Another approach that has been successfully used is to develop the area factors by using the RESRAD–BUILD computer code and adjusting these derived area factors to account for the fact that RESRAD–BUILD typically gives less conservative dose estimates. With this approach, the screening DCGL values are converted into the appropriate concentration unit for RESRAD–BUILD [i.e., from (dpm per 100 cm^2) to (pCi/m^2)]. Area factors calculated by RESRAD–BUILD can then be adjusted by the ratio of the dose from RESRAD–BUILD to 0.25 mSv/y (25 mrem/y) (i.e., the equivalent dose from DandD).

I.3.3.3.6 Offsite Scenarios

In rare situations, the scenario resulting in the highest exposures from the residual radioactivity will be offsite use scenarios. For these evaluations, the dose limit remains that of 10 CFR 20, Subpart E, even though the situation may seem similar to the clearance of materials prior to license termination (see also Section G.1.1 of this volume). In these scenarios, the exposure to the radioactive material will occur because it has been removed from the current location, and this results in either new or enhanced exposure pathways. For example, a site has poor ground water characteristics (thereby, allowing the licensee to remove the ground water pathway from any applicable scenarios) and the reasonably foreseeable land use is either commercial or industrial. The primary contaminant is Tc-99, which primarily results in dose through either the ground water or vegetable pathways, both of which are not applicable to the physical characteristics of the site or land use assumptions. The residual radioactivity is present in the site's top soil. A possible offsite scenario is that during construction of any commercial interest on the site after license termination, the removed topsoil is sold for use in a residential setting. In this case, it is likely that the topsoil with residual radioactivity will be unintentionally mixed with other topsoil at the offsite location. Licensees can use generic analyses to screen the importance of offsite uses with such sources as NUREG-1640, "Radiological Assessments for Clearance of Materials from Nuclear Facilities." (NRC 2003)

Even if offsite use is not considered reasonably foreseeable, offsite scenarios may be less likely but plausible scenarios and should be analyzed as scenarios, to understand the robustness of the analysis.

I.3.3.3.7   Determining the Compliance Scenario

In many situations a licensee will be faced with selecting a compliance scenario from potentially a large suite of scenarios and exposure groups. The licensee is expected to base their demonstration of compliance on the exposure to the highest group, consistent with the definition of the critical group. Licensees may find it advantageous to use an iterative approach to screen all the potential scenarios. This will allow the licensees to focus their more detailed analyses on the important scenarios. Licensees may be able to use information from NUREG/CR-5512, NUREG-1640, and NUREG-1717, as well as other licensees' analyses to screen their potential scenarios with quantitative methods. Licensees also may be able to provide qualitative arguments to demonstrate that the dose from certain scenarios are bounded by the dose of higher level scenarios (e.g., a residential gardening scenario will bound the dose for the residential non-gardening scenario). The licensees should provide justifications on the basis, method, and results of their scenario screening in their DP.

Even after screening the scenarios, a licensee will likely be left with a few scenarios that may require detailed analyses to determine which will result in the critical group. For licensees with multiple radionuclides, commonly, determining the compliance scenario depends on the final mixture of radionuclides. This can provide a dilemma for licensees creating DCGLs. The licensee must show that the final concentrations at the site meet the dose criteria of 10 CFR Part 20, Subpart E. Three possible approaches that the licensee may use to show compliance are, but are not limited to the following:

1. Use the most limiting DCGL for each radionuclide, regardless of the scenario, and use the sum of fractions, ignoring the scenario basis for each DCGL. This approach requires limited justification. It will always either estimate the same dose as the individual scenarios or overestimate the dose. Generally, it will greatly overestimate the dose for the individual scenarios.

2. Use a surrogate approach to limit the number of radionuclides of importance. A surrogate approach relies on different radionuclides having relatively fixed ratios. For example, assume that at a site with cesium-137 and strontium-90 can show that for every 37 Bq/kg (1 pCi/g) of Cs-137, there will be a 18 Bq/kg (0.5 pCi/g) of Sr-90. By using this relationship, an effective DCGL for combined Cs-137 and Sr-90 can be created. The licensee may be able to reduce the number of critical scenarios, specifically those driven by exposure to Sr-90. This approach requires that the licensee have the necessary information on relative ratios of the radionuclides.

3. Commit to demonstrating the final dose for each of the important scenarios in the final status survey reports. This approach will require the licensee to establish operational DCGLs to fully utilize MARSSIM (see Section 2.5).

The licensee needs to provide either quantitative analysis of or a qualitative argument discounting the need to analyze all the scenarios generated from the less likely but plausible land uses. The results of these analyses will be used by the staff to evaluate the degree of sensitivity of dose to overall scenario assumptions (and the associated parameter assumptions). Analyses of less likely but plausible scenarios are not meant to be 'worst-case' analyses and should not utilize a set of 'worst-case' parameters. Selection of parameters for less likely but plausible scenarios should be consistent with the guidance in this Appendix. The reviewer will consider both the magnitude and time of the peak dose from these scenarios. If peak dose from the less likely but plausible land use scenarios is significant, the licensee would need to provide greater assurance that the scenario is unlikely to occur, especially during the period of peak dose. The licensee may be able to show that the compliance scenario bounds the results of all or many of the scenarios associated with the less likely but plausible land uses.

I.3.4   Generic Examples

The following examples are provided as situations where the default pathways may be removed or modified. Note, the examples assume that an adequate level of justification has been provided by the licensee.

I.3.4.1   Removal of Ground Water Pathways

A licensee has extensive contamination of the upper soil horizon and the upper aquifer, which is unconsolidated and the licensee wishes to remove the ground water pathway because the upper aquifer would not be used as a water source. The aquifer shows relatively high levels of microbial activity, turbidity, and nitrates. In addition, adjacent to the site is a small patch of wetlands that shows a great deal of communication with the upper aquifer. The potential yield rate of the upper aquifer is sufficient for domestic use, but there is a better-quality, confined aquifer, whose horizon is at a depth of approximately 30 meters (100 feet). The licensee has also demonstrated that the deeper aquifer will not become contaminated from the upper aquifer. Considering all of these reasons in combination, it is questionable whether the upper aquifer would actually be used. Although it may be possible for someone to treat the contaminants and use the aquifer, there are better sources of water easily available. After consultation with the EPA and the State, it is agreed that it would be unreasonable to assume someone would use the upper aquifer as a water source. Therefore, the licensee is allowed to remove the ground water pathway from the scenario.

I.3.4.2   Scenario Development for Buried Residual Radioactivity

I.3.4.2.1   Example 1: Subsurface Soil

A site has residual radioactivity buried at a few feet below the surface and the licensee is requesting unrestricted release. The residual radioactivity does not have enough highly energetic gamma-emitters to result in an external dose in the current configuration. Two exposure scenarios can be developed (without any other site-specific information): (1) leaching of the radionuclides to the ground water, which is then used by a residential farmer; and (2) inadvertent intrusion into the buried residual radioactivity by house construction for a resident farmer with the displaced soil, which includes part of the residual radioactivity, spread across the surface. Exposure scenario 2 encompasses all the exposure pathways and, although not all of the source term is in the original position, leaching may occur both from the remaining buried residual radioactivity and the surface soil. Except for cases where an additional 0.6 m (2 ft) of unsaturated zone may make a tremendous difference in travel time to the aquifer, the ground water concentrations should be similar and, therefore, analysis of the second exposure scenario appears to be the appropriate scenario for the critical group exposure. This example is described in greater detail and integrated with the other guidance in Appendix J of this volume.

I.3.4.2.2   Example 2: Embedded Piping

At another site, the licensee is requesting unrestricted release of its site. It is removing the buildings, but is evaluating the need to remove the concrete pads, which have embedded piping that contains the residual radioactivity. Two scenarios can be reasonably envisioned. The first scenario involves a resident farmer onsite. The farmer builds a house on the concrete pad, without disturbing the embedded piping. Possible exposure pathways would be external dose from the piping and exposure to leached materials from the piping through ground water use (e.g., drinking, irrigation, etc.). The second scenario is similar to the building renovation scenario, where the concrete pad and piping are removed from the site. The licensee should investigate both to find the limiting scenario.

I.3.4.3   Scenario Development for Restricted Release

The site restrictions planned for an alternate site include a restriction, for this example, on the deed, on the use of the property for only parkland, and an engineered cover is placed over the residual radioactivity. The engineered cover is contoured for use as parkland and has a vegetative cover (i.e., not a mound covered in rip-rap). Three scenarios are easily envisioned for the restricted release analysis. The first is recreational use of the property as a city park or golf course, which would limit exposure scenarios to possible external exposure. The second would involve offsite use of ground water that contains radionuclides leached from the buried residual radioactivity. The default offsite user would be a resident farmer using the ground water for all water needs. The third scenario would be a worker maintaining the park.

The doses assuming the loss of the institutional control (i.e., the deed restriction) and degradation of the engineered cover also must be evaluated. Again, two main scenarios can be envisioned.

The first scenario is similar to the default exposure and would involve a residential farmer that uses ground water from the aquifer under the site. The engineered cover may have been compromised by the placement of the buildings, but the cover may still work in some degraded function (e.g., the water infiltration rate would increase from the design rate to some higher rate, but probably not as high as the infiltration rate would have been if the cover had never been constructed). Whether buried residual radioactivity had been transported to the surface by the construction of a basement under the resident farmer's house would depend on the thickness of the engineered cover. If typical basement depth were deeper than the engineered cover's thickness, some portion of residual radioactivity would be transported to the surface, mixed with the "clean" cover material, and spread over the site.

The second scenario would involve possible erosion of the cover and subsequent exposure of an onsite resident to the buried radionuclides or radionuclides redistributed by surface water. The exposure scenario would still be a resident farmer. The reasonableness of this scenario would depend on the thickness and erosion-resistance of the engineered cover.

I.4   Criteria to Establish Conceptual Models

I.4.1   Introduction

Analyzing the release and migration of radionuclides through the natural environment and/or engineered systems, at a specific site, requires the licensee to interpret the nature and features of the site so that the site can be represented by mathematical equations (i.e., mathematical models). This simplified representation of the site, including the associated mathematical models, is commonly referred to as the conceptual model of the site.

Figure I.1 depicts the process of conceptual model development. In dose assessments, developing a conceptual model involves making an abstraction of site data into a form that is capable of being modeled. This development should generally involve making simplifying assumptions, including simplification of the appropriate governing equations, to reflect the physical setting. These simplifying assumptions are usually made in describing the geometry of the system, spatial and temporal variability of parameters, isotropy of the system, and the influence of the surrounding. The conceptual model should provide an illustration, or description, of site conditions, which shows, or explains, contaminant distributions, release mechanisms, exposure pathways and migration routes, and potential receptors. In other words, the conceptual model should explain or illustrate how radionuclides enter, move through, and/or are retained in, and leave, the environment.

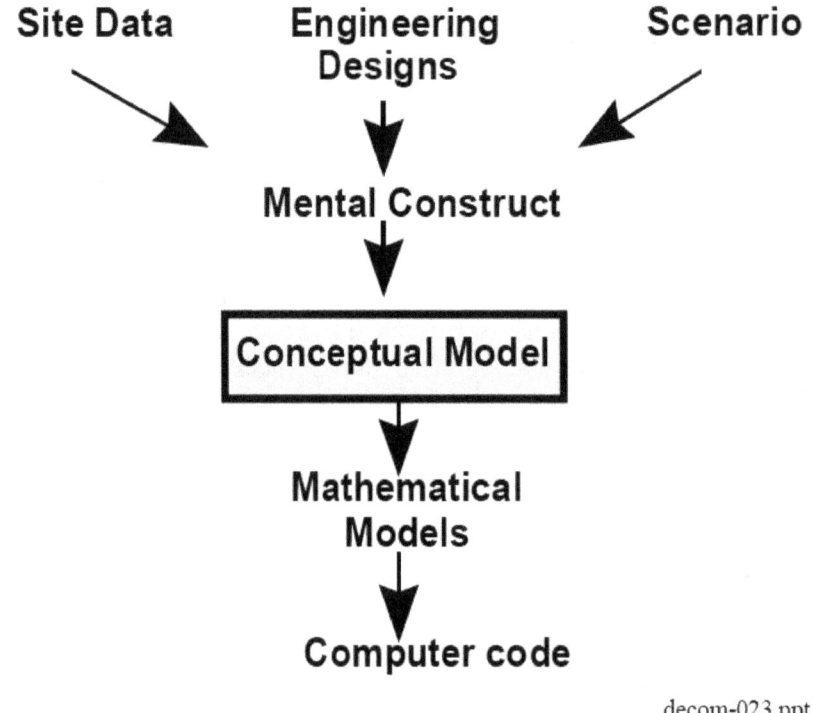

decom-023.ppt
031102

Figure I.1   Conceptual Model Development.

As shown in Figure I.2, developing a conceptual model at a site is Step 3 of the Decommissioning Decision Framework (see Section 2.6 of this volume). Conceptual model development follows after assimilation of site data (Step 1) and definition of scenarios (Step 2), because information from these two steps feeds into its development. In other words, the conceptual model should be based on what is known about the site from data and information gathered as part of Step 1, and how the site evolves during the period covered by the analysis based on the assumed land-use defined under Step 2.

Mathematical models are a quantitative representation of the conceptual model. Because the conceptual model provides the linkage between site conditions and features (Steps 1 and 2) and the computer code(s) (with its associated mathematical models) used in the dose analysis (Step 4 of the Decommissioning Framework), it is a key step in a dose assessment and should not be taken lightly.

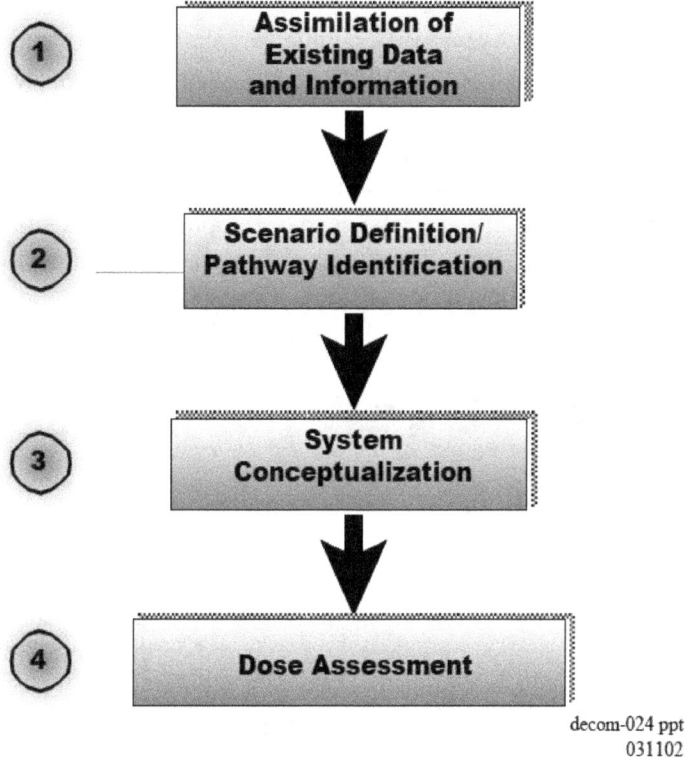

decom-024 ppt
031102

Figure I.2   Decommissioning Decision Framework.

I.4.2   Issues

Uncertainties in conceptual models can be large, and possibly even larger than uncertainties in parameters used in the analysis (James and Oldenburg 1997). Thus, conceptual model uncertainties can be a significant source of uncertainty in the overall dose assessment. Uncertainties in the conceptual model(s) are generally caused by incomplete knowledge about the natural system being analyzed and differing views about how to interpret data representing the system.

Development of conceptual models is a subjective process based on interpretation of limited (or in most cases, sparse) site data. From these limited data, the licensee should determine the key processes and features at the site and how they are likely to affect the movement of radionuclides through the environment. Because the conceptual of the site is based on incomplete information, it is possible that multiple interpretations of the same data can be derived. A licensee should also determine the appropriate level of simplification acceptable for representing the site. An overly simplified conceptual model may leave out key site features or conditions that are important in estimating where radionuclides are likely to be transported (thus, where people might be exposed) and when they might get there (thus, the radionuclide concentration when it arrives). On the other hand, an overly complex conceptual model may introduce unnecessary uncertainty and costs into the analyses. As a broad example, simple models contained in screening codes may oversimplify features and processes at a specific site. The licensee also should ensure that the appropriate level of detail is provided in the conceptual model. It is important that the conceptual model have sufficient detail and scope for a license reviewer to be able to assess the appropriateness of the computer codes used in the analysis and the defensibility of the assumptions made. In summary, key issues in developing and presenting the conceptual model are: (a) identifying the important site features and processes that need to be included in the conceptual model; (b) deciding among possible competing interpretations of the site data; and (c) determining the level of detail needed to describe those features and processes.

I.4.3   Recommended Approach

I.4.3.1   Screening

An acceptable dose assessment analysis need not incorporate all the physical, chemical, and biological processes at the site. The scope of the analysis, and accordingly the level of sophistication of the conceptual model, should be based on the overall objective of the analysis. A performance assessment conceptual model can be simple if it still provides satisfactory confidence in site performance. For an initial screening analysis, little may be known about the site from which to develop a conceptual model. Computer codes used for screening analyses are generally intended to provide a generic and conservative representation of processes and conditions expected for a wide array of sites. Accordingly, the generic conceptual model in such codes may not provide a close representation of conditions and processes at a specific site. Such a generic representation is still acceptable as long as it provides a conservative assessment of the performance of the site.

The DandD code has two default land-use scenarios; a building occupancy and a resident farmer scenario. The building occupancy scenario is intended to account for exposure to both fixed and removable residual radioactivity within a building. Exposure pathways included in the building occupancy scenario include external exposure to penetrating radiation, inhalation of resuspended surface residual radioactivity, and inadvertent ingestion of surface residual radioactivity. The resident farmer scenario is intended to account for exposure to residual radioactivity in soil. Exposure pathways included in the resident farmer scenario include: external exposure to penetrating radiation; inhalation exposure to resuspended soil; ingestion of soil; and ingestion of contaminated drinking water, plant products, animal products, and fish. The predefined conceptual models within DandD are geared at assessing releases of radioactivity, transport to, and exposure along, these pathways.

For the building occupancy scenario, DandD models external exposure to penetrating radiation as an infinite area source, using surface source dose rate factors from Federal Guidance Report No. 12 (EPA, 1993). Exposure to inhalation of resuspended surface residual radioactivity is modeled as a linear static relationship between surface residual radioactivity and airborne concentrations. The model accounts for ingrowth and decay. Exposure to incidental ingestion of surface residual radioactivity is modeled with a constant transfer rate.

The generic conceptual models for the resident farmer scenario are more complicated because of the large number of exposure pathways and considerations of release of radioactivity from the source area and transport of radionuclides in the environment. DandD models external exposure from volume soil sources when the person is outside as an infinite slab of residual radioactivity 15 cm (6 in) thick, using dose rate factors from Federal Guidance Report No. 12, for volume residual radioactivity. When the person is indoors, exposure from external radiation is modeled in a similar manner, except the exposure is assumed to be attenuated through the use of a shielding factor (note: the higher the shielding factor, the lower the assumed attenuation). Exposure through ingestion of contaminated animal and plant products is modeled simply through the use of transfer factors. Instantaneous equilibrium is assumed to occur between radionuclide concentration in the soil and the concentration in plants, and between animal feed and animal products.

The generic source-term conceptual model in DandD assumes a constant release rate of radionuclides into the water and air pathways. Release of radionuclides by water is assumed to be downward and a function of a constant infiltration rate, constant contaminant zone thickness, constant moisture content, and equilibrium adsorption. DandD assumes that there are no radioactive gas or vapor releases. Release of radioactive particulates is assumed to be upward, instantaneous, uniform, and a function of a constant particulate concentration in the air and the radioactivity within the soil. Radionuclides in the contaminant zone are assumed to be uniformly distributed in a single soil layer, 15 cm (6 in) thick. No transport is assumed to occur within the source zone, but radioactive decay is taken into account. In terms of containment, DandD assumes that there are no containers (or that they have failed), and that there is no cover over the contaminated zone.

The DandD generic conceptual model for the ground water pathway assumes a single hydrostratigraphic layer for each of the unsaturated and saturated zones. The unsaturated zone (vadose zone) can be broken into multiple layers within DandD; however, each layer is assumed to have the same properties. For radionuclides entering the vadose zone, DandD accounts for adsorption-limited leaching by considering the vadose zone to behave as a well-mixed chemical reactor with a constant water inlet and outlet rate set at the infiltration rate. Accordingly, it is assumed that the vertical saturated hydraulic conductivity of the unsaturated zone is greater than or equal to the infiltration rate (i.e., there is neither ponding nor runoff on the surface). The outlet concentration from one unsaturated zone layer to another is assumed to be a function of the constant infiltration rate, equilibrium partitioning, the thickness of the layer, a constant moisture content, and radioactive decay. Radionuclides entering the saturated zone are assumed to be instantaneously and uniformly distributed over a constant volume of water equivalent to the larger of either the volume of infiltrating water (i.e., the infiltration rate times the contaminated area) or the sum of the water assumed to be removed for domestic use and irrigation. Based on the default parameters in DandD, dilution in the ground water pathway is based on the water use.

No retardation is assumed to occur in the aquifer; however, radioactive decay is taken into account. A volume of contaminated water equivalent to the irrigation volume is assumed to be returned annually to the source zone. The concentration of radionuclides in the irrigation water is assumed to remain constant during the year. Radionuclides deposited on the vegetation are assumed to be removed at a constant rate. The DandD ground-water model should generally provide a conservative representation of the ground water system because it allows very little dilution and nominal attenuation.

The generic surface-water conceptual model in DandD assumes that radionuclides are uniformly mixed within a finite volume of water representing a pond. Radionuclides are assumed to enter the pond at the same time and concentration as they enter the ground water. Accordingly, there is assumed to be no transport of radionuclides through the ground water to the pond and thus no additional attenuation (besides the initial ground water dilution) is assumed for transport in the ground water. The surface-water model within DandD should provide a conservative dose estimate as long as a small volume is assumed for the surface-water pond. Because the parameters in DandD are selected to provide a conservative dose estimate, the generic conceptualization of the surface-water pathway should generally provide a conservative representation of transport of radionuclides through the surface-water pathway. Figure I.3 shows the generic ground-water and surface-water conceptual model within DandD.

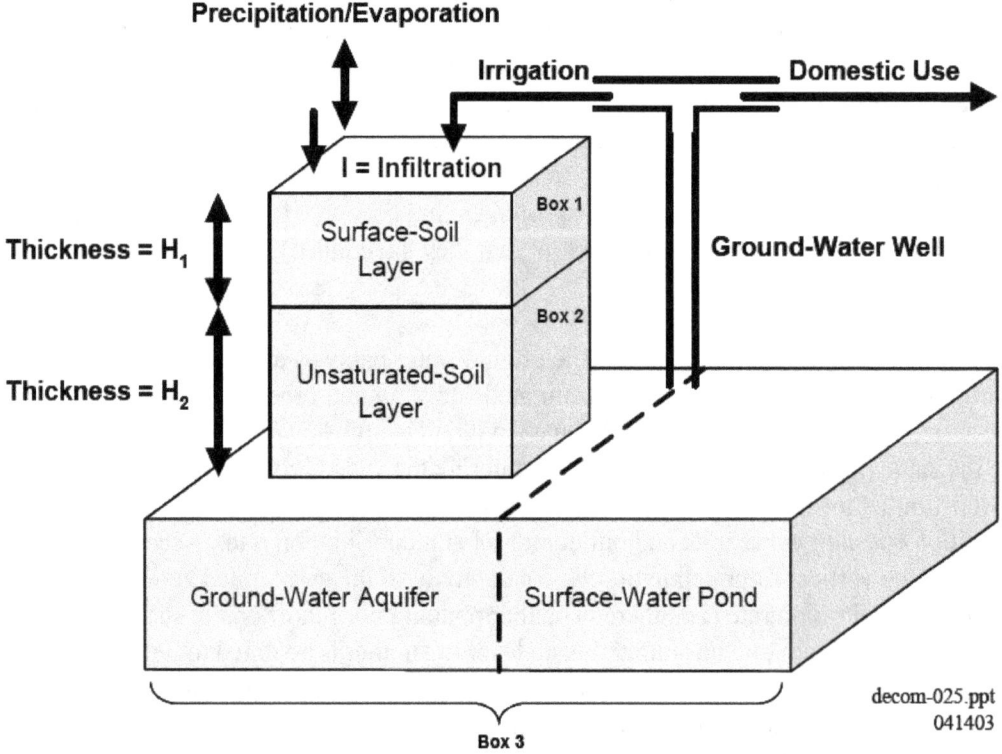

Figure I.3   DandD Conceptual Model of the Ground-Water and Surface-Water Systems (from NUREG/CR–5621).

The generic conceptual model of the air pathway in DandD assumes an equilibrium distribution between radionuclides in the air and soil. The concentration in air is assumed to be a function of the soil concentration and a constant dust loading in the air. Accordingly, all radionuclides in the air are assumed to be in a particulate form. The air pathway model within DandD is very simple and should generally allow a conservative dose estimate as long as a conservative particulate concentration is assumed. Because the default parameters in DandD are geared to be conservative, in general the air pathway in DandD should allow a conservative dose estimate.

In general, the conceptual models within DandD are expected to provide a conservative representation of site features and conditions. Therefore, for screening analyses, NRC staff should consider such generic conceptual models to be acceptable provided it is acceptable to assume that the initial radioactivity is contained in the top layer (building surface or soil) and the remainder of the unsaturated zone and ground water are initially free of residual radioactivity. In using DandD for site-specific analyses, it is important to ensure that a more realistic representation of the site that is consistent with what is known about the site would not lead to higher doses. Some site features and conditions that may be incompatible with the generic conceptual models within DandD are listed in Table I.4. The relative importance of the incompatibilities varies with the scenario and radionuclides involved. More information on the assumptions of the model available in the development documentation (e.g., the NUREG/CR–5512 series).

For any site where it is known that one or more of these conditions or features are present, the licensee should provide an appropriate rationale on why the use of the DandD should not result in an underestimation of potential doses at the specific site.

Table I.4 Site Features and Conditions that May be Incompatible with Those Assumed in DandD

- Sites with highly heterogeneous radioactivity

- Sites with wastes other than soils (e.g., slags and equipment)

- Sites that have multiple source areas

- Sites that have contaminated zones thicker than 15 cm (6 in.)

- Sites with chemicals or a chemical environment that could facilitate radionuclide releases (e.g., colloids)

- Sites with soils that have preferential flow conditions that could lead to enhanced infiltration

- Sites with a perched water table, surface ponding, or no unsaturated zone

- Sites where the ground water discharges to springs or surface seeps

- Sites with existing ground water contamination

- Sites where the potential ground water use is not expected to be located immediately below the contaminated zone

- Sites with significant transient flow conditions

- Sites with significant heterogeneity in subsurface properties

- Sites with fractured or karst formations

- Sites where the ground water dilution would be less than 2000 m^3 (70,000 ft^3)

- Sites where overland transport of contaminants is of potential concern

- Sites with radionuclides that may generate gases

- Sites with stacks or other features that could transport radionuclides to result in a higher concentration offsite than onsite

As example, it may be possible to demonstrate the acceptable use of DandD for analyzing sites that contain H-3 and C-14, although both radionuclides may be occur as a gas. The following approach can be used to demonstrate the acceptable use of DandD for analyzing sites that contain either H-3 or C-14 (NRC 1999a): (1) determine the area of the contaminated zone; (2) run DandD for the site with only H-3 or C-14; (3) read the associated activity ratio factor for the given area from Figure I.4; and (4) estimate the potential missed dose by multiplying the inhalation dose calculated from DandD by the activity ratio factor.

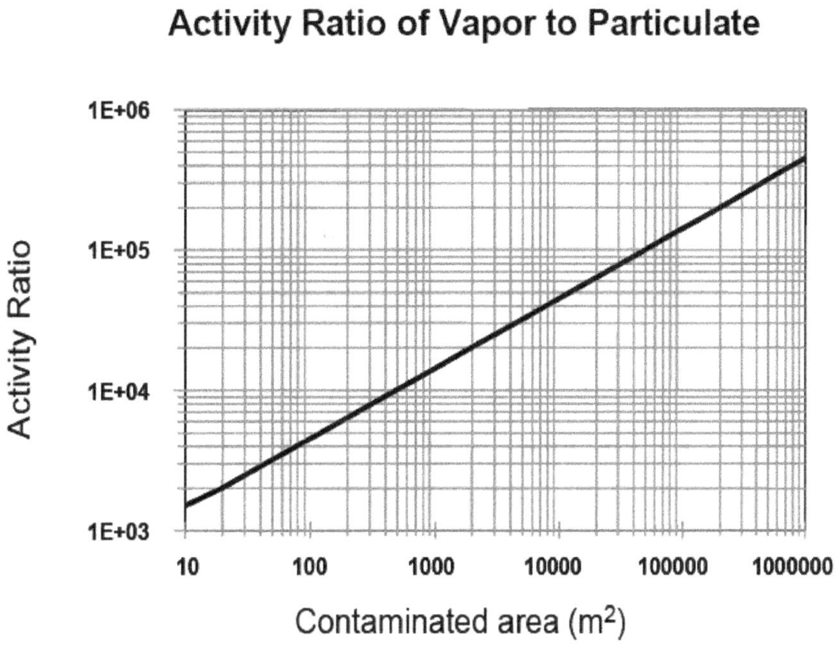

Activity Ratio of Vapor to Particulate

Figure I.4   Activity Ratio of Vapor to Particulate as a Function of Contaminated Area.

I.4.3.2   Site-Specific

For site-specific analyses, the intent is to provide a more realistic assessment of doses based on more site-specific information and/or data. Presumably for such analyses, more is known about the site from which to develop a conceptual model. For site-specific analyses, the licensee should provide a schematic or verbal description of the problem that it is attempting to analyze. Even when using a computer code that has a predefined conceptual model, it is important for the licensee to identify any site features or conditions that may differ from those assumed in the code. In developing a site-specific conceptual model or identifying potential limitations with a predefined conceptual model, the issues listed in Table I.5 should be considered.

Because conceptual models are developed based on limited data, in most cases more than one possible interpretation of the site can be justified based on the existing data. This uncertainty should be addressed by developing multiple alternative conceptual models and proceeding forward with the conceptual model(s) that provides the most conservative estimate of the dose and *yet is consistent with the available data*. Consideration of unrealistic and highly speculative conceptual models should be avoided. Consistent with the overall dose modeling framework of starting with simple analyses and progressing to more complex modeling, as warranted, it may be advisable for the analyst to begin with a simple, conservative analysis that incorporates the key site features and processes and progress to more complexity only as merited by site data. It is important to stress that a simple representation, of the site, in itself does not mean that the analysis is conservative. It is incumbent on the licensee to demonstrate that its simplification is justified, based on what is known about the site and the likelihood that alternative representations of the site would not lead to higher calculated doses.

Table I.5 Issues To Be Considered in Developing a Site-Specific Conceptual Model

• Whether a more realistic representation of the site would lead to higher doses
• Whether the conceptual model accounts for the most important physical, chemical, and biological processes at the site
• Whether the conceptual model adequately represents responses to changes in stresses
• Whether the conceptual model includes consistent and defensible assumptions

In general, there are two primary areas of the dose analysis where the conceptual model is expected to change from one site to another; these are related to the source term and environmental transport. Aspects of the analysis related to the exposure pathways in the biosphere and dosimetry are largely determined by the scenario and the assumed behavior of the critical group. Accordingly, models related to the exposure pathways in the biosphere and dosimetry should not change from one site to another unless there is a significant change in the scenario and associated critical group. The principal environmental transport pathways that should have to be considered in a dose assessment are ground water (including transport through the unsaturated zone), surface water, and air.

The conceptual model of the source area should describe the contaminants and how they are likely to be released into the environment. Specifically, it should describe key features and processes such as the infiltration of water into the source area, the geometry of the source zone, the distribution of contaminants, release mechanisms, the physical form of the contaminants, near-field transport processes, and containment failure. If the contaminants are assumed to be uniformly distributed, this is an important assumption that needs to be justified because in general contaminants may not be uniformly distributed (see discussion under Section I.2 of this appendix). The source description should clearly identify how the contaminants are assumed to be released from the media. Common release mechanisms are diffusion, dissolution, surface release, and gas generation. The source description should also identify key processes and features that may retain or limit the release of contaminants from the source area (e.g., solubility and sorption). In addition, the description of near-field transport should state assumptions made regarding the dimensionality. In general, the assumption of one-dimensional vertical flow should be appropriate, unless there is some type of barrier present that may hinder flow in the vertical direction. The description of the source term should also describe failure mechanisms for any containment (e.g., corrosion, concrete degradation, or cover degradation) if containers or other forms of containment are present.

The conceptual model of the ground water pathway should describe how contaminants could migrate through the unsaturated and saturated zones to potential receptors (e.g., a well, spring, or surface-water bodies). Essential features that should be included in the conceptual model include hydrostratigraphic units; information on the geometry of the pathway (i.e., boundaries and boundary conditions); the physical form of the contaminants (i.e., dissolved, suspended sediment, gas, etc.); structural features of the geology (i.e., those that influence contaminant transport such as fractures, faults, and intrusions); and physical and chemical properties. Important processes that should be characterized include the dimensions and state conditions (e.g., steady-state) of flow; dimensions and state conditions of transport (e.g., dispersion); chemical and mass transfer processes (e.g., sorption, precipitation, complexation); and transformation processes (e.g., radioactive ingrowth and decay). Although contaminant migration through both the unsaturated and saturated zones is best represented in three dimensions, it may be appropriate to assume only one or two dimensions, if this provides a more conservative representation of contaminant migration, and/or if it can be demonstrated that migration in one or more other directions is not expected to result in exposure to potential receptors.

The conceptual model of the surface-water pathway should describe potential contaminant migration through surface-water bodies such as lakes, streams, channels, or ponds to potential receptors. Essential features that should be included in the conceptual model include: the geometry of the surface-water body (i.e., boundaries and boundary conditions); the physical form of the contaminants (e.g., dissolved or solid); and physical and chemical properties. Key processes that should be described include: the dimensions and state conditions of flow and transport; chemical and mass transfer processes (e.g., sorption, precipitation, volatization); and transformation. One key boundary condition that should be described is how the contaminants are expected to initially mix or interact with the surface water.

The conceptual model of the air pathway should describe potential contaminant migration through the air to potential receptors. Essential features that should be included in the conceptual model are similar to those for the other environmental pathways—namely, the geometry (i.e., boundaries and boundary conditions); form of contaminants (e.g., particulates or gases); and physical and chemical properties. Key processes that should be described include the dimensions and state conditions of flow and transport, and transformation processes.

I.4.3.2.1   Site-Specific Computer Codes

Two common computer codes used for site-specific analyses are RESRAD and RESRAD–BUILD. Both these computer codes have predefined conceptual models. Therefore, in using these codes, it is important for the licensee to demonstrate that key site features and conditions are consistent with the modeling assumptions within the codes or, where they are not consistent, the analysis may not result in an underestimation of potential doses.

APPENDIX I

I.4.3.2.1.1 RESRAD–BUILD

The RESRAD–BUILD code can be used to evaluate doses for the building occupancy scenario.

It considers exposure from external radiation at the source and air submersion, inhalation of airborne material, and inadvertent ingestion of radioactive material. Exposure to direct radiation at the source is calculated using surface source dose rate factors from Federal Guidance Report No. 12. RESRAD–BUILD incorporates correction factors to account for a finite area source, for any offset of the receptor from the axis of the disk of residual radioactivity, and for shielding by material covering the residual radioactivity. Exposure to external radiation from air submersion is calculated as an infinite cloud of material using dose rate conversion factors for an infinite cloud. RESRAD–BUILD models airborne concentration of radionuclides using a dynamic model that accounts for the kinetic introduction and removal of radioactive material to and from indoor air. Exposure to incidental ingestion of radioactive material is modeled using a constant transfer rate.

I.4.3.2.1.2 RESRAD

RESRAD can be used for analyzing the resident farmer scenario. As with the generic conceptual models used by DandD for analyzing the resident farmer scenario, the conceptual models in RESRAD (see Figure I.5) are more complex than those in RESRAD–BUILD. RESRAD models external exposure from volume soil sources when the person is outside, using volume dose rate factors from Federal Guidance Report No. 12. Correction factors are used to account for soil density, areal extent of residual radioactivity, thickness of residual radioactivity, and cover attenuation. When the person is indoors, exposure from external radiation is modeled in a similar manner except that additional attenuation is included to account for the building. Exposure through ingestion of contaminated animal and plant products is modeled simply through the use of transfer factors.

The generic source-term conceptual model in RESRAD assumes a time-varying release rate of radionuclides into the water and air pathways. Radionuclides in the contaminant zone are assumed to be uniformly distributed. No transport is assumed to occur within the source zone, but radioactive decay is accounted for. In terms of containment, the radioactive material is not assumed to be contained (or containers are assumed to have failed). RESRAD does allow inclusion of a cover over the contaminated area. However, the cover is not assumed to limit infiltration of water, and is assumed to function only in terms of providing shielding from gamma radiation. Release of radionuclides by water is assumed to be a function of a constant infiltration rate, time-varying contaminant zone thickness, constant moisture content, and equilibrium adsorption. The contaminant zone is assumed to decrease over time from a constant erosion rate. RESRAD assumes a uniform release of tritium and C-14 gases, based on a constant evasion loss rate. Particulates are assumed to be instantaneously and uniformly released into the air as a function of the concentration of particulates in the air, based on a constant mass loading rate.

NUREG-1757, Vol. 2, Rev. 1   I-42

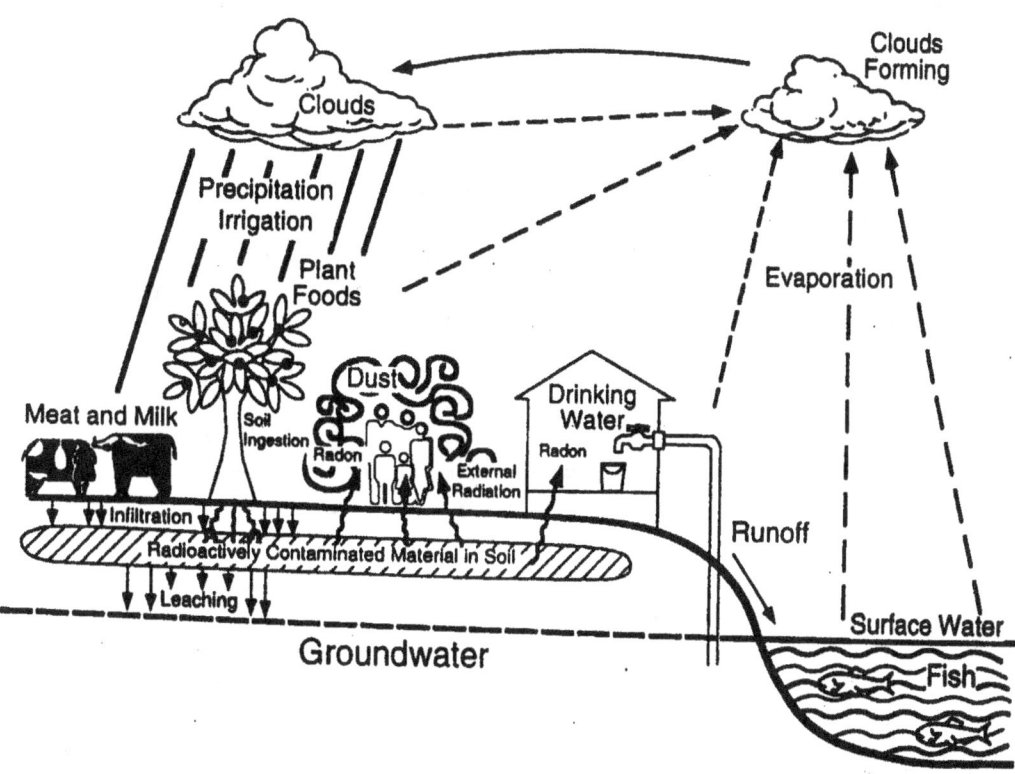

Figure I.5   Conceptualization Modeled by RESRAD (from ANL/EAD/LD–2).

The RESRAD generic conceptual ground-water model assumes one or more horizontal homogeneous strata for the unsaturated zone. Transport in the unsaturated zone is assumed to result from steady-state, constant vertical flow, with equilibrium adsorption, and decay, but no dispersion. RESRAD has two different ways of modeling radionuclides once they reach the saturated zone. In the mass-balance approach, radionuclides entering the saturated zone are assumed to be instantaneously and uniformly distributed over a constant volume equivalent to the volume of water removed by the hypothetical well (as long as the pumping rate is larger than the rate of leachate entering the ground water—if not, no dilution is assumed to occur in the ground water). For the mass-balance approach, radionuclides are assumed to enter a well pumping immediately beneath the contamination zone. The mass-balance approach is very similar to the ground-water modeling approach in DandD. In the nondispersion approach, transport in the saturated zone is assumed to occur in a single homogeneous stratum, under steady-state, unidirectional flow, with a constant velocity, equilibrium adsorption, and decay. It assumes no dispersion; however, radionuclides are assumed to be diluted by clean water as a function of the assumed capture zone of the hypothetical well, in relation to the width of residual radioactivity and the depth of residual radioactivity, in relation to the depth of the hypothetical well. Radioactive decay and equilibrium adsorption are assumed to occur for the nondispersion approach. Radionuclides are assumed to enter a well located at the immediate downgradient edge of the contamination zone. For the nondispersion model, the calculated width of the effective pumping zone could be a factor of 2 larger than what one would predict from a steady-state capture zone analysis; this could lead to a slight overestimation in the amount of dilution (NRC 1999a).

In determining which of these two conceptual models to use, consideration should be given to where the hypothetical well may be located (i.e., either at the center of the residual radioactivity or at the edge of the residual radioactivity); the relative half-life of the radioactivity; and the potential capture zone of the hypothetical well. Use of the nondispersion model will generally result in lower estimated doses. Both models assume that radionuclides enter the well as soon as they reach the water table. However, the nondispersion model, unlike the mass-balance model, calculates the time it takes for the peak concentration to occur after the initial breakthrough. Accordingly, the nondispersion model accounts for radioactive decay during the interval between the initial breakthrough and arrival of the peak concentration. Generally, the amount of decay should be small unless the radionuclides have short half-lives and are retarded. In addition, unlike with the mass-balance model, for the nondispersion model no assumption is made that all radionuclides released from the contaminated zone are withdrawn through the well. Therefore, the nondispersion model may include dilution. The only way that dilution is not considered is if the expected capture zone of the hypothetical well is small in relation to the width and thickness of the residual radioactivity. Because the nondispersion model will generally give a lower estimated dose than the mass-balance model, it is important for the licensee to justify the use of this model for the specific analysis. Use of the mass-balance approach should always be acceptable. In Equations I–1 and I–2, use of the nondispersion model should be acceptable, without additional justification, for modeling long-lived radionuclides (i.e., where radioactive decay is not important) when either one of the following conditions are met:

$$\frac{U_w}{v \bullet d_w} > \frac{A}{len} \quad \text{and}$$

$$\left(\frac{I}{v}\right) len < d_w \tag{I–1}$$

or

$$\frac{U_w}{v \cdot d_w} \leq \frac{A}{len} \quad \text{and}$$

$$\left(\frac{I}{v}\right) len \geq d_w \tag{I–2}$$

where U_w =   pumpage rate from the well (m^3/y);
      v =   ground-water darcy velocity (m/y);
      A =   area of residual radioactivity (m^2);
      d_W =   depth of well intake below water table (m);
      len =   length of residual radioactivity parallel to ground water flow (m); and
      I =   infiltration rate (m/y).

As a general rule, use of the nondispersion approach should be acceptable when the area of residual radioactivity is known to be larger than the assumed capture area of the hypothetical well. Assuming an essentially flat water table and steady-state conditions, the capture area of the hypothetical well can be calculated in Equation I–3 as follows:

$$A_w = \left(\frac{U_w}{I} \right) \qquad \textbf{(I–3)}$$

where A_W =   area of well capture (m^2);
      U_W =   pumpage rate from the well (m^3/y); and
      I =   infiltration rate (m/y).

The generic conceptual model of the surface-water pathway in RESRAD assumes that radionuclides are uniformly distributed in a finite volume of water within a watershed. For example, the default watershed area in RESRAD Version 5.91 is 1x10^6 m^2 (250 ac). Radionuclides are assumed to enter the watershed at the same time and concentration as in the ground water. Accordingly, no additional attenuation is considered as radionuclides are transported to the watershed. In the surface water, radionuclides are assumed to be diluted as a function of the size of the contaminated area in relation to the size of the watershed. The RESRAD surface-water conceptual model assumes that all radionuclides reaching the surface water are derived from the ground water pathway. Thus, transport of radionuclides overland from runoff is not considered. In addition, additional dilution from overland runoff is not considered.

The generic conceptual model of the air pathway in RESRAD uses a constant mass loading factor and area factor to model radionuclide transport. The area factor, which is used to estimate the amount of dilution, relates the concentration of radionuclides from a finite area source to the concentration of radionuclides from an infinite area source. It is calculated as a function of particle diameter, wind speed, and the side length of a square-area source. The conceptual model assumes a fixed particle density, constant annual rainfall rate, and constant atmospheric stability. No radioactive decay is considered. See Chang, et al., (1998) for more detail. Tritium and C-14 gases are assumed to be uniformly mixed in a constant volume of air above the contaminated zone. RESRAD does not model the transport of tritium and C-14 as particulates in the air.

I.4.3.2.2 Limitations of Site-Specific Computer Codes

In general, the conceptual models within RESRAD and RESRAD–BUILD are expected to provide an acceptable generic representation of site features and conditions. Some specific site features and conditions that may be incompatible with this generic representation are listed in Table I.6. At any site where it is known that one or more of these conditions or features are present, the licensee should provide appropriate justification for use of the computer code.

Table I.6 Site Features and Conditions that May be Incompatible with the Assumptions in RESRAD

- Sites with highly heterogeneous radioactivity

- Sites with wastes other than soils (e.g., slags and equipment)

- Sites with multiple source areas

- Sites that have chemicals or a chemical environment that could facilitate radionuclide releases

- Sites with soils that have preferential flow conditions that could lead to enhanced infiltration

- Sites where the ground water discharges to springs or surface seeps

- Sites where the potential ground water use is not expected to be located in the immediate vicinity of the contaminated zone

- Sites with significant transient flow conditions

- Sites with significant heterogeneity in subsurface properties

- Sites with fractured or karst formations

- Sites where overland transport of contaminants is of potential concern

- Sites with stacks or other features that could transport radionuclides off the site at a higher concentration than onsite

---

I.4.4   Generic Examples

I.4.4.1   Screening

A hypothetical research and development (R&D) facility is authorized to use radiological chemicals through an NRC license. Because the R&D facility plans to discontinue its use of radioactive material, it wants to decommission the facility and terminate its license. A historical site assessment (HSA) reveals that use of radioactive material were limited to a single building within the facility. The floor area of the facility is estimated to be 560 m² (6000 ft²). The wall area is 430 m² (4600 ft²). In addition, an outside area of roughly 930 m² (10,000 ft²) was used for dry storage of chemicals. A preliminary characterization program has determined that approximately 10 % of the building floor area and 5 % of the wall area are contaminated with Cs-137 and Co-60. Surficial soils covering an area of approximately 2500 m² (27,000 ft²) are contaminated from windblown dust and runoff from spills in the storage area. The soils are also contaminated with Cs-137 and Co-60.

The licensee proposes to use a screening analysis, using DandD, to demonstrate compliance with the LTR. A building occupancy scenario is assumed for the building and a residential farmer scenario is assumed for the contaminated soils. Based on what is known about the site, the licensee certifies that the use of the generic conceptual models within DandD is appropriate for the analysis.

I.4.4.2   Site-Specific

A hypothetical manufacturing facility has a former radioactive waste burial area that may be decommissioned for unrestricted release. Radioactively contaminated trash was previously buried in 0.2 cubic meter (55-gallon) drums, in trenches covering an area of roughly 2000 m² (22,000 ft²). The trenches, which are roughly 0.9 m (3 ft) deep are covered with 1.2 m (4 ft) of native soil. A review of site operating records show that the radionuclides of concern are natural uranium, enriched uranium, and natural thorium.

Based on information from the local county agricultural extension office and published reports, the geology and hydrogeology at the site are described as follows. This description shows that none of the site features or conditions in Table I.6 are present at this site.

> "The surface geology at the site contains 14 to 27 m (46 to 89 ft) of till consisting primarily of fine, silty sand to sandy silt with narrow, discontinuous sand lenses. Sandstone bedrock underlies the unconsolidated till. A shallow unconfined aquifer occurs in the unconsolidated till. The average depth to the water table ranges between three to four meters below the land surface. The mean horizontal hydraulic conductivity is roughly 60 m/y (197 ft/y). The average vertical hydraulic conductivity of the till is estimated to be an order of magnitude less. The hydraulic gradient is estimated to range between 0.006 to 0.021. The mean precipitation at the site is roughly 0.8 m/y (30 in/y). The site is located in the reach of a surface water drainage basin that has a drainage area of approximately 500,000 m² (5.4 million ft²)."

A residential farmer scenario is assumed as a reasonable future land use. The licensee proposes to use the RESRAD computer code for the dose analysis. Because the contaminated media is trash, an assumption is made that the trash degrades and becomes indistinguishable from soil. In addition, the metal drums are assumed to have degraded away. Given the relative short lifespan for metal drums and the long half-life of the radionuclides, this should be a reasonable assumption. The cover is also assumed to be breached through the construction of a basement for the house. The contaminated soil is assumed to be uniformly mixed with the excavated cover. Because the trash is assumed to be indistinguishable from soil, it is also assumed that once the cover is breached the future hypothetical farmer may not recognize the contaminated material as contaminated. The licensee also assumes that the hypothetical future well is located at the center of the residual radioactivity because of limited bases for assuming otherwise.

The licensee determines that the other aspects of conceptual models within RESRAD are acceptable for analyzing the problem.

I.5   Criteria for Selecting Computer Codes/Models

I.5.1   Introduction

Dose assessment commonly involves the execution of numerical model(s) that mathematically represent the conceptual model of the contaminated site. The numerical models used to implement the mathematical equations are usually linked via the conceptual model and codified in a software package known as "the code." The words "code" and "model" are frequently used to express the software package, including the embedded numerical models or the specific models contained in the code. For example, "DandD code" may refer to the software package, including the associated exposure models (e.g., the water-use model, food-ingestion pathway model, inhalation-exposure model, etc.) embedded in the code. The "DandD model" may also refer to DandD software, the DandD conceptual model, or to any of the numerical models, or the group of models used in the code (e.g., DandD ground water model). Within the context of this volume, the word "code" will refer to the software package and the associated numerical models. However, the word "model" will refer to the mathematical representation of the conceptual model, including representation of the specific exposure scenario and pathways. This section describes the process and criteria used in selection of codes and models for the dose assessment.

The codes and models used in the dose assessment can be either generic screening codes/models or site-specific codes/models. Regardless of the intent of the use of the code/model (e.g., for screening or site-specific analysis), NRC staff should ensure that the dose assessment codes/models and the associated databases are properly documented and verified in accordance with a rigorous QA/QC criteria which is acceptable to NRC staff. Currently, the only acceptable generic screening code is DandD Version 2. If site-specific models/codes are used, a justification of the conceptual model should be provided (see Section I.4.3.2 of this appendix). NRC staff should also review the source-term model(s), the transport models, the exposure models, and the overall dose models. NRC staff should assess the QA/QC documentation and the level of conservatism of any alternate code/model.

This section describes the generic issues associated with the selection of the screening and site-specific codes/models that NRC staff may encounter, and recommends approaches and criteria, for NRC staff acceptance of the codes/models. In addition, this section presents as generic description of the two common dose assessment codes, DandD Screen and RESRAD/RESRAD–BUILD. NRC staff developed or modified these codes. In addition, these codes have been used by NRC staff and licensees for demonstrating compliance with the dose criteria in Subpart E.

I.5.2   Issues in Selection of Computer Codes/Models

The major issues associated with the selection of computer codes/models include:

1. **Generic criteria for the selection of computer codes/models**: This issue pertains to NRC staff's review criteria of code aspects related to QA/QC requirements, specifications, testing, verification, documentation, interfacing, and other features related to uncertainty treatment approaches.

2. **Acceptance criteria for selection of site-specific codes/models**: This issue pertains to NRC staff's review of additional specific requirements for the justification of the use of the conceptual model, the numerical mathematical models, the source-term model and its abstraction, and the transport and exposure pathway models.

3. **Options for selection of deterministic or probabilistic site-specific codes**: This issue pertains to NRC staff's review of the justification to support the decision to use either of these two approaches.

A generic description of the DandD Version 2 is presented below to familiarize users with this code. Further, the rationale for development of DandD Version 2 and the issue of excessive conservatism in DandD Version 1 are also addressed. A description of the inherent excessive conservatism in DandD model and approaches to minimize such excessive conservatism, using DandD Version 2, site-specific input data, or use of other models/codes is included.

For site-specific analysis, NRC staff should accept any model or code that meets the criteria described below in "Generic Criteria for Selection of Codes/Models." However, NRC staff is expected to conduct a more detailed and thorough review of less common codes/models (e.g., codes other than DandD, and RESRAD), specifically those developed by licensees. NRC sponsored development of the probabilistic RESRAD (Version 6) and RESRAD–BUILD (Version 3) codes for site-specific analysis. These have already been reviewed for QA/QC and are acceptable.

Selection of appropriate models/codes for complex sites may also present challenges. For example, sites with multiple source terms, with significant ground water/surface water contamination, or sites with existing offsite releases, may require more advanced codes/models than common codes such as DandD or RESRAD. Complex sites may also include sites with engineered barrier(s), or with complex geological conditions like highly fractured geologic formations. Because of site complexity and variability, there is no standard dose analysis review criteria for these sites.

I.5.3   Recommended Approach

I.5.3.1   Generic Criteria for Selection of Codes/Models

The generic criteria under this subsection pertain to NRC staff review of codes/models other than commonly used codes, specifically, those developed or modified by NRC staff (i.e., other than DandD and RESRAD/RESRAD–BUILD). NRC staff should use the generic criteria when the codes/models have no readily available documentation of testing, verification, and QA/QC review. In this context, NRC staff should use the following generic criteria in reviewing the codes/models selected for the dose assessment:

- NRC staff should review the adequacy and completeness of the database available regarding QA/QC aspects of the code/model. The QA/QC database should be comparable to NRC's QA/QC requirements [NUREG/BR–0167 (Douglas 1993) and NUREG–0856 (NRC 1983)]. The QA/QC should include information regarding mathematical formulation, code/model assumptions, consistency of the pathways with the assumed conceptual model(s) used in the code, and accuracy of the software to reflect the model's mathematical formulation and correct representation of the process or system for which it is intended.

- NRC staff should ensure that the software used for the code are in conformance with the recommendations of IEEE Standard 830–1984, IEEE Guide for Software Requirement Specifications.

- NRC staff should review the adequacy and appropriateness of the code/model documentation with regard to: (a) software requirements and intended use; (b) software design and development; (c) software design verification; (d) software installation and testing; (e) configuration control; (f) software problems and resolution; and (g) software validation.

- For uncommon codes/models, NRC staff should review code data including: (a) a software summary form; (b) a software problem/change form; (c) a software release notice form; and (d) a code/model user's manual, which covers code technical description, software source code, functional requirements, and external interface requirements (e.g., user interface, hardware interface, software interface, and communication interface), if necessary.

- NRC staff should review the conceptual model of the selected code to ensure compatibility with the specific site conceptual model, including the pathways and the exposure scenario. The source-term assumptions of the selected code should also be compatible with site-specific source term. NRC staff may accommodate minor modifications in the source-term conceptual model, as long as the basic model assumptions are not violated.

- NRC staff should review the selected code to verify that the exposure scenario of the selected code is compatible with the intended scenario for the site. For example, models/codes designed for the onsite exposure scenario may not be appropriate for assessment of an offsite receptor scenario or a scenario to estimate an offsite collective public dose.

- NRC staff should review the selected model/code formulation to account for radionuclide decay and progenies. The code should have proper and timely formulation, as well as linkages of decay products with the receptor location and the transport pathways, via corresponding environmental media;

- NRC staff should examine documentation of the selected code/model performance; specifically, test and evaluation, as well as code comparison with commonly used (accepted) codes and models (e.g., DandD and RESRAD codes). NRC staff should also review documentation on code/model verification, if available, to support decisions for code acceptance.

- NRC staff should review code/model features regarding sensitivity/uncertainty analysis to account for variability in selection of input parameters and uncertainty in the conceptual model and multiple options for interpretation of the system.

I.5.3.2   Acceptance Criteria for Selection of Site-Specific Codes/Models

This issue involves NRC staff's review of additional requirements supporting the justification for using the conceptual model, the numerical mathematical models, the source-term model and its abstraction, and the transport- and exposure-pathway models.

CONCEPTUAL MODELS

NRC staff review should compare the conceptual model for the site with the conceptual model(s) in the selected code, to ensure compatibility with site-specific physical conditions and pathway assumptions for the critical group receptor.

NUMERICAL MATHEMATICAL MODELS

NRC staff should review the equations used in the code to implement the conceptual model and the numerical links between mathematical models to ensure correctness and consistency. For codes developed or modified by NRC staff (e.g., DandD, RESRAD & RESRAD–BUILD), NRC staff review would be minimal because these codes were revised by NRC staff and examined early for consistency with NRC's QA/QC requirements. For less commonly used codes, or codes developed locally by user(s), NRC staff should verify the numerical mathematical models, including the numerical links between these models. In this context, NRC staff may examine, if necessary, each mathematical model used for the specific transport-exposure pathway, to ensure that the code is designed for its intended use.

SOURCE-TERM MODELS

NRC staff should review the source-term model(s) used for the specific site. In this context, NRC staff review should include the following source-term aspects:

- **Building Occupancy Scenario Source Term**: NRC staff should review the HSA and other relevant data regarding extent of the source term and its depth [e.g., within 1 to 10 mm (0.04 to 0.39 in) deep into the building surface or more]. Based on this review, NRC staff should identify the source term as surficial or volumetric source. In addition, NRC staff should examine assumptions made for the loose/fixed fractions of the source. Sources of residual radioactivity on surfaces that are not integral parts of the building (e.g., equipment, pipes, and sewer lines) should be addressed separately, because the applicable model and exposure scenario could be different. Therefore, source-term model assumptions for such surfaces should be reviewed on a case-by-case basis.

  NRC staff should also review the source term regarding radionuclide mixture and if a constant ratio is assumed in the dose analysis. NRC staff should determine if surrogate radionuclides are used in the source-term model assumption. The latter two situations may require additional NRC staff verification of the source-term model and review of consistency with the intended final survey methodology.

  NRC staff should also review the use of multiple sources (e.g., multiple rooms). Certain codes may use advanced source-term assumptions, such as two to three rooms, with multiple-story buildings. The source term under these conditions allows for source depletion due open air circulation and common ventilation. For example, the RESRAD–BUILD code model uses two- or three-room models with two- or three-story buildings, allowing for air exchange within the rooms, and source-term depletion. The indoor air-quality model (e.g., building ventilation and infiltration), and the indoor air-concentration model, as well as the adaptation of the air-quality model in RESRAD–BUILD code should be reviewed, to ensure consistency with the site-specific condition. Input parameters associated with these models should be verified. NRC staff may accept such site-specific source-term models after an assessment of the compatibility of the source-term model with the conceptual model of the site. NRC staff should also review the physical parameters defining the source term, to ensure consistency with site-specific conditions, and the occupancy parameters, to ensure consistency with the exposure scenario.

- **Resident Farmer Scenario Source Term**: NRC staff should examine the source-term information to identify the source as surficial or volumetric, to ensure consistency with the model in the selected code. NRC staff should also review the vertical and horizontal extent of residual radioactivity, to verify the model assumed for the contaminated zone (CZ), and to determine if there is subsurface and/or ground water contamination at the site. For surficial source terms, DandD model and other codes like RESRAD (assuming appropriate thickness) may be used. For volumetric sources, DandD cannot be used directly before simulation of the volumetric source into a surficial source. The source-term model should also be reviewed, to examine the contaminated area and its shape, to check for possible correction for the area and/or for geometry of the source. NRC staff should also determine if a cover or a barrier is assumed at the top of the CZ, and the justification for such an assumption. The cover and/or barrier issue should be examined within the context of the institutional control assumptions, if appropriate, and the physical performance of the cover or the barrier within the compliance period (e.g., 1000 years).

NRC staff should also review the physical and chemical form of the source term to evaluate the soil leaching model assumption and the two components, sorbed mass and leached mass of the source. This review should help assess the source mass-balance model and the transport model within the concerned environmental media. In addition, NRC staff review of these source-term aspects would help establish consistencies for the selection of relevant parameters. NRC staff should also review source-term horizontal distribution and homogeneity, and variation of source concentration with depth. NRC staff should use either an upper-bounding value for modeling the thickness or an area-weighted approach to calculate the representative thickness. In certain cases, NRC staff may evaluate the need for modeling of multiple sources and the need for more advanced subsurface source-term modeling.

TRANSPORT MODELS

The transport models simulate transport mechanisms of contaminants from the source to the receptor. NRC staff should review transport models for consistency and compatibility with respect to (a) the source term; (b) the exposure scenario defined for the critical group receptor; and (c) the simplified conceptual model, which describes site-specific physical conditions. The transport models may include diffusive and advective transport of contaminants via air, surface water, and ground water. The transport models can be overly simplified, using simple conservative assumptions such that minimal characterization data would be required to execute the model(s). Transport models can also be very complex, requiring advanced mathematical derivation and extensive site-specific, or surrogate, data about the site.

For the building occupancy scenario, the associated transport models (e.g., transport models for ingestion, inhalation, and direct exposure pathways) of DandD code are simple and conservative. For example, the ingestion pathway depends on the effective transfer rate of the removable surface residual radioactivity from surfaces to hands and from hands to mouth. The inhalation transport model depends largely on mechanical disturbance of the contaminated surface, resuspension of residual radioactivity in the air, and subsequent breathing of contaminated air. The external dose formulation assumes exposure from a nonuniform source of residual radioactivity on the walls, ceiling, and floor of a room. This model was found to be comparable to the infinite plane source for the building occupancy scenario (NRC 1992).

For the resident farmer scenario, the associated DandD transport models include models of contaminants transport to ground water, to surface water (e.g., three-box model that relies on transfer of contaminate through leaching), and to air (e.g., through dust mass loading and indoor resuspension). Transport models of contaminants via the air include dust loading, resuspension of contaminated soil, and use of mass loading factor for deposition. Transfer of contaminants from the soil/water to plants, fish, animals, and animal products are calculated using a water-use model, along with transfer factors, translocation factors, and bio-accumulation factors. For carbon and tritium, separate models were used, as described in NUREG/CR–5512, Volumes 1, 2, and 3. The RESRAD model assumes a volumetric source, with an idealized cylindrical shape of the contaminated zone, and allows for a cover at the top of the contaminated zone, if appropriate.

In general, NRC staff should conduct a review of the selected code, with respect to transport models and appropriateness of such models with respect to the site-specific conditions (e.g., area, source, unsaturated zone, and aquifer conditions). In addition, NRC staff should review, for compatibility and consistency, the transport model assumptions and the generic formulation pertaining to the applicable pathways of the critical group exposure scenario. The extent of transport model review depends on the familiarity of NRC staff with these models. Because certain codes/models were commonly used and were developed or modified by NRC staff (e.g., DandD, RESRAD, and RESRAD–BUILD), NRC staff is more familiar with such common codes. Therefore, NRC staff review of these common codes/models, would be less than NRC staff review of a less common codes/model developed by users or other parties. NRC staff review should also include updated new models or code versions and studies regarding code/models testing, comparison, and verifications.

RESRAD–BUILD is a more advanced code than DandD, because it employs multiple sources and more advanced particulate air transport models. In other words, each contaminated location may be considered a distinct source. Depending on its geometric appearance, the source can be defined either as a volume, area, or as a point source. RESRAD–BUILD depends on erosion of the source and transport of part of its mass into the indoor air environment, resulting in airborne residual radioactivity. The RESRAD–BUILD model differs from DandD because it assumes air exchange among all compartments of the building. In other words, the model assumes that the airborne particulates are being loaded into the indoor air of the compartment and then transported to the indoor air of all compartments of the building. In addition to air exchange between compartments, the indoor air model also simulates air exchange between compartments and the outdoor air. Descriptions of models pertaining to indoor air quality, air particulate deposition, inhalation of airborne dust, and ingestion of removable materials and deposited dust, were documented in Argonne National Laboratory report "ANL/EAD/LD–3," (ANL 1994). The exposure pathways in the RESRAD–BUILD code include (a) the external exposure to radiation emitted directly from the source and from radioactive particulates deposited on the floors, and exposure caused by submersion from radioactive particulates; (b) inhalation of airborne radioactive particulates; and (c) ingestion of contaminated material directly from the source, and airborne particulates deposited onto the surface of the building.

EXPOSURE PATHWAY MODELS

The exposure pathway models pertain to the formulation of the links between the radiological source, the transport of contaminants within environmental media, the critical group receptor location, and behaviors of the receptor that lead to its exposure to residual radioactivity through direct exposure, inhalation, and ingestion of contaminated water, soil, plants, crops, fish, meat, milk, and other dairy products. NRC staff should review the conceptual model(s) that describe the human behaviors that lead, or control, the amount of receptor exposure. Therefore, the occupational, behavioral, and metabolic parameters describing these models should be reviewed and compared with the default model scenarios and associated parameters. NRC staff should review exposure model(s) and associated parameters to ensure conservatism, consistency, and comparability with site-specific conditions and scenario assumptions. NUREG/CR–5512, Volumes 1, 2, and 3 provide detailed information regarding default parameters and approaches for changing parameters in dose modeling analysis.

INTERNAL AND DIRECT EXPOSURE DOSE CONVERSION FACTORS

In general, NRC staff should review the dose conversion factors for inhalation and ingestion, to ensure that the factors used are those developed by EPA, published in Federal Guidance Report No. 11 (EPA, 1988). Similarly, NRC staff review should ensure that EPA's external dose factors, although they may correct for actual area, published in Federal Guidance Report No. 12 (EPA, 1993) were used or another appropriate code such as Microshield. These dose factors were selected to ensure consistency of the dosimetry models used in deriving these factors with NRC's regulations in Part 20. The default parameter sets for DandD and the RESRAD family of codes are based on Federal Guidance Reports No. 11 and No. 12.

Licensees may request an exemption from Part 20 to use the latest dose conversion factors (e.g., ICRP 72). Scenarios and critical group assumptions should be revisited to look at age-based considerations. Licensees may not "pick and choose" dosimetry methods for radionuclides (e.g., Federal Guidance Report No.11 for six radionuclides and current International dose conversion factors for three radionuclides).

I.5.3.3   Option for Selection of Deterministic or Probabilistic Site-Specific Codes

Licensees may select either a deterministic analysis approach or a probabilistic approach for demonstrating compliance with the dose criteria in 10 CFR Part 20, Subpart E. A deterministic analysis uses single parameter values for every variable in the code. By contrast, a probabilistic approach assigns parameter ranges to certain variables, and the code randomly selects the values for each variable from the parameter range each time it calculates the dose. While a deterministic analysis will calculate the results from a single solution of the equations each time the user runs the code, a probabilistic analysis will calculate hundreds of solutions to the equations, using different values for the parameters from the parameter ranges. The deterministic analysis gives no indication of the sensitivity to certain parameters or of the uncertainty in the value of the parameters. Therefore, the deterministic approach may require more elaborate justification of code input parameter values and may require further analysis of doses using upper or lower bounding conditions.

NRC–approved data sets for both DandD and RESRAD are for the probabilistic calculation and not the deterministic mode.

Section I.7.3.2 of this appendix provides a detailed description of a NRC staff review for deterministic and probabilistic analysis.

I.5.3.4   Modeling of Subsurface Source-Term Residual Radioactivity

For subsurface residual radioactivity (residual radioactivity at depths >15–30 cm (6–12 in)), NRC staff should review existing historical site data (including previous processes or practices) and site characterization data, to establish an adequate conceptual model of the subsurface source, specifically the horizontal and vertical extent of residual radioactivity. Section I.2.3.1 describes approaches for subsurface source-term abstraction for dose modeling analysis.

I.5.3.5   Generic Description and Development of DandD

Two scenarios are implemented in DandD, the building occupancy and the residential scenario. The building occupancy scenario relates volume and surface residual radioactivity levels in existing buildings (presumably released after decommissioning for unrestricted commercial or light-industrial use) to estimates of the TEDE received during a year of exposure, with the conditions defined in the scenario. The exposure pathways for this scenario include external exposure, inhalation exposure, and secondary ingestion pathways.

The more complex and generalized residential scenario is meant to address sites with residual radioactivity in soils and ground water. The exposure pathways include external exposure, inhalation, and ingestion of contaminated crops, meat, soil, plants, fish, and drinking water (NRC 1992). A generic water-use model was developed to permit the evaluation of the annual TEDE from drinking water from wells and from multiple pathways associated with contaminated soil. Section I.4.3.1 describes the three-box water-use model of the DandD code.

I.5.3.5.1   Excessive Conservatism in DandD Version 1 Methodology and Parameters

DandD, Version 1.0, was a deterministic screening code, with a single set of default parameters, which is an acceptable screening tool to calculate the screening values to demonstrate compliance with the dose limit in Part 20, Subpart E. NRC staff used this code to develop the screening numbers published in the *Federal Register* (see Section 5.1 and Appendix H of this volume). NRC staff examined several areas where the DandD code may be overly conservative. These areas include (a) reevaluation of the resuspension factor (RF); (b) reevaluation of default parameter selection; (c) model comparison study (NRC 1999a); and (d) ground water model comparison study (NRC 1998a). A technical basis document for revision of the RF is still under review and development.

Version 1.0 of the DandD code used a deterministic set of default parameters. These deterministic values, however, were selected from a range of possible values, rather than by establishing single bounding values. A probability density function (PDF) was established for the range of values for each parameter in the DandD code. A single set of default parameters was selected by probabilistically sampling the PDFs for each of the parameters, to maintain a 90% confidence level that doses would not exceed the dose limit for a combination of all radionuclides. A detailed discussion of the way the default parameters were selected is contained in NUREG/CR–5512, Volume 3.

This method of selecting the default parameter set tends to overestimate the dose. That is, if the default parameter set were selected for a single radionuclide rather than for all radionuclides, the dose calculated using DandD with the single radionuclide default parameter set would, in most cases, be lower than with the "all radionuclides" default parameter set in DandD, Version 1.0. For example, the $DCGL_W$ corresponding to 0.25 mSv/y (25 mrem/y) for Cs-137 using the "all radionuclide" default parameter set is approximately 37 Bq/kg (1 pCi/g); while the $DCGL_W$ using the "single radionuclide" default parameter set is approximately 407 Bq/kg (11 pCi/g). The results from DandD, Version 1.0 using the two default parameter sets are discussed in a Letter Report from Sandia National Laboratories, dated January 30, 1998. To improve this area, Version 2 of the code was developed (and replaced Version 1) to calculate a unique default parameter set based on the specific radionuclides in the source term.

To evaluate the overall conservatism in DandD, a study was conducted to compare the DandD code with the RESRAD and RESRAD–BUILD codes for both the residential and building occupancy scenarios, respectively. This comparison is documented in NUREG/CR–5512, Volume 4 (NRC 1999a). In summary, the models in the DandD codes appeared appropriate for screening (e.g., simplistic, and defensible with minimal data). The default soil mass loading factor for foliar deposition for DandD appears to be too high. The soil-to-plant transfer factors, distribution coefficients, and bio-accumulation factors for certain radionuclides appear to be too conservative. This conservatism is mainly caused by the DandD Version 1.0 approach for selection of the solution vector, to generate a single set of default parameters for all radionuclides. Therefore, the deterministic DandD code in Version 1.0 has been revised as a probabilistic code, DandD, Version 2. An arithmetic error was also found in the default parameter value of the S-35 radionuclide. Also, the code did not model tritium and carbon-14 realistically. This could lead to an underestimation of doses where ground water is not a predominate pathway. It was also determined that RESRAD and RESRAD–BUILD may be better suited to deal with "hot spots."

Another area where NRC staff evaluated the excess conservatism in the DandD code was the ground water model. The basic conceptual ground water model in DandD was described in NUREG/CR–5512, Volume 1. This ground water model was compared to two more realistic ground water models in NUREG/CR–5621 (NRC 1998a). These two models are the STOMP code, as the realistic vadose zone model, and the CFEST code, as the realistic aquifer compartment model. The study concluded that the maximum ground water concentration increased with the number of vadose zone compartments for the DandD model, and that it exaggerated vadose zone dispersion. The study recommended that the maximum vadose zone compartment (layer) thickness in the DandD code should be set to 1 m (3.3 ft). This could be a problem where the vadose zone is thicker than 10 m (33 ft), because the DandD code only allows 10 vadose zone compartments. In general, the study concluded that the DandD model described realistic and conservative representations, of an aquifer and vadose zone, that are appropriate for site assessment. However, it was indicated that, for radionuclides with short half-lives compared to the vadose zone transit time, the DandD model may not be adequate.

I.5.3.5.2   Probabilistic DandD Version 2

Because of the overly conservative approach resulting from the artifact in the way the single default parameter set was selected in DandD Version 1.0, NRC staff has developed a probabilistic DandD, Version 2. DandD, Version 2, updates, improves, replaces and significantly enhances the capabilities of Version 1.0. In particular, Version 2 allows full probabilistic treatment of dose assessments, whereas Version 1.0 embodied constant default parameter values and only allowed deterministic analyses. DandD implements the methodology and information contained in NUREG/CR–5512, Volume 1, as well as the parameter analysis in Volume 3, that established the probability distribution functions (PDFs) for all of the parameters associated with the scenarios, exposure pathways, and models embodied in DandD.

Finally, DandD Version 2 includes a sensitivity analysis module that assists licensees and NRC staff to identify those parameters in the screening analysis that have the greatest impact on the results of the dose assessment. Armed with this information and the guidance available in NUREG–1549, licensees are able to make informed decisions regarding the allocation of resources needed to gather site-specific information related to the sensitive parameters. When cost and the likelihood of success associated with acquisition of this new knowledge are considered, licensees are better able to optimize the costs to acquire site data that allow more realistic dose assessments that, in turn, may lead to demonstrated and defensible compliance with the dose criteria for license termination.

I.5.3.6   Generic Description of RESRAD/RESRAD–BUILD Codes

The RESRAD and RESRAD–BUILD computer codes were developed by Argonne National Laboratory under the sponsorship of the U.S. Department of Energy, and other Agencies, such as NRC. These two codes are pathway analysis models designed to evaluate potential radiological doses to an average member of the specific critical group. RESRAD code uses a residential farmer scenario (ANL 1993a) with nearly identical exposure pathways as the DandD residential scenario described in NUREG/CR–5512, Volume 1 (NRC 1992). The RESRAD–BUILD code uses a building occupancy scenario that covers all exposure pathways in the DandD building occupancy scenario, plus pathways corresponding to external exposures from air submersion and deposited material, and to ingestion of deposited material. Brief descriptions of RESRAD and RESRAD–BUILD codes and conceptual models were presented in previous sections (see Section I.4.3.2 in this appendix). For detailed descriptions of these two codes, the user is referred to ANL/EAD/LD–2 (ANL 1993a), ANL/EAD/LD–3 (ANL 1994), and NUREG/CR–6697 (NRC 2000b). The deterministic versions of these codes were widely used by NRC staff and licensees, prior to the LTR, to estimate doses from radioactively contaminated sites and structures. NRC sponsored development of the probabilistic versions (RESRAD Version 6 and RESRAD–Build Version 3) and their default probabilistic data sets. These two codes were selected because they possess all three of the following attributes:

1. The software has been widely accepted and there is already a large user base among NRC staff and licensees.

2. The models in the software were designed, and have been applied successfully, to more complex physical and residual radioactivity conditions than DandD code.

3. Verification and validation of these two codes are well-documented (Yu 1999; NRC 1998a).

It should be noted that the RESRAD code has been widely used and tested by national and international agencies and has gone through verification (HNUS 1994), dose model comparison (NRC 1999a; EPRI 1999) and benchmarking (DOE 1995). Therefore, RESRAD and RESRAD–BUILD codes are continuously developed and updated with new code versions. Licensees should strive to use the latest version of the RESRAD and RESRAD–BUILD codes and should document in their DP the version used.

I.5.4   Use of Codes and Models Other than DandD and RESRAD

NRC staff should provide flexibility for possible use of other codes and models selected by licensees. However, less common codes, specifically those developed by users, may require more extensive NRC staff review and verifications. In this context, NRC staff may review the following pertinent aspects when using other less common codes:

- scope of code application and applicability to the concerned site;

- extensive review of the generic code selection criteria listed previously;

- review of the mathematical formulation of the associated models and the selected dose conversion factors;

- review of the conceptual model, including the source-term model, used in the code, and compatibility with site conditions;

- review of code performance and comparison with commonly used and verified codes;

- review of code capability regarding handling of default pathways and consistency in selection of default parameters (e.g., occupancy, behavioral, and metabolic parameters); and

- detailed review of codes/models documentation and updates for code/model modifications, including QA/QC reviews.

I.6   Criteria for Selecting or Modifying Input Parameter Values

I.6.1   Introduction

Any analytical approach to dose assessment should involve the selection of appropriate values for input parameters. Each computer modeling code or other analytical methods that a licensee may use should have its own suite of input parameters. Also, unless the licensee is performing a screening analysis, each site should likely have its own defining characteristics that should be incorporated into the dose assessment through the selection of input parameter values.

This section provides general guidelines for NRC staff to consider in evaluating a licensee's selection of values for input parameters. This section addresses three aspects of parameter value selection:

- selection of parameter values or range of values;

- technical justification to support value selection; and

- evaluation of the impact of parameter selection on dose assessment results.

NUREG/CR–5512, Volume 1, and the deterministic parameter set from DandD, Version 1, have been superseded by NUREG/CR–5512, Volume 3, and DandD, Version 2, respectively. Therefore, a licensee should not refer to NUREG/CR–5512, Volume 1, as a primary source for a default deterministic parameter set. Similarly, DandD Version 1, which did not support probabilistic analyses, provided a default deterministic input parameter set. DandD Version 2 has replaced Version 1 and the DandD, Version 1 default deterministic parameter set should not be used as a reference data set for any parameters. This is especially important for the Version 1 defaults, as all the defaults in the code were selected by a method that made them highly interdependent. Each single value in the default deterministic data set was selected based on the values of the other parameters. Thus, if a single parameter is changed in DandD Version 1, the appropriateness of all of the other parameters in the code may be questionable.

I.6.2   Issues in Modifying Parameters

In addressing the three aspects of parameter value selection identified above, several issues should be discussed. First is the distinction between screening analysis and site-specific analysis, with respect to parameter value modification. Second is the appropriateness of accepting default input parameter values in site-specific analyses. Third is the level of justification expected to support the selection of site-specific input parameter values. NRC staff should consider these issues in evaluating a licensee's dose assessment.

I.6.2.1   Screening Analyses Versus Site-Specific Analyses

A licensee may perform a screening analysis to demonstrate compliance with the radiological criteria for license termination specified in Part 20, Subpart E. The screening analysis described in Chapter 16 of this volume requires that the licensee either (a) refer to radionuclide-specific screening values listed in the *Federal Register* (63 FR 64132 and 64 FR 68395) or (b) use the latest DandD computer code. A licensee pursuing the screening option may find that implementation of the DandD code is necessary if radionuclides not included in the *Federal Register* listings should be considered.

NRC staff should ensure that a licensee performing a screening analysis using the DandD code limits parameter modification to identifying radionuclides of interest and specifying the radionuclide concentrations. NRC staff should verify that the licensee has not modified any other input parameter values. The output file generated by DandD identifies all parameter values that have been modified. Modifying any input parameter value from a default value will constitute a site-specific analysis.

I.6.2.2   Default Values Versus Site-Specific Values

DandD and many other computer codes used for dose assessment provide the user with default values for the input parameters. Often, the user only needs to select radionuclides to execute the code. This allows the user to quickly obtain results with very little time expended in developing input data sets. This is basically how DandD, Version 2 was envisioned to be used for screening analyses.

Codes with default parameters, while developed to be run with little user-input or thought, require several considerations that should be made and justified to NRC staff. In actuality, they may be inappropriate for site conditions, scenario, time period, etc. Basically, in using an off-the-shelf computer code and its default parameters, the user agrees with (a) the conceptual model used by the computer code, (b) the exposure scenario, and (c) the process used to select the default parameters so that they are appropriate for the site being modeled.

Users of computer codes should have an understanding of the conceptual and numerical modeling approaches of the code through the process of developing or justifying data input sets. If default parameter values are unavailable or inappropriate, the user should address each and every input parameter by (a) determining what characteristics of the modeled system the parameter represents and how the parameter is used in the code and (b) developing a value for the input parameter that is appropriate for both the system being modeled and for the conceptual and numerical models implemented by the code. In fact, many default data values in the computer code may be simply "placeholders" for site data.

NRC staff realizes that the theoretical approach is quite intensive and probably inappropriate, based on the risk from some sites. Experience has shown that the availability of default values for input parameters can result in the user performing a "site-specific" analysis to modify values for parameters for which site data are readily available and accept the default values as appropriate for the remaining parameters, without an adequate understanding of the parameters and the implications of accepting the default values. Therefore, for site-specific analyses, NRC staff requests that the licensee provide justification for using both the model and the default parameters, along with any justification for site-specific modifications. The level of justification appropriate for the parameter value is not, necessarily, constant for all parameters. This is why Section I.7 of this appendix discusses uncertainty and sensitivity analyses to provide a means to focus both licensees and NRC staff resources on the important parameters.

NRC staff have reviewed, and considered appropriate for dose assessments using these codes, default parameter ranges for both DandD, Version 2 and RESRAD, Version 6. This supports decommissioning by (a) promoting consistency among analyses (where appropriate); (b) focusing licensee and NRC staff resources on parameters considered significant with respect to the dose assessment results; and (c) facilitating review of the licensee's dose assessment by NRC staff. Therefore, most licensees could use the code and its default parameter ranges with little justification. If parameters have been modified, the licensee may need to provide some more justification for default parameters associated with the site-specific parameters. While these are default data for the associated computer code, that does not mean that they can be transferred to another computer code for use in it without justification.

To benefit from the advantages while minimizing the disadvantages, NRC staff should ensure that the licensee employs default parameter values or ranges in a manner consistent with the guidance provided in this section.

I.6.2.3   Justifying Site-Specific Parameter Values

The NRC reviewer should evaluate whether a licensee submitting a site-specific dose assessment has demonstrated that all parameter input values are appropriate for the site being modeled. However, this does not require the licensee to submit a detailed analysis to support the values selected for each and every input parameter. Instead, the level of justification required should be based on the parameter classification and should be commensurate with the significance of the parameter relative to the dose assessment results, as evaluated through sensitivity analyses. The sensitivity analyses should reflect the relative significance of exposure pathways. Note that the relative significance of exposure pathways may change as parameters are modified.

Dose assessment input parameters may be generally classified as behavioral, metabolic, or physical. Behavioral parameters (B) collectively describe the receptor—the exposed individual for whom the dose received is being assessed. The values selected for these input parameters should depend on the behavior hypothesized for the exposed individual. Metabolic parameters (M) also describe the exposed individual, but generally address involuntary characteristics of the individual. Physical parameters (P) collectively describe the physical characteristics of the site being modeled. These would include the geohydrological, geochemical, and meteorological characteristics of the site. The characteristics of atmospheric and biospheric transport up to, but not including, uptake by, or exposure of, the dose receptor, would also be considered physical input parameters.

There is always uncertainty associated with the behavior of a hypothetical receptor. For this reason, the licensee may accept a generically defined receptor for its analysis. The generically defined receptor is the "average member of the critical group." The characteristics of this exposed individual and the criteria for modifying the characteristics for a site-specific analysis are discussed in Section I.3 of this appendix. The licensee may use default values for the behavioral and metabolic parameters, with limited justification, if the values are consistent with the generic definition of the average member of the critical group, and the screening group is reflective of the scenario.

In site-specific analyses, all efforts should be made by the licensee to use site-specific information for important physical parameters. "Site-specific" in this context includes (a) information directly related to the site; (b) information, characterizing the region, that is consistent with site conditions; and (c) generic information that is consistent with the specific geohydrologic conditions at the site (e.g., consistent with the surface-soil unsaturated-zone soil classifications). The justification for site-specific physical parameter values should demonstrate that the site-specific values selected are not inconsistent with the known or expected characteristics of the physical site being modeled. The level of justification should be based on the significance of the parameter to the results of the dose assessment. The licensee should evaluate the significance through sensitivity analyses (see Section I.7 of this appendix). Because of the importance of groundwater, NRC staff should verify that the licensee used site-specific values for all physical parameters (or parameter ranges) related to geohydrologic conditions. If a licensee relies on the DandD default parameter ranges for the physical parameters describing other geochemical conditions (i.e., partition coefficients) and biosphere transport (e.g., crop yields, soil-to-plant concentration factors), NRC staff should evaluate whether the default parameter ranges are inconsistent with known or expected conditions at the site.

I.6.3   Input Parameter Data Sets

I.6.3.1   DandD Default Probabilistic Parameter Set

Probabilistic analyses using the DandD computer code were performed to establish the screening values for building and surface-soil residual radioactivity that were published in the *Federal Register* in November 1998 and December 1999 (63 FR 64132 and 64 FR 68395). In performing these screening analyses, data were compiled for over 600 input parameters and reviewed by NRC staff. These data are discussed in great detail in NUREG/CR–5512, Volume 3, and are directly incorporated into DandD. These data form the reference input parameter set for probabilistic analyses using DandD. The user is referred to NUREG/CR–5512, Volume 3, and the current version of the DandD computer code for the current default parameter ranges and basis.

The DandD computer code may be used to evaluate radiological doses for two exposure scenarios: (1) the building occupancy scenario and (2) the residential scenario. These exposure scenarios and the associated exposure pathways are discussed in detail in NUREG–1549 and NUREG–CR/5512, Volume 1.

A licensee may use the default deterministic behavioral and metabolic parameters from NUREG/CR–5512, Volume 3, or the current version of the DandD computer code, with limited justification. The justification should examine how the licensee's scenario is consistent with the generic scenario from DandD. Similarly, a licensee may use the parameter range for a physical parameter, provided they justify why the parameter range is consistent with the site conditions.

Note that deterministic physical parameter values may not be used without substantial justification (including sensitivity and uncertainty analyses).

I.6.3.2   DandD Default Deterministic Parameter Set

Several default parameter sets have been developed to support deterministic analyses with the DandD code. NUREG/CR–5512, Volume 1, initially presented the conceptual and mathematical foundation of the DandD code, and deterministic values for many input parameters were presented in the document. Volume 3 of NUREG/CR–5512 incorporated much of the parameter information from Volume 1 in developing the default probabilistic input parameter set, making corrections and updating values as necessary. Therefore, a licensee should not refer to NUREG/CR–5512, Volume 1, as a primary source for a default deterministic parameter set.

Similarly, DandD Version 1, which did not support probabilistic analyses, provided a default deterministic input parameter set. DandD Version 2 has replaced Version 1, the DandD Version 1 default parameter set should not be used as a reference data set for any parameters.

Licensees may perform deterministic analyses using DandD (Version 2 or later). This would require licensees change all parameter distribution types to "constant" and specify a single value for each parameter. However, NRC staff does not intend to provide a default deterministic input parameter set to be used in conjunction with DandD. Also, a licensee intending to support decommissioning activities with deterministic dose assessments should ensure that the deterministic approach should provide the information necessary to demonstrate compliance (e.g., support necessary sensitivity analyses, as described in Section I.7 of this appendix).

I.6.3.3   RESRAD Default Probabilistic Parameter Set

The most recent versions of the RESRAD and RESRAD–BUILD computer codes include the option to perform probabilistic dose assessments. The RESRAD team at Argonne National Laboratory worked with NRC staff to develop a default input parameter set that may be used to perform probabilistic dose assessments with the RESRAD and RESRAD–BUILD codes. These default probabilistic input parameter sets are documented in NUREG/CR–6697, "Development of Probabilistic RESRAD 6.0 and RESRAD–BUILD 3.0 Computer Codes" (NRC 2000b).

I.6.3.4   RESRAD Default Deterministic Parameter Set

Versions of RESRAD (e.g., Versions 5.82, 6.0, 6.1) and RESRAD–BUILD (Version 2.37) include default parameter values that support the RESRAD and RESRAD–BUILD deterministic analyses. Many of these default parameters are documented in "Data Collection Handbook to Support Modeling the Impacts of Radioactive Material in Soil" (ANL 1993b). As a set, these are not considered to be acceptable default input parameter values for performing dose assessments in support of decommissioning. Instead, a licensee may use the parameter set described in the preceding section as a starting point for its analyses. NRC staff should ensure that a licensee justifies the selected values and that the values are consistent with existing or expected conditions at the site.

I.6.3.5   Input Data Sets for Other Computer Codes

A licensee may choose to use a computer code or analytical approach other than DandD or RESRAD/RESRAD–BUILD to perform the dose assessment in support of decommissioning. Each code or analytical approach should have a unique set of input parameters. However, there will likely be some input parameters that are also included in the DandD input parameter set.

NRC staff should verify that a licensee provides a listing of all input parameters required in its analysis. For each parameter, the licensee should provide a discussion similar to that provided in NUREG/CR–5512, Volume 3, Chapters 5 and 6. The discussion should include the parameter name, a description of the parameter, a discussion of how the parameter is used in the dose assessment model, and the licensee's classification of the input parameter (i.e., behavioral, metabolic or physical). For the parameters being represented by constant values, the licensee should provide the range of appropriate values for the parameter, the single value selected for the parameter, and the basis for the range and selected value, including references. The level of justification to be provided in the basis should be based on the classification of the parameter (i.e., behavioral, metabolic or physical) and the relative significance of the parameter in the dose assessment.

For input parameters classified as "behavioral" or "metabolic," NRC staff should verify that the licensee specifies values that are consistent with the default screening values specified for the DandD behavioral and metabolic parameters, as long as the definition of the critical group has not been modified. Consistency may depend on the conceptual and numerical models underlying the code being used and the manner in which the parameters are used in the models. Using consistent behavioral and metabolic parameter values for the default critical group may support a relatively standardized definition of the average member of the critical group among analyses. The basis the licensee provides for these parameters should identify the comparable

DandD parameters and discuss any adjustments necessary to accommodate differences between DandD and the code or analytical method being used.

For the input parameters the licensee classifies as physical, other than those related to geochemical conditions and atmospheric and biospheric transport, NRC staff should verify that the licensee uses site-specific values whenever available. The licensee should provide the soil classification for all soil units and specify consistent values for all geohydrologic parameters. For geochemical parameters, such as partition coefficients, the licensee may rely on DandD default probabilistic ranges, as long as justification is provided to demonstrate that the ranges are consistent with geochemical conditions at the site. Site conditions may require that the licensee modify the default parameters to ensure consistency. Additionally, it is important to note that the distributions may not be applicable to codes other than DandD. For meteorological parameters, the licensee should use values that are based on applicable site or regional data. For physical parameters related to atmospheric and biospheric transport, the licensee may accept DandD default parameter ranges with minimal justification, using NUREG/CR–5512, Volume 3, as a starting reference point. Physical parameters related to biosphere transport would include parameters such as crop yields, animal ingestion rates, transfer factors, and crop growing times. NRC staff should evaluate whether the justification provided by the licensee demonstrates that the default values are consistent with conditions at the site.

I.6.3.6   Internal and Direct Exposure Dose Conversion Factors

NRC staff should review the dose conversion factors for inhalation and ingestion, to ensure that the factors used are those developed by EPA, published in Federal Guidance Report No. 11 (EPA 1988). Similarly, NRC staff review should ensure that EPA's external dose factors, although they may correct for actual area, published in Federal Guidance Report No. 12 (EPA 1993) were used or another appropriate code such as Microshield. These dose factors were selected to ensure consistency of the dosimetry models used in deriving these factors with NRC regulations in Part 20.

Licensees may request an exemption from Part 20 to use the latest dose conversion factors. Scenarios and critical group assumptions should be revisited, and justified, to explore at age-based considerations. Licensees may not "pick and choose" dosimetry methods for radionuclides (e.g., Federal Guidance Report No.11 for six radionuclides and current International dose conversion factors for three radionuclides).

I.6.4   Recommended Approach to Parameter Modification

Any analysis that does not meet the conditions of a screening analysis may be considered a site-specific analysis. This will include all analyses using the DandD computer code where one or more input parameters values have been modified from default ranges (or values for behavioral and metabolic parameters), as well as analyses using analytical methods or computer codes other than DandD.

I.6.4.1   Modifying the DandD Default Probabilistic Parameter Set

A reviewer should expect that a licensee who is modifying parameter values for a site-specific analysis using DandD is cognizant of the following:

- what the parameter represents,

- how the parameter is used in the DandD code,

- the basis for the default parameter value, and

- which parameters are physically or numerically correlated.

NUREG/CR–5512, Volumes 1–3, describes in detail what each parameter is intended to represent. Volume 1 provides the original parameter definitions but has been superceded by Volume 3 for parameter values. Volume 1 also provides the mathematical formulations, underlying the DandD code, that should allow the user to (a) understand how each parameter is used and the implication of parameter modification on the resulting calculated dose; and (b) identify numerical correlations among parameters. Volume 2 (the DandD user's manual) redefines several of the input parameters and mathematical formulations based on implementation of the Volume 1 methodology in the DandD computer code. Finally, Volume 3 provides a detailed discussion of most input parameters, allowing the user to fully understand the basis for the default ranges. Volume 3 provides a parameter description and a discussion of how parameters are used in the code, a review of the information sources on which the default values are based, a discussion of uncertainty in the default parameter values, and insight into the selection of alternative parameter values. The DandD user performing site-specific analyses with DandD should be cognizant of the information provided in the three volumes of NUREG/CR–5512.

A licensee may modify DandD behavioral (B) and metabolic (M) input parameter values for the building occupancy and residential scenarios to reflect the characteristics of the average member of a *site-specific* critical group. NUREG/CR–5512, Volume 3, provides the basis for the default value for each behavioral and metabolic parameter. If the licensee modifies the values for these parameters, NRC staff should verify that the licensee has defined a *site-specific* critical group. The licensee may provide site-specific parameter distributions that reflect the variability of the behavior of the average member of the site-specific critical group, or the licensee may use the mean of the site-specific information as a constant-value input for these parameters, consistent with the concept of the "average member" of the critical group. The level of justification required to support modification of behavioral and metabolic parameter values should be consistent with the sensitivity of the parameter.

For the DandD building occupancy scenario, there are only three physical parameters: the resuspension factor (Rfo*), which is derived from the loose fraction (Fl) and the loose resuspension factor (Rfo). Unless the licensee has site-specific information to indicate that the default values are inconsistent with the default values, NRC staff should verify that the licensee has used the default values for these physical parameters in its calculations.

There are many more physical parameters for the DandD residential scenario. The physical parameters may be considered in several groups. The following physical parameters address the geohydrologic conditions:

Unsaturated Zone Thickness (H2)
Soil Classification (SCSST)
Porosity Probability (NDEV)
Permeability Probability (KSDEV)
Parameter "b" Probability (BDEV)
Water Application Rate (AP)
Surface Soil Porosity (N1)
Unsaturated Zone Porosity (N2)
Surface Soil Saturation (F1)
Unsaturated Zone Saturation (F2)
Infiltration Rate (INFIL)
Surface Soil Density (RHO1)
Unsaturated Zone Density (RHO2)
Surface Soil Permeability (Ksat1)
Soil Moisture Content (sh)

For these physical parameters, the licensee should use site-specific distributions and values. [As stated previously, "site-specific" in this context includes (a) information directly related to the site; (b) information characterizing the region that is consistent with site conditions; and (c) generic information that is consistent with the specific geohydrologic conditions at the site (e.g., consistent with the unsaturated zone soil classification)].

NRC staff should verify that the licensee has provided site-specific information for the thickness of the unsaturated zone and the soil classification. In addition, the licensee should ensure that the water application rate is consistent with the irrigation rate (behavioral parameter) if the licensee modifies the irrigation rate. Alternatively, the licensee may demonstrate, through sensitivity analyses, that the dose assessment results are insensitive to these parameters, and use the default ranges.

Values for the derived parameters will be generated internally according to the soil classification indicated and the uniform distributions defined for the porosity probability (NDEV), the permeability probability (KSDEV), and the parameter "b" probability (BDEV). NRC staff should verify that the licensee has not modified the uniform distributions for these three parameters. If site-specific data are available, the licensee may proceed to modify the derived geohydrologic parameters, consistent with the information presented in NUREG/CR–5512, Volume 3.

The only geochemical parameter used in DandD is the element-specific partition coefficient. As documented in NUREG/CR–5512, Volume 3, the partition coefficients at a site are generally dependent on geochemical conditions and are generally independent of soil classification. If the licensee has used the default distributions, NRC staff should evaluate whether the defaults are inconsistent with known or expected conditions at the site.

The following physical parameters address radionuclide transport through the atmosphere and exposure to direct radiation:

Outdoor Shielding Factor (SFO)
Flood dust loading (PD)
Indoor Resuspension Factor (RFR)
Outdoor Dust Loading (CDO)
Indoor Dust Loading (CDI)
Indoor/Outdoor Penetration Factor (PF)
Gardening Dust Loading (CDG)

The remaining physical parameters address characteristics of transport through the biosphere:

Growing Periods (produce, forage, grain, hay) [TG_(#)]
Animal Product Specific Activity (SATac)
Livestock Feeding Periods [TF_(#)]
Animal Product Yields [YA(#)]
Interception Fractions [R_(#)]
Translocation factors [T_(#)]
Contaminated Fractions [x_(#)]
Crop Yields [Y_(#)]
Wet-to-dry conversion factors [W_(#)]
Animal Ingestion Rates [Q_(#)]
Mass-Loading factors [ML_(#)]
Carbon Fractions [fc_(#)]
Hydrogen Fractions [fh_(#)]
Hydrogen Fraction: Soil (fhd016)
Tritium Equivalence: Plant/Soil (sasvh)
Tritium Equivalence: Plant/Water (sawvh)
Tritium Equivalence: Animal Products (satah)

These two groups of physical parameters describe characteristics of the transport of radionuclides through the atmosphere or biosphere up to the point of ingestion or inhalation by, or external exposure to, the receptor. The licensee may accept the default distributions for these parameters as long as the default distributions are consistent with conditions that may exist at the site in the future. The licensee should review the basis given in NUREG/CR–5512, Volume 3, for the default distributions, to determine whether the basis is inconsistent with conditions hypothesized for the site. If so, the licensee should modify the input values accordingly. NRC staff should ensure that the licensee documents this assessment for each of the physical parameters. Note that modifying several of these parameters (e.g., crop yields, animal product yields) should affect the derived behavioral parameters (e.g., area of land cultivated).

For the physical parameters, the licensee may use representative distributions or values. A representative distribution should take into account spatial and temporal variation of the parameter at the site. A representative distribution, for example, would be a precipitation rate based on the historical precipitation data for the site, if available, or from surrounding defensibly relevant monitoring locations. The arithmetic or geometric mean value is often used in defining a representative value. However, the calculation of a mean value should be weighted to account for nonuniform sampling or other nonuniform parameters (e.g., material volume) and parameter sensitivity and uncertainty. The licensee is not required to routinely adopt worst-case, bounding, upper- or lower-percentile, or other overly conservative values in defining distributions.

NRC staff review of this information should be facilitated if the licensee presents the information in a tabular or list format. NRC staff should verify that the licensee has listed every DandD input parameter with the default screening distributions or value (for behavioral or metabolic parameters). For those parameters for which the licensee is using site-specific values (e.g., the physical parameters), the licensee should provide the range of plausible values for the site, the selected distribution or value, and supporting justification, including references.

I.6.4.2 Modifying the RESRAD Default Probabilistic Parameter Set

A licensee using the RESRAD or RESRAD–BUILD codes may change parameters from the default values to reflect a site-specific critical group or site-specific conditions, or to incorporate site-specific data. As discussed in the preceding section, NRC staff should expect that a licensee who is modifying parameter values for a site-specific analysis using RESRAD or RESRAD–BUILD is cognizant of the following:

- what the parameter represents;

- how the parameter is used in the code;

- the basis for the default parameter value; and

- which parameters are physically or numerically correlated.

The licensee should refer to the current code documentation to determine the basis for and how the parameter distributions are used in the code. References to the documentation should be provided. With respect to the basis for the default parameter distributions and values, the licensee should refer to Yu, et al. (2000).

When modifying parameter distributions and values, the licensee should consider whether the parameters are classified as behavioral, metabolic or physical. For behavioral and metabolic parameters for which probability distributions have been developed, the licensee may adopt the DandD default distribution, or the mean of the DandD default distribution, as long as the licensee has not modified the definition of the critical group. For behavioral and metabolic parameters for which distributions have not been developed, the licensee should use values or distributions that are consistent with the DandD default distributions, as applicable.

A licensee may modify behavioral and metabolic default input parameter values to reflect the characteristics of the average member of a *site-specific* critical group. The licensee may modify the values for these parameters if the licensee has defined a *site-specific* critical group. The licensee may provide site-specific parameter distributions that reflect the variability of the behavior of the average member of the site-specific critical group, or the licensee may use the mean of the site-specific information as a constant-value input for these parameters, consistent with the concept of the "average member" of the critical group. The level of justification required to support modification of behavioral and metabolic parameter values should be consistent with the sensitivity of the parameter.

For the physical parameters, the licensee should use site-specific information for the physical parameters addressing geohydrologic and meteorologic conditions. The level of justification for the parameter values should be based on sensitivity analyses. Alternatively, sensitivity analyses may be used to support the use of default distributions or representative values.

For the physical parameters describing geochemical conditions (i.e., distribution coefficients), the licensee should use values that are consistent with the RESRAD default distributions, as long as the values are consistent with known or expected site conditions. Justification supporting the values should be based on sensitivity analyses.

For the remaining physical parameters (atmospheric and biospheric transport), the licensee may use distributions or representative values that are consistent with the RESRAD default distributions, as applicable, as long as the default distributions are consistent with known or expected site conditions.

I.6.4.3   Sensitivity Analyses

The level of justification required to support site-specific parameter values should be commensurate with the sensitivity of the results of the dose assessment to the selected values. Sensitivity analyses are discussed in detail in Section I.7 of this appendix.

I.6.4.4   Site-Specific Distribution Coefficients for Soil or Concrete

The following describes an acceptable approach for the developing input distribution coefficient (K_d) values for soil or concrete for use in site-specific dose modeling codes. This guidance is from Question 4 of Appendix O.

It is noted that K_d values commonly reported in the literature may vary by as much as six orders of magnitude for a specific radionuclide. Generally, no single set of ancillary parameters, such as pH and soil texture, is universally appropriate in all cases for determining appropriate K_d values. Although K_d values are intended to represent adsorption, they are in most cases a lumped parameter representing a myriad of processes. Given the above, the proper selection of a range of K_d values, for either soils or concrete, from the literature will require judicious selection.

APPENDIX I

The licensee is encouraged to use sensitivity analyses to identify the importance of the K_d parameter on the resulting dose either (1) to demonstrate that a specific value used in the analysis is conservative or (2) to identify whether site-specific data should be obtained (if the licensee feels K_d is overly conservative). The sensitivity analysis should encompass an appropriate range of K_d values. As noted above, the input range for the sensitivity analysis may be obtained from literature, DandD default distribution, or RESRAD probabilistic default distribution.

The licensee should use sensitivity analyses, which include an appropriate range of K_d values, to identify the importance of the K_d to the dose assessment and how the change in K_d impacts the dose (i.e., how dose changes as K_d increases or decreases). The range of K_d values that bound the sensitivity analysis may be obtained from (a) the literature, (b) the default distribution in DandD, or (c) the default distribution in the probabilistic code of RESRAD.

Using the results of the sensitivity analysis, the licensee can choose a conservative K_d value, depending on how it affects the dose. For example, if higher K_d values result in the larger dose, an input K_d value should be selected from the upper quartile of the distribution, or if lower K_d values result in the larger dose, an input K_d value should be selected from the lower quartile of the distribution. For those isotopes where the K_d does not have a significant impact on the dose assessment (i.e., K_d is not a sensitive parameter), the median value within the range is an acceptable input parameter.

If the licensee feels that the K_d value is overly conservative, the licensee is encouraged to perform a site-specific K_d determination, so that the dose assessment reflects true site conditions.

DandD

The use of the default K_d values from DandD Version 1.0 outside of the scope of DandD may not be justified, since the single set of default parameters derived for DandD was developed assuming a specific set of exposure pathways and a specific source term. Any single parameter value taken from the default set of parameters outside of the context of the given exposure scenario, source term, and other parameters will have no meaning in terms of the original prescribed probability; therefore there is no basis to conclude that any default K_d value will give a conservative result. However, the distribution of K_d values, used in DandD (which can be found in NUREG/CR–5512, Volume 3, "Residual Radioactive Contamination From Decommissioning —Parameter Analysis," Table 6.86), can be used as the range of K_d values for the sensitivity analysis.

RESRAD

RESRAD default parameter values (including K_d values) should not be used. The defaults were included in the code primarily as place holders that enable the code to be run; it was assumed that site-specific values would be developed. However, it is appropriate to use the default parameter distribution, developed for RESRAD Version 6.0, as the range for use in the sensitivity analysis.

After performing sensitivity analysis with the appropriate K_d ranges, the K_d value at the upper or lower quartile of the distribution, resulting in the highest derived dose, is an acceptable value to input into the dose code, and no further justification is required. For those K_d values that are overly conservative, a site-specific K_d value may be determined by the direct measurement of site samples. Appropriate techniques for K_d determination include American Society for Testing and Materials (ASTM) and U.S. Environmental Protection Agency (EPA) Methods 9–83, "Distribution Ratios by the Short-Term Batch Method"; ASTM D 4646–87, "24-h Batch-Type Measurement of Contaminant Sorption by Soils and Sediments"; and "Understanding Variation in Partition Coefficient, K_d Values, Volumes I and II, EPA 402–R–99–004A, 8/99" available at http://www.epa.gov/radiation/technology/partition.htm#voli.

I.7   Uncertainty/Sensitivity Analyses

I.7.1   Introduction

Uncertainty is inherent in all dose assessment calculations and should be considered in regulatory decision making. In general, there are three primary sources of uncertainty in a dose assessment; (1) uncertainty in the models, (2) uncertainty in scenarios, and (3) uncertainty in the parameters (NRC 1988a, DOE 1991). As stated in Section I.4 of this appendix, models are simplifications of reality and, in general, several alternative models may be consistent with available data. Uncertainty in scenarios is the result of our lack of knowledge about the future of the site. Parameter uncertainty results from incomplete knowledge of the model coefficients.

NRC's risk-informed approach to regulatory decision making suggests that an assessment of uncertainty be included in estimating doses. Specifically, the Probabilistic Risk Assessment (PRA) Policy Statement (60 FR 42622, August 16, 1995) states, in part, "The use of PRA technology should be increased in all regulatory matters to the extent supported by the state of the art in PRA methods and data, and in a manner that complements NRC's deterministic approach...." In the past, dose assessments in support of NRC decommissioning requirements have primarily included the use of deterministic analyses. The deterministic approach has the advantage of being simple to implement and easy to communicate to a nonspecialist audience. However, it has a significant drawback in not allowing consideration of the effects of unusual combinations of input parameters and by not providing information on uncertainty in the results, which would be helpful to the decision-maker. Furthermore, a deterministic analysis that had a high assurance of not being exceeded would have to rely on the use of pessimistic estimates of each parameter of the model, often leading to overly conservative evaluations. Even with the use of probabilistic analyses, it is generally recognized that not all sources of uncertainty can be considered in a dose assessment, nor need to be considered. The primary emphasis in uncertainty analysis should be to identify the important assumptions and parameter values that, when altered, could change the decision.

Sensitivity analysis performed in conjunction with the uncertainty analysis can be used to identify parameters and assumptions that have the largest effect on the result. Sensitivity analysis provides a tool for understanding and explaining the influence of these key assumptions and parameter values on the variability of the estimated dose.

I.7.2   Issues in Uncertainty/Sensitivity Analyses

Uncertainty analysis imparts more information to the decision-maker than deterministic analysis. It characterizes a range of potential doses and the likelihood that a particular dose may be exceeded.

An important issue in uncertainty and sensitivity analysis is that not all sources of uncertainty can be easily quantified. Of the three primary sources of uncertainty in dose assessment analyses, parameter uncertainty analysis is most mature. However, approaches for quantifying conceptual model and scenario uncertainty are less well-developed. Difficulties in predicting the characteristics of future society, especially those influencing exposure, can lead to large uncertainties. At most, one is able to assert that an acceptably complete suite of scenarios has been considered in the assessment (Flavelle 1992). For these reasons, we make no attempt to quantify formally model or scenario uncertainty, although to a certain extent, these are captured in parameter uncertainty analyses. Choices of the scenarios and conceptual model(s) to be used for the site are discussed in Sections I.3 and I.4, of this appendix, respectively.

Uncertainty analyses frequently use the Monte Carlo method. Input variables for the models are selected randomly from probability distribution functions, which may be either independent or correlated to other input variable distributions. Critics of formal uncertainty analysis have often pointed out that limitations of knowledge about the nature and extent of correlation among variables fundamentally limit our ability to make meaningful statements about the degree of uncertainty in dose assessments (Smith et al., 1992).

Because the results of an uncertainty analysis provide a distribution of doses, it should be recognized that some percentage of the calculated doses may exceed the regulatory limit. A key issue that should be addressed in the treatment of uncertainty is specifying how to interpret the results from an uncertainty analysis in the context of a deterministic regulatory limit. Agency practice has not been to require absolute assurance that the regulatory limit will be met, so regulatory compliance could be stated in terms of a metric of the distribution such as the mean, or a percentage of calculated doses allowed to exceed the limit. Even for a deterministic analysis, it is recognized that the reported dose is simply one of a range of possible doses that could be calculated for the site; therefore, there is still an issue of where this calculated dose should lie in terms of the unquantified spectrum of possible doses.

In summary, the key issues in addressing uncertainty are (a) incorporating alternative conceptual models and scenarios to identify a complete suite of possibilities; (b) determining how to select appropriate parameter distribution and ranges, along with the associated correlation between parameters for the analysis; and (c) specifying the metric of the dose distribution to use in determining compliance with the dose limit.

I.7.3   Recommended Approach

I.7.3.1   Screening Analyses

Often the first step in evaluating site compliance should be a screening analysis. At preliminary stages of the evaluation, there may be little information available about the site. Therefore, NRC's screening approach is designed to ensure that there is high confidence that the dose should not be underestimated. As discussed in Sections I.3 and I.4 of this appendix, the models and scenarios used in screening were selected to represent generic conditions and are intended to be "prudently conservative." The screening analysis assumes that all that is known about a site is the source term. Accordingly, the default parameters were selected to make it unlikely for the dose that would be calculated using site-specific information to exceed the screening dose.

NRC published a screening table for building-surface residual radioactivity and surface soil (see Appendix J). NRC staff performed a Monte Carlo analysis, using the DandD code, with values of the input parameters sampled from wide ranges selected to represent the variability in those parameters across the United States. The default values of input parameters for the DandD code (i.e., the values that the code would use without specification by the user) were then chosen from distributions of those parameters that would never cause the 90th percentile of the output dose distribution from the Monte Carlo analysis to be exceeded for any radionuclide, as illustrated in Figure I.6 (NRC 1999a).

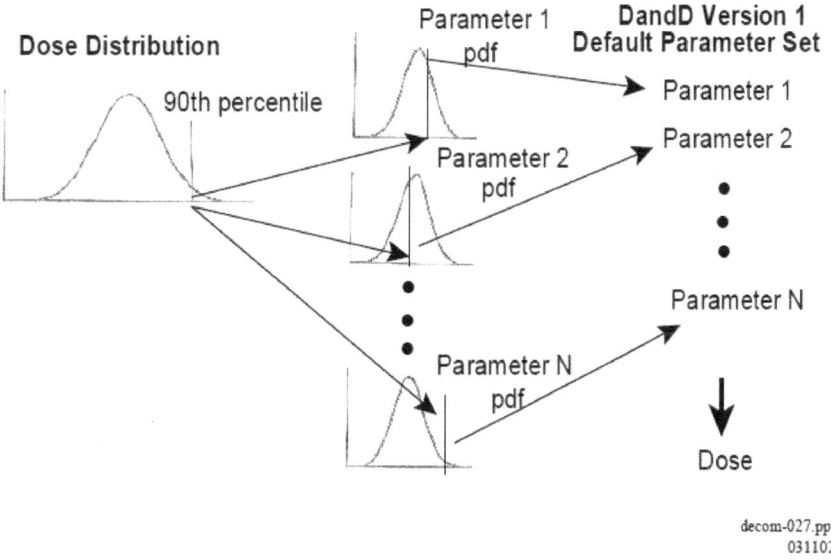

decom-027.ppt
031102

Figure I.6   Treatment of Parameter Uncertainty in DandD Version 1.

The intent of the specification of default parameter values, scenario, and conceptual models in the DandD code was to ensure that there should be less than a 10 % probability that the calculated dose using site-specific information may exceed the dose limit. Because the default parameters, scenarios, and conceptual models in DandD Version 1.0 were designed to provide high confidence that the dose should not be underestimated, an licensee using the screening criteria does not need to quantify the uncertainty in the dose analysis. The calculated results may be considered to represent a "prudently conservative" estimate of the dose (i.e., the calculated dose is likely an overestimation of the true dose). In many cases, however, the default parameter values chosen were highly conservative, making the outcome of the deterministic analysis overly stringent.

DandD Version 2 is designed to allow Monte Carlo analyses which give a distribution of doses as illustrated in Figure I.7. The code automatically performs the probabilistic analyses and aggregates the results for the user. To maintain consistency in approaches used for Versions 1 and 2, and previously published screening tables, the 90th percentile of the dose distribution should be used to determine compliance with the Subpart E when used for screening analysis. Default parameter probability density functions have been incorporated into the code for screening analyses; therefore, for screening analyses, the license reviewer may only need to ensure that these aforementioned default parameters were used.

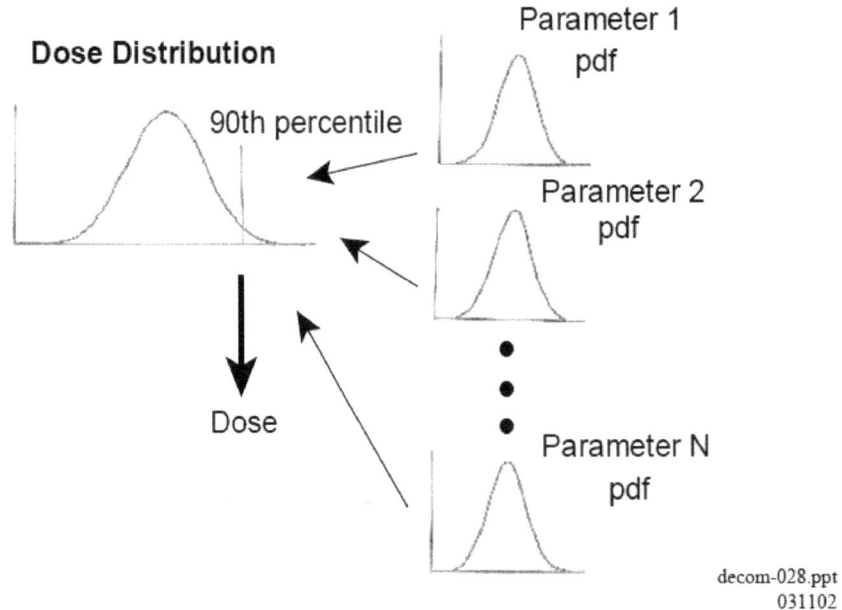

Figure I.7   Treatment of Parameter Uncertainty in DandD Version 2.

I.7.3.2   Site-Specific Analyses

I.7.3.2.1   Deterministic Analysis

For site-specific analyses, the treatment of uncertainty in deterministic and probabilistic analyses should be handled differently. NRC's risk-informed approach to regulatory decision making suggests that an assessment of uncertainty should be included in dose analyses. However, in some cases such analyses may not be needed (e.g., bounding type analyses). Because no information is provided on the uncertainty in bounding analyses, it is important for the licensee to demonstrate that the single reported estimate of the peak dose is likely to be an overestimation of the actual peak dose. Use of conservatism in only some aspects of the analysis may not necessarily result in a conservative estimate of the dose. Uncertainties in the conceptual model may be larger than uncertainties in parameters used in the analysis; therefore, use of conservative parameter values do not necessarily ensure a conservative estimate of the dose. To ensure that the results from a deterministic analysis are unlikely to underestimate the dose, it is recommended that the licensee use the approaches discussed in Sections I.3 and I.4, of this appendix, for developing land-use scenarios and conceptual models. In addition, the licensee should use conservative values for key parameters. The approaches discussed below on performing sensitivity analyses should be used in identifying key parameters in the analysis.

I.7.3.2.2   Probabilistic Analysis

Although bounding analyses are a good starting point for determining regulatory compliance, the demonstration that a single, deterministic result is bounding may be too difficult to prove. For site-specific probabilistic analysis, it is only necessary to demonstrate that the mean dose does not exceed the regulatory criterion.

> A single deterministic calculation using the mean values of parameters is unlikely to result in the mean dose.

Parameter uncertainty analysis provides a quantitative method for estimating the uncertainty in calculated doses, assuming the structure of the model is an adequate representation of the real world, and the exposure scenario is an appropriate reflection of potential future land-use at the site. Several methods have been developed for quantifying parameter uncertainty, including (a) analytical methods, (b) Monte Carlo methods, (c) response surface methods, and (d) differential methods (DOE 1990). In addition, alternative approaches, such as the first-order reliability method, have recently been applied on a wide variety of environmental problems (DOE 1998). Of these methods, the Monte Carlo methods are recommended because they are easy to implement and provide significant versatility.

Monte Carlo methods can be applied to either linear or nonlinear models, and analytical or numerical models. Input parameter uncertainties are represented as probability density functions. Parameter values randomly sampled from probability density functions are used as inputs to multiple runs or "realizations" of the model.

For probabilistic analyses, the peak of the plot of mean dose over time should be compared with the regulatory standard to determine compliance. Equation I–4 shows how the mean dose as a function of time can be derived. For Monte Carlo Runs:

$$Mean(t_i) = \frac{\sum_{k=1}^{N} Dose_k(t_i)}{N} \qquad \textbf{(I–4)}$$

where $Mean(t_i)$ = mean dose at time t_i
$Dose_k(t_i)$ = dose at time t_i for run k
t_i = time in years
i = time steps (1 to 1000)

Essentially, a mean dose is determined at each discrete time in the analysis. A plot is then made of these means over time. The mean dose provides the "best estimate" of dose at each discrete time. The overall peak of these best estimates is then used to determine compliance with the rule. Figure I.8 shows how such a plot would be used to determine compliance with the regulations.

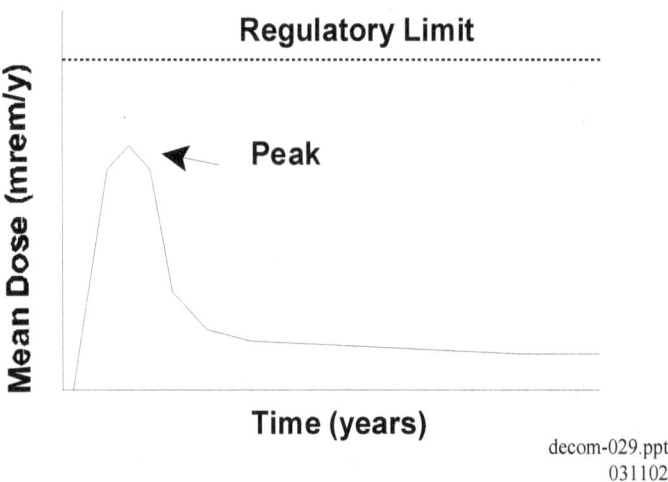

decom-029.ppt
031102

Figure I.8   Application of "Peak of the Mean" Dose.

Licensees using probabilistic dose modeling should use the "peak of the mean" dose distribution for demonstrating compliance with 10 CFR Part 20, Subpart E. The "peak of the means" approach is one method for determining compliance with the regulations using probabilistic analyses. Other probabilistic approaches, such as "mean of the peaks," if justified, may also be acceptable for demonstrating compliance. If the licensee intends to use any probabilistic approach to calculate DCGLs, the licensee should discuss their planned approach with NRC staff.

I.7.4   Input Parameter Distributions for Monte Carlo Analysis

A key aspect of any Monte Carlo analysis is defining the ranges and statistical distribution of parameters treated as uncertain in the analysis. It is important for the licensee to avoid assigning overly restrictive ranges that suggest an unwarranted precision in the state of knowledge. On the other hand, the specification of unreasonably large ranges may not account for what is known about a parameter and also may lead to "risk dilution." The distributions used in the analysis should characterize the degree of belief that the true but unknown value of a parameter lie within a specified range of values for that parameter.

Sensitivity results are generally less dependent on the actual distributions assigned to the input parameters than they are on the ranges chosen for the parameters. However, distributional assumptions can have a large impact on the dose distribution (SNL 1993). Resources can often be used most effectively by performing a Monte Carlo analysis in an iterative manner. Initially, rather crude ranges and distribution assumptions can be used to determine which input variables dominate the behavior of the calculated dose. Often, most of the variation in the calculated dose is caused by a relatively small subset of input parameters. Once the most important input parameters are identified, resources can be concentrated on characterizing their uncertainty. This avoids spending a large effort characterizing the uncertainty in parameters that have little impact on the dose (SNL 1993).

A reasonable strategy for assigning distributions for parameters used in Monte Carlo analyses is summarized below (NRC 2000):

- **Select parameters to be assigned distributions**—Not all parameters of the system under study require specification of a distribution. Those parameters that may well be distributed, but have little impact ultimately on the results, can be assigned constant values. Even if a parameter is known to have a significant effect on the results, its value may be specified at a constant value if it can be demonstrated that the choice leads to a conservative result.

- **Assign distributions for important parameters**—The assignment of parameter distributions usually is a matter of the quantity of available data.

- **Ample data available**—Where there are ample data, empirical distributions of a parameter can be generated directly.

- **Sufficient data available**—Data plotted as histograms or in probability coordinates can be used to identify standard distributional forms (e.g., normal, lognormal, and uniform).

- **Parameters with some data**—Where there are insufficient data to estimate the shape of an empirical distribution, data may be supplemented by other soft information. For example, if there were a mechanistic basis for assigning a given distribution, or if a distribution were well-known for the parameter, on a regional basis, this information could be used to estimate the likely shape of the distribution. Alternatively, the new data can be used to supplement a prior, non-site-specific parameter distribution (e.g., Bayesian updating).

- **Parameters with insufficient information**—If sufficient data are not available, but there were other kinds of data that imply the likely behavior of a parameter, then it may be possible to supplement the desired data indirectly. An example of such a procedure is the use of root uptake factors to infer distribution coefficients in soil (ORNL 1984). If only incomplete information is known about the parameter (e.g., its mean, or its range), and no correlations to other types of data are available, then the choice of the parameter distribution should reflect the uncertainty. The distribution should have the least-biased value, which is generally a wide distribution encompassing all the possible values. One procedure to assure that the distribution has the least bias is known as the "maximum entropy formalism," based on Shannon's informational entropy (Harr 1987). This formalism allows the investigator to pick the distribution based on the kinds of information available on the parameter to assure that the result is least-biased; for example, if only the range of the data is known, a uniform distribution between the range is least-biased. Table I.7 describes the maximum entropy solutions for several classes of data (Harr 1987). Other, empirical sources of guidance for choosing parameter distributions can be found in several other references (IAEA 1989; NCRP 1996a).

- **Parameter correlations**—Many of the parameters used in the probabilistic analyses may be correlated to other parameters. Some parameter distributions may in fact be used to derive other distributions (e.g., root uptake factors may be used to derive soil distribution coefficients). Also, correlations are expected on physical grounds, such as the relationship between hydraulic gradient and permeability. Where available, these correlation coefficients can then be used to generate correlated values of distributed parameters. This may help to avoid the situation where two correlated quantities are treated as uncorrelated, leading to unlikely combinations of parameters (e.g., high gradient and high-hydraulic conductivity). The effects of assumed minimum versus assumed maximum levels of correlation can be investigated to evaluate the importance of including an explicit estimate of dependency between model parameters. In some cases, explicit modeling of the dependency between model parameters is possible, based on knowledge about the explicit mechanistic reasons for the dependencies. In general, it is more important to consider the effect of dependency when correlations are strong among the model's most sensitive parameters (see discussion below on identifying sensitive parameters); weak correlations between sensitive parameters and strong correlations among insensitive parameters will generally have very little impact on the overall calculated dose (NCRP 1996a).

Table I.7 Maximum Entropy Probability Distributions (Adapted from Harr 1987)

Given Constraints on Data	Assigned Probability Density
Minimum and maximum only	Uniform
Expected value only	Exponential
Expected value and standard deviation	Normal
Expected value, standard deviation, minimum and maximum	Beta
Mean occurrence rate between arrival of independent events	Poisson

I.7.5   Sensitivity Analysis

Uncertainty and sensitivity analyses are closely linked, and ideally, they should be considered together. The primary aim of a sensitivity analysis is to identify the input parameters that are the major contributors to the variation or uncertainty in the calculated dose. Identifying these key parameters is essential for building a defensible case in support of the assessment. It is very important for the licensee to justify the value or range of values used in the assessment to represent these key parameters. Several of the more-popular sensitivity methods used in other performance assessments conducted at NRC are presented, very briefly, below (NRC 1999). It may be necessary for the licensee to use more than one approach in identifying the key parameters.

The licensee should focus on the pathways and radionuclides that are providing the greatest dose. If these pathways are modified or eliminated, a re-evaluation of the sensitivity analysis should be done to verify the important parameters for the analysis, consistent with the iterative nature of the "Decommissioning and License Termination Framework" (see Section 1.5 of this volume). For sites with a suite of radionuclides, the licensee may use expected concentrations or relative ratios of radionuclides to focus resources on the overall critical pathways and parameters. In addition, the licensee also should evaluate the effects of uncertainty on the relative ratios.

I.7.5.1   Deterministic Sensitivity Analysis

Two types of sensitivity analysis techniques are widely used: deterministic and Monte Carlo. The first, deterministic sensitivity analysis, calculates the change in the output result (i.e., peak dose) with respect to a small change in the independent variables, one at a time. The following formula illustrates the normalized sensitivity coefficient calculated from a deterministic analysis.

$$S_i = \left[\frac{\overline{X}_i}{d(\overline{X}_i)}\right]\left(\frac{\partial d}{\partial \overline{X}_i}\right) \qquad\qquad \textbf{(I–5)}$$

where
- S_i = sensitivity coefficient
- $\overline{X}_i$ = baseline value of the ith parameter
- $d(\overline{X}_i)$ = peak dose for the baseline case
- ∂d = change in peak dose
- ∂X_i = change in i^{th} parameter

Variable transformations, such as *normalization,* used in this example, are described further below.

The advantage of the deterministic technique is that it is unambiguous in terms of demonstrating a cause and effect for the given conceptual model. The disadvantages are that at least one evaluation of the model should be performed for every independent variable, and the sensitivity result applies only locally (i.e., for one location in the space of all of the independent variables).

I.7.5.2   Statistical Sensitivity Analysis Techniques

The techniques used herein (except deterministic analysis) rely on the use of the Monte Carlo method for probabilistically determining system performance. Statistical analyses of Monte Carlo results starts with a large pool of realizations (hundreds to thousands). These techniques determine sensitivities of the dependent variable (dose) to changes in the independent variables. The main advantage of these techniques is that they allow sensitivity to be determined over wide ranges of the independent variables, as opposed to the deterministic techniques that apply to only one point within the ranges. The disadvantage of statistical techniques is that it is often difficult to extract useful information on sensitivity except for a small set of the most important variables, because smaller sensitivities are obscured. A compilation of some of the more popular techniques for analyzing sensitivity from Monte Carlo results is presented below.

Usually, statistical sensitivity techniques have been applied to the set of peak doses drawn from the realizations. Sensitivity information from the ensemble of the peak doses provides useful information, and would be the correct approach if one were pursuing the "mean of the peaks" dose. However, this approach is not as meaningful for the "peak of the mean" dose. For the latter, the statistical techniques should be applied to the set of doses drawn from the Monte Carlo runs at the time of the "peak of the mean" dose.

I.7.5.2.1   Scatter Plot and Linear Regression on One Variable

In the scatter plot/single linear regression technique, peak TEDE is plotted versus each of the sampled input variables. This is often a good starting point for examining Monte Carlo results because strong relationships between peak dose and the independent variables are often obvious. Single linear regression of Monte Carlo results may fail to show unambiguous correlation since other sampled parameters that affect the output are varying at the same time.

I.7.5.2.2   Use of the T-Statistic to Determine Significance of Single Linear Regression Parameters

The t-test estimates the confidence that an estimated parameter value differs from another value. In this case, it is used to determine if there is a specified (e.g., 95-%) confidence that the slope (m_i) of a single linear regression is different from zero (Benjamin and Cornell 1970).

The t statistic of the slope of the regression line is defined:

$$t_i = m_i \sqrt{n \frac{S_{i,x}^2}{S^2}} \qquad \text{(I–6)}$$

where t_i = t-statistic for regression coefficient i
m_i = estimated value of regression coefficient (i.e., slope of the best-fit line for dose versus the independent variable i)
S = estimated standard deviation of dose
$S_{i,x}$ = estimated standard deviation of independent variable x_i
n = number of samples

When the number of realizations is large, the t distribution may be represented by the normal distribution. The critical value to ensure 95-% confidence that m_i differs from zero under these conditions is 1.96. Equation I–6 is used therefore to determine whether the absolute value of the t statistic for each independent variable is greater than 1.96. If not, then the hypothesis that the independent variable is significant is rejected.

I.7.5.2.3 Partial Rank Correlation

The partial rank correlation coefficient measures the strength of the relationship between variables after any confounding influences of other variables have been removed. The partial rank correlation coefficient between X_1 and Y, with the influence of X_2 removed, is given by:

$$\rho(X_1 Y X_2) = \frac{\rho_{X_1 Y} - (\rho_{X_1 X_2})(\rho_{Y X_2})}{\left[\left(1 - \rho_{X_1 X_2}^{\ 2}\right)\left(1 - \rho_{Y X_2}^{\ 2}\right)\right]^{\frac{1}{2}}} \qquad \text{(I–7)}$$

where $\rho(X_1 Y X_2)$ = partial rank correlation coefficient between X_1 and Y, with the influence of X_2 removed

$\rho_{X_1 Y}$ = rank correlation coefficient between X_1 and Y

$\rho_{X_1 X_2}$ = rank correlation coefficient between X_1 and X_2

$\rho_{Y X_2}$ = rank correlation coefficient between Y and X_2

I.7.5.2.4 Stepwise Multiple Linear Regression

Stepwise multiple linear regression (stepwise regression) determines the most influential independent variables on output uncertainty according to how much each reduces the residual sum of squares (RSS) (SNL 1991). The form of the regression equation is:

$$y = m_1 x_1 + m_2 x_2 + \ldots + m_n x_n + b \qquad \text{(I–8)}$$

where y = dependent variable (i.e., peak dose)

x_i = independent variables

m_i = regression coefficients

b = intercept

The variables may be the raw variables, transformed variables (e.g., logarithms), or ranks (see Section I.7.5.3.2 of this appendix). The stepwise algorithm calculates the reduction in RSS for the independent variables in the order that gives the greatest reduction first. The regression coefficients m_i are the partial derivatives of the dependent variable with respect to each of the independent variables; therefore, m_i provides a measure of the relative change in output with respect to a change in the input variable, given that the other input variables are held constant.

I.7.5.2.5   Nonparametric Tests

Nonparametric tests differ from regression and differential analyses in that they do not require fitting the data to prespecified functional form. The Kolmogorov–Smirnov (KS) test is one such test that determines whether a set of samples has been drawn from a specific distribution (NRC 1988). It is used to determine whether an independent variable is important by comparing a subset of the independent variable composed of the values from the highest category (e.g., 10 %) of the peak TEDE realizations to the theoretical distribution of that independent variable. If the distributions are equivalent, then peak TEDE is not sensitive to the variable in question. Conversely, if the distributions are different, then the variable in question does have an effect on peak TEDE.

I.7.5.3   Variable Transformations and Their Attributes

Demonstrating the relationship among input and output variables can be enhanced by transforming the variables. This section describes some common variable transformations used in sensitivity analysis.

I.7.5.3.1   Normalization

In normalization, the input variable x_i is transformed by dividing by its mean value (or another baseline such as the median, 90th percentile, etc.):

$$x_i^* = \frac{x_i}{\overline{x}_i} \qquad\qquad\text{(I–9)}$$

Normalized variables are dimensionless and are scalar multiples of their baseline values. Dimensionless variables allow the comparison of sensitivities to other independent variables with different dimensions. Normalized variables are a natural outcome of sensitivity derived from regression of log-transformed variables. Such sensitivity measures describe only the relative change in the dependent variable (peak TEDE) to changes in the independent variables. Sensitivities calculated from normalized variables do not take into account the uncertainty in the independent variables.

I.7.5.3.2   Rank Transformation

Rank transformation, a dimensionless transform, replaces the value of a variable by its rank (i.e., the position in a list that has been sorted from largest to smallest values) (Iman and Conover 1979). Analyses with ranks tend to show a greater sensitivity than results with untransformed variables, and diminish the influence of the tails in highly skewed distributions.

I.7.5.3.3 Logarithmic Transformation

For situations in which input and output variables range over many orders of magnitude, it may be advantageous or even necessary to perform analyses on the logarithm of the variables instead of the variable values themselves. The log transformation is also valuable for creating regression equations, where the subprocesses of the model multiply each other to form the output variable. For the present situation, in which the dose calculation results from radionuclide releases from the waste form, transport through the geosphere, and uptake by humans, the processes are indeed largely multiplicative rather than additive. Log transforms therefore tend to give better fits to the Monte Carlo results than untransformed variables. The log transformation is generally used in conjunction with normalization.

I.7.5.3.4 Standardization

The independent and dependent variables can be standardized by subtracting the mean and dividing by the standard deviation, that is,

$$x_i^* = \frac{x_i - \bar{x}}{\sigma_x} \tag{I–10}$$

The advantage of standardization over normalization is that it inserts the approximate range of the variables into the sensitivity analyses. Therefore a variable that has a large per-unit sensitivity, but is well-known and has a narrow range, will have an increased sensitivity when standardized. Conversely, independent variables with wide ranges may show a reduced sensitivity when standardized.

Sensitivity measures based on standardized variables (standardized sensitivities) have the advantage of taking into account the uncertainty (in terms of the standard deviation) of the independent variable. This technique decreases the sensitivity if the range of the independent variable is large. Furthermore, the standardized sensitivities preserve the absolute values of peak TEDE since the derivatives are divided by the standard deviation for the entire set of calculations, rather than the mean peak TEDE at the evaluation point.

I.7.6 Conclusions

Sensitivity analyses should be used to identify parameters of the models and assumptions that have the largest effect on the results. These sensitivity results should be used to determine if more information on key parameters is warranted to make a convincing case for the acceptability of the site. The sensitivity techniques discussed here portray sensitivity in different ways, and all have their strengths and weaknesses. A useful way to use sensitivity results is to employ several different techniques, and then to determine if a common set of parameters regularly turns out to be important.

I.8   References

Argonne National Laboratory (ANL). ANL/EAD/LD–2, "Manual for Implementing Residual Radioactive Material Guidelines Using RESRAD 5.0." Working Draft for Comment. ANL: Argonne, IL. September 1993(a).

— — — — —. ANL/EAD/LD–3, "RESRAD–BUILD: A Computer Model for Analyzing the Radiological Doses Resulting from the Remediation and Occupancy of Buildings Contaminated with Radioactive Material." ANL: Argonne, IL. November 1994.

— — — — —. ANL/EAD/TM–82, "Evaluation of the Area Factor Used in the RESRAD Code for the Estimation of Airborne Contaminant Concentrations of Finite Area Sources." ANL: Argonne, IL. July 1998.

— — — — —. ANL/EAIS–8, "Data Collection Handbook to Support Modeling the Impacts of Radioactive Material in Soil." ANL: Argonne, IL. April 1993(b).

— — — — —. Letter Report, Rev. 1, for NRC Project Y6112, "Selection of RESRAD and RESRAD–BUILD Input Parameters for Detailed Distribution Analysis." ANL: Argonne, IL. October 1999.

Benjamin, J.R. and C.A. Cornell. *Probability, Statistics, and Decision for Civil Engineers.* McGraw–Hill: New York, NY. 1970.

Department of Energy (U.S.) (DOE), Washington DC. DOE/LLW–100,"Guidelines for Sensitivity and Uncertainty Analysis of Performance Assessment Computer Codes, National Low-Level Waste Management Program." DOE: Washington, DC. September 1990.

— — — — —. DOE/ORO–2033, "Benchmarking Analysis of Three Multimedia Models: RESRAD, NMMSOILS and MEPAS." DOE: Washington, DC. November 1995.

— — — — —. Proceedings of a Low-Level Waste Conference. "Treatment of Uncertainty in Low-Level Waste Performance Assessment." DOE 13th Annual Low-Level Waste Conference: November 19–21, 1991.

Environmental Protection Agency (U.S.) (EPA), Washington DC. EPA–402–R–93–081, "Federal Guidance Report No. 12: External Exposure to Radionuclides in Air, Water and Soil." EPA: Washington, DC. September 1993.

— — — — —. EPA–520/1–88–020, "Federal Guidance Report No. 11: Limiting Values of Radionuclide Intake and Air Concentration and Dose Conversion Factors for Inhalation, Submersion, and Ingestion." EPA: Washington, DC. September 1988.

Electric Power Research Institute. TR–112874, "Comparison of Decommissioning Dose Modeling Codes for Nuclear Power Plant Use: RESRAD and DandD." Electric Power Research Institute: Palo Alto, CA. November 1999.

Flavelle, P., 1992. "A Quantitative Measure of Model Validation and Its Potential Use for Regulatory Purposes." *Advances in Water Resources*. Vol. 15, pp. 5–13. 1992.

Halliburton NUS Corporation. HNUS–ARPD–94–174, Contract No. 32532403, "Verification of RESRAD – A Case for Implementing Residual Radioactive Material Guidelines, Version 5.03." June 1994.

Harr, M.E., 1987. *Reliability Based Design in Civil Engineering*. McGraw–Hill: New York, NY. 1987.

Helton, J.C., 1993. "Uncertainty and Sensitivity Analysis Techniques for Use in Performance Assessment for Radioactive Waste Disposal." *Reliability Engineering and System Safety*, Vol. 42, Nos. 2–3, pp. 327–367. 1993.

International Atomic Energy Agency (IAEA). BIOMASS, 2001. *BIOsphere Modelling and ASSessment Programme Version β2* [sic]. Compact Disc. IAEA: Vienna, Austria. 2001.

— — — — —. IAEA BIOMASS/T1/WD01, "Long-Term Releases From Solid Waste Disposal Facilities: The Reference Biosphere Concept." IAEA: Vienna, Austria. 1999(a).

— — — — —. IAEA BIOMASS/T1/WD03, "Guidance on the Definition of Critical and Other Hypothetical Exposed Groups for Solid Radioactive Waste Disposal." IAEA: Vienna, Austria. 1999(b).

— — — — —. IAEA–SS–100, "Evaluating the Reliability of Predictions Made Using Environmental Transfer Models." IAEA: Vienna, Austria. 1989.

Iman, R.L., and W.J. Conover. "The Use of Rank Transform in Regression." *Technometrics*. Vol. 21, No. 4, pp. 499–509: 1979.

James, A.L., and C.M. Oldenburg. "Linear and Monte Carlo Uncertainty Analysis for Subsurface Contaminant Transport Simulation." *Water Resources Research*. Vol. 33, No. 11: November 1997.

Mirshra, S. "A Methodology for Probabilistic Risk-Based Performance Assessments Using the First-Order Reliability Method." *Proceedings of the Topical Meeting on Risk-Based Performance Assessment and Decision Making*. American Nuclear Society: April 5–8, 1998.

National Council on Radiation Protection and Measurements (NCRPM). NCRP Commentary No. 14, "A Guide for Uncertainty Analysis in Dose and Risk Assessments Related to Environmental Contamination." NCRPM: Bethesda, MD. May 10, 1996(a).

— — — — —. NCRP Report No. 123, "Screening Models for Releases of Radionuclides to Atmosphere, Surface Water, and Ground." NCRPM: Bethesda, MD. January 22, 1996(b).

— — — — —. NCRP Report No. 129, "Recommended Screening Limits for Contaminated Surface Soil and Review of Factors Relevant to Site-Specific Studies." NCRPM: Bethesda, MD. January 29, 1999.

Nuclear Regulatory Commission (U.S.) (NRC), Washington, DC. NUREG–0856, "Final Technical Position on Documentation of Computer Codes for High Level Waste Management." NRC: Washington, DC. June 1983.

— — — — —. NUREG–1549, "Decision Methods for Dose Assessment to Comply with Radiological Criteria for License Termination." Draft Report for Comment. NRC: Washington, DC. July 1998(a).

— — — — —. NUREG–1573, "A Performance Assessment Method For Low-level Waste Disposal Facilities: Recommendations of NRC's Performance Assessment Working Group." NRC: Washington, DC. October 2000(a).

— — — — —. NUREG–1575, "Multi-Agency Radiation Survey and Site Investigation Manual (MARSSIM)." NRC: Washington, DC. December 1997.

— — — — —. NUREG–1620, Rev. 1, "Standard Review Plan for the Review of a Reclamation Plan for Mill Tailings Sites Under Title II of the Uranium Mill Tailings Radiation Control Act of 1978." NRC: Washington, DC. June 2003.

— — — — —. NUREG–1668, Volume 2, "NRC Sensitivity and Uncertainty Analyses for a Proposed HLW Repository at Yucca Mountain, Nevada, Using TPA 3.1." NRC: Washington, DC. March 1999(a).

— — — — —. NUREG/BR–0058, Rev. 2, "Regulatory Analysis Guidelines of the U. S. Nuclear Regulatory Commission." NRC: Washington, DC. November 1995(b).

— — — — —. NUREG/BR–0167, "Software Quality Assurance Program and Guideline." NRC: Washington, DC. February 1993(c).

— — — — —. NUREG/CP–0163, "Proceedings of the Workshop on Review of Dose Modeling Methods for Demonstration of Compliance with the Radiological Criteria for License Termination." NRC: Washington, DC. May 1998(a).

—————. NUREG/CR–4604, "Statistical Methods for Nuclear Material Management." NRC: Washington, DC. December 1988(a).

—————. NUREG/CR–5211, "A Review of Uncertainties Relevant in Performance Assessment of High-Level Radioactive Waste Repositories." NRC: Washington, DC. September 1988(b).

—————. NUREG/CR–5453, Volumes 1 and 2, "Background Information for the Development of a Low-Level Waste Performance Assessment Methodology: Identification of Potential Exposure Pathways." December 1989.

—————. NUREG/CR–5512, Volume 1, "Residual Radioactive Contamination From Decommissioning: Technical Basis for Translating Contamination Levels to Annual Total Effective Dose Equivalent." NRC: Washington, DC. October 1992.

—————. NUREG/CR–5512, Volume 2, "Residual Radioactive Contamination from Decommissioning, User's Manual, Draft Report." NRC: Washington D.C. May 1999(b).

—————. NUREG/CR–5512, Volume 3, "Residual Radioactive Contamination from Decommissioning, Parameter Analysis, Draft Report for Comment." NRC: Washington, DC. October 1999(d).

—————. NUREG/CR–5512, Volume 4, "Comparison of the Models and Assumptions used in the DandD 1.0, RESRAD 5.61, and RESRAD–Build Computer Codes with Respect to the Residential Farmer and Industrial Occupant Scenarios Provided in NUREG/CR–5512, Draft Report for Comment." NRC: Washington, DC. October 1999(a).

—————. NUREG/CR–5621, "Groundwater Models in Support of NUREG/CR–5512." NRC: Washington, DC. December 1998(b).

—————. NUREG/CR–6697, "Development of Probabilistic RESRAD 6.0 and RESRAD–BUILD 3.0 Computer Codes." NRC: Washington, DC. December 2000(b).

—————. Regulatory Guide DG–4006, "Demonstrating Compliance With the Radiological Criteria for License Termination, Draft." NRC: Washington, DC. August 1998(c).

—————. Staff Requirements SECY–98–051, "Staff Requirements – SECY–98–051 – Guidance in Support of Final Rule on Radiological Criteria for License Termination." Staff Requirements Memorandum, from John C. Hoyle, U.S. Nuclear Regulatory Commission, to L. Joseph Callan. NRC: Washington, DC. July 8, 1998(d).

Oak Ridge National Laboratory (ORNL). ORNL–5786, "A Review and Analysis of Parameters for Assessing Transport of Environmentally Released Radionuclides Through Agriculture." ORNL: Oak Ridge, Tennessee. September 1984.

Office of Management and Budget (U.S.) (OMB), Washington, DC. "Economic Analysis of Federal Regulations under Executive Order 12866." OMB: Washington, DC. January 11, 1996. (Available on the Web at http://www.whitehouse.gov/omb/inforeg/riaguide.html.)

Sandia National Laboratories. SAND90–7103, "Sensitivity Analysis Techniques and Results for Performance Assessment at the Waste Isolation Pilot Plant." SNL: 1991.

Smith, A.E., P.B. Ryan and J.S. Evans, 1992. "The Effect of Neglecting Correlations When Propagating Uncertainty and Estimating the Population Distribution of Risk." *Risk Analysis*, Vol. 12, No. 4, 1992.

Swedish Radiation Protection Institute (SSI). Technical Report No. 6, "Development of a Reference Biospheres Methodology for Radioactive Waste Disposal, Final Report of the Reference Biospheres Working Group of the BIOMOVS II Study." SSI: Stockholm, Sweden. September 1996.

Thomas, D., W. Hareland, W. Beyeler, D. Brosseau, J. Cochran, T. Feeney, T. Brown, P. Pohl, R. Haaker and B. Fogleman, 2000. *Process for Developing Alternate Scenarios at NRC Sites Involved in D&D and License Termination*, Sandia Letter Report, Sandia National Laboratories, January 2000. (ADAMS Accession Number ML003698994.)

Yu, C., 1999. "RESRAD Family of Codes and Comparison with Other Codes for Decontamination and Restoration of Nuclear Facilities." *Decommissioning and Restoration of Nuclear Facilities*, M.J. Slobodien (ed.). Medical Physics Publishing: Madison, Wisconsin. 1999.

Appendix J

Assessment Strategy for Buried Material

J.1   Generic Description of Situation

For purposes of this appendix, a licensed site has the following characteristics:

- It has buried radioactive material from the late 1960s and early 1970s. Older sites may have 10 CFR 20.304 burial units that were in use prior to the early 1980s, when Section 20.304 was removed from the regulations.

- The site has no other sources of residual radioactivity. Note that if it did, the licensee could use a similar approach but the total dose from all sources would still have to below the dose limit (see Section 2.7 of this NUREG report).

- Information on the inventory (radionuclide concentrations, disposal dates, form, etc.) is limited. However, the licensee has enough information to estimate or to bound the total activity or concentrations of radioactive material present (see Section 4.0 for information on HSA, Section 4.2 for information on characterization surveys, and Section I.2 of Appendix I for information about source-term abstraction).

- It is known that the material is buried deep enough that an external dose is not possible in the current configuration.

- The site also does not have any of the physical limitations that would preclude use of DandD or RESRAD (see Appendix I, Section 4, Tables I–4 and I–6).

- The site is underlain by a unsaturated layer and aquifer. The aquifer is potable and there is enough land for a residential farmer.

- The soil at the site is assumed to be capable of growing crops without significant soil engineering.

Figure J.1 shows a simple conceptual figure of the site.

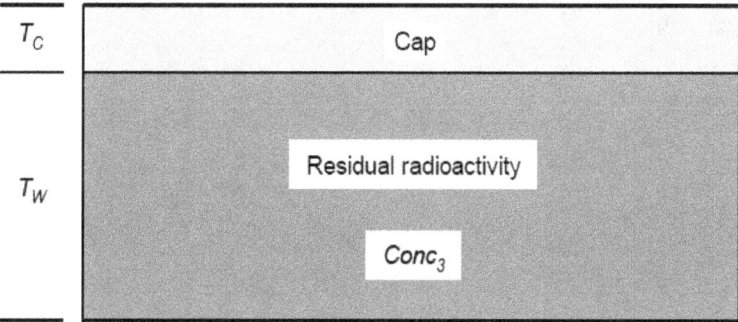

Figure J.1   Conceptual Figure of Soil at the Site.

J.2   Scenario, Exposure Pathways, and Critical Group

To develop the exposure scenario(s) for the critical group, the analyst should to address the following questions:

- How does the residual radioactivity move through the environment?

- Where can humans be exposed to the environmental concentrations?

- What are the exposure group's habits that will determine exposure? (What do they eat and where does it come from? How much? Where do they get water and how much? How much time do they spend on various activities?)

Again, these are not in a strict order to answer and are highly interdependent. In this example, since there is no site information that could preclude the use of a residential farmer, and this is a generic situation, the members of the critical group will be assumed to be residential farmers as described in the default exposure scenarios. It may be possible, even at fairly generic sites, to modify this assumption. For more information, see Section I.3 of Appendix I and Appendix M of this volume.

A conservative analysis could just assume all of the material was spread on the surface (Figure J.2). But by considering the other two questions, two alternate exposure scenarios can be developed: (1) leaching of the radionuclides from their buried position to the ground water, which is then used by a residential farmer; and (2) inadvertent intrusion into the buried residual radioactivity by house construction for a resident farmer with the displaced soil, which includes part of the residual radioactivity, spread across the surface (Figure J.3). The second alternative exposure scenario encompasses all the exposure pathways and, although not all of the source term is in the original position, leaching will occur both from the remaining buried residual radioactivity (if there is any) and the surface soil. Unless differences in the thickness of the unsaturated zone will make a tremendous difference in travel time to the aquifer, the ground water concentrations should be similar and, therefore, will generally result in higher doses than the first alternate scenario.

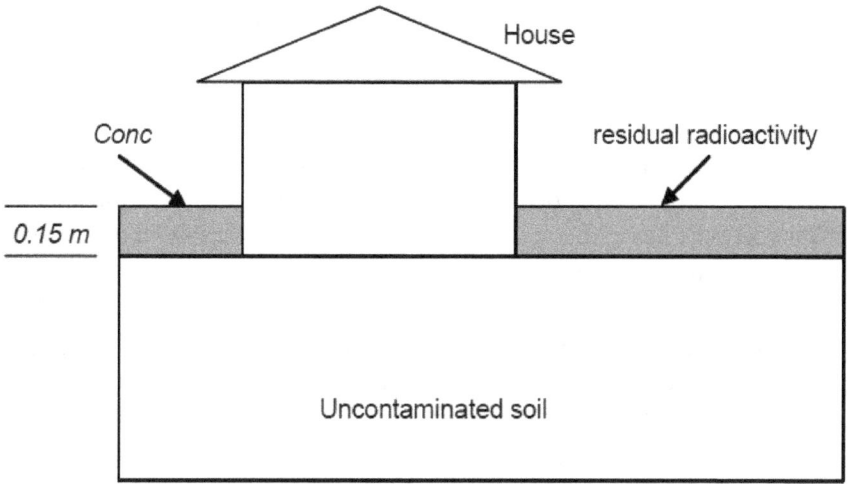

Figure J.2   Alternative Conceptual Disposal.

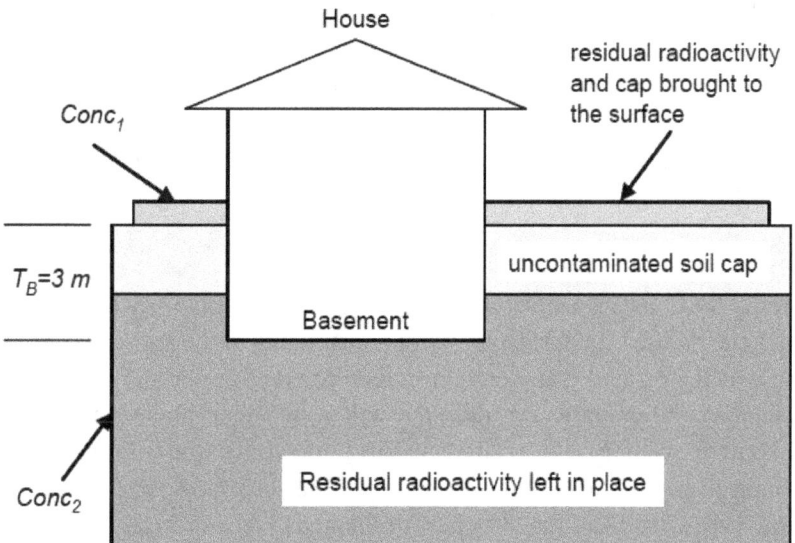

Figure J.3   Conceptual Disposal Problem.

J.3 Analyses with DandD

The DandD code is designed for two-levels of analyses, generic screening and limited site-specific analyses. Licensees can obtain a copy of the DandD code from their NRC reviewer or from NRC staff in the Decommissioning Directorate in NMSS. Screening analyses with DandD relies on the use of default parameter ranges, predefined models, and predefined scenarios. The result is expected to provide a prudently conservative estimate of the dose; that is, an overestimation of the actual dose that individuals might receive. Site-specific analyses with DandD involve the use of some site-specific parameter values with predefined models and scenarios. In following the iterative approach outlined in Section 1.5 of this volume, an analyst is encouraged to start the assessment using the generic screening approach. If the generic screening approach shows that the dose limit can be met, the analysis is done. If license termination or partial site release is being pursued, the analyst would then move onto looking at demonstrating ALARA (Step 6). If the generic screening analysis gives doses above the dose limit, the analyst will need to do some type of sensitivity analysis (Step 8 and 9) to identify parameters where more site-specific data would be helpful in refining the parameter and analysis. In going through the framework a second or subsequent time, the analyst will then use DandD with the site-specific parameter value(s). More details on going through the framework are provided in Section 1.4 of this volume.

DandD is designed to perform screening analyses using only the source inventory or concentration. The NRC reviewer should ensure (1) that the source inventory used is appropriate, (2) that the default parameter values have not been changed, and (3) that there is no known existing ground-water contamination at the site or other features not appropriately represented by the DandD conceptual model. Because the source inventory is the only input parameter in a screening analysis, it is important that there be appropriate justification through the use of: (1) measured data, (2) operational and burial records, or (3) possession limits in the license.[5]

Assuming that the NRC staff reviewer has determined the acceptability of using DandD (in general, DandD can be used for screening, with adjustments to the source term, unless the site is known to have existing ground-water contamination or other important pathways not included in the generic scenario), the primary consideration will be whether the licensee has appropriately converted the source inventory (i.e., source activity) into concentrations and also whether the licensee has changed any of the default parameters. The scenario (i.e., a resident farmer) and conceptual model are already assumed as part of the code. Accordingly, an analyst following the Decommissioning Framework would establish their source concentration (Step 1) and then move directly to calculating the dose (Step 5).

---

[5] For sites with old burials under 10 CFR 20.304, the maximum quantity that was allowed to be buried in trenches should be used with care because NRC staff have identified instances where disposal limits have been exceeded.

The DandD code requires that the source inventory (i.e., activities) be input as a source concentration (i.e., in pCi/g or Bq/g). Accordingly, the inventory should be averaged over some volume. There are three acceptable approaches to calculating the source concentration. These three approaches move from conservative to more realistic ways of dealing with the source concentration.

J.3.1   Mass Balance

Assume that the source activity is distributed uniformly over a default volume of 360 m^3 (12,700 ft^3) through the following relationship:

$$Conc(i) = \frac{Activity(i)}{\left(\rho * Ar * T * CF\right)} \qquad \textbf{(J-1)}$$

where:

$Conc(i)$ = concentration of radionuclide i (pCi / g)

$Activity(i)$ = total activity of radionuclide i (pCi)

Ar = cultivation area in DandD (m^2) = 2400 m^2

ρ = *waste* density (kg / m^3) = 1431 kg / m^3 in DandD

CF = conversion factor (g / kg) = 1000 g / kg

T = thickness of the residual radioactivity (m) = 0.15 m

$\therefore$

$$Conc(i) = \frac{Activity(i)}{\left(5.15x10^8\right)}$$

This approach should be used if the thickness of the residual radioactivity is unknown and it can be safely assumed the volume of residual radioactivity is greater than or equal to 360 m³ (12,700 ft³). Because of the small volume, it will always provide a conservative source concentration. The 360 m³ (12,700 ft³) volume is based on a 2400 m² (25,800 ft²) cultivation area multiplied by a residual radioactivity thickness of 0.15 m (6 in). The activity should be adjusted to account for radioactive decay since waste emplacement through the following relationship:

$$A_t = A_0 e^{-(\lambda t)} \quad \Rightarrow \quad A_t \approx A_0 \left(\frac{1}{2^n} \right) \qquad \textbf{(J-2)}$$

where:

A_t = activity (Ci)

A_0 = initial activity (Ci)

λ = decay constant (year^{-1})

  = $0.693/T_{1/2}$

$T_{1/2}$ = half-life (years)

t = time (years)

n = number of half-lives

DandD accounts for the ingrowth of some progeny by assuming that the parent and daughter radionuclides are at secular equilibrium when the progeny has a half-life less than nine hours and a half-life less than ten percent of the parent half-life. In DandD, an analyst can also assume secular equilibrium for an entire chain by selecting radionuclides that have a "+C" designation.

J.3.2   Single Simulation

A single simulation can be used by assuming that the contaminants are distributed uniformly within the volume of contaminated soil and interspersing clean soil, and assuming that the soil is distributed over a surface to a depth of 0.15 m (6 in). Figure J.2 shows a conceptualization of this alternative. The following relationship can be used to calculate the source concentration:

$$Conc_1(i) = \frac{Activity(i)}{SA * T_w * 1.431x10^6} \qquad \textbf{(J-3)}$$

where:

$Conc_1(i)$ = Concentration of radionuclide i

T_W = True thickness of the residual radioactivity (m)

The equivalent cultivation area (A_r) that should be used in DandD would be:

$$Ar_1 = \frac{SA * T_w}{0.15} - 200 \qquad\qquad\qquad \textbf{(J - 4)}$$

where:

SA = Surface area of residual radioactivity (m^2)

This assumes that the area of the hypothetical house is 200 m^2 (2150 ft^2). It should be noted that the average waste concentration can be used if concentration measurements have been made.

For this approach, the hypothetical individual is assumed to be exposed through all pathways. This second approach requires that the depth of residual radioactivity (T_w) be known. This approach should in general provide comparable results to the dual simulation approach (described below) especially if the ground water is expected to be an important environmental pathway. It should be noted that no credit is taken for an existing cover in order to evaluate the impacts from gamma exposure and because DandD assumes no cover over the residual radioactivity. This approach may not be appropriate for large areas of residual radioactivity because the activity is diluted more as the area is increased. It is recommended that this approach not be used for areas of residual radioactivity larger than 2400 m^2 (25,800 ft^2). For burials larger than 2400 m^2 (25,800 ft^2), the analyst should consider using some other method for calculating the source term. The surface area represents the area of residual radioactivity plus any interspersing clean soil.

J.3.3　Dual Simulation

Assume that the activity is uniformly distributed over the volume of contaminated soil and interspersing clean soil. Further assume that a volume equivalent to the size of a single-family-house basement is excavated and spread out over the land surface to a depth of 0.15 m (6 in). Figure J.3 shows a schematic conceptualization of the problem. Note that there will be two different concentrations, $Conc_1$ and $Conc_2$. $Conc_1$ represents radionuclides mixed with the cover material and spread out over the land. $Conc_2$ represents the concentration of the remaining radionuclides left in place (i.e., in the waste but not excavated). The two zones of residual radioactivity will not represent the same exposure to the hypothetical farmer. The farmer can be exposed through all pathways from the top zone (at concentration $Conc_1$); however, the farmer's exposure to the second zone will be limited primarily through what is leached out and reaches the ground water. Because of the two concentrations and different exposure pathways associated with each, this conceptual problem will require two simulations with the DandD code. The first simulation is used to evaluate exposure from contaminants spread out over the land surface. For this first simulation all exposure pathways are considered with the exception of drinking water and irrigation (these will be covered in the second simulation). To exclude the drinking water and irrigation pathways set the following parameters to zero: water ingestion, domestic use, infiltration rate, and irrigation rate. If the total activity within the waste area is known, the following approach can be used to calculate source concentrations for this first simulation:

$$If\ T_c + T_w > 3,$$

$$Conc_1(i) = \frac{Activity(i)(3 - T_c)}{SA * T_w * 4.293x10^6}$$

$$If\ T_c + T_w < 3,$$

$$Conc_1(i) = \frac{Activity(i)}{SA * 4.293x10^6}$$

(J - 5)

where:

$Conc_1(i)$ = concentration of material on the surface (pCi / g)

SA = surface area of residual radioactivity (m^2)

T_c = thickness of cap (m)

T_w = thickness of residual radioactivity (m)

Derivation of the above equations is provided in Section J.7 of this appendix. In the above formulas, the cap and waste are both assumed to be represented by soil at a density of 1.43 g/cm^3 (the DandD V1.0 default). In addition, the basement height is assumed to be three meters. The surface area represent the area of residual radioactivity and any interspersing clean soil. The cultivation area (Ar) parameter in DandD should be set to 4000 m^2 (43,100 ft^2), i.e., 600 m^3 (21,200 ft^3) divided by 0.15 m (6 in). The area of the hypothetical house is assumed to be 200 m^2 (2150 ft^2).

The second simulation is used to evaluate exposure from the remaining inventory, which could leach into the ground water. The second simulation can be eliminated entirely if the licensee can demonstrate conclusively that the ground water will not be used at the site. Furthermore, the second simulation can be eliminated if the contaminated volume is ≤600 m^3 (21,200 ft^3) which represents excavation of the entire source term. If the second simulation is eliminated, then all pathways including drinking water and irrigation should be evaluated in assessing the material brought to the surface, for the first simulation.

Because the second simulation is used to evaluate exposure from contaminated ground water, several parameters will have to be set to zero in order to eliminate or reduce the calculated exposure from the other pathways (i.e., external, inhalation, plant ingestion, and resuspension). Accordingly, the following parameters will have to be set to zero for the second simulation: floor dust, resuspension factor, indoor dust, outdoor dust, gardening dust, indoor breathing, outdoor breathing, gardening breathing, time spent gardening, time spent outdoors, and soil ingestion rate. In addition, the indoor shielding factor should be set to 1.0 and the plant mass loading factor should be set to 0.0011 (the smallest value allowed in DandD). As with the first simulation, the surface area represents the area of residual radioactivity plus interspersing clean soil. Source concentrations for the second simulation can be obtained using the following functional relationship:

$$Conc_2(i) = \frac{Activity(i)}{SA * T_w * 1.431x10^6}$$ (J - 6)

where:

$Conc_2(i) = concentration$ in waste area for second simulation (pCi / g)

For this second simulation, the activity removed for the first simulation is not accounted for, because irrigation and drinking water are excluded in the first simulation. Accordingly, the whole activity is used in evaluating impacts from exposure from these pathways in the second simulation. The cultivation area (Ar) parameter in DandD should be calculated as follows:

$$Ar_2 = SA - 200$$ (J - 7)

Again, the area of the hypothetical house is assumed to be 200 m² (2150 ft²).

The total dose can be obtained by summing the dose from the two simulations. If the peak doses for both simulations occur at roughly the same time, the reported doses from each simulation can be simply added together. However, if the two peaks occur at vastly different times, some type of integration of the two dose curves will be needed. In any event, it will be always conservative to simply sum the two peak doses.

The activities for both Equations J-5 and J-6 should be adjusted to account for radioactive decay since waste burial. This third approach (i.e., the dual simulation approach) also requires that the depth of residual radioactivity be known. In addition, it accounts for the presence of an existing cover over the burial. If there is no cover over the burial area, the formulations are still valid, the analyst only has to set T_c to zero. Although less conservative than the mass balance and single simulation approaches, the dual simulation approach should be appropriate in most cases because it is consistent with the assumed resident farmer scenario. That is, the resident farmer scenario assumes that an individual's activities take place over the whole area and is not limited to exposures from isolated spots; thus, the concentration contacted over time is best represented by a spatially averaged concentration. However, for large areas this approach is not appropriate because the activity becomes more diluted as the surface area gets larger. As a cut off, it is recommended that this approach not be used for areas of residual radioactivity larger than the 2400 m² (25,800 ft²) area assumed in DandD. For burials larger than this, the analyst will need to consider using some other method to devise the source term.

The above formulas can be used if the analyst knows the total activity in the waste area. If concentration measurements have been made, the average concentration can be used. For the first simulation, the average concentration can be used in the following relationship:

$$If\ T_c\ +\ T_w\ >\ 3,$$

$$Conc_1(i) = \frac{\left[\overline{Conc(i)}(3 - T_c)\right]}{3}$$

$$If\ T_c\ +\ T_w\ <\ 3,$$

$$Conc_1(i) = \frac{\overline{Conc(i)} * T_w}{3}$$

(J - 8)

where:

$\overline{Conc(i)}$ = average concentration of rationuclide i

from measurements (pCi / g)

For the second simulation, the arithmetic average concentration from the measurements can be used directly in the analysis.

For all three of these approaches, it is assumed that the activity is uniformly distributed over some defined volume. In using either of the last two approaches it is important to assess the appropriateness of assuming that the activity is uniformly distributed over the waste volume. This assumption may not be appropriate for situations where the waste is very heterogeneous or if there are isolated large areas of elevated concentrations. Using sensitivity analyses or other methods, licensees should demonstrate the appropriateness of assuming an uniform distribution.

No credit is assumed to be taken for any waste containers (e.g., metal drums or boxes); that is, containers are assumed to have failed or decayed. In general, this assumption should be appropriate because of the expected lifespan of most waste containers are expected to be short relative to the time frame of the dose assessment. The equations described in these three approaches can be easily evaluated, especially for a large number of radionuclides, in a spreadsheet.

After evaluating the source concentration, the NRC staff reviewer should evaluate the licensee's DandD output report. Any changes to default parameters are echoed in the output. Accordingly, it is important that the NRC staff reviewer request a copy of the licensee's output report. NRC staff can also determine that the default parameter set has not been altered by running DandD using the licensee's source concentration as input.

J.4 Analyses with RESRAD

RESRAD is a computer code developed by Argonne National Laboratory (ANL) for the Department of Energy (DOE) to calculate site-specific residual radiation guidelines and radiation dose to future hypothetical on-site individuals at sites contaminated with residual radioactive material. The RESRAD code was adopted by DOE in Order 5400.5 for derivation of soil cleanup criteria and dose calculations, and it is widely used by DOE, other Federal agencies, and industry.

The RESRAD code is continuously updated. NRC staff reviewers will need to ensure that the latest version has been used in assessments that they are reviewing. If an earlier version has been used, the analyst should be required to document that the earlier version is not expected to give significantly different results from the latest version. The RESRAD Web site http://www.ead.anl.gov/~resrad/reshstry.html provides information on all the updates from one version to another.

RESRAD, like DandD, has an assumed conceptual model; therefore, the analyst only has to determine if the assumed conceptual model is appropriate for the problem. However, unlike DandD, RESRAD does not have prescribed exposure scenarios. The analyst should develop the exposure scenario by switching on or off various exposure pathways. For the standard resident farmer scenario used by the NRC staff, all of the exposure pathways should be switched on with the exception of the radon pathway. The analyst needs to provide justification for excluding any of the other pathways. For example, if it can be shown that the ground water at the site cannot be used because of either poor ambient water quality (e.g., salinity) or low yields, it should be justifiable to exclude the ground-water pathway. A finding that the ground water is unsuitable is typically made in coordination with State agencies. For more information on removing pathways, see Appendix I, Section 3.

RESRAD, like DandD, requires that the radioactive inventory be input as a source concentration. Because RESRAD is designed for conducting site-specific analyses, it is expected that for most analyses, the analyst will have data on radionuclide concentrations at the site.[6] Given that we are assuming a resident farmer scenario, it should be appropriate to use the arithmetic average of the radionuclide concentration in the analysis (note this also includes any interspersing clean soil). RESRAD allows the user to input information on the area and thickness of the residual radioactivity (i.e., these are not fixed, although defaults are provided). For surface residual radioactivity [$\leq$0.9 m (3 ft), the default rooting depth in RESRAD], the site-specific mean concentration, area of residual radioactivity, and thickness of the residual radioactivity can be used directly in the code. For deeper residual radioactivity or if the residual radioactivity is capped (such as with burials) some assumptions should be made about how much waste may be brought to the surface and how it may be mixed with uncontaminated soil. In general, the schematic in Figure J.3 should apply. Analyzing this conceptual model, as with DandD, requires two simulations. During the first simulation it is assumed that a small volume of waste [600 m^3 (21,200 ft^3)] is brought to the surface and spread out over an area to a depth of 0.9 m (3 ft). For the first simulation, the important result is the dose from exposure to the material brought to the

---

[6]   RESRAD is primarily designed to look at radioactively contaminated soils; therefore, for analyses involving other types of wastes, the analyst will have to make some assumptions about the waste form and how the radionuclides will be released from this waste form. These assumptions should be clearly laid out.

surface, such as, direct gamma radiation, inhalation, soil ingestion, and plant ingestion (excluding irrigation with contaminated water). Exposure from ground water, irrigation, and aquatic use should be considered in the second simulation. Accordingly, the drinking water and aquatic pathways should be switched off for the first simulation. In addition, the irrigation rate should be set to zero. The source concentration for this first simulation would be derived using Equation J-8 as previously defined.

The concentrations should be adjusted to account for radioactive decay. The area that should be used in the first simulation should be 700 m^2 (7550 ft^2), i.e., 600 m^3 (21,200 ft^3) divided by 0.9 m (3 ft). The assumed contaminated thickness would be 0.9 m (3 ft) (note: T_w that should be used in the above formulation represents the true residual radioactivity thickness in its current configuration). The second simulation looks at effects from exposure from the remaining waste. The primary environmental transport pathway for this remaining waste should be ground water. For the second simulation the external gamma, inhalation, and soil ingestion pathways should be switched off. In addition, the mass loading for foliar deposition parameter should be set to zero. Furthermore, if the residual radioactivity is presently capped, the residual radioactivity can be assumed to be covered for the second simulation, unless there are reasons to model the situation with the cover removed (e.g., through a high soil erosion rate). The source concentration for the second simulation should be the mean concentration for the waste area. This includes interspersing clean soil. The area and thickness of the residual radioactivity used in the second simulation would be based upon the true existing waste zone configuration. Accordingly, to use this approach the analyst will have to know something about the waste zone configuration.

An alternative to using the dual simulation approach is to simply assume that the waste is uniformly distributed over the source volume, taking no credit for the cover (i.e., by assuming that the cap is not present). This should provide comparable, but conservative results to the dual simulation approach especially if the ground water is an important pathway. Using this simpler approach, the analyst would use the mean concentration as the source concentration.

In using either of these approaches it is important for the NRC staff reviewer to assess the appropriateness of assuming that the activity is uniformly distributed over the waste volume. This assumption may not be appropriate for situations where the waste is very heterogeneous or if there are isolated large areas of elevated concentrations.

If all that is known is the source inventory (activities), such as at some old burial sites, the source concentration can be calculated with Equations J-5 and J-6. It should be noted that the density for the residual radioactivity should be set to 1.431 or the concentration should be calculated with the same density assumed in the analysis.

RESRAD can be executed both in deterministic and probabilistic modes. Analysts should use the probabilistic mode because a single default parameter set has not been established for performing generic analyses. Although RESRAD has default parameters, these parameters may or may not be suitable or provide a conservative estimate of the dose for any given site. The probabilistic data set of parameter ranges is approved for use for generic site-specific analyses. See Section I.6 from Appendix I for more information.

J.5 Derivation of Equations J-4 through J-8

J.5.1 Equation J-4

For DandD, the cultivation area needs to be equivalent to the area of residual radioactivity. Therefore for the single simulation approach, the size of the area of residual radioactivity is an equivalent volume of the waste limited to a depth of 15 cm (6 in).

$$Ar_2 = \frac{Vol_{waste}}{0.15} \qquad \textbf{(J - 4)}$$

where:

$Ar_2 = $ cultivation area for DandD (m^2)

$Vol_{waste} = $ volume of residual radioactivity (m^3)

$\qquad = SA * T_w$

$SA = $ surface area of residual radioactivity (m^2)

$T_w = $ thickness of residual radioactivity (m)

The area taken up by the house is subtracted; therefore, the equivalent cultivation area is:

$$Ar_2 = \frac{SA * T_w}{0.15} - 200$$

J.5.2 Equation J-5

The initial concentration in the waste or residual radioactivity can be derived as follows:

$$Conc_0(i) = \frac{Activity(i)}{Vol_{waste} * \rho_{waste} * CF} \qquad\qquad \textbf{(J - 5)}$$

where:

$Conc_0(i) =$ initial concentration of radionuclide i in the
residual radioactivity (pCi / g)

$Vol_{waste} = volume$ of residual radioactivity (m^3) $= SA * T_w$

$SA = surface$ area of residual radioactivity (m^2)

$T_w = thickness$ of residual radioactivity (m)

$\rho_{waste} = density$ of residual radioactive materials $=$ 1431 kg / m^3 (DandD default)

CF = conversion factor $=$ 1000 g / kg

$\therefore$

$$Conc_0(i) = \frac{Activity(i)}{SA * T_w * 1.431x10^6}$$

The concentration in the material brought to the surface, for the first simulation should depend upon how much of the basement extends into the waste or residual radioactivity. This concentration can be represented as a fraction of the volume of material excavated to the total volume of material in the basement.

$$Conc_1(i) = Conc_0(i) * Fraction_1$$

where:

$Conc_1(i)$ = concentration of radionuclide i in the material brought to the surface (pCi / g)

$$Fraction_1 = \frac{Vol_e}{Vol_b}$$

Vol_e = volume excavated (m^3)

$$= A_b(T_b - T_c) \qquad\qquad T_b < T_c + T_w$$

$$= A_b * T_w \qquad\qquad T_b > T_c + T_w$$

where:

A_b = area of house (m^2)

T_b = thickness of the basement (m)

T_c = thickness of the cap (m)

T_w = thickness of residual radioactivity (m)

Vol_b = volume of the basement (m^3)

$$= A_b * T_b$$

If we assume a basement thickness of 3 meters,

$$Vol_e = A_b(3 - T_c) \qquad\qquad 3 < T_c + T_w$$

$$= A_b * T_w \qquad\qquad 3 > T_c + T_w$$

$\therefore$

$$Conc_1(i) = Conc_0(i)\frac{A_b(3 - T_c)}{A_b * 3} \qquad\qquad 3 < T_c + T_w$$

$$= Conc_0(i)\frac{A_b * T_w}{A_b * 3} \qquad\qquad 3 > T_c + T_w$$

Canceling terms and substituting in $Conc_0(i)$:

$$Conc_1(i) = \frac{Activity(i)(3 - T_c)}{SA * T_w * 4.293x10^6} \qquad\qquad 3 < T_c + T_w$$

$$= \frac{Activity(i)}{SA * 4.293x10^6} \qquad\qquad 3 > T_c + T_w$$

J.5.3   Equation J-6

For the second simulation, the impacts from gamma radiation or plant uptake are not of concern; therefore, the 0.15 m (6 in) residual radioactivity thickness is not important. Therefore, concentrations can be determined based upon the existing geometry of the residual radioactivity. The concentration in the waste or residual radioactivity is simply:

$$Conc_2(i) = \frac{Activity(i)}{Vol_{waste} * \rho_{waste} * CF} \qquad \textbf{(J - 6)}$$

where:

$Conc_2(i) = concentration$ of radionuclide i in residual radioactivity (pCi / g)

Activity(i) = total activity of radionuclide i in residual radioactivity (pCi)

$Vol_{waste} = volume$ of residual radioactivity (m^3) $= T_w * SA$

$T_w = thickness$ of residual radioactivity (m)

SA = surface area of residual radioactivity (m^2)

$\rho_{waste} = density$ of residual radioactive material $=$ 1431 kg / m^3

$CF = conversion$ factor $= 1000$ g / kg

$\therefore$

$$Conc_2(i) = \frac{Activity(i)}{SA * T_w * 1.431x10^6}$$

J.5.4   Equation J-7

The cultivation area should be equivalent to the area of residual radioactivity. The default cultivation area cannot be used if the residual radioactivity is assumed to spread out over an area different than the default of 2400 m^2 (25,800 ft^2). In this case, the size of the area of residual radioactivity is *SA*. In addition, the assumed area of the house is subtracted; accordingly,

$$Ar_2 = SA - 200 \qquad \textbf{(J - 7)}$$

J.5.5   Equation J-8

Derivation of Equation J-8 is the same as Equation J-5; however, the initial concentration ($Conc_0(i)$) is assumed to be the average from the measurements.

J.6   References

Bonano, E.J., P.A. Davis, and R.M. Cranwell, "A Review of Uncertainties Relevant in Performance Assessment of High-Level Radioactive Waste Repositories," U.S. Nuclear Regulatory Commission: NUREG/CR–5211, September 1988.

Kozak, M.W., N.E. Olague, D.P. Gallegos, and R.R. Rao, "Treatment of Uncertainty in Low-Level Waste Performance Assessment," 13th Annual DOE Low-Level Waste Conference, November 19-21, 1991.

Maheras, S.J. and M.R. Kotecki, "Guidelines for Sensitivity and Uncertainty Analyses of Performance Assessment Computer Codes," National Low Level Waste Management Program: DOE/LLW–100, September 1990.

NCRP, "A Guideline for Uncertainty Analysis in Dose and Risk Assessments Related to Environmental Contamination," National Council on Radiation Protection and Measurements: NCRP Commentary No. 14, May 10, 1996.

Appendix K

Dose Modeling for Partial Site Release

K.1 Dose Modeling Considerations for Partial Site Release

This appendix consists of the technical guidance, for review of the release, under 10 CFR Part 20, Subpart E, of a portion of a site before final termination of the entire site; a process called partial site release. This is generally applicable to Decommissioning Groups 2–5.

The guidance in this appendix has been developed to encompass the needs of the most complex situations, but the specific informational needs for a partial site release request should be tailored to the complexity and safety significance of the proposed action. This appendix is split into three sections. The first section, which complements Chapter 5 of this volume, details the review criteria to be used in assessing compliance with Subpart E. The second section provides technical information, which supplements the guidance in Appendices H and I. The third section contains two hypothetical simple examples of partial site release considerations.

The guidance is focussed on partial site release requests that occur prior to the decommissioning plan (DP) being approved, but it is also applicable to those requested for phased release of areas after DP approval (see Section K.1.8).

K.1.1 Partial Site Release Reviews

For a partial site release (PSR), dose modeling is not necessarily limited to the dose caused by areas with residual radioactivity on the partial site, but, also, residual radioactivity outside of the partial site. Areas around the PSR may contribute direct radiation or have natural processes that may move residual radioactivity to the partial site release. For purposes of this volume, "offsite sources" means potential sources of exposure that are not on the partial site, but still on impacted areas under (or previously under) the control of the licensee. For example, a licensee may have impacted groundwater under the site. At the present time, the groundwater under the partial site release is not impacted. Possible movement of this impacted groundwater (an "offsite source" of residual radioactivity) from the remaining site to the PSR must be considered.

In addition to compliance analyses for the PSR, there should be evaluations of potential prospective analyses. These analyses should evaluate how the PSR could impact the license termination of the licensed site, including any additional PSRs. For example, releasing an area of the site at higher DCGLs than is likely for the rest of the site could constrain the future decommissioning, forcing the licensee to use DCGLs for the rest of the site that are below what they could have been if the PSR never occurred.

K.1.2 Incorporation into Review Process

The licensee may use either a screening or site-specific dose assessment to show compliance with Subpart E. Although they may use generic screening analyses to create the DCGLs of PSRs, the overall review should be a site-specific review. The NRC reviewer should use the appropriate section of Chapter 5 for the review of the assessments and the additional considerations for source terms, scenarios, and pathway identification detailed below.

K.1.3   Evaluation Information

The difference between a dose assessment for license termination of the entire site license and a dose assessment for a PSR is that other sources under control of the licensee may affect the potential dose on the PSR. In license termination of the entire site, when the site is released for unrestricted use, there are no offsite sources remaining under the control of the licensee to affect the projected dose for residents or workers using the site. In contrast, after a PSR, the remaining licensed site may still be operating and thus have dose contributions to the critical group receiving dose from the PSR, such as from surface water run-off or ground water migration. In addition, sources on the remaining licensed site may result in dose to the public after unrestricted release of the remaining site in the future, such that members of the PSR critical group receive doses from sources on both the PSR and on the remaining (now terminated) site.

NRC staff should review the licensee's assessment of offsite sources that may influence the dose analysis, and NRC staff would evaluate these sources similar to a source on the PSR. The development of the appropriate scenarios for compliance evaluation should identify which sources NRC staff should focus on. The primary areas of additional consideration given to PSR cases in developing reasonable scenarios are:

- How does the licensed site or a previous PSR influence the dose on the PSR (e.g., effluent releases, ground-water plumes, future combined use, etc.)?

- How could the PSR influence dose estimates for the licensed site during its decommissioning?

- How does the PSR influence previous PSRs (e.g., possible effects on the PSR's final DCGLs to limit the impacts on the previous PSR, so that the potential dose on the previous PSR does not exceed Subpart E)?

K.1.4   Development and Identification of Partial Site Release Scenarios

Based on the questions above, scenarios can be divided for purposes of analysis into two categories: compliance and prospective. Analysis of both of these categories of scenarios should assist in establishing the finality of the decision regarding the PSR.

Compliance scenarios involve assessing the compliance of the proposed PSR, or the continued compliance of a previous PSR affected by the proposed PSR, with the Subpart E dose limit. Compliance scenarios involve current or future exposure routes between the PSR and the previous PSR or the licensed site [e.g., see Section K.3.1's gamma radiation from the low-level waste storage area]. Compliance scenarios that calculate exposures in excess of the regulatory limit or a licencee self-imposed limit (e.g., from a previous PSR's approval) should then entail remedial actions on the proposed PSR [not the previous PSR(s)] or more realistic dose assessments.

Prospective scenarios involve assessing possible interactions between the PSR and any future decommissioning actions on the licensed site, including another PSR. The purpose of prospective analyses is to scope out the potential interactions in the future and address them either by additional remediation of the PSR or by placing or acknowledging possible the constraints on future decommissioning of the other sources.

K.1.4.1   Screening of Features, Events, and Processes

NRC staff should review the licensee's analysis using the worksheet in Appendix L to guide reviews of potential sources of interaction between the PSR and offsite sources. The purpose of this screening is to answer the questions from above, by identifying any potential interaction and evaluating the impact(s) on the dose calculations.

The licensee should have adequate justification for excluding each of the potential sources, transport processes, or exposure pathways not evaluated in the dose assessment analyses. Justification can be quantitative or qualitative.

There are three acceptable methods of handling the offsite impact related to interactions that have not been screened out: (1) incorporate the source, transport mechanisms, and pathways into the conceptual model and the dose analyses; (2) remediate those sources; or (3) apply constraints on the PSR's DCGLs, to accommodate potential exposures from offsite sources, or to previous PSRs.

Therefore, NRC staff should evaluate the information to verify that:

- the licensee screened potential interactions with the licensed site and previous PSRs

- the screening arguments are justified; and

- the licensee properly addressed the remaining potential exposure pathways.

Section K.3.1 illustrates some of these considerations.

K.1.4.2   Screening the Use of the Partial Site and Other Areas by the Critical Group

A member of the critical group could be potentially exposed to higher doses than those resulting from the PSR alone. This would be through the use of other impacted areas, after they have been released (including previous PSRs), in addition to continuing the use of the PSR.

Three general situations can result in doses to individuals that are higher than that for the PSR alone:

- One of the land area's DCGLs took into account the small size of the area.

- Use of more than one exposure area would result in the dose receptor receiving exposure from radionuclides or sources not present on the PSR.

- Use of more than one exposure area would result in the dose receptor receiving exposure from new exposure pathways or would increase the degree of exposure to a current exposure pathway.

Section K.3.2 illustrates a hypothetical review of a situation involving multiple land use.

If the licensee has used the same DCGLs for a previous PSR, or commits to use the same DCGLs for areas surrounding the partial site, multiple use of the areas is not likely to result in a higher dose, as long as none of the above situations is present, and the scenarios and assumptions used in the calculations are appropriate for all areas.

If the licensee has (1) used different DCGLs, (2) has at least one of the above situations present; (3) found that the scenarios and assumptions regarding the proposed PSR used for a previous PSR are no longer appropriate, or (4) has not committed to use the same analyses for surrounding areas (as long as it would be valid for the other areas, too), then NRC staff should evaluate the licensee's analyses of potential multiple use scenarios. For example, for interactions with a previous PSR, NRC staff needs to look at: (1) any prospective analyses and associated constraints, if established, done for the previous PSR; (2) the estimated dose from the residual radioactivity on both the previous PSR and the proposed PSR; and (3) any new or updated analyses performed by the licensee.

K.1.5 Partial Site Release Evaluation Criteria

NRC staff should verify the following points regarding PSR considerations:

- For PSR and previous PSR interactions:
  — the scenarios used in the prospective analyses for the previous PSR, that analyzed the interactions, between the previous PSR and the area encompassed by the proposed PSR, continue to be appropriate, or have been updated appropriately;

  — the licensee did incorporate any constraints, imposed by the previous PSR, that remain appropriate in determining the DCGLs for the proposed PSR;

  — the licensee appropriately identified those sources, that may affect the dose to the average member of the critical group, on either the previous PSR or the proposed PSR;

  — the licensee provided adequate justification for each excluded potential source, transport mechanism, and pathway;

— the licensee incorporated, or addressed by other appropriate means, any sources, transport mechanisms, or pathways that could not be screened out;

— the licensee evaluated (either quantitatively or qualitatively) reasonable scenarios to account for interactions between the previous PSR and proposed PSR. This includes the prospective analyses for the previous PSR, as well as any new scenarios that needed to be evaluated based on new information; and

— the DCGLs for the proposed PSR should not result in exposures exceeding the dose limit at either the previous PSR or the proposed PSR. The dose assessment for the proposed PSR should also include any appropriate contributions from the licensed site.

- For PSR and interactions with the licensed site, considering both current and future sources (e.g., potential impacts from other decommissioning activities, or potential future parallel use of impacted areas on the licensed site and the PSR):

— the licensee appropriately identified those current and potential future offsite sources that may affect the dose calculated for the partial site;

— the licensee provided adequate justification for each excluded potential source, transport mechanism, and exposure pathway;

— the licensee incorporated, or addressed by other appropriate means, any sources, transport mechanisms, and exposure pathways that could not be screened out;

— the licensee evaluated reasonable scenarios to account for interactions between the proposed PSR and the licensed site—this includes any prospective analyses that estimate exposures after the licensed site is decommissioned;

— the DCGLs should not result in exposures exceeding the dose limit at the proposed PSR. The dose assessment for the proposed PSR should also include any appropriate contributions from previous PSRs; and

— the licensee has clearly documented any constraints placed on current and potential future sources of exposure on the licensed site.

K.1.6   Dose Modeling Approaches

Licensees proposing PSRs may still be able to use either dose modeling option: screening numbers or site-specific analyses.

- If a licensee proposes to use the screening criteria, the following should be by NRC staff:

— Interactions with the licensed site or previous PSRs have been appropriately evaluated;

— Any sources of potential exposure from the licensed site have been either constrained or remediated;

— Any sources of potential exposure increasing either the dose to residents or workers on the proposed PSR or a previous PSR have been either constrained or remediated;

— The screening criteria have been appropriately scaled by all the considerations associated with the PSR. For example, in Section K.3.1, the licensee limited the ground water dose to 0.05 mSv (5 mrem). Therefore, the screening values for the PSR's surface soil would need to be scaled to 80 % [0.2 mSv (20 mrem)/0.25 mSv (25 mrem)] of the published values or those received by using the current version of the DandD computer code; and

— The PSR and its analysis meet the other requirements of Sections 5.1.1 and 5.1.2, as appropriate.

• If a licensee uses site-specific modeling, the following have to be verified by NRC staff:

— All sources from the licensed site or previous PSRs have been incorporated, as necessary, into the analyses;

— Any constraints used by the licensee have been properly reflected in calculating the DCGLs; and

— The modeling meets all the other review criteria of Section 5.2, as appropriate.

K.1.7 License Termination: The Effect of Previous Partial Site Releases

At the time of final license termination, NRC staff should take into consideration any previous PSRs. The entire site (including the previous PSRs) should meet the Subpart E dose limit. Reviewing the impact on the license termination is exactly the same as that discussed under "For PSR and previous PSR interactions" above. In this case, it is necessary to consider the rest of the licensed site as the PSR.

K.1.8 Use of Partial Site Release During Decommissioning

Reviewers can use this guidance when licensees request release of portions of their site(s), either as part of a DP submittal or after the DP has been approved. After the DP has been approved, some of the issues are not as relevant. If the licensee has prepared a DP for the entire site, more information may likely be available at the time of the PSR. Importantly, the NRC reviewer may be able to review the PSR's DCGLs, as well as those for other areas of the site, and any plans on continued remediation of other areas of the site. Prospective analyses of critical group behavior after the entire site is released may still need to be completed, but these scenarios are likely to be easier to define and evaluate.

K.2 Partial Site Release Technical Basis for Dose Modeling

K.2.1 Considerations for Partial Site Release Dose Assessments

Although the license termination requirements in Subpart E provide options of unrestricted and restricted release, normally, PSRs would be used for unrestricted release. PSR has many aspects in common with the existing approach for unrestricted release, and the available guidance is generally applicable. One key difference is that PSR does not occur concurrent with license termination. As a result, continuation of licensed activities outside of the PSR represents a potential source of exposure. In turn, the residual radioactivity on the PSR may impact dose analyses for other areas of the facility during subsequent PSR requests and/or eventual license termination.

Because this volume's guidance requires that the dose assessment include all significant exposure pathways, the need to consider the potential for accumulation processes resulting in increased radionuclide concentrations over time is not a new concept for PSR. Nonetheless, the importance of accumulation is increased under PSR because the license will not be terminated. The existing site areas outside the PSR are not required to be remediated at the time of PSR. Therefore, the potential for accumulation, on the partial site, that could impact the dose assessment, is increased.

One of the most important concepts of the guidance for PSR is the finality of the decision. The purpose of the guidance is to establish the scope of the review and focus NRC staff attention and resources to early identification of aspects important to compliance. The primary objective of the PSR guidance is to ensure that any PSR meets Subpart E requirements, even if potentially impacted by later PSRs and/or license termination. The secondary objective of the PSR guidance is designed to ensure, at the time of license termination, that all prior released areas are considered and included, as necessary, in dose assessments to provide assurance that the entire site meets Subpart E requirements. To meet these two objectives, the licensee is requested to perform both compliance calculations for current conditions at the PSR (or the effect on a previous PSR) and prospective calculations to estimate the impact on other decommissioning activities by a licensee. This set of analyses should help ensure that the DCGLs chosen for the PSR should not result in the need for future remediation of the PSR or unduly constrain the decommissioning of the entire site.

The existence of a PSR may place constraints on future activities that occur nearby, including currently licensed activities to limit the potential for exposures to a critical group residing or working (non-radiation workers) on the PSR from exceeding other public dose limits (see Section 3.4 of this volume). For example, 10 CFR 20.1301(d) requires that a fuel cycle facility comply with the requirements of 40 CFR Part 190. This standard (40 CFR Part 190) limits the total dose that a member of the public may receive from all fuel cycle activities. For the remaining site to show compliance with 40 CFR Part 190, the dose from the partial site and any effluents from the remaining site's operations will need to be combined. The constraint, then, is not on the compliance of the PSR with Subpart E limits but on maintaining compliance with 40 CFR Part 190 for the remaining activities (e.g., an independent spent fuel storage installation). Existing NRC effluent control and operational dose limits and their associated guidance should generally limit operational releases to acceptable levels. Adjustments may need to be made to effluent compliance calculations or environmental sampling areas to account for the removal of the PSR from licensee's control (e.g., because of changing the site boundary). Additional guidance addressing these constraints (other than 10 CFR Part 20, Subpart E) may be developed in the future.

The dose criterion of 40 CFR Part 190 is based on actual annual doses. Thus, it should be noted that a licensee does not have to use their Subpart E compliance calculation (generally a prospective calculation) as the dose contribution from the PSR for the 40 CFR Part 190 compliance determination. Information will be available to the licensee to estimate the actual annual dose based on the actual activities that occurred on the PSR. For example, a PSR is released using a scenario similar to a residence. The next year the licensee is performing calculations to show compliance with 40 CFR Part 190. In the year since the PSR was approved, the land was used as a public parking lot. For compliance with 40 CFR Part 190, the licensee can evaluate the PSR dose to a member of the public using the parking lot, instead of the residence scenario used in approving the PSR for 10 CFR Part 20, Subpart E. The licensee may also account for decay since the FSS.

For non-impacted areas, the NRC technical review should normally be limited to the sufficiency of bases in the site characterization. PSR, effectively, narrows the definition of non-impacted because of the possibility of future licensee actions resulting in impacting the PSR. For example, in some cases, close proximity to existing operations, contaminated areas, future remediation sites, or potential storage areas, may not allow a licensee to designate an area as non-impacted and release it without a dose assessment. For impacted areas, PSR requests can involve a more complicated compliance demonstration and review effort. The site characterization should include areas of the site outside of the PSR to the extent necessary to provide assurance that residual radioactivity from the licensed site (or previous PSR) is unlikely to transport material, to the PSR, that would result in potential exposures, to users of the property (including subsurface water and ground water), in the future.

K.2.2   Partial Site Release and Decommissioning Guidance

This NUREG report was written, in large part, to address decommissioning of sites as part of the license termination process. As a result, termination of the license is often discussed as the end result of decommissioning. When applying this guidance to a PSR, most of the references to license termination should be regarded to imply the completion of the partial site decommissioning effort. Despite the frequent use of the term, "license termination," licensees should be aware that PSR will not result in license termination, as the entire licensed site (including any PSRs) should meet the Subpart E requirements at the time of license termination. Therefore, true license termination issues only need to be considered in PSR reviews when assessing prospective analyses that may raise issues that need to be considered or analyzed at the time of license termination (e.g., creation of new license conditions that identify pathways that should be included in DCGL calculations). Licensees should also be aware that the existence of a PSR adjacent to impacted areas could place limitations on future decommissioning methods and actions related to the license termination (e.g., to minimize the potential for decommissioning to re-contaminate previously PSRs).

The terms "site" and "facility" are used interchangeably in this NUREG report. Under PSR, most of the references to site or facility will apply to the boundaries of the area proposed for PSR (i.e., the area to be decommissioned). Exceptions to this would be when the consolidated guidance discusses the need to collect site characterization information, in which case the terms "site" or "facility" can include areas beyond the boundary of the PSR and potentially encompass the entire site and any previous PSR(s), as necessary to establish contaminant source, transport, and exposure pathways for DCGL calculations. NRC staff is expected to use pre-submittal meetings with the licensee to develop the amount of information needed on the licensed site for specific areas of such safety concerns.

K.2.3   MARSSIM and Partial Site Release

The MARSSIM approach involves demonstration of compliance on a survey unit by survey unit basis. Survey units are determined based on the expected level of residual radioactivity in areas across the site as well as spatial and topographical considerations. By allowing compliance demonstration by survey unit, the current approach is congruent with a PSR concept. As a result, in general, the MARSSIM approach can be directly applied to a PSR without significant problems.

To limit the potential for interactive dose effects, any impacted areas, identified to exist on the PSR, which continue across the proposed PSR boundary, should be fully included in the proposed PSR final surveys. If buildings are intended for PSR, the building should be included in the PSR unless a licensee can provide information to demonstrate a low potential for future exposure of individuals in the PSR portions of the building from other impacted areas of the building and that any significant dose contributions from areas outside the PSR are included in determination of the DCGLs for the PSR.

K.2.4   Dose Modeling Specific Issues

The compliance methodology in this NUREG report emphasizes dose modeling to derive DCGLs that should be used as input to the MARSSIM process (see Section 2.5). Simple sites that only involve surface contamination and low potential for migration of residual radioactivity should require straightforward dose calculations to derive DCGLs. Sites with both subsurface and ground-water residual radioactivity or migration of radioactive material from one area to another may generally require more complex modeling and compliance demonstration methods.

Because areas of the site outside a PSR may not be remediated at the same time as the PSR, a primary concern with the calculation of DCGLs is that the dose calculation includes all applicable transport and exposure pathways. For PSR, special consideration needs to be given to any potential for significant transport of material into the PSR from outside the boundary, or from the PSR to other areas of the licensed site or a previous PSR. DCGLs should account for movement of radioactive material under circumstances where accumulation processes could lead to media concentrations significantly increasing, if the transport were included, or would add new radionuclides or exposure pathways for the PSR dose assessment.

For example, an area designated for PSR may not have impacted ground water, but an impacted area on the licensed site up-gradient has impacted ground water that is expected to migrate into the PSR in the future. The future ground-water residual radioactivity should be included in the dose calculation for the PSR (unless its contribution would not be significant) or addressed by other methods. The surface DCGLs for the PSR may need to be limited to ensure the total PSR, including any future ground water dose, complies with the 0.25 mSv/y (25 mrem/y) dose limit for the 1000-year compliance period. Similar situations could exist with up-gradient surface or subsurface contamination (e.g., leaching and transport from the sources to the PSR).

Similarly, residual radioactivity sources on the PSR should be evaluated for potential transport to the licensed site or other previous PSRs. Most licensees should be able to assess the potential for transport pathway communication between site areas using available site characterization information. Complex sites may require collection of additional site characterization information (inside and outside of the PSR) to support evaluation of transport pathways. The scope of site characterization work should be consistent with the expected level of residual radioactivity and potential dose consequences.

Records of the PSR are needed to ensure the residual radioactivity at the PSR can be included in subsequent PSR analyses and in the overall site license termination process. Residual radioactivity at PSRs may be a concern for site license termination when the potential for migration and accumulation of radioactive materials to the licensed site exists. This circumstance may only be significant when a number of PSRs exist, in close proximity, that share common transport pathways with the licensed site, such that accumulation of transported material is possible. Another situation is when the critical group may use multiple PSRs. In addition, the PSR's approval may have involved licensee agreed-upon limits on the dose contributions from decommissioning activities or residual radioactivity sources that may remain on the licensed site.

For each subsequent PSR request and at license termination, all prior PSRs need to be considered for potential contributions to dose that may need to be included in partial or site DCGLs that are calculated. For some sites, this could mean that prior PSRs could constrain the amount of residual radioactivity allowed in a PSR, or at the remainder of a site, at license termination. If the review of a PSR identifies important features, events, or processes (FEPs) that need to be considered at license termination, NRC staff may develop a license condition to ensure the matter is addressed.

Review of impacts on previous PSRs is very important because the previous PSR was approved based on the calculations and evaluations done to show compliance with the Subpart E limit. If another portion of the site is decommissioned and released at a later date, the possible impacts on the dose estimates at the previous PSR need to be reviewed. The first area to investigate is to review the previous approval and look at the prospective analyses done at that time. If they remain valid and bound any impacts that could be caused by the proposed PSR, then the impact of the proposed PSR can be considered acceptable. If the proposed PSR may result in impacts that may cause the previous PSR to exceed the Subpart E limit, the DCGLs for the proposed PSR should be constrained to limit the impact so that the dose on the previous PSR remains below the Subpart E limit.

Scenario development, especially for prospective analyses, does involve some speculation. NRC staff should focus on reasonable scenarios to limit the degree of speculation. Both human behavior and FEPs should be based on present knowledge. Speculation regarding activities that are not present in the region, not reasonably likely to occur, or would change the behavior of the FEPs, should be avoided. For example, a scenario that involves modifying the local topography, unless that is part of a remediation option, so that surface water would then transport radioactive material from an impacted area to the PSR is generally too speculative and not a reasonable scenario.

An important part of the detailed technical review may be determining if a licensee has included all applicable exposure pathways in the DCGL calculations and provided sufficient bases for exclusion of exposure pathways. Applicable exposure pathways are determined by considering all three of the following:

- The means by which the critical group can be exposed to localized residual radioactivity;

- The potential for sources and transport of radioactive materials (from the PSR, the licensed site, or previous PSRs) to the location of the applicable critical group; and

- Concurrent use, if appropriate, of the PSR and previous PSRs by the critical group.

This NUREG report generally addresses exposure pathways for localized residual radioactivity, and the methods are relatively straightforward. This section focuses on analyzing sources and transport pathways because the potential risk of additional sources of exposure impacting a PSR (or a previous PSR) is increased when the entire site is not decommissioned at the same time. Scenario definition and pathway identification are therefore key aspects of DCGL dose modeling that are impacted by the unique circumstances possible under PSR.

K.2.5 Features, Events, and Processes

Applicable source, transport, and exposure pathways comprise the exposure scenario for DCGL calculations. DCGL calculations can be done using all-pathway models or pathways can be decoupled from the modeling and their results allocated to pathway-specific DCGLs that can be combined to generate a survey unit or PSR $DCGL_W$ that equates to the Subpart E dose limit. A number of options for calculating DCGLs exists, and the specific option, chosen by the licensee, for a site, may be determined by the site conditions, complexity, and level of risk involved.

K.2.5.1 Screening Methods

The purpose of screening various sources, transport mechanisms, and exposure pathways is to evaluate whether the PSR may have processes that could result in radioactive material being transferred between the PSR and either the licensed site or previous PSRs. The first goal of the screening criteria is to eliminate various FEPs from consideration, while minimizing the amount of information needed by NRC staff to make a decision. A second goal is for the screening criteria to factor in the availability and cost of information (i.e., the first criterion should not require the need to develop a complex site-specific three-dimensional ground-water model). The screening of these criteria should not only focus on the effect of the licensed site, or a previous PSR, on the PSR, but, also, the potential contribution of the PSR on the dose assessment for the entire site at the time of decommissioning, or the current compliance of a previous PSR with Subpart E. The following are the general categories of screening criteria:

- The presence of residual radioactivity in various media, including effluent releases from the operating site (e.g., soil, ground water, air).

- The availability of mechanisms to either move material from one location to another, (e.g., ground water movement) or project exposure from one area to another (e.g., direct radiation).

- The availability of exposure pathways to cause dose in humans after it is moved or projected to the area.

After a medium, such as ground water, is found to contain residual radioactivity, it may be screened out if it has minimal levels of residual radioactivity (compared with the residual radioactivity currently present in the media at the critical group location). If the source is not screened, then the transport mechanism(s) is (are) screened to evaluate the capability of the process to move material to the area of interest. This can then be compared to the residual radioactivity levels for each radionuclide currently present on the area of interest or other processes moving material. Finally, the potential exposure pathways can be screened to remove those pathways that would result in insignificant doses or are not present at the location where the material is being deposited.

In formulating a complete exposure scenario for a proposed PSR, initial consideration should be given to available information (from the PSR area, the site, or any prior PSR) that can rule out further consideration of specific sources or transport pathways. Although investigation of potential sources and transport pathways can become complicated, a number of potential sources and transport pathways can be ruled out with relatively simple and available information. Appendix L provides a worksheet of source, transport, and exposure pathways with questions that can be used for screening. Use of a "top-down" approach to screening can avoid unnecessary and costly investigation into details that may not have a significant impact on DCGL calculations. It is expected that once a potential release or transport pathway has been identified, licensees may provide simple, yet reasonably conservative, screening-type calculations to assess importance. Pathways may be excluded because of only a small dose contribution, if the pathway results in less than 10 % of the dose limit, and the sum total of all pathway exclusions does not exceed 10 % of the dose limit (see Section 3.3 of this volume). The licensee should clearly identify all screened pathways, and should show sufficient bases for exclusion.

Example of Screening Process for FEPS

In Appendix L, a worksheet has been provided as one method of screening FEPs for PSR. The purpose of the worksheet is to provide some general topics that can, in most cases, be considered with generally available information, to minimize unnecessary site characterization, modeling, and review. The worksheet can be used to develop both compliance and prospective analyses. Ultimately, if radionuclides cannot be released or transported to the critical group location, there is no point for further consideration of the FEP(s) in the dose assessment.

Site-specific conditions and available information may make it desirable for a licensee to initially focus on source, transport, or both, when trying to screen FEPs. In some cases, it may be necessary to conduct limited dose calculations to provide information to justify the exclusion of a source, transport mechanism, or exposure pathway. If a source cannot be screened out, then it should be considered for transport screening. If pathways cannot be excluded using this worksheet, they should be considered in initial dose calculations, by either inclusion in the analysis or in modifying the dose limit through the use of an agreed-upon limit. Results of the initial dose calculations can provide additional insights to the significance of pathways with respect to dose and may provide additional means for further refinement of the calculations to address only the important features and processes. All source, transport, and exposure pathway exclusions from modeling should be identified and accompanied by an appropriate justification for exclusion.

The worksheet is split into three parts: (1) Sources; (2) Transport Processes; and (3) Exposure Pathways. The method is to start with the source questions and follow the directions under each item as necessary. The user should follow the path down until the item is screened out or needs to be considered in the analyses. After reaching the end of a path, the user should go back to where the branching occurred and continue with the questions, if applicable. For example, a site has some residual radioactivity in soil and the licensee reviews the questions under L.2.2.2 ("Soil Transport: Leaching"). The questions lead the user on to L.2.4 ("Ground Water") and the user follows that path to its conclusion. The user then needs to go back and still evaluate L.2.2.3–L.2.2.6 for that residual radioactivity in soil.

In general, for each "yes" the analysis continues to more detailed questions on that source and media type. Each "no" on a black bulleted question means no further evaluation of that area (and its related questions) is necessary for that specific source or media combination. For a black-bulleted question with a list of more detailed questions (i.e., with the empty bullets) to be excluded, all of the detailed questions need to be "no." Some instructions may provide exceptions to this general rule.

For example, the last question of L.2.3.1 ("Deep Soil Transport: Leaching") includes three specific transport mechanisms from deep soil. If the answers to all three were "no," the leaching of the source would be screened out of the dose assessment. If the answer was "no" for surface water and other, but "yes" for ground water, potential leaching of the deep soil source would need to be addressed unless the ground water transport or related exposure pathways were subsequently screened out.

K.2.5.2   Human-Induced Scenarios

Another source of exposure that may lead to interactive dose effects between the PSR and another impacted area under (or previously under) control of the licensees is individuals using both the PSR and the impacted area(s) after the licensee no longer controls those areas. The concern is that a critical group could use the PSR, such that it still receives a large fraction of the Subpart E dose limit, and reasonably use another impacted area that would lead to the critical group receiving, in total, doses in excess of the Subpart E dose limit.

Most of the human-induced scenarios are prospective scenarios to evaluate the human-induced scenarios after the other impacted area is released for unrestricted use. In cases where the human-induced scenario involves a previous PSR, the analysis is one of compliance for the PSR, which may also verify that the human-induced scenario may not result in exposures to the previous PSR, above the limit. The licensee can use self-identified and agreed-upon limits to address the exposures from future use of areas that are on the licensed site (see the example in Section K.3.2).

Three situations can result in dose assessments higher than that for the PSR alone:

*   One of the DCGLs for the land areas took into account the small size of the area.

*   Use of more than one impacted areas would result in the dose receptor receiving exposure from radionuclides not present on the PSR.

*   Use of more than one impacted areas would result in the dose receptor receiving exposure from new exposure pathways or would increase the degree of exposure to a current exposure pathway.

Taking an area's size into account when developing the DCGLs is a special case of the aforementioned third bullet. This is because usually size-related modifications for dose modeling result in reducing the number of pathways or amount of exposure, but these changes may not be obvious especially if the code itself (like RESRAD does) modifies the dose calculations.

K.2.6   Subsurface Residual Radioactivity

Subsurface residual radioactivity can exist in soils and deeper geologic strata. Common sources of subsurface residual radioactivity include material leached from surface soils, buried waste, and impacted ground water. Impacted areas can be either saturated with ground water or unsaturated (where water may percolate through but does not fill all pore spaces). Currently, this volume suggests applying the MARSSIM (surface-based) methodology to subsurface, with a few modifications to address volume sources. Guidance is expected to be updated in the future to improve statistical methods for subsurface residual radioactivity. This section discusses special considerations for addressing subsurface residual radioactivity under the PSR scenario(s), with an emphasis on pathway identification for DCGL calculations. Because addressing subsurface residual radioactivity is merely a component of the same dose modeling discussed in the previous sections, the same framework for DCGL calculations applies. For the purpose of discussion, surface water is included in some examples because of the interconnection between surface water and ground water systems.

If a site is classified as impacted by the MARSSIM methodology, this volume suggests that surface water surveys and ground water surveys should be designed on a site-specific basis. If important information necessary to understand subsurface characteristics (including extent and amount/type of residual radioactivity) is not immediately available when PSR is requested, some characterization of surface water flow, sediment movement, and ground water flow for both the PSR and adjacent areas, as necessary, on the licensed site may be needed to support the amendment request. The source locations in conjunction with the site complexity determine the surface and ground-water characterization needed at the time of PSR. The level of surveys for surface and ground-water residual radioactivity should factor in all three of the following:

- The extent of existing residual radioactivity of soil on the PSR.

- The proximity of the PSR to existing and potential impacted areas on the licensed site.

- The complexity of the surface and ground water hydrology.

As noted previously, dose modeling is required for a PSR that has been classified as impacted. Subsurface residual radioactivity, once identified, should be assessed for inclusion or exclusion in dose modeling to derive DCGLs. Residual subsurface radioactivity that contributes less than 10 % of the dose limit does not need to be included in the DCGL calculations, as long as all exclusions do not consist of more than 10 % of the dose limit, but its exclusion should be documented for future consideration at license termination.

Simple situations that need to include subsurface residual radioactivity in dose modeling may involve only radioactive material originating from the PSR or only one offsite source of impacted ground water in a relatively simple hydrology system. More complex PSR can involve numerous additional sources of residual radioactivity migrating from areas outside the PSR or migrating off the PSR onto a previous PSR or the licensed site. An important aspect is the possibility of multiple sources coalescing in the surface or ground water systems (i.e., the additive effect of multiple sources from the licensed site, the PSR, or other previous PSRs).

All potential processes for migration of material need to be considered; however, some pathways can be easily excluded with available information (see Section 3 of Appendix I and Appendix L). There is a large dilution effect when radionuclides migrate into bodies of water, such as streams, rivers, lakes, and ponds. Sediment movement and ground water flow are commonly slow processes, relative to surface water flow. Reduction of residual concentrations in ground water (caused by mechanical mixing and sorption) and radioactive decay effects are associated with the longer time factor in the transport legs for sediment movement and ground water transport. A clear example of pathway exclusion would be a PSR in a watershed that is isolated from the licensed site operations and impacted areas, and the PSR is located upstream of all other offsite sources. It is reasonable in this case that the residual radioactivity at the PSR area can be neglected in dose modeling at the time of license termination if it can be demonstrated that there is no significant dose contribution to the site DCGLs. Another simple example is the exclusion of drinking water pathways, given the absence of a drinking water aquifer accessible to the critical group.

Types of surface and ground water features that could lead to a focusing of residual radioactivity from multiple, spatially separated source areas can be separated into two categories—common features and site-specific features. To determine if focusing occurs, site characterization data are needed to identify spatially convergent ground-water flow directions or convergent surface water flow and sediment movement. The level of site characterization needed should be determined by the potential for these features to occur at the particular site. Examples of each are described below.

The most common feature leading to convergent mass movement is a river, stream, or pond in a watershed. Multiple radionuclide sources at various locations around a watershed could all potentially migrate in the surface and subsurface towards the main stream channel or pond. All surface water in the watershed could be routed into the main channel or pond. Whereas most watersheds have an outlet, some lakes, ponds, or bogs may be the terminal point in a transport pathway where residual radioactivity may accumulate. Changing chemistry of the transport path (e.g., the reducing environment of a swamp) can also impact the deposition or dissolution/mobilization of specific contaminants.

In that same watershed, the uppermost aquifer may also focus ground water flow into the stream since gradients in the unconfined aquifers typically follow the topography and commonly seep into stream and river channels. The exception is for uppermost aquifers with water tables that lie below the stream elevation; these aquifers would not, necessarily, seep into the stream channels or ponds and would not lead to a convergence of ground water flow directions unless dictated by another feature.

Site-specific features, such as faults, karst terrains, and alluvial channel deposits, have the potential to focus water from diverse locations into single transport pathways. These features may lead to a channelization of flow in the subsurface. Licensees should first determine if such subsurface features exist at a site. If present, the candidates can then be analyzed for the potential to focus transport pathways from impacted areas of the licensed site or a previous PSR.

Facilities and PSRs with the potential for multiple sources of residual radioactivity that could migrate to surface or ground water should use or obtain sufficient site characterization data to ascertain if there is a potential for convergent features to exist on the site. This site characterization data may have to be obtained at the time of a PSR if they are not already available. The potential for overlapping transport pathways needs to be assessed from multiple source areas, where those sources could be on the PSR, the licensed site, or previous PSRs.

K.2.7 Records and Documentation

Maintaining complete records of PSRs is important because the information may likely be needed for any subsequent PSRs and at the time of license termination. NRC staff should consider all prior licensing actions in the reviews for a license termination, of which PSR is only one example. Similarly, the framework for PSR involves consideration of all prior PSRs and consideration of whether the residual radioactivity needs to be included in DCGL calculations for license termination. Because considerable time may elapse between a PSR and the eventual license termination, maintenance of complete records is an important aid to the licensee, as well as NRC staff. Incomplete records may result in the need for additional site characterization at the time of license termination. Records should include identification of impacted areas, and information describing the MARSSIM RSSI methods used and results obtained, including all site characterization information, applicable to the PSR, that supports DCGL calculations. Any information supporting source, transport, or exposure pathway exclusions at the PSR, in common with the licensed site, is of particular importance, as are any licensee agreed-upon limits used to simplify the previous dose assessments. This information may be used to support a determination of whether PSRs may have to be included in DCGL calculations at the time of license termination. See NUREG–1757, Volume 3, for more information on record-keeping requirements.

K.3 Hypothetical Examples

K.3.1 Contributions from the Remaining Licensed Site

A licensee requests to release a portion of a site, 10 years before the DP is estimated to be provided to NRC. The PSR has surface soil residual radioactivity of cobalt-60 (Co-60) and cesium-137. Adjacent to the PSR, on the licensed site, is the low-level waste storage area, which is a potential source of gamma exposure to individuals on the PSR. The only other potential offsite source is a ground water plume from the licensed site. The licensee evaluates the two offsite sources and eliminates all other offsite sources because of the absence of valid transport mechanisms to allow significant impact on dose analyses. The licensee then takes the two following actions to address the remaining potential exposure sources:

- A berm is going to be built between the low-level waste storage area and the PSR, on the licensed site, to reduce the external gamma exposure. The berm is estimated to reduce the potential dose from 0.05 mSv (5 mrem) to less than 0.001 mSv (0.1 mrem). The contribution is now insignificant and the source can be eliminated from further consideration in estimating the dose for the compliance calculations for the PSR's DCGLs. The presence of the berm would then likely become a license condition. Removal of the berm in the future may require re-analysis of the total dose to the critical group, to verify that exposures on the

PSR should not exceed the Subpart E limit, after unrestricted release of the low-level waste storage area.

> Note that if a previous PSR is impacted by a proposed PSR, or the decommissioning of the site, such that doses on the previous PSR may exceed the Subpart E dose limit, constraints are to be placed on the current action(s) and do not require that the previous PSR be remediated [except as noted by 10 CFR 20.1401(c)].

- The ground water plume is estimated to reach the PSR in approximately 15 years. The licensee has currently no final plans on the level of remediation that may be done to the plume. Current conservative dose modeling estimates the annual peak exposure to be approximately 0.05 mSv (5 mrem) from an all-pathway analysis, using the ground water concentration at its current location. The licensee proposes partitioning the unrestricted dose limit for the PSR. The licensee proposes constraining the annual peak dose from the surficial soil residual radioactivity on the PSR to 0.2 mSv (20 mrem) and the ground water dose to 0.05 mSv (5 mrem). To ensure the ground water concentrations do not exceed specific concentrations associated with the 0.05-millisievert (5-millirem) licensee agreed-upon limit, the licensee may install monitoring wells between the plume and the new licensed site boundary and develop a corrective action plan to use in case the concentrations raise above some specified fraction of the licensee agreed-upon limit.

Although the PSR should meet the Subpart E dose limits at the time of approval, these actions may result in further impact on the final decommissioning of the licensed site. For example, assume, at the time of license termination, residual amounts of Co-60 and niobium-94 are in the surface soil around the low-level waste storage area, and a building code requires the berm to be removed before the entire site is released for unrestricted use. When the berm is removed, external exposure would result in 0.04 mSv/y (4 mrem/y) to the average member of the critical group on the PSR. Based on the concentrations from the FSSR of the PSR and the ground water licensee agreed-upon limit, the total dose estimate for the PSR is now estimated as 0.29 mSv/y (29 mrem/y), which does not meet the Subpart E dose limit for unrestricted release. The final DCGLs of the low-level waste storage area may be limited because of the effect on the PSR. Other options available to the licensee would be to re-evaluate the PSR dose assessment to account for decay and new information on the dose from the ground-water plume or additional remediation of the ground-water plume.

K.3.2   Use of Multiple Areas

A second licensee requests to release an impacted portion of a licensed site. The area has residual radioactivity of uranium and thorium. The PSR is rocky, with poor soil, and the licensee can provide adequate justification that the critical group would not plant extensive gardens nor use the ground water under the PSR. No offsite sources or transport mechanisms could affect the dose if the critical group used only the PSR.

The closest other radioactive source, under the control of the licensee, is some ground water concentrations of Hydrogen-3 and Chlorine-36, present nearby, on the licensed site, from old tracer tests. The land over the ground-water residual radioactivity is suitable for extensive gardening, or farming and the aquifer is potable. As part of the dose assessment, the licensee evaluates a prospective scenario where the critical group may use both the PSR and portions of the licensed site after it is decommissioned. After review of the sources, impacted areas, and routes of exposure, it is decided that a reasonable scenario would involve the person living on the PSR, and using the offsite area and its impacted aquifer for drinking water and growing an extensive garden. The licensee, believing it should be easy to remediate the ground water, addresses the offsite pathways in this scenario by proposing an aggressive limit (i.e., a small fraction of the current dose estimate) on the dose from the waterborne pathways (which are all from the offsite area). Accounting for the licensee agreed-upon ground water limit, the licensee calculates DCGLs for the PSR, based only on the radioactivity on the PSR; performs an FSS; and gains NRC approval to release the PSR for unrestricted use.

At the time of site decommissioning, years later, the licensee, having better characterized the ground water plume and having run some well pumping tests, finds that it may be difficult to meet the limit it established on the ground water dose, without extensive remediation of the ground water. Therefore, the NRC staff's perspective is that at license termination, the licensee is effectively left with three options:

1. Remediate the ground water down to the licensee's agreed-upon limit.

2. Revise the licensee's agreed-upon limit based on additional modeling (e.g., taking into account actual FSS results for the PSR, decay of the sources, new information known about the ground water system, and associated residual radioactivity, or more realistic models of ground-water dispersion and transport).

3. Perform a combination of the two options.

NRC views remediating a previous PSR as the option of last resort, consistent with NRC's policy on intervention of terminated licenses. Obviously, if the licensee desires to remediate the previous PSR, NRC would not necessarily stop the licensee. However, the situation could involve a number of issues related to regulatory authority and require a need for the current site owner's approval.

Note that if the ground water with residual radioactivity had been under a previous PSR instead of the licensed site, the options would have been different. The options would be to remediate the uranium and thorium residual radioactivity on the proposed PSR, do more complex modeling, or a combination of the two.

Appendix L

Worksheet for Identifying Potential Pathways for Partial Site Release

This worksheet is provided to assist the staff and licensee in screening potential sources and transport pathways from consideration in dose modeling for Derived Concentration Guideline Levels for a partial site release (hereafter, partial release). It is intended that the results of this worksheet summarize the exclusion or inclusion of each item and the screening argument, as well as reference the more complete screening argument, if necessary. The questions should be considered for all sources of residual radioactivity and potential critical group locations. Although this worksheet has been designed for use in identifying features, events, or processes that could result in additional sources of exposure for the critical group on the partial release, it can also be used for general scenario and pathway development.

> **NOTE: The worksheet focuses on physical features, events, and processes that may transport radioactive material to the partial site. Additionally, it covers situations where offsite radioactive material may directly expose critical group members using the partial release. It does not explicitly address sources or routes of exposure that result from the critical group using more than the partial release.**

INSTRUCTIONS

The worksheet is split into three parts: (1) "Screening Sources," (2) "Screening Transport Processes," and (3) "Exposure Pathways." The method is to start with the source questions and follow the directions under each item as necessary. The licensees should follow the path down until the item is screened out or needs to be considered in the analyses. After reaching the end of a path, the licensees should go back to where the branching occurred and continue with the questions, if applicable. For example, a site has some residual radioactivity in soil and the licensee reviews the questions under Subsection L.2.2.2 ("Soil Transport: Leaching"). The questions lead the licensee on to Section L.2.4 "Ground Water" and the licensee follows that path to its conclusion. The licensee then needs to go back and still evaluate Subsections L.2.2.3–L.2.2.6 for that source of residual radioactivity in soil.

L.1   Screening Sources (Yes/No)

Do the following section that is appropriate for each possible source of residual radioactivity.

L.1.1   Existing/Historical Residual Radioactivity (Yes/No)

- Is there residual radioactivity present in media? (yes/no)

    — Surface soil [less than 30 cm (1 ft)]?

    — Deep soil [greater than 30 cm (1 ft)]?

    — Ground water?

- — Surface water?

- — Structures?

- — Other?

- Evaluate, for each media type: is there a sufficient amount of residual radioactivity to include in dose calculations? (yes/no)

  - — Surface soil [less than 30 cm (1 ft)]? If "yes," go to Section L.2.2.

  - — Deep soil [greater than 30 cm (1 ft)]? If "yes," go to Section L.2.3.

  - — Ground water? If "yes," go to Section L.2.4.

  - — Surface water? If "yes," go to Section L.2.5.

  - — Structures? If "yes," go to Section L.2.6.

  - — Other? If "yes," follow the process for the most similar media.

L.1.2   Current Operational Releases (Yes/No)

- Are there current effluents or 10 CFR 20.2002 ongoing disposals from the operating facility in the media? (yes/no)

  - — Gaseous or particulate release? If "yes," got to Section L.2.1.

  - — Surface soil [less than 30 cm (1 ft)]? If "yes," go to Section L.2.2.

  - — Deep soil [greater than 30 cm (1 ft)]? If "yes," go to Section L.2.3.

  - — Ground water? If "yes," go to Section L.2.4.

  - — Surface water? If "yes," go to Section L.2.5.

  - — Other? If "yes," follow the process for the most similar media.

- Are there ongoing or planned decommissioning activities involved with media containing residual radioactivity? (yes/no)

  - — Gaseous or Particulate Release? If "yes," got to Section L.2.1

  - — Surface Soil [less than 30 cm (1 ft)]? If "yes," go to Section L.2.2.

  - — Ground water? If "yes," go to Section L.2.4.

  - — Surface Water? If "yes," go to Section L.2.5.

  - — Structures? If "yes," go to Section L.2.6.

  - — Other? If "yes," follow the process for the most similar media.

L.2   Screening Transport Processes (Yes/No)

Do the following appropriate section(s) for the media type/source combination.

L.2.1   Air Transport (Yes/No)

• Does the wind travel a significant portion of the year from the source to the critical group location? (yes/no)

• Is the source location near enough to the critical group location to avoid significant dilution of suspended or gaseous residual radioactivity? (yes/no)

• Do the structures, topography, and vegetation between the source and critical group locations provide only small amounts of dispersion? (yes/no)

If the answer to any one of the above questions in this section is "no," answer the following question. If all are "yes," go to Section L.3.1.

• Is there the potential for this source's air-transported residual radioactivity to accumulate with other source/air transport combinations that have been screened out, so that the combined effect of all sources would result in a significant source of exposure? (yes/no)

If the answer is "yes," air transport for this source and for any other sources identified by this question are not screened out. Go to Section L.3.1. If "no," air transport for the source is screened out.

L.2.2   Surface Soil Transport (Yes/No)

Each source should go through all subsections. Screening out one subsection does not mean all subsections are screened out, necessarily. To screen out the entire surface soil transport mechanism for a source, Subsections L.2.2.1–L.2.2.5 all need to be screened out individually.

L.2.2.1   Erosion (Yes/No)

• Is the residual radioactivity chemical/structural form erodible within analysis time frame? (yes/no)

• Is the rainfall, runoff, or wind speed sufficient to erode source contaminants? (yes/no)

• Is the proximity of the source location to the critical group location sufficient for erosion to transport contaminants to the critical group location? (yes/no)

• Do the structures, topography, and vegetation between the source location and the critical group favor transport of material to the critical group location? (yes/no)

If the answer to any one of the above questions in this section is "no," answer the following question. If all are "yes," skip the next question, and then answer the last question of this subsection.

- Is there the potential for this source's eroded residual radioactivity to accumulate with other source/erosion transport combinations that have been screened out so that the combined effect of all sources would result in a significant source of exposure? (yes/no)

If the answer is "yes," the erosion subsection for this source and any other sources identified by this question are not screened out. Answer the following question. If "no," the erosion subsection is screened out. Go to Section L.2.2.2.

- If erosion were to occur, where would the material end up so that it can be transported to the critical group location?

    — Direct overland flow? (yes/no) If "yes," go to Section L.2.5 and answer surface water questions for potential overland flow.

    — Surface water body? (yes/no) If "yes," go to Section L.2.5.

    — Other? (yes/no) If "yes," go to the appropriate similar transport mechanism.

If the answer to any one of these questions is "yes," the erosion subsection for this source is not screened out. Proceed as directed by the specific question. When complete with that pathway, return, and proceed through Section L.2.2.2. If "no," the erosion subsection is screened out. Go to Section L.2.2.2.

L.2.2.2   Leaching (Yes/No)

- Is the rainfall or infiltration amount sufficient for leaching of residual radioactivity to occur to a significant degree? (yes/no)

- Will the residual radioactivity leach within the analysis time frame? (yes/no)

- Does the geochemistry of the soil and radionuclides [e.g., distribution coefficients (K_d)] allow leached residual radioactivity to reach the ultimate transport mechanism within the analysis time frame (e.g., will the residual radioactivity be able to move through the unsaturated zone and enter into the ground water aquifer)? (yes/no)

If the answer to any one of the above questions in this section is "no," answer the following question. If all are "yes," skip the next question, and then answer the last question of this subsection.

- Is there the potential for this source's leached residual radioactivity to accumulate with other source/leach transport combinations that have been screened out so that the combined effect of all sources would result in a significant source of exposure? (yes/no)

If the answer is "yes," the leaching subsection for this source and for any other sources identified by this question are not screened out. Answer the following question. If "no," the leaching subsection is screened out. Go to Section L.2.2.3.

- If leaching were to occur, where would the material end up so that it can be transported to the critical group location?

    — Ground water aquifer? (yes/no) If "yes," go to Section L.2.4.

    — Surface water body? (yes/no) If "yes," go to Section L.2.5.

    — Other? (yes/no) If "yes," go to the appropriate similar transport mechanism.

If the answer to any one of these above questions is "yes," the leaching subsection for this source is not screened out. Proceed as directed by the specific question. When complete with that pathway, return, and proceed through Section L.2.2.3. If the answers to all of these empty bullets are "no," the leaching subsection is screened out. Go to Section L.2.2.3.

L.2.2.3   Resuspension (Yes/No)

- Does the wind travel a significant portion of the year from the source to the critical group location? (yes/no)

- Is the source location near enough to the critical group location to avoid significant dilution of suspended or gaseous residual radioactivity? (yes/no)

- Do the structures, topography, and vegetation between the source location and the critical group favor transport of material to the critical group location? (yes/no)

- Can enough of the residual radioactivity be resuspended to affect the dose to the critical group? (yes/no)

If the answer to any one of the above questions in this section is "no," answer the following question. If all are "yes," skip the next question and go to Section L.3.1. When complete with that pathway, return and proceed through Section L.2.2.4.

- Is there the potential for this source's resuspended residual radioactivity to accumulate with other source/resuspension or air-transport combinations that have been screened out so that the combined effect of all sources would result in a significant source of exposure? (yes/no)

If the answer is "yes," the resuspension subsection for this source and for any other sources identified by this question are not screened out. Go to Section L.3.1. When complete with that pathway, return, and proceed through Section L.2.2.4. If "no," the resuspension subsection is screened out. Go to Section L.2.2.4.

L.2.2.4   Manual Redistribution: Excavation and Fill (Yes/No)

- Do source area characteristics allow future excavation and reuse? (yes/no)

- Would reuse be reasonable for use on or near the partial site? (yes/no) A "no" on this question does not screen this subsection out.

- Would the source be able to become airborne as part of fugitive dust emissions? (yes/no) If "yes," go to Section L.2.2.3. A "no" on this question does not screen this subsection out.

If the answer to the first bullet is "no," or the second and third bullets are "no," the manual redistribution subsection is screened out. Go to Section L.2.2.5. If manual redistribution is not screened out, go to Section L.3.2. When complete with that pathway, return, and proceed through Section L.2.2.5.

L.2.2.5   Direct Radiation (Yes/No)

- Are the radionuclides significant external hazards? (yes/no)

- Is the source location close enough to the critical group location to avoid significant reduction in dose rate? (yes/no)

- Do the structures, topography, and vegetation between the source and critical group locations provide inadequate shielding to minimize the external exposure? (yes/no)

If the answer to any one of the above questions in this section is "no," the direct radiation subsection is screened out. If all are "yes," go to Section L.3.2

L.2.3   Deep Soil Transport (Yes/No)

Each source should go through both subsections. Screening out one subsection does not mean both subsections are screened out, necessarily. To screen out the entire deep soil transport mechanism for a source, Subsections L.2.3.1 and L.2.3.2 need to be each screened out individually.

L.2.3.1   Leaching (Yes/No)

- Is the rainfall or infiltration amount sufficient for leaching of residual radioactivity to occur to a significant degree? (yes/no)

- Will the residual radioactivity leach within the analysis time frame? (yes/no)

- Does the geochemistry of the soil and radionuclides (e.g., K_d) allow leached residual radioactivity to reach the ultimate transport mechanism within the analysis time frame (e.g., will the residual radioactivity be able to move through the unsaturated zone and enter into the ground water aquifer)? (yes/no)

<u>If the answer to any one of the above questions in this section is "no," answer the following question. If all are "yes," skip the next question, and then answer the last question of this subsection.</u>

- Is there the potential for this source's leached residual radioactivity to accumulate with other source/leach transport combinations that have been screened out so that the combined effect of all sources would result in a significant source of exposure? (yes/no)

<u>If the answer is "yes," the leaching subsection for this source and for any other sources identified by this question are not screened out. Answer the following question. If "no," the leaching subsection is screened out. Go to Section L.2.3.2.</u>

- If leaching were to occur, where would the material end up so that it can be transported to the critical group location?

  — Ground water aquifer? (yes/no) <u>If "yes," go to Section L.2.4.</u>

  — Surface water body? (yes/no) <u>If "yes," go to Section L.2.5.</u>

  — Other? (yes/no) <u>If "yes," go to the appropriate similar transport mechanism.</u>

<u>If the answer to anyone of these is "yes," the leaching subsection for this source is not screened out. Proceed as directed by the specific question. When complete with that pathway, return, and proceed through Section L.2.3.2. If the answers to all these empty bullets are "no," the leaching subsection is screened out. Go to Section L.2.3.2.</u>

L.2.3.2 Manual Redistribution: Excavation and Fill (Yes/No)

- Do source area characteristics allow future excavation and reuse? (yes/no)

- Would reuse be reasonable for use on or near the partial site? (yes/no) <u>A "no" on this question does not screen this subsection out.</u>

- Would the source be able to become airborne as part of fugitive dust emissions? (yes/no) <u>If "yes," go to L.2.2.3. A "no" on this question does not screen this subsection out.</u>

<u>If the answer to the first bullet is "no," or all bullets are "no," the manual redistribution subsection is screened out. If manual redistribution is not screened out, go to Section L.3.2.</u>

L.2.4 Ground Water Transport (Yes/No)

- Does saturated ground water exist that is in hydraulic connection with the radioactive source? (yes/no)

- Does the ground water (including unconfined or confined aquifers, as necessary) flow from the source to the location of the critical group? (yes/no)

- Is the aquifer fit for use? (yes/no)

- Potable? (yes/no)

- Irrigation? (yes/no)

- Can the residual radioactivity enter the ground water aquifer in significant amounts [e.g., is the aquifer not protected from all potential migrating contaminants by low-permeability geologic strata (e.g., clay layer)]? (yes/no)

- Is the yield rate of the aquifer sufficient? (yes/no)

  — Household and drinking water? (yes/no)

  — Irrigation? (yes/no)

- Is the distance traveled from source to the critical group location close enough to avoid significant dilution and sorption of migrating radionuclides? (yes/no)

If the answer to any one of the above questions in this section is "no," the ground water transport mechanism is screened out. If all are "yes," go to Section L.3.3

L.2.5   Surface Water Transport (Yes/No)

- Does surface water flow from the source of residual radioactivity (or from zones of mobilized radionuclides) to the critical group location? (yes/no)

- Does the volume of surface water allow transport of significant concentrations of either dissolved or suspended radioactive solids? (yes/no)

If the answer to either of the above questions in this section is "no," the surface water transport mechanism is screened out. If both are "yes," answer the following question.

- Is significant sediment buildup possible at the critical group location? (yes/no)

If the answer is "yes," go to Section L.3.5. When complete with that pathway, return and go to 3.4. If the answer is "no," go to Section L.3.4.

L.2.6   Structures (Yes/No)

L.2.6.1   Direct Radiation (Yes/No)

- Are the radionuclides significant external hazards? (yes/no)

- Is the source location close enough to the critical group location to avoid significant reduction in dose rate? (yes/no)

- Do the structures, topography, and vegetation between the source and critical group locations provide inadequate shielding to minimize the external exposure? (yes/no)

If the answer to any one of the above questions in this section is "no," the direct radiation subsection is screened out. Go to Section L.2.6.2. If all are "yes," go to Section L.3.2. When complete with that pathway, return and proceed through Section L.2.6.2.

L.2.6.2   Leaching (Yes/No)

- Is the rainfall or infiltration amount sufficient for leaching of residual radioactivity to occur to a significant degree? (yes/no)

- Will the residual radioactivity leach from the structure within the analysis time frame? (yes/no)

- Does the geochemistry of the soil and radionuclides (e.g., K_d) allow leached residual radioactivity to reach the ultimate transport mechanism within the analysis time frame (e.g., will the residual radioactivity be able to move through the unsaturated zone and enter into the ground water aquifer)? (yes/no)

If the answer to any one of the above questions in this section is "no," answer the following question. If all are "yes," skip the next question, and then answer the last question of this subsection.

- Is there the potential for this source's leached residual radioactivity to accumulate with other source/leach transport combinations that have been screened out so that the combined effect of all sources would result in a significant source of exposure? (yes/no)

If the answer is "yes," the leaching subsection for this source and for any other sources identified by this question are not screened out. Answer the next question. If "no," the leaching subsection is screened out.

- If leaching were to occur, where would the material end up so that it can be transported to the critical group location?

    — Ground water aquifer? (yes/no) If "yes," go to Section L.2.4.

    — Surface water body? (yes/no) If "yes," go to Section L.2.5.

    — Other? (yes/no) If "yes," go to the appropriate similar transport mechanism.

If the answer to any one of these questions is "yes," the leaching subsection for this source is not screened out. Proceed as directed by the specific question. If the answers to all these empty bullets are "no," the leaching subsection is screened out.

L.3   Exposure Pathways

"No" on the black bullets will not eliminate the entire section.

APPENDIX L

L.3.1   Air Pathways (Yes/No)

- Based on the critical group habits and activities, are the following viable? (yes/no)
  — Inhalation? (yes/no)
  — Submersion External Dose? (yes/no)
- Is significant deposition viable? (yes/no) <u>If "yes," go to Section L.3.2 and consider the potential soil pathways at the deposition area.</u>

L.3.2   Soil Pathways (Yes/No)

- Is external exposure viable? (yes/no)
- Is exposure through ingestion viable? (yes/no)
  — Direct soil ingestion? (yes/no)
  — Garden or crops? (yes/no)
  — Leafy vegetables? (yes/no)
  — Non-leafy vegetables? (yes/no)
  — Fruits? (yes/no)
  — Grain? (yes/no)
  — Animal husbandry? (yes/no)
  — Meat? (yes/no)
  — Milk? (yes/no)
  — Eggs? (yes/no)
- Is exposure through inhalation viable? (yes/no)
  — Indoors? (yes/no)
  — Outdoors? (yes/no)

L.3.3   Ground Water Pathways (Yes/No)

- Is exposure via drinking water viable? (yes/no)
- Is exposure via irrigation viable? (yes/no)
  — Garden or crops? (yes/no)
  — Animal husbandry? (yes/no)
  — Fish farming? (yes/no)

If irrigation is viable, go to Section L.3.2. Consider the soil pathways appropriate for the soil impacted by the irrigation.

- Is water used for purposes other than household uses (including drinking water) or irrigation? Examples would include evaporative coolers, dust suppression, etc. (yes/no)

If "yes," go to Section L.3.2. Consider the soil pathways appropriate for the impacts of the activity.

L.3.4   Surface Water Pathways (Yes/No)

- Is internal exposure viable? (yes/no)
    — Fish? (yes/no)
    — Drinking water? (yes/no)
    — Inadvertent intakes? (yes/no)
    — Is exposure via irrigation viable? (yes/no)
    — Garden or crops? (yes/no)
    — Animal husbandry? (yes/no)

If irrigation is viable, go to Section L.3.2. Consider the soil pathways appropriate for the soil impacted by the irrigation.

- Is water used for purposes other than household uses (including drinking water) or irrigation? Examples include evaporative coolers, dust suppression, etc. (yes/no)

If "yes," go to Section L.3.2. Consider the soil pathways appropriate for the impacts of the activity.

- Are recreational activities viable? (yes/no)

If recreational activities are viable, go to Section L.3.2. Consider the exposure pathways appropriate for recreational activities in the water (e.g., incidental ingestion during swimming)

L.3.5   Sediments (Yes/No)

- Are recreational activities viable? (yes/no)

If recreational activities are viable, go to Section L.3.2. Consider the exposure pathways appropriate for recreational activities on the shore, or involving sediments (e.g., incidental ingestion from making sand castles).

• Is use of sediments for land-based activities viable (e.g., fill or crops, etc)?

<u>If use of sediments is viable, go to Section L.3.2. Consider the soil pathways appropriate for the impacts of the activity.</u>

DOCUMENTATION

The information from the worksheet should be summarized in tables. The tables should summarize (1) the source, (2) whether it is included or excluded, (3) the FEPs screened, (4) the screening argument, and (5) the reference for the screening argument. For example, one format is below, and it uses the example in Section 3.1 of Appendix K as a basis. The level of detail is only needed for the question being used to screen out the source, transport mechanism, or pathway. Common pathways using the same or similar screening arguments can be grouped (e.g., fourth row of example table).

Table L.1   Example of Summary Format

Source	Status	Screening Pathway[a]	Screening Argument	Reference
Ground Water (GW) Plume–Area 4-10	Incl	GW (1.1)–GW (2.4)–GW (3.3)–Soil (3.2)	N/A	N/A
GW Plume–Area 4-12	Excl	GW (1.1)–GW (2.4/YIELD)	Yield of Pico Aquifer <30 L/day.	DP Chapter 3.7.3
Low-Level Waste (LLW) Storage Area	Incl	Other (1.2)–Soil (2.6)–Soil (3.2/direct)	N/A	N/A
	Excl	Other (1.2)–Soil (2.1–2.2)	No significant erosion or leaching of LLW Area within 1000 years.	DP Chapter 4.1.5
Note: a Numbers in this column indicate the appropriate sections in Appendix L.				

Appendix M

Process for Developing Alternate Scenarios at NRC Sites Involved in DandD and License Termination

Note that some of the Web addresses may no longer be valid due to both the fluid nature of the Internet and the age of the document (the document was produced in the spring and summer of 1998).

Acronyms

BRAC	Base Realignment and Closure
CERCLA	Comprehensive Environmental Response, Compensation, and Liability Act
D&D	Decontamination & Decommissioning
DoD	Department of Defense
GIS	Geographic Information Systems
HLW	High Level Waste
IHI	Inadvertent Human Intruder
LLW	Low Level Waste
MOP	Member of Public
NEA	Nuclear Energy Agency
NRCS	Natural Resources Conservation Service
NWPA	Nuclear Waste Policy Act
OSWER	Office of Solid Waste and Emergency Response
RCRA	Resource Conservation and Recovery Act
SCS	Soil Conservation Service
SNL	Sandia National Laboratories
USDA	U.S. Department of Agriculture
WIPP	Waste Isolation Pilot Plant

M.1   Introduction

M.1.1   Purpose

The process for developing alternate scenarios complements the Decisionmaking Framework and is meant to be used in conjunction with the methodology discussed in Section 1.2 of this volume and in the guidance on scenarios, exposure pathways, and critical groups discussed in Section I.3 from Appendix I.

As noted in Section I.3, this Appendix is primarily focused on users that have chosen to create site-specific scenarios by modifying the basic screening scenarios based on physical considerations of the site. Licensees also are free to construct scenarios by other methods, and, depending on the reasonably foreseeable, and less likely but plausible use land uses, may not need to address the screening scenarios. See Section I.3 for more details. This Appendix, while constructed as a method of modifying the screening scenarios, may still be useful to other scenario generation methods. Specifically, a licensee may try to find a scenario on a later panel that approximates the site-specific land use assumptions and start there. In addition, most licensees may find the discussions on sources of data for various aspects that relate to their site useful.

Two basic screening scenarios are used; the residential farmer and the building occupancy scenario. The residential farmer scenario is meant to be applied to sites with land and water residual radioactivity and the building occupancy scenario is to be applied to sites with contaminated structures. A generic critical group, with acceptable default parameter values to represent the average member of each group is associated with each scenario. The default pathways, models and parameter values for the critical group combine to form exposure scenarios.

This appendix steps through the process for eliminating the various pathways from the residential farmer scenario and describes the associated information needed to justify the actions. This is a simple illustration of the process. Following the illustration methodically will not always identify the scenario containing the critical group for a site-specific scenario. As illustrated, the process assumes that the combination of radionuclides and remaining pathways will result in the resident still getting the highest dose. However, in some cases, other scenarios, including offsite use of materials, could become the critical group's scenario. For example, a construction scenario may become more important than a residential scenario for certain radionuclides.

M.1.2   Background

There is significant variability among decommissioning sites with respect to geography and site residual radioactivity. The original purpose of the site, historical development, and the resulting processes that generated the site residual radioactivity vary widely. Residual radioactivity has occurred in buildings, process equipment and other site structures, soils (surface and subsurface), ponds, lagoons, surface waters, and ground water. Sites are located in urban and suburban, residential, commercial and industrial, rural, and agricultural areas, and many are located on or directly adjacent to rivers, lakes, oceans, estuaries, wetlands, flood plains, or wildlife areas. The waste form is highly variable as slag, general soil or sediment residual radioactivity, sludge, debris, dust or sand piles, packaged (drums, crates, etc.), and dispersed in liquid media.

In general, scenarios represent possible realizations of the future state of the system (Cranwell *et al.*, 1990). Scenarios are needed to establish potential future conditions which might lead to human exposure.

M.2   Process Schematic

The process for developing alternate scenarios is presented in this report as a twelve-panel schematic flow diagram (Figures M.1 through M.12). This diagram is supported by text in Sections M.3, M.4, and M.5.

The schematic begins with the definition of the source and describes a step-by-step procedure of using site-specific information to alter the resident farmer scenario by removing pathways. The supporting text should be referred to for specific details about steps, standards, and data needed to defend the removal of a pathway.

Although this step-by-step process provides an efficient way to introduce site-specific data to rule out pathways, shortcuts can be and should be taken at specific points in the process when data developed by the decision analysis warrants it. For example, if the decision analysis shows the aquatic pathway to be primary in the computation of the TEDE, then the licensee when using this process should skip other pathways and focus on evidence that could either rule out the aquatic pathway or make the dose estimate more realistic.

If the residual radioactivity at a site is fully contained within a building (and would reasonably be expected to remain there throughout the period when it could cause a TEDE greater than the threshold), the default resident farmer scenario would not be applicable and the licensee should use the building occupant scenario.

M.2.1   Panel 1: Beginning the Process

The first panel (Figure M.1) begins with a more detailed version of the Decision Framework (Section 1.2 of this volume) and shows the context of this process in relation to that framework. This panel takes the licensee from defining the source through initial and iterative dose assessments, to sensitivity and decision analyses, and finally to the use of site-specific information to develop alternate scenarios.

While this schematic shows other actions that can be taken subsequent to the sensitivity analysis, the schematic (and this report) concentrate solely on those actions associated with the process for developing alternate scenarios through the introduction of specific information. Other actions include using site-specific information to modify pathway parameters, changing or altering the pathway models, releasing a license for restricted use, and cleaning up the site.

Section M.3 of this appendix (Initial Computation) provides descriptions of processes shown in Panel 1 (Figure M.1) with regard to the source definition and the initial and iterative dose assessments. Section M.4 (Sensitivity Analysis) describes a sensitivity analysis process and presents an example of both text and graphics reports developed using the NRC software DandD 1.0. This example shows how a sensitivity analysis can help the licensee understand which specific pathway and radionuclides dominate the computed dose.

If the initial computation results in a TEDE to an average member of the critical group that exceeds 0.25 mSv/y (25 mrem/y), the licensee would proceed to Panel 2 (Figure M.2) to consider land use. The projected use of the land is critical to beginning of this process. If the future use of the land is shown to be urban or industrial, rather than the default resident farmer, the starting scenarios contain significantly fewer pathways than the resident farmer scenario and the TEDE should always be significantly lower than the initial TEDE.

M.2.2   Panel 2: Land Use Data

The second panel (Figure M.2) illustrates the decisions necessary to determine if there is sufficient evidence to bypass the resident farmer scenario and go directly to an urban or industrial worker scenario. These decisions are based on the persistence of the TEDE over the 0.25 mSv/y (25 mrem/y) threshold and on the current and projected land use at the site. Future land use should be projected for the time period that the TEDE is expected to be greater than the 0.25 mSv/y (25 mrem/y) threshold.

One hundred years is considered a reasonable cutoff point for future land use projections to provide the basis for scenario assumptions. For example, if the future land use can reasonably be predicted to be either urban or industrial, the resident farmer scenario may potentially be bypassed, allowing the licensee to concentrate on these land uses. See Section I.3 for more information on justifying and selecting reasonably foreseeable land uses.

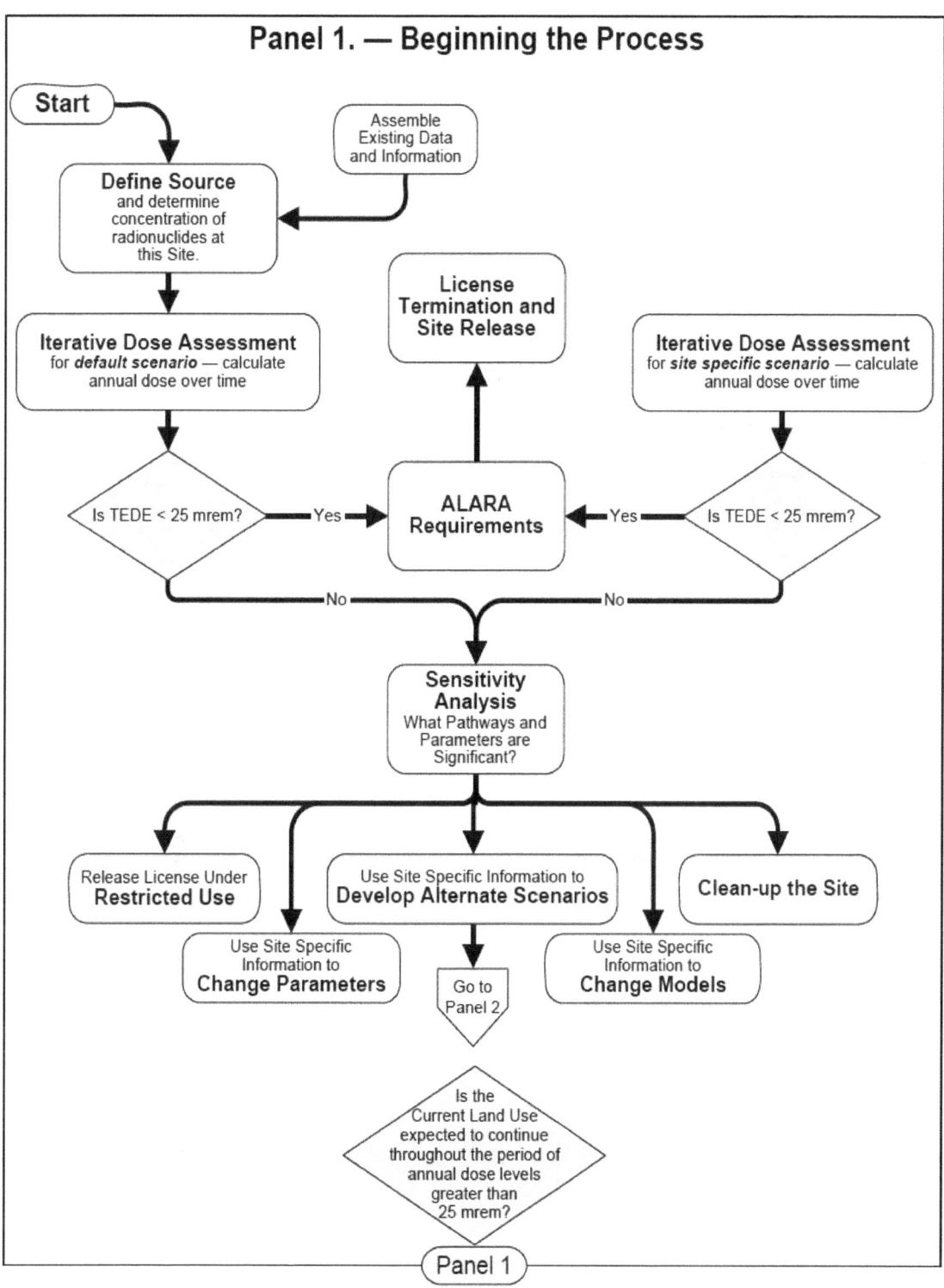

Figure M.1 Panel 1: Beginning the Process.

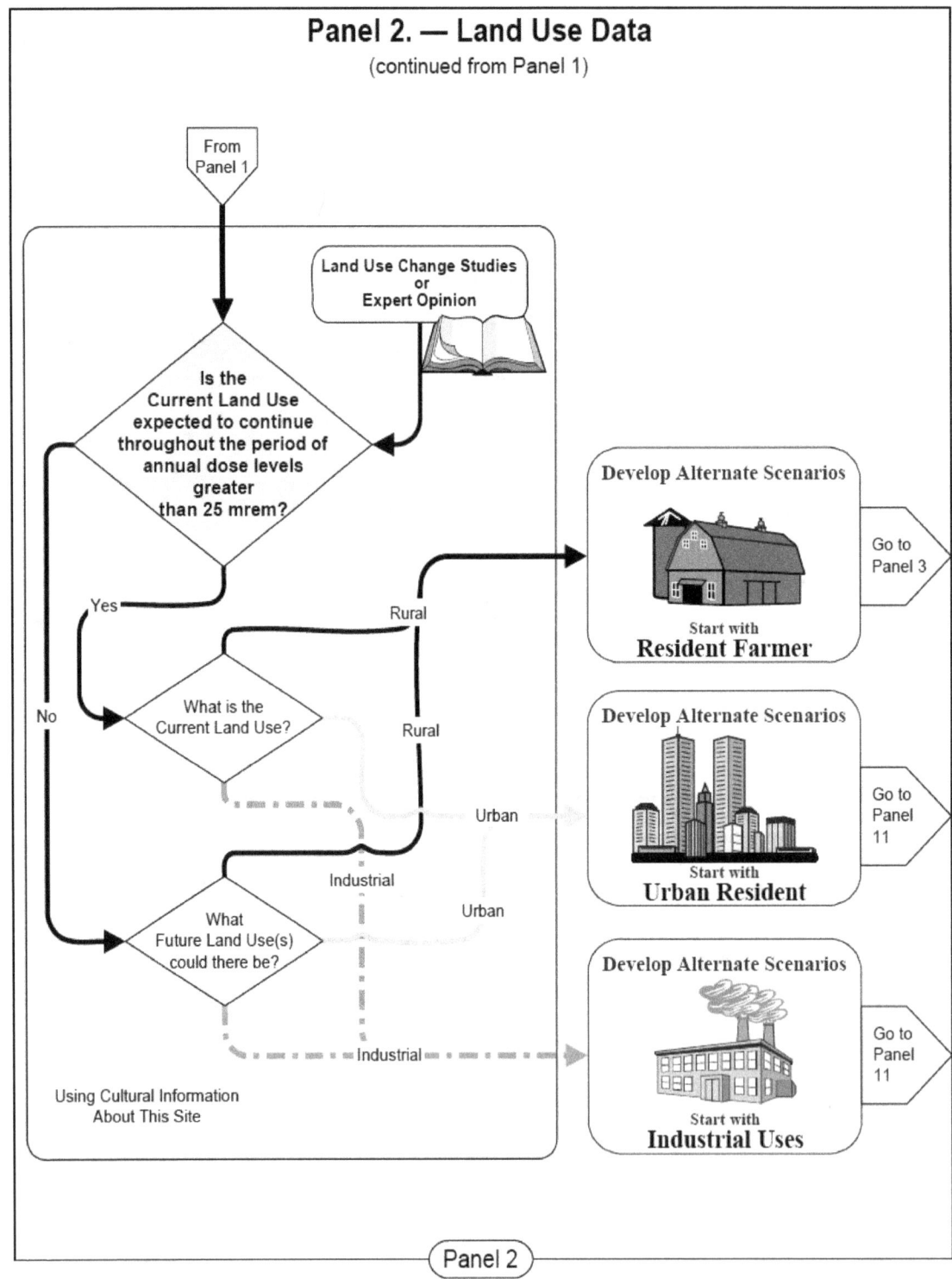

Figure M.2 Panel 2: Land Use Data.

Section M.5.1 presents procedures for determining current land use and for estimating future land use. It also presents tables of Web sites where land-use information might be obtained.

Panel 2 (Figure M.2) will direct the licensee to Panel 3 (Figure M.3), to begin the process of devolving the resident farmer scenario by removing pathways, to Panel 11 (Figure M.11) where the urban resident and the industrial worker scenarios are considered, or to both if multiple land uses are viable.

M.2.3   Panel 3: Start with Resident Farmer — Ground Water — Limited Use

The third panel (Figure M.3) is a continuation of Panel 2 and begins the process of introducing physical information about the site. The starting point here is the resident farmer scenario with all pathways (the default scenario). Since water is critical to the key pathways in this scenario, the first question to ask is "Is ground water available?"

If the answer to this question is "Yes," it may be available for only limited uses, depending on the ground-water yield. Section M.5.2.1 addresses the availability of ground water and the documentation that would need to be submitted to NRC if the licencee wants to remove or limit the ground water pathway on the basis of ground water unavailability. If the yield is great enough to satisfy all uses, Panel 3 (Figure M.3) will direct the licensee to Panel 5 (Figure M.5), to consider the suitability of ground water as an environment for the resident farmers' fishery. If the yield can only sustain irrigation and drinking water, Panel 3 will direct the licensee to Panel 6 (Figure M.6), to consider the suitability of ground water for agriculture. If the yield would limit the use to household and drinking water only, Panel 3 will direct the licensee to Panel 8 (Figure M.8), to consider the potability of ground water.

If ground water is not available, the ground water pathway (and all pathways that depend on ground water) would be removed from the resident farmer scenario, resulting in a resident farmer scenario where all water needs are assumed to be met through the use of an outside, uncontaminated water source. The licensee would review Panel 4 (Figure M.4) to consider the resident farmer scenario without ground water.

M.2.4   Panel 4: Agriculture — No Pond

The fourth panel (Figure M.4) is a continuation of Panel 3 and continues the process of introducing physical information about the site. If ground water has been documented to be unavailable, an iterative dose assessment should be done to see if the TEDE to an average member of the critical group still exceeds 0.25 mSv/y (25 mrem/y). If it still exceeds this threshold value, the next logical question to ask is "Are soil and topography at this site suitable for agriculture?" The details of this issue are addressed in Section M.5.2.2.

If the answer to this question is "No," that either soil or topography at this site are determined to be unsuitable for agriculture, and the agricultural pathway would be removed resulting in a scenario that has a rural resident with no agriculture, pond, or drinking water, since the ground water pathway had already been removed. The resident farmer scenario has now devolved into a what is essentially a building occupancy scenario combined with modified external exposure and inhalation pathways. Section M.5.2.2 describes the documentation that should be submitted to NRC if the licencee wants to remove the agricultural pathway on the basis of either topography or soil being unsuitable to agriculture.

After the agricultural pathway is removed, another dose assessment would be done for a scenario that includes only the building occupancy scenario and external exposure and inhalation pathways. These pathways should be modified to reflect that the resident is no longer working on the "farm." If the TEDE to an average member of the critical group still exceeds 0.25 mSv/y (25 mrem/y), the licensee should begin a more critical analysis of the pathway parameters for the pathways in this scenario, but there is no need, at this point, to continue with alternate scenario development.

M.2.5   Panel 5: Aquatic Life

The fifth panel (Figure M.5) is a continuation of Panel 3. This panel starts with a resident farmer and all pathways. Ground water is available, but is it suitable for aquatic life?

Section M.5.2.1.2 considers the suitability of ground water as an environment for the resident farmers' fishery and presents the standards for this water to be considered acceptable for this use. If the water is unsuitable for aquatic life, the aquatic pathway would be removed, resulting in a resident farmer scenario with no pond. An iterative dose assessment would be performed, and if the TEDE to an average member of the critical group still exceeds 0.25 mSv/y (25 mrem/y), the licensee would go to Panel 5 (Figure M.5) to consider the suitability of ground water for agricultural use.

If the answer to the first question is "Yes," the ground water is suitable for a pond, cultural data for the area should be introduced to answer the question, "Do residents of this area use ponds as fisheries?" See Section M.5.1.2 for more details on information sources and documentation needed. If the answer is "No," the licensee would proceed as in the previous paragraph for the removal of the aquatic pathway and subsequent analyses, including iterative dose assessment. If the answer is "Yes," the licensee would go to Panel 7 (Figure M.7) to consider the suitability of ground water for agriculture.

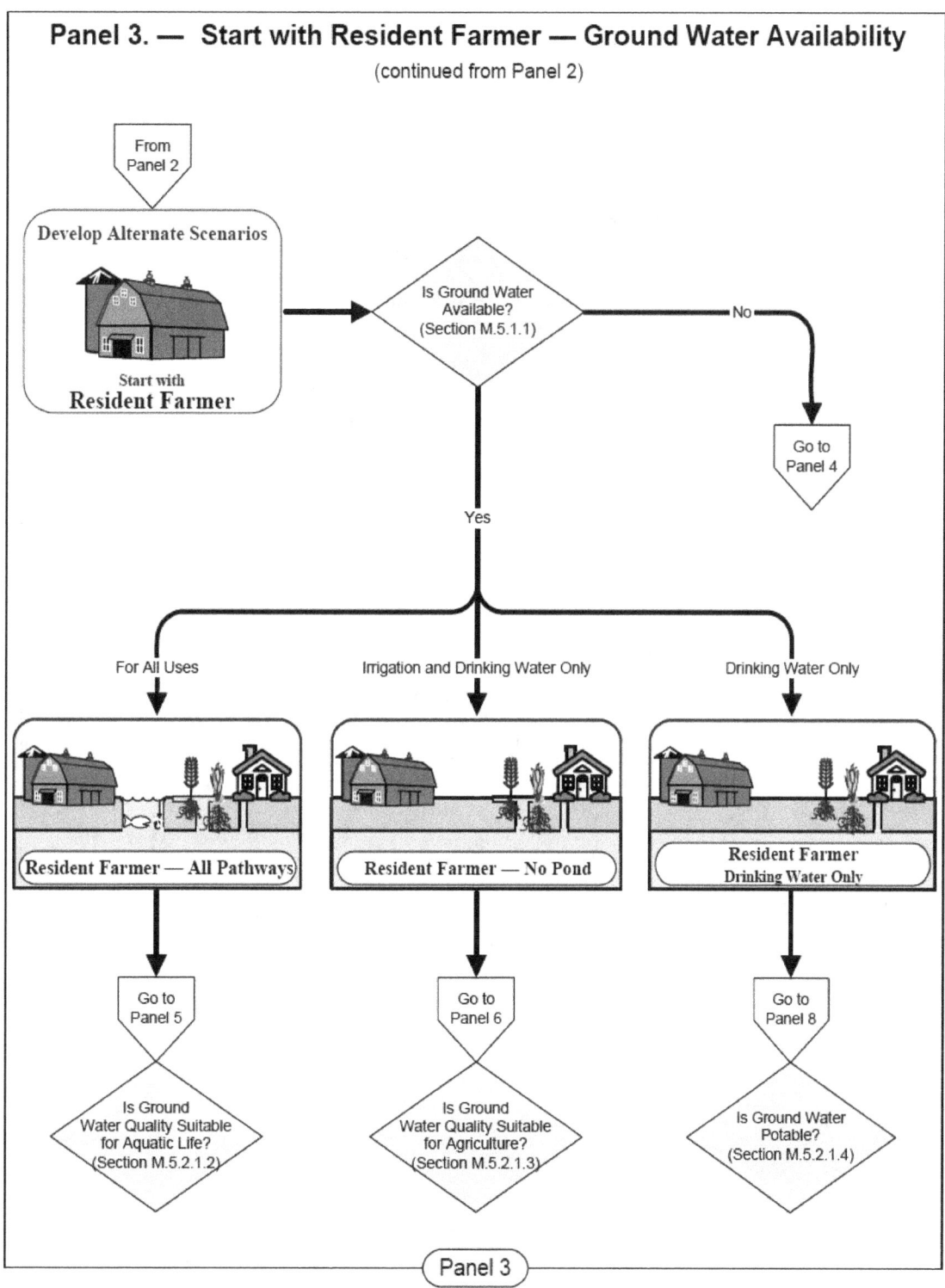

Figure M.3 Panel 3: Start with Resident Farmer — Ground Water Availability.

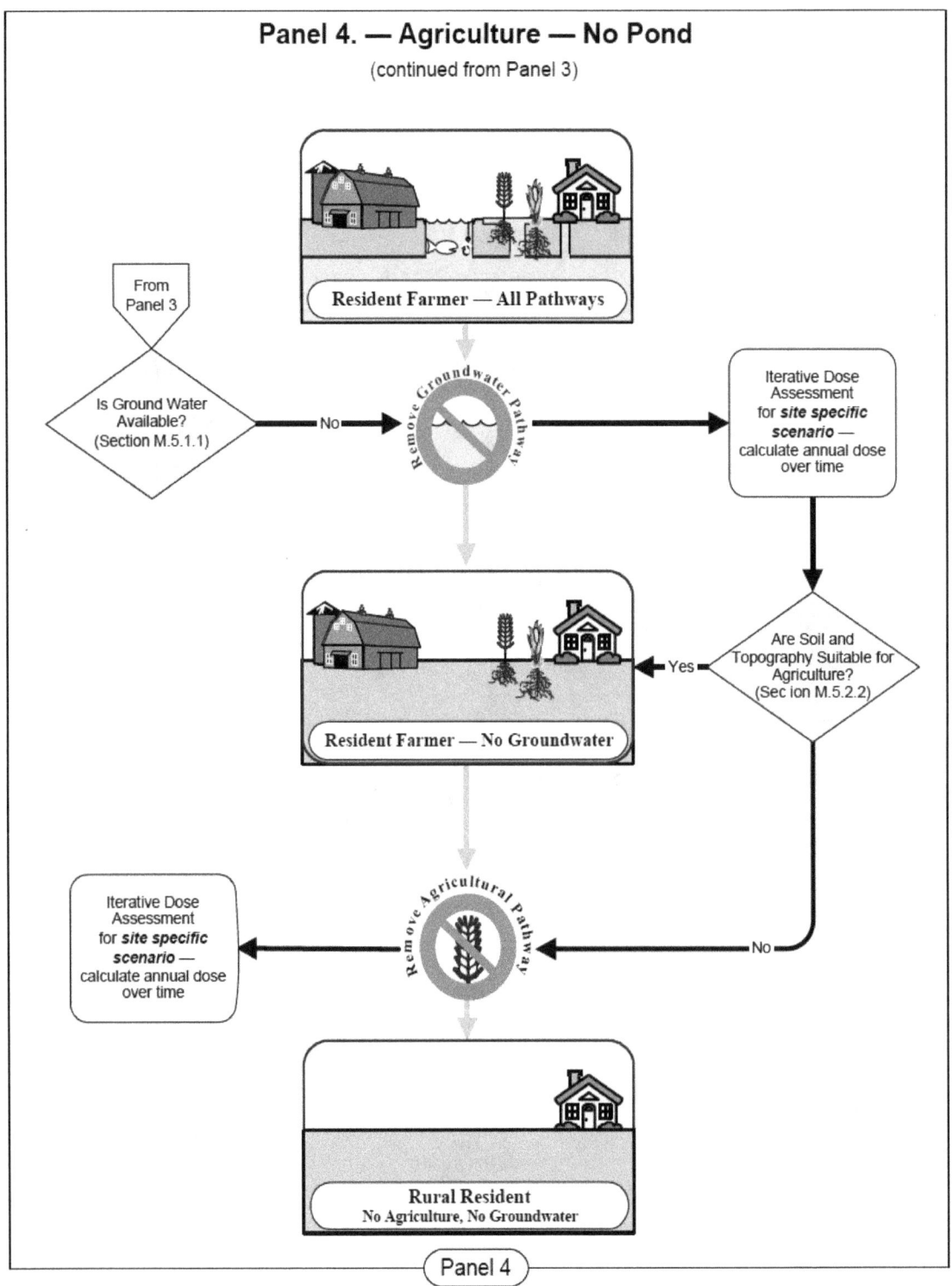

Figure M.4 Panel 4: Agriculture — No Pond.

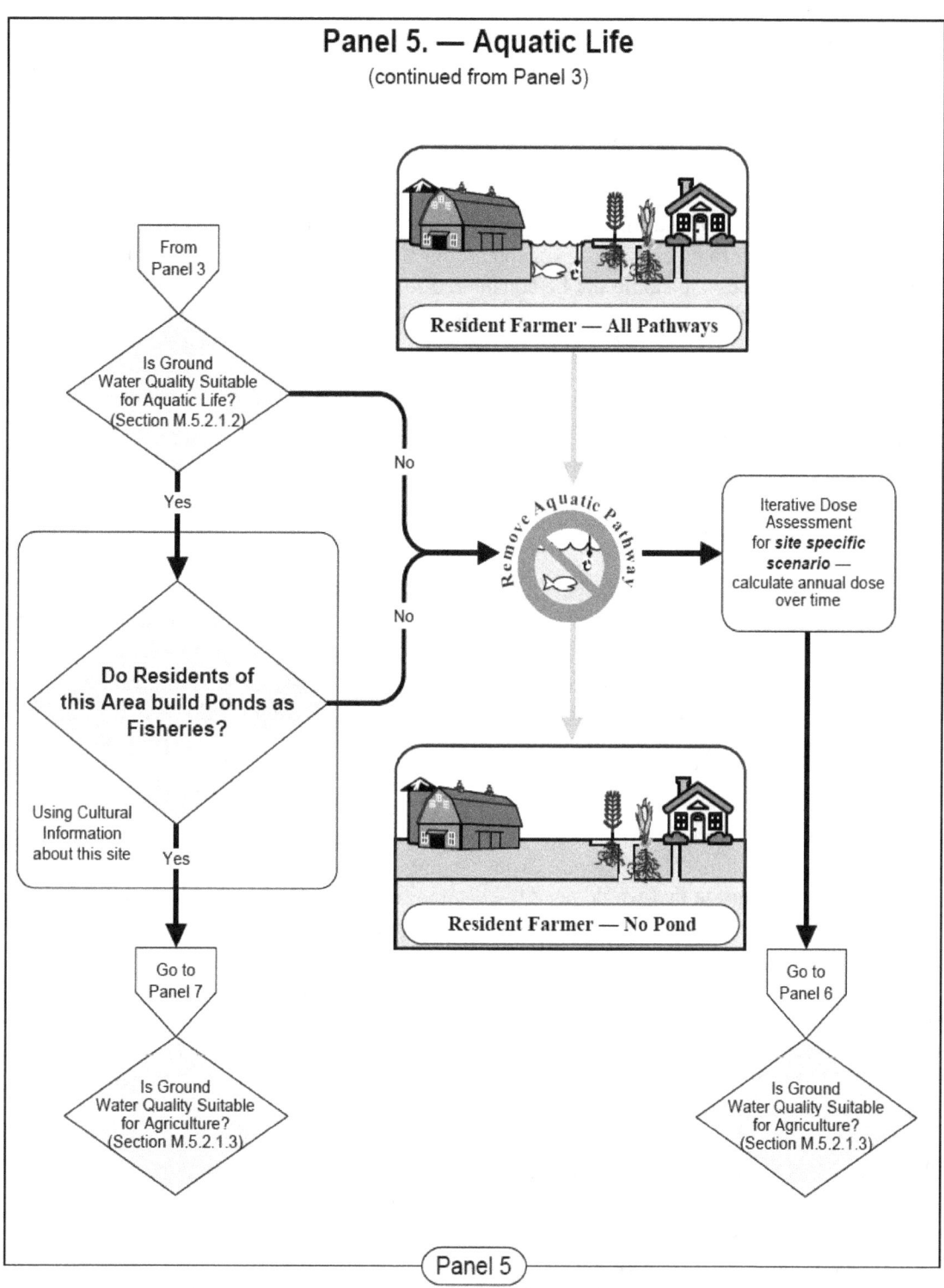

Figure M.5 Panel 5: Aquatic Life.

M.2.6   Panel 6: Agriculture — No Pond

The sixth panel (Figure M.6) is a continuation of Panel 5. This panel starts with a resident farmer without a pond. Ground water is available, but it is not suitable for a pond. The question asked here: "Is the ground water quality suitable for agricultural use?"

Section M.5.2.1.3 considers the suitability of ground water for agriculture and present the standards for this water to be considered acceptable for this use. If the water is unsuitable for irrigation (growing crops), it should not be considered suitable as drinking water for the farmer or for his animals. In this case, the following pathways would be removed: the irrigation pathway, the drinking water pathway, and any pathways associated with farm animals drinking water.

The resultant scenario would be a resident farmer scenario with no ground water use. All water needs would be met by uncontaminated water from an outside source, but the farmer would still be growing crops in soil with residual radioactivity, and his animals would still be ingesting food and soil with residual radioactivity.

If an iterative dose assessment shows the TEDE to an average member of the critical group still exceeds 0.25 mSv/y (25 mrem/y), the next logical question to ask is "Are soil and topography at this site suitable for agriculture?" Additional details concerning this issue can be found in Section M.5.2.2. If the answer to this question is "No," and either soil or topography at this site are determined to be unsuitable for agriculture, the agricultural pathway would be removed, leaving a rural resident with no agriculture, pond, or drinking water, since the these pathway have already been removed.

The resident farmer scenario has now devolved into a what is essentially a building occupancy scenario combined with modified external exposure and inhalation pathways. Section M.5.2.2 presents the documentation that would need to be submitted to NRC if the licencee wants to remove the agricultural pathway on the basis of either topography or soil being unsuitable to agriculture.

After the agricultural pathway is removed, another dose assessment would be done. If the TEDE to an average member of the critical group still exceeds 0.25 mSv/y (25 mrem/y), the licensee should begin to analyze the critical parameters for the this scenario, but there is no need, at this point, to continue with alternate scenario development.

If the answer to the question regarding the suitability of the soil and topography at this site for agriculture is "Yes," the scenario returns to that of a resident farmer with no ground water use getting all his water needs met by uncontaminated water from an outside source. For this situation, the scenario has been defined and there is no need to introduce additional site data. An iterative dose assessment should be done after critical parameters have been modified.

If the answer to the first question in this panel, "Is ground water quality suitable for agriculture?" is "Yes," the licensee would go to Panel 8 (Figure M.8) to consider the potability of the ground water.

M.2.7   Panel 7: Agriculture — All Pathways

The seventh panel (Figure M.7) is also a continuation of Panel 5, but it starts with a resident farmer and all pathways. Ground water is available and it is suitable for a pond. The question asked here is the same as in Panel 6: "Is the ground water quality suitable for agricultural use?"

The procedure here is identical to Panel 6, except that in each resultant scenario, the farmer still has a pond. In the final situation, where both questions have been answered with a "No," the scenario is of a rural resident with a pond — the building occupancy scenario combined with the aquatic scenario and modified versions of the external exposure and inhalation pathways.

If the answer to the first question in this panel, "Is ground water quality suitable for agriculture?" is "Yes," the licensee would go Panel 9 (Figure M.9) to consider the potability of the ground water.

M.2.8   Panel 8: Potability — No Pond

The eighth panel (Figure M.8) is a continuation of Panel 6, a resident farmer without a pond. Ground water is available and is suitable for agriculture, but it is not suitable for a pond. The questions asked here are "Is the ground water potable?" and "Can the farmer drink the water?".

Section M.5.2.1.4 considers the potability of ground water, drinking water standards, and documentation needed for the NRC. If the ground water does not meet drinking water standards, the drinking water pathway would be removed, and an iterative dose assessment would be done. If TEDE to an average member of the critical group still exceeds 0.25 mSv/y (25 mrem/y), the licensee would consider the suitability of the soil and topography for agricultural use.

This suitability of the topography and soil for agriculture would be considered in the same manner as it was in Panel 6 (Figure M.6). If either the soil or topography is determined to be unsuitable, the agricultural pathway would be removed and the scenario would devolve to the building occupancy scenario of a rural resident with no pond, no agriculture, and no drinking water.

After the agricultural pathway is removed, another dose assessment would be done, and if TEDE to an average member of the critical group still exceeds 0.25 mSv/y (25 mrem/y), the licensee should begin analysis of the critical parameters for the building occupancy scenario. If the TEDE is still above the threshold value, the licensee would need to consider modifications to the critical parameters, but there would be no need, at this point, to continue with alternate scenario development.

If the answer to the first question in this panel, "Is ground water potable?" is "Yes," the licensee would go to Panel 10 (Figure M.10) to consider the potability of the ground water.

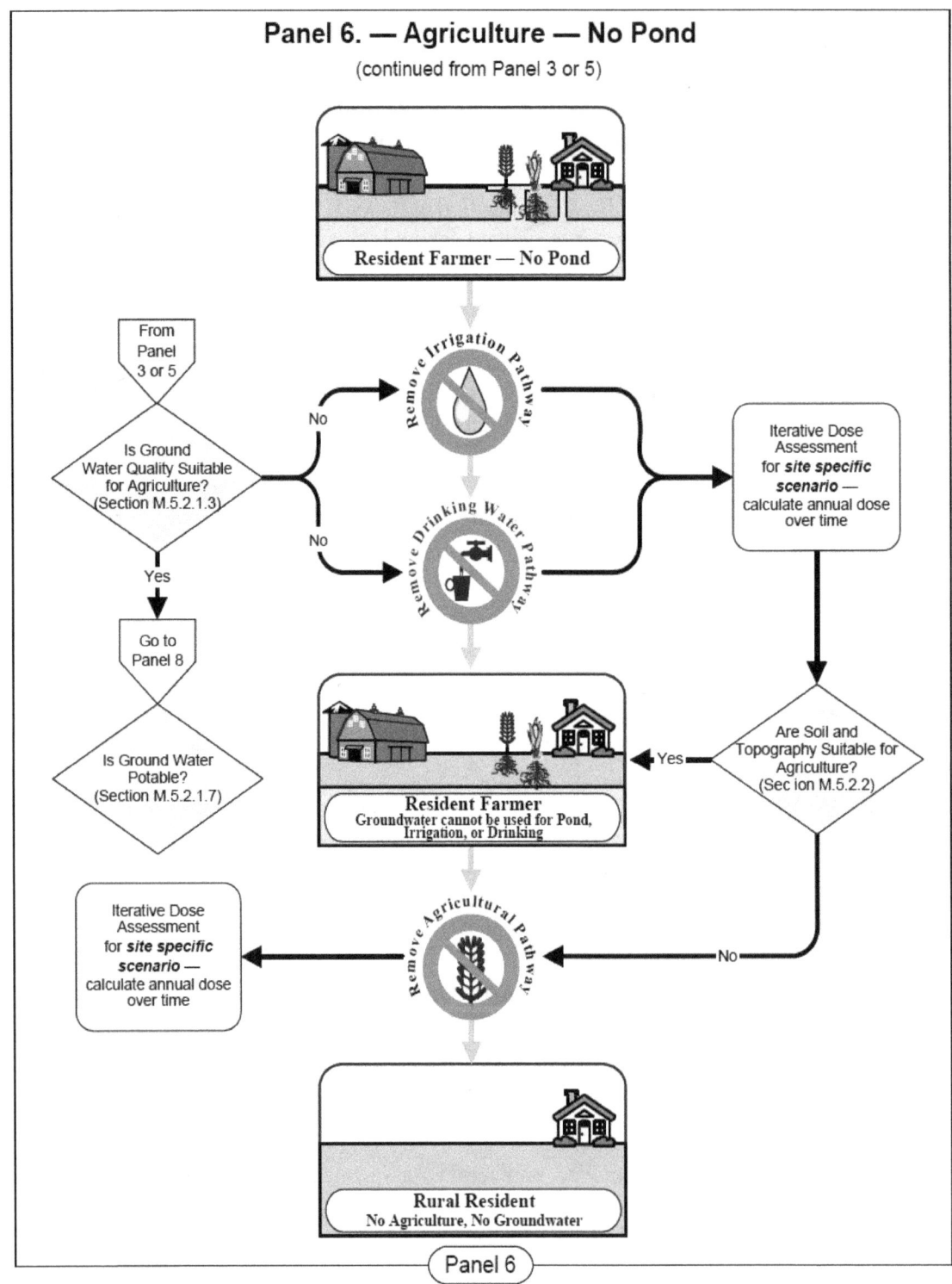

Figure M.6 Panel 6: Agriculture — No Pond.

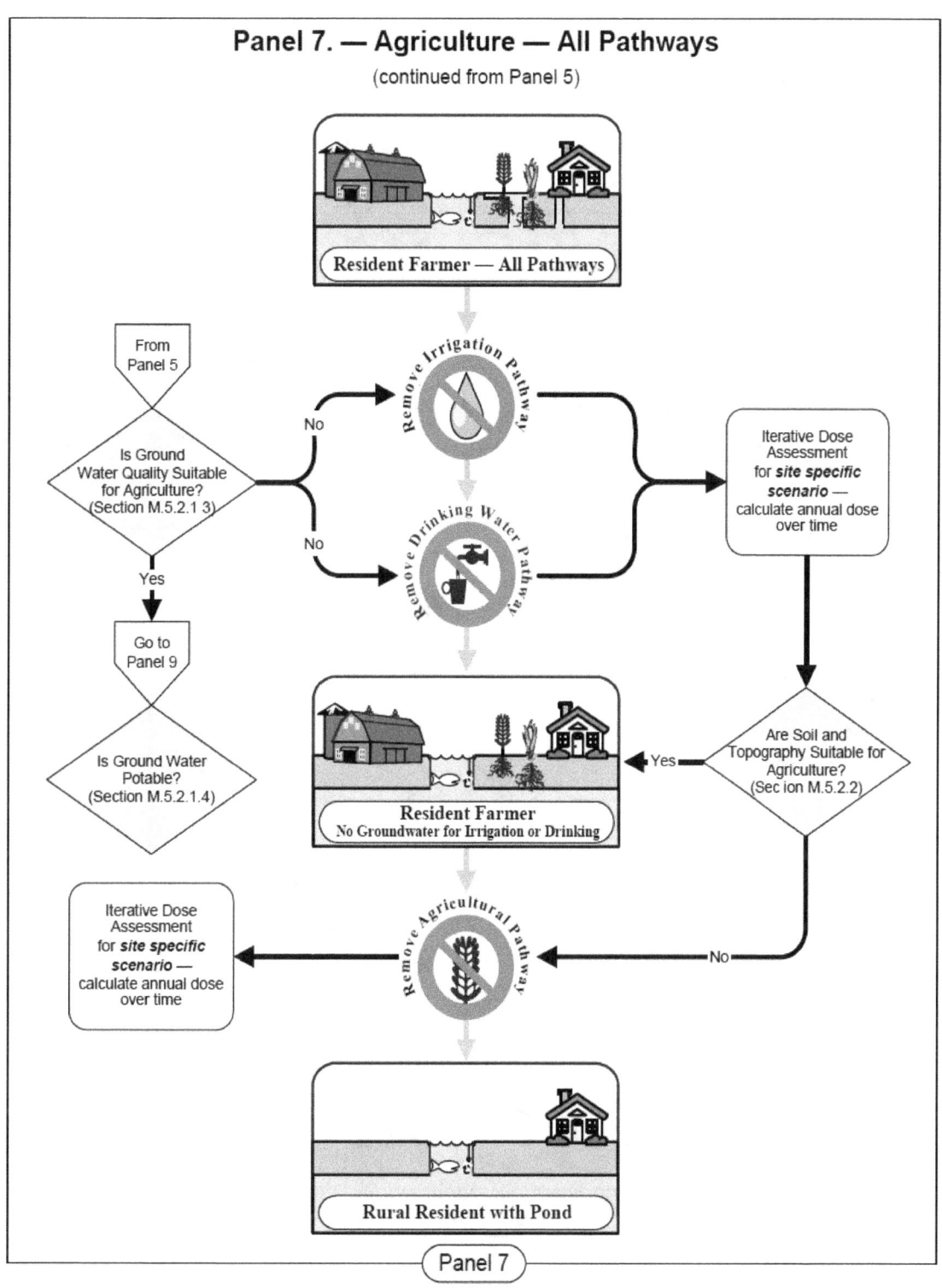

Figure M.7 Panel 7: Agriculture — All Pathways.

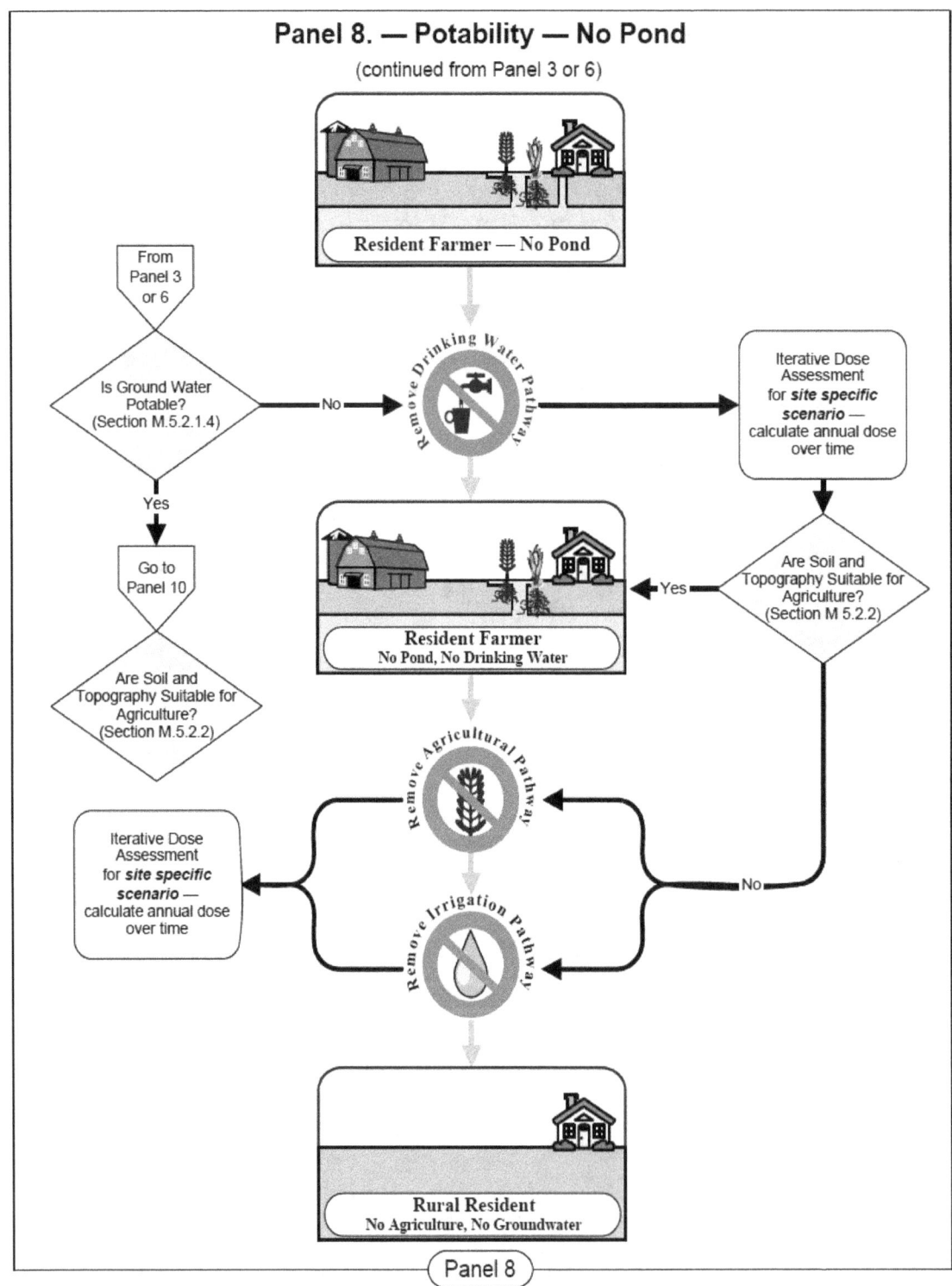

Figure M.8 Panel 8: Potability — No Pond.

M.2.9   Panel 9: Potability — All Pathways

The ninth panel (Figure M.9) is the continuation of Panel 7 and is almost the same as Panel 8, except that it starts with a resident farmer and all pathways. Ground water is available and is suitable for both pond and agriculture. As with Panel 8, the question asked here is, "Can the farmer drink the water?"

The procedure here is identical to Panel 8, except that in each resultant scenario the farmer still has a pond. In the final situation, where both questions have been answered with a "No," the scenario would be that of a rural resident with a pond — the building occupancy scenario combined with the aquatic scenario and modified versions of the external exposure and inhalation pathways.

As with Panel 8, if the answer to the first question in this panel, "Is ground water quality suitable for agriculture?", is "Yes," the licensee would go Panel 11 (Figure M.11) to consider the suitability of topography and soil for agriculture.

M.2.10   Panel 10: Topography and Soil — No Pond

The tenth panel (Figure M.10) is the continuation of Panel 8; it starts with a resident farmer with no pond. Ground water is available, and although not suitable for a pond, it is suitable for both agriculture and drinking. The question here is the suitability of topography and soil for agriculture.

The question of the suitability of topography and soil for agriculture is considered in the same manner as it was for Panel 6. If either the soil or topography is determined to be unsuitable, the scenario would devolve to a rural resident with drinking water but no pond. This would essentially be the building occupancy scenario combined with the drinking water scenario, and modified versions of the external exposure and inhalation pathways.

After the agricultural pathway is removed, another dose assessment would be done, and if the TEDE to an average member of the critical group still exceeds 0.25 mSv/y (25 mrem/y), the licensee should begin analyzing of the critical parameters for this scenario, but there would no need, at this point, to continue with alternate scenario development.

If the answer to the first question in this panel regarding the suitability of topography and soil for agriculture is "Yes," the licensee would assume that the correct scenario is the resident farmer with all pathways except a pond, and would begin examining critical parameters for that scenario using information from a sensitivity analysis.

M.2.11 Panel 11: Topography and Soil — All Pathways

The eleventh panel (Figure M.11) is the continuation of Panel 9; it starts with a resident farmer and all pathways. Ground water is available and has been determined to be suitable for a pond, for agriculture, and for drinking. The question now is the suitability of topography and soil for agriculture. This suitability of topography and soil for agriculture is considered here in the same manner as it was in Panel 6. If either the soil or topography is determined to be unsuitable, the scenario would devolve to a rural resident with drinking water and a pond. This would essentially be the building occupancy scenario combined with the drinking water scenario, the aquatic scenario, and modified versions of the external exposure and inhalation pathways.

After the agricultural pathway is removed, another dose assessment would be done and if TEDE to an average member of the critical group still exceeds 0.25 mSv/y (25 mrem/y), the licensee should begin analysis of the critical parameters for this scenario, but there would be no need, at this point, to continue with alternate scenario development.

If the answer to the first question in this panel regarding the suitability of topography and soil for agriculture is "Yes," the licensee would assume that the correct scenario is the resident farmer with all pathways and would begin examining critical parameters based on a sensitivity analysis.

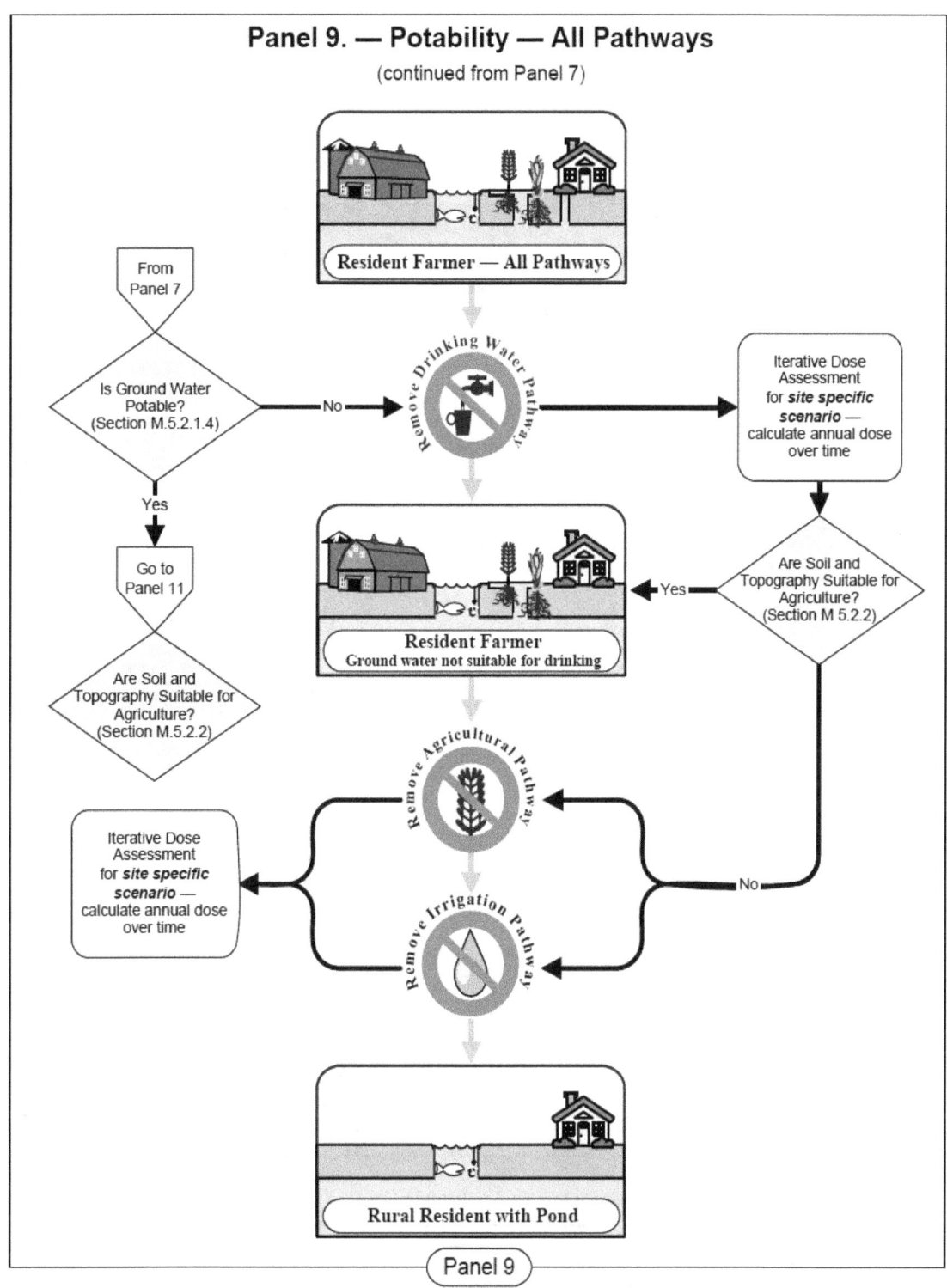

Figure M.9 Panel 9: Potability — All Pathways.

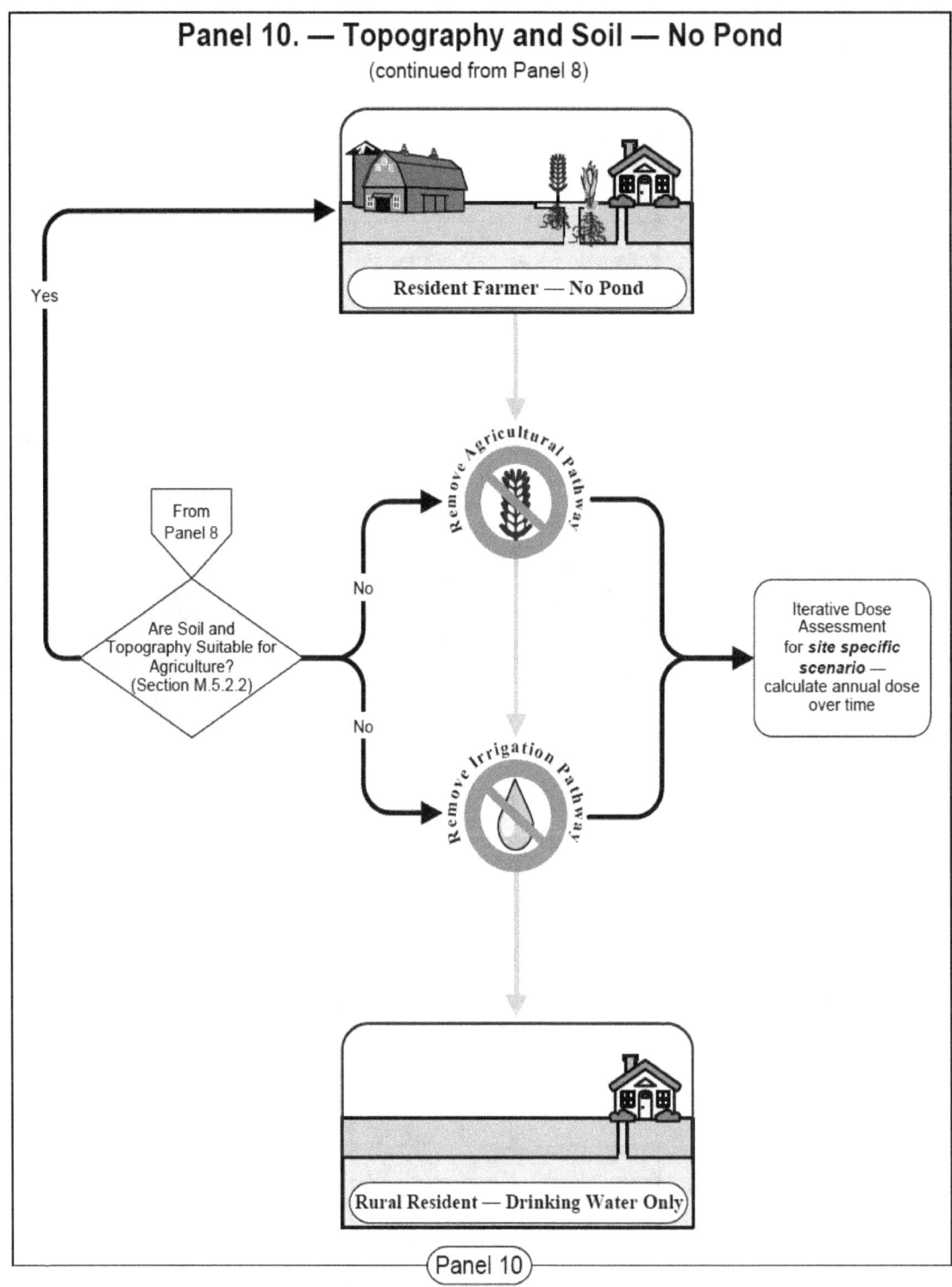

Figure M.10   Panel 10: Topography and Soil — No Pond.

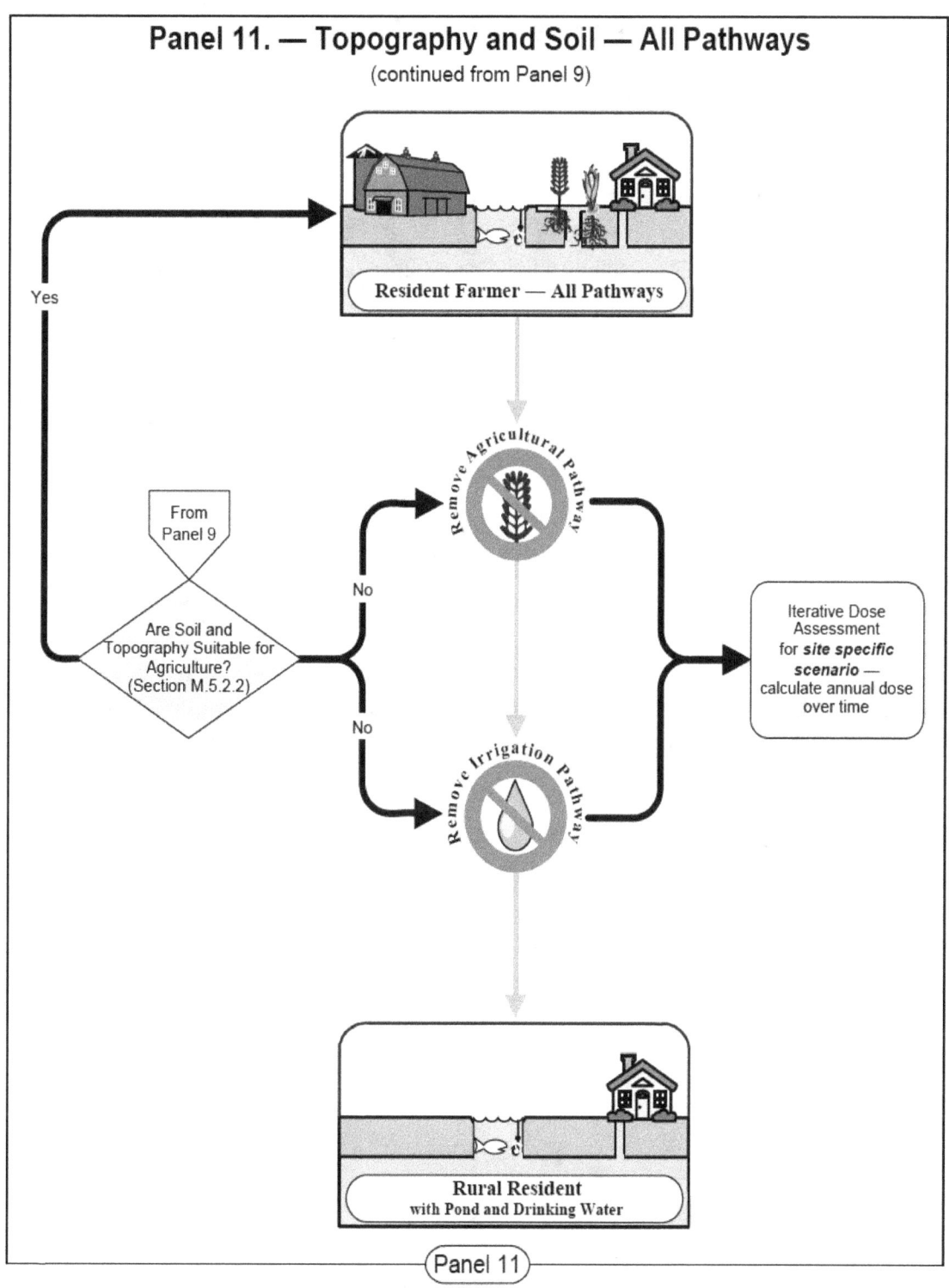

Figure M.11   **Panel 11: Topography and Soil — All Pathways.**

M.2.12 Panel 12: Urban Resident and Industrial Uses

The twelfth panel (Figure M.12) is the continuation of Panel 2; it starts with the urban resident scenario and shows the industrial uses scenario.

M.2.12.1 Urban Resident

The urban resident scenario is essentially a building occupancy scenario that includes a garden scenario (modified from the resident farmer scenario) and modified versions of the external exposure and inhalation pathways. Cultural information regarding future land use needs to be introduced here to answer the question, "Is this urban resident likely to have a garden?" The information presented in Section M.5.1 and specifically M.5.1.2.2 can be used to help answer this question and determine the documentation that would need to be submitted to the NRC on this issue.

If the urban resident is likely to have a garden, the licensee should begin analyzing of the critical parameters for this scenario, but there would be no need, at this point, to continue with alternate scenario development.

If it is considered unlikely for the urban resident to have a garden, the garden pathway would be removed, an iterative dose assessment would be done. If TEDE to an average member of the critical group still exceeds 0.25 mSv/y (25 mrem/y), the licensee should begin analyzing of the critical parameters for the urban resident scenario, but there would be no need, at this point, to continue with alternate scenario development.

M.2.12.2 Industrial Worker

The industrial worker scenario includes the building occupancy scenario and modified versions of the external exposure and inhalation pathways. While there is no additional site-specific information to further devolve this scenario, site-specific information can be used to modify the pathway parameters for this scenario.

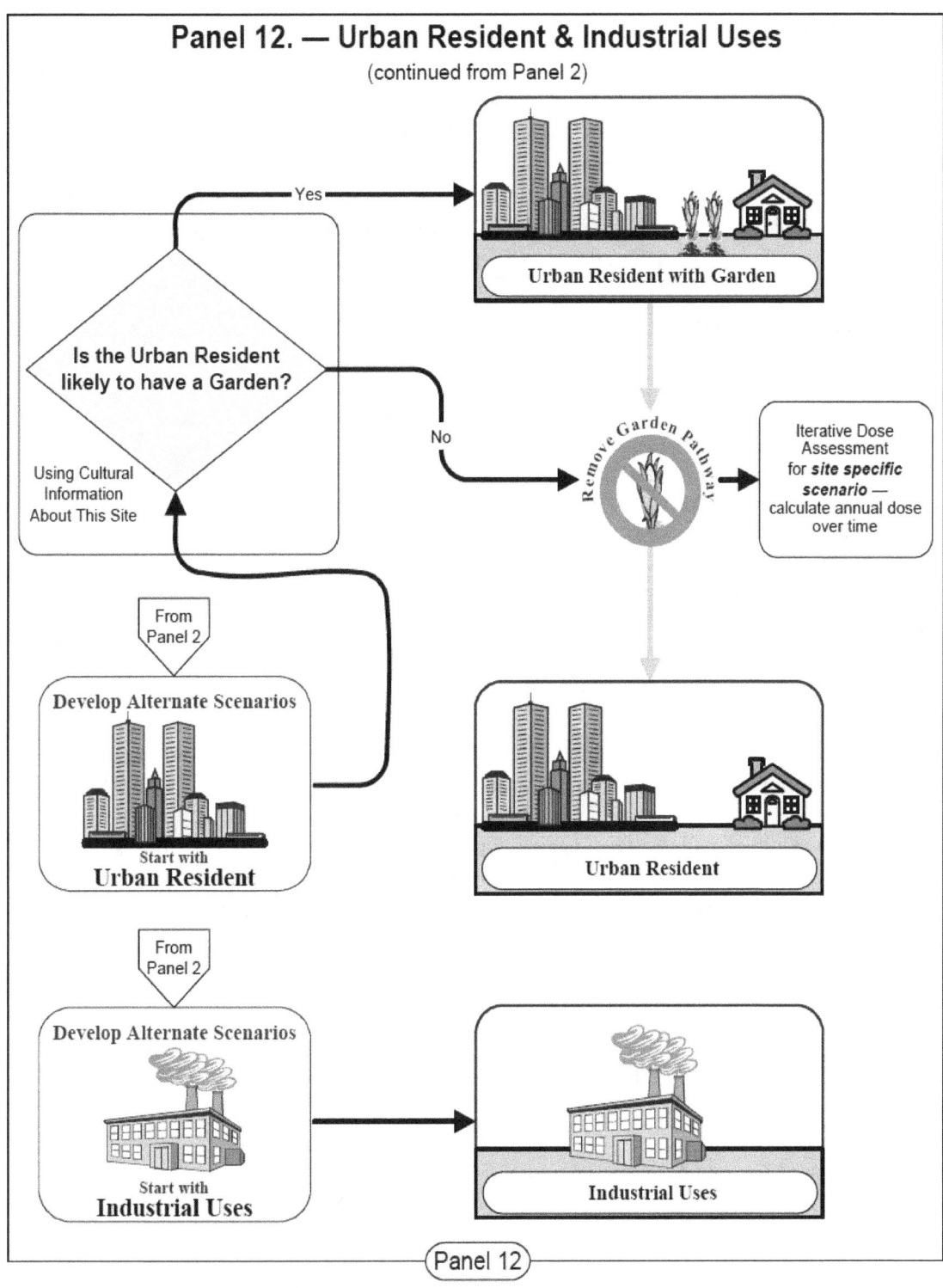

Figure M.12   **Panel 12: Urban Resident and Industrial Uses.**

M.3   Initial Computation

The process for developing alternate scenarios begins with the decommissioning and license termination framework as described in Section 1.2 of this volume. The process described in this report integrates with the Decision Framework and expands upon the introduction of site-specific information, the revision of pathways, and iterative dose assessment. Figure M.1 shows a more detailed framework diagram that includes the sensitivity analysis step and highlights the process that uses site-specific information to develop alternate scenarios.

M.3.1   Define Source

M.3.1.1 Assemble Existing Data

Existing data for the site should be gathered, assembled, and evaluated. The first step is to determine the types and amounts of radioactive material possessed by the licensee at this site; this information is needed to perform the initial dose assessment.

Information about any surveys and leak tests that have been performed, as well as any records important to decommissioning as described in 10 CFR Parts 30.35, 40.36, 50.75, 70.25, and 72.30, need to be assembled as appropriate. This information may be needed to quantify the amount of residual radioactivity present at the site.

Information regarding ground water depth and quality, soil type, and local cultural practices may be needed to develop alternate scenarios, to evaluate models, or to modify model parameters, but an initial dose assessment can be performed before expending resources to gather this data. If the initial dose assessment, using site-specific source concentrations, default pathways, and default pathway parameters, shows TEDE to an average member of the critical group to not exceed 0.25 mSv/y (25 mrem/y), there is no reason to gather and evaluate this site-specific information.

M.3.1.2 Calculate Source Concentration

The calculation of source concentration should be done according to the guidance in Appendix I.2 of this volume.

M.3.2   Initial Dose Assessment

Since the process for alternate scenario development set forth in this document is essentially a dissection of the resident farmer scenario, the initial dose assessment should be done using the default resident farmer scenario with its associated default pathways and parameters. Within this process of dissection, pathways should be removed as appropriate site-specific information is introduced.

The exception to this would be where the residual radioactivity at a site is fully contained within a building. If the case can be made that the contaminant would remain in the building throughout the period when it could cause a TEDE to an average member of the critical group to exceed the 0.25 mSv/y (25 mrem/y) threshold, the default resident farmer scenario would not be applicable and the building occupant scenario should be used.

Whenever pathways are removed, the licensee is expected to perform an iterative dose assessment that reflects the new scenario. These recurring computations are best done using software which has (1) the built-in NRC–approved default parameters and pathways, (2) procedures for removing entire pathways, and (3) procedures for modifying pathway parameters.

If the initial dose assessment results in an TEDE to an average member of the critical group that does not exceed 0.25 mSv/y (25 mrem/y), there is no need to collect more data or to develop alternate scenarios for this site. After ALARA concerns have been met, the site would be considered a candidate for unrestricted use.

If this dose assessment results in an TEDE to an average member of the critical group that is greater than 0.25 mSv/y (25 mrem/y), one of the options is to use site-specific information to modify the resident farmer scenario by eliminating pathways that are inappropriate for the site in question. There are other options at this point, but this appendix concentrates on the development of alternate scenarios.

If the process presented in this report is followed, the amount of data that needs to be gathered and the level of analyses that need to be done should be kept as low as possible. The first step in this process is to perform a sensitivity analysis by examining the results of the initial dose assessment to determine the pathways and radionuclides that significantly influence the TEDE. Section M.4 provides greater detail on this procedure and gives a specific example of a sensitivity analysis.

M.3.3   Iterative Dose Assessment

Iterative dose assessments should be performed whenever a pathway is eliminated or parameters are modified. Since the process began with the resident farmer scenario and default pathways and parameters, the introduction of site-specific data should reduce the TEDE.

If at any point in the process, the iterative dose assessment shows the TEDE to an average member of the critical group does not exceed 0.25 mSv/y (25 mrem/y), there is no need to introduce more data nor to continue developing alternate scenarios. After ALARA concerns have been met, the site should be acceptable for release.

M.4　Sensitivity Analysis

If the dose assessment shows that the results exceed 0.25 mSv/y (25 mrem/y) TEDE to the average member of the critical group, the results of initial or iterative dose assessments need to be examined to determine which of pathways and parameters are significant. This sensitivity analysis should help the licensee concentrate subsequent analyses on those pathways or parameters that are major contributors to the TEDE. As the licensee moves through this process, the licensee can take shortcuts, jumping to those pathways that are significant and ignoring those that are not.

A simple sensitivity analysis can be done following the initial dose assessment and following each iterative dose assessments as pathways are eliminated or parameters are modified. The results of the dose assessment should show the percentage of the TEDE attributable to each major pathway and to each of the radionuclides.

M.5　Introducing Site-Specific Information

Site-specific information can be divided into two broad categories: cultural information and physical information. Physical information includes the location, climate, topography, geology, soil types, water availably, etc. of the site. Cultural information is essentially how the land is used by the human population. Physical properties of land are essentially unchanging, while cultural properties are constantly changing. In reality, physical properties change (sometimes as a result of cultural activities), but the change is slow compared to the cultural use of the land.

Since the initial dose assessment for this process was done using the resident farmer scenario with NRC–approved default pathways and parameters, the introduction of either cultural or physical information about a D&D site is likely to reduce the TEDE.

M.5.1　Cultural Information

For developing alternate scenarios, the most important element of cultural information about any site is the future land use, because radionuclides can persist over long periods of time. The future is assessed on the basis of the past and the present. Experience has shown that while this is an inexact science, the near future can be estimated with some degree of accuracy. What is the near future? It depends on the location, the culture, and what is being estimated. In this time and space, and for what is being predicted, it is probably substantially less than 100 years, but the line should be drawn somewhere, and in this case, it is drawn at 100 years. The key to the assessment of future land use is the current and past use of the land, at both the site and in the region. The licensee will need to address both the dose from reasonably foreseeable land use (for the compliance calculation) and less likely but plausible land uses (for informational purposes).

M.5.1.1   Current Land Use

The determination of current land use is the initial step in the process of estimating future land use. Land use should be determined not only for the site, but also for the land within a 80-km (50-mile) radius surrounding the site. This assessment of land use does not need to be complicated or detailed; it should be fairly simple, dividing the land into only three categories: urban, rural, or industrial.

Current land use can be determined through one or more of the following information sources:

- site description,

- topographic maps,

- planning agencies,

- zoning maps,

- aerial photographs, or

- site visits.

The majority of the US has codified land use/zoning, and many administrative areas have developed land use master plans. For this reason, the primary source of information on current land use should be the planning agencies of the State, county, and/or municipality in which the site resides. In most cases, the easiest way to find these planning agencies is in the government section of the local phone book.

A large amount of data also is available on the Internet at Web sites maintained by government agencies. Tables M.1, M.3, and M.4 list current Web sites for every State in the U.S. These Web sites contain indices to all types of data about each specific State. Examples of the types of information available are given in Table M.2. Land use planning information is often available at these sites. (Note: Web sites are volatile. Addresses, the amount, and type of information at any Web site may change at any time.)

Table M.1   State Web Sites

STATE NAME	WEB ADDRESS	STATE NAME	WEB ADDRESS
Alabama	*http://www.state.al.us/*	Montana	*http://www.state.mt.us/*
Alaska	*http://www.state.ak.us/*	Nebraska	*http://www.state.ne.us/*
Arizona	*http://www.state.az.us/*	Nevada	*http://www.state.nv.us/*
Arkansas	*http://www.state.ar.us/*	New Hampshire	*http://www.state.nh.us/*
California	*http://www.state.ca.us/*	New Jersey	*http://www.state.nj.us/*
Colorado	*http://www.state.co.us/*	New Mexico	*http://www.state.nm.us/*
Connecticut	*http://www.state.cn.us/*	New York	*http://www.state.ny.us/*
Delaware	*http://www.state.de.us/*	North Carolina	*http://www.state.nc.us/*
Florida	*http://www.state.fl.us/*	North Dakota	*http://www.state.nd.us/*
Georgia	*http://www.state.ga.us/*	Ohio	*http://www.state.oh.us/*
Hawaii	*http://www.state.hi.us/*	Oklahoma	*http://www.state.ok.us/*
Idaho	*http://www.state.id.us/*	Oregon	*http://www.state.or.us/*
Illinois	*http://www.state.il.us/*	Pennsylvania	*http://www.state.pa.us/*
Indiana	*http://www.state.in.us/*	Rhode Island	*http://www.state.ri.us/*
Iowa	*http://www.state.ia.us/*	South Carolina	*http://www.state.sc.us/*
Kansas	*http://www.state.ks.us/*	South Dakota	*http://www.state.sd.us/*
Kentucky	*http://www.state.ky.us/*	Tennessee	*http://www.state.tn.us/*
Louisiana	*http://www.state.la.us/*	Texas	*http://www.state.tx.us/*
Maine	*http://www.state.me.us/*	Utah	*http://www.state.ut.us/*
Maryland	*http://www.state.md.us/*	Vermont	*http://www.state.vt.us/*
Massachusetts	*http://www.state.ms.us/*	Virginia	*http://www.state.va.us/*
Michigan	*http://www.state.mi.us/*	Washington	*http://www.state.wa.us/*
Minnesota	*http://www.state.mn.us/*	West Virginia	*http://www.state.wv.us/*
Mississippi	*http://www.state.ms.us/*	Wisconsin	*http://www.state.wi.us/*
Missouri	*http://www.state.mo.us/*	Wyoming	*http://www.state.wy.us/*

Table M.2   Land Use Information Types

Current land use plans	Cultural resources in the area
Zoning laws	Threatened and endangered species
Zoning maps	Natural resource inventory information
Community comprehensive master plan	Floodplain/wetlands designations
State comprehensive master plan	Local/regional geologic information
Demographics	Wellhead projection areas/aquifer recharge areas
Historical population growth patterns	State comprehensive ground water protection program
Current site location relative to other land uses	Historical aerial photography
Federal/State land designations for surrounding lands	Environmental justice issues

Table M.3   Federal Sites Containing Data Relevant to Land Use

Sites (National)	Web Address
USGS National Biological Information Infrastructure	http://www.nbii.gov/index html
USGS National Mapping Information	http://mapping.usgs.gov/
USGS Water Resources of the United States	http://water.usgs.gov/
USDA Natural Resource Conservation Service	http://www nrcs.usda.gov/
EPA Spatial Data Library System	http://www.epa.gov/enviro/html/esdls/esdls_over ht ml
USDOC National Oceanic and Atmospheric Administration (NOAA)	http://www.noaa.gov/
USDOC Census Bureau TIGER Data	http://www.census.gov/geo/www/tiger/index html
USDOC Census Bureau Population Topics & Household Economic Topics	http://www.census.gov/population/www/index.html
USDOC Census Bureau Economic Programs	http://www.census.gov/econ/www/index html
USGS EROS Data Center	http://edcwww.cr.usgs.gov/eros-home.html
USGS Mapping Applications Center	http://mapping.usgs.gov/mac/
USGS Mid-Continent Mapping Center	http://mcmcweb.er.usgs.gov/
USGS Rocky Mountain Mapping Center	http://rmmcweb.cr.usgs.gov/
USGS Western Mapping Center	http://www-wmc.wr.usgs.gov/

Table M.3   Federal Sites Containing Data Relevant to Land Use (continued)

Sites (National)	Web Address
USDA NRCS: East	http://www.ea nrcs.usda.gov/
USDA NRCS: Mid-West	http://www.mw nrcs.usda.gov/
USDA NRCS: Northern-Plains	http://www.np nrcs.usda.gov/
USDA NRCS: South Central	http://www ftw nrcs.usda.gov/regional/sc_reg.html
USDA NRCS: Southeast	http://www.ga.nrcs.usda.gov/index html
USDA NRCS: West	http://www.rcw nrcs.usda.gov/
EPA Region 1	http://www.epa.gov/region01/
EPA Region 2	http://www.epa.gov/region02/
EPA Region 3	http://www.epa.gov/region03/
EPA Region 4	http://www.epa.gov/region04/
EPA Region 5	http://www.epa.gov/region05/
EPA Region 6	http://www.epa.gov/region06/
EPA Region 7	http://www.epa.gov/region07/
EPA Region 8	http://www.epa.gov/region08/
EPA Region 9	http://www.epa.gov/region09/
EPA Region 10	http://www.epa.gov/region10/

Table M.4   State Sites Containing Data Relevant to Land Use

State	Organization	Web Address
Alabama (GSA)	Geologic Survey of Alabama	http://www.gsa.tuscaloosa.al.us/gsa/gsa html
Alaska (ASGDC)	Alaska State Geo-Spatial Data Clearinghouse	http://www.asgdc.state.ak.us/homehtml/intro.html
Arizona (AGIC)	Arizona Geographic Information Council	http://www.land.state.az.us/agic/agichome html
Arkansas (ASLIB)	Arkansas State Land Information Board	http://www.dis.state.ar.us/LIB/Lib_Home htm
California (CGIA)	California Geographic Information Association	http://www.cgia.org/

Table M.4 State Sites Containing Data Relevant to Land Use (continued)

State	Organization	Web Address
Colorado (CGICC)	Colorado Geographic Information Coordinating Committee	http://www-gis.cudenver.edu/~gicc/
Connecticut (CEGIC)	Connecticut Environmental and Geographic Information Center	http://dep.state.ct.us/cgnhs/
Delaware (DGDC)	Delaware Geographic Data Committee	http://www.state.de.us/planning/coord/dgdc htm
Florida (FGIB)	Florida Geographic Information Board	http://als.dms.state.fl.us/
Georgia (GITPC)	Georgia Information Technology Planning Council	http://www.state.ga.us/itpc/
Hawaii (HSGISP)	Hawaii Statewide GIS Program	http://www.hawaii.gov/dbedt/gis/index html
Idaho (IGDC)	Idaho Geographic Data Center	http://geolibrary.uidaho.edu/
Illinois (ISGS)	Illinois State Geologic Survey	http://www.isgs.uiuc.edu/
Indiana	**No GIS site found; this is the State Web Page	http://www.state.in.us/
Iowa (IGIC)	Iowa Geographic Information Council	http://www.gis.state.ia.us/default htm
Kansas (KDASC)	Kansas Data Access and Support Center	http://gisdasc kgs.ukans.edu/dasc html
Kentucky (KOGIS)	Kentucky Office of Geographic Information Systems	http://ogis.state ky.us/
Louisiana (LGISC)	Louisiana Geographic Information Systems Council	http://www.doa.state.la.us/lgisc/
Maine (MOGIS)	Maine Office of GIS	http://apollo.ogis.state.me.us/homepage.htm
Maryland (MSGIC)	Maryland State Government Geographic Information Coordinating Committee	http://www.dnr.state md.us/MSGIC/index htm
Massachusetts (MassGIS)	Massachusetts Geographic Information System	http://www.magnet.state ma.us/mgis/massgis.htm
Michigan (MIC)	Michigan Information Center	http://www.state.mi.us/dmb/mic/

Table M.4   State Sites Containing Data Relevant to Land Use (continued)

State	Organization	Web Address
Minnesota (GCGI)	Minnesota Governor's Council on Geographic Information	http://www.lmic.state mn.us/gc/gc htm
Mississippi (MARIS)	Mississippi Automated Resource Information System	http://www.maris.state ms.us/
Missouri (MSDIS)	Missouri Spatial Data Information Service	http://msdis missouri.edu/
Montana (MGIC)	Montana Geographic Information Council	http://www.mt.gov/isd/groups/mgic/index htm
Nebrasksa (NGISSC)	Nebraska Geographic Information Systems Steering Committee	http://www.calmit.unl.edu/gis/
Nevada (NSMAC)	Nevada State Mapping Advisory Committee	http://www.nbmg.unr.edu/smac/smac htm
New Hampshire (NHOSP)	New Hampshire Office of State Planning	http://www.state.nh.us/osp/ospweb.htm
New Jersey (CIO–GIS)	New Jersey GIS	http://www.state.nj.us/cio/gis/index.html
New Mexico (NMGIC)	New Mexico Geographic Information Council	http://nmgic.unm.edu/
New York (NYSGISC)	New York State GIS Clearinghouse	http://www.nysl nysed.gov/gis/clhs_new htm
North Carolina (NCGICC)	North Carolina Geographic Information Coordinating Committee	http://cgia.cgia.state nc.us:80/gicc/
North Dakota (NDSMAC)	North Dakota State Mapping Advisory Committee	http://www.state.nd.us/ndgs/SMAC html
Ohio (OGRIP)	Ohio Geographically Referenced Information Program	http://www.state.oh.us/ogrip/
Oklahoma (OSEIC)	Spatial and Environmental Information Clearinghouse	http://www.seic.okstate.edu/seic html
Oregon (OSSCGIS)	Oregon State Service Center for Geographic Information Systems	http://www.sscgis.state.or.us/index html

Table M.4   State Sites Containing Data Relevant to Land Use (continued)

State	Organization	Web Address
Pennsylvania (PASDA)	Pennsylvania Spatial Data Access	http://www.pasda.psu.edu/
Rhode Island (RIGIS)	Rhode Island Geographic Information System	http://www.edc.uri.edu/rigis/
South Carolina (SCDNRGDC)	South Carolina Department of Natural Resources GIS Data Clearinghouse	http://www.dnr.state.sc.us/gisdata/index html
South Dakota	**No GIS site found; this is the State Web Page	http://www.state.sd.us/
Tennessee (TGIS)	Tennessee Geographic Information System	http://www.state.tn.us/finance/oir/admin/gishome.html
Texas (TNRIS)	Texas Natural Resource Information System	http://www.tnris.state.tx.us/digital htm
Utah (UGISAC)	Utah Geographic Information Systems Advisory Council	http://www.its.state.ut.us/agrc/html/gisac2 html
Vermont (VGIS)	Vermont Geographic Information System	http://geo-vt.uvm.edu/
Virginia (UVALGIC)	University of Virginia Library Geographic Information Center	http://www.lib.virginia.edu/gic/services.html
Washington (WAGIC)	Washington Geographic Information Council	http://www.wa.gov/gic/
West Virginia (WVGIST)	West Virginia Geographic Information System Techweb	http://wvgis.wvu.edu/
Wisconsin (WISCLINC)	Wisconsin Land Information Clearinghouse	http://badger.state.wi.us/agencies/wlib/sco/pages/wisclinc html
Wyoming (WGIAC)	Wyoming Geographic Information Advisory Council	http://wgiac.state.wy.us/index.html

Assumptions and predictions regarding future land uses are important considerations in the development of scenario definitions and descriptions for analysis. If the site currently exists in a highly populated urban area, a residential farmer scenario is very unlikely. Exposure scenarios for certain sites may exclude exposures via agricultural pathways if agricultural land uses are clearly incompatible with existing and anticipated future conditions at the sites. Exposures via ingestion of contaminated ground water may be discounted if the affected ground water is of such poor quality as to preclude human consumption.

M.5.1.1.1 Use of Ponds as Fisheries

In addition to physical limitations on the likelihood of a farmer using a pond as a fishery, local cultural information should be used to determine if local residents currently engage in this practice. This question might be answered by the USDA county extension agent nearest to the D&D site. Contact information for county extension agents can be found at: http://www.reeusda.gov/.

M.5.1.2 Future Land Use

Specific local conditions should be taken into account when deciding how far into the future land use can be estimated. The general range for estimation is within 100 years. In areas where rapid change has occurred in the past, this cutoff might be considerably less than 100 years, whereas in other areas, such as the heart of New York City, it may be reasonable to argue that urban conditions should prevail for more than 100 years.

The first step in estimating future land use is to determine the current land use at the site. The past use of the land should also be ascertained because it is the combination of past and present uses that should indicate what changes have occurred and the rate of those changes. This information should be used in a documented process that a reviewer would be able to follow. This documentation should include the types and sources of material that were used and how the final projected use was determined. Tables M.1, M.3 and M.4 list possible Web site sources that may contain useful information.

Land use and changes in land use within the 80-km (50-mile) radius of the site should be considered as part of this process. For example, a site that is currently located in a rural area within 16–32 km (10–20 miles) of a growing metropolitan area should likely be in the suburbs of the metropolitan area within a decade or two, depending on population growth.

The 80-km (50-mile) radius is only a suggestion for determining the size of the area to consider. There may be valid reasons for increasing or decreasing the area of consideration, depending on local conditions and the length of time that a TEDE greater than the 0.25 mSv/y (25 mrem/y) threshold is expected to occur. Other factors that may influence this decision are critical pathways and the estimated distribution of residual radioactivity.

M.5.1.2.1 Sources of Information for Determining Future Land Use

The primary document referenced for information types was EPA OSWER Directive No. 9355.7–04: Land Use in the CERCLA Remedy Selection Process, dated May 25, 1995. This directive by the EPA's Office of Solid Waste and Emergency Response (OSWER) is also referenced by the Department of Defense (DoD) for use in Base Realignment and Closure (BRAC) installations.

Based on the OSWER directive, and on personal experience, Tables M.1, M.3, and M.4 contain information types which were used in determining possible data/information sources.

The are many sources at the Federal, State, and local levels for the information types listed in Table M.2. The list of sources provided here is not definitive, but the sources listed should, in most cases, be able to point the licensee to additional sources of information.

Because so much of the information used to describe current land uses and to determine possible future land uses is geographic in nature, the sources provided are for government geographic information system (GIS) providers at both the national and State levels. State GIS organizations should be able to direct the licensee to local sources for much of this information and, in many cases, may have links to that information directly from their data sites.

Table M.3 lists Federal government sources for data useful for determining possible future land use, and Table M.4 lists each State and the corresponding location for digital data.

M.5.1.2.2 Urban Gardens

The subsistence farm associated with the resident farmer is unlikely to exist in an urban situation, but gardens are very likely in urban and suburban settings. The "Victory Gardens" of World War II demonstrate this possibility. Exceptions would be places like the concrete and steel core of large cities like New York, where gardens using locally obtained soil and water would be highly unlikely.

Documentation to be Submitted to NRC

Current Land Use should be documented by maps, descriptions, or information from one of the other sources listed in Tables M.1, M.3, and M.4.

Estimates of Future Land Use should be supported by the documented process described in Section M.5.1.2.

M.5.2   Physical Information about Site

Physical information about the site includes climate, topography, vegetation, and, most importantly, water. Since water is a key factor in many of the pathways, its availability and proximity are very important.

M.5.2.1 Ground Water and Surface Water

Ground water is present at some depth at most every site. If ground water is only found at great depths, surface water may be ephemeral and may exist only in response to rainfall or snowmelt. Surface water for the resident farmer is a fish pond that is connected to the ground water.

There are several key questions about ground water that should be answered using site-specific information. The most important question regards the availability of water. Subsequent questions regard its quality and suitability for use.

M.5.2.1.1 Is Ground Water Available?

The first question that should be answered is "Is ground water available as a resource for the scenario resident?" More specific questions include the following:

1. Is it shallow enough and does it have sufficient yield such that it can reasonably be pumped by the resident to irrigate a small farm and provide domestic drinking water?

2. Is it shallow enough to intercept and connect to a fish pond, and does it have sufficient yield to sustain the pond?

With regards to the first question, the resident would need to drill a well into a permanent aquifer that has water sufficient for his needs and then be able pump that water into his house and onto his crops. Under the assumption that the well drilling and pumping technology available to the resident is similar to what exists today, it would not be unreasonable for the farmer to drill a well to and pump from a depth of 120 m (400 ft), but this depth should be considered somewhat subjective. Specific local conditions should be considered when deciding how deep an aquifer a subsistence farmer would be able to use. A commercial farmer would be likely to drill much deeper than a subsistence farmer would.

Local trends in ground water decline should be taken into account. In areas where ground water is being withdrawn at an unsustainable rate, water levels may be dropping. If it can be reasonably assumed that this trend may continue into the future, this should be taken into account when assessing the availability of ground water for the resident farmer.

If ground water is not available at a reasonable depth for drinking water or irrigation, it may also not be available for a pond. Under these circumstances, the resident farmer scenario can be devolved to exclude all three of the major pathways based on ground water usage: irrigation, drinking water, and aquatic (pond). If ground water is unavailable, it is also reasonable to exclude the use of surface water, since the aquatic scenario considers the concentration of radionuclides in the surface water to be related to the concentration in the ground water aquifer [Kennedy, 1992].

Even if ground water were available at reasonable depth, the yield may be insufficient for all uses. Under these circumstances, the resident farmer scenario can be devolved to exclude major pathways that cannot be satisfied by the yield.

Documentation to be Submitted to NRC

Ground Water Unavailable: USGS or independent consultant report showing that either ground water does not exist, or that it is too deep >120 m (>400 ft) to reasonably be used by a subsistence farmer.

If ground water is available for drinking or irrigation, it may not be available for a fishery pond. It would not be reasonable to expect that the farmer would continually pump water into a pond to maintain it as a fishery. The ground water would have to be shallow enough that a sufficient pond level would be maintained through its connection to the pond. This would mean the ground water would have to be no deeper than about 5 m (15 feet). Information about local topography and specific conditions at each site could be used to adjust this number up and down. If ground water is not available for the pond, the aquatic pathway should be removed from the resident farmer scenario.

Documentation to be Submitted to NRC

Ground Water Unavailable for Fish Pond: USGS or independent consultant report showing that either ground water does not exist, or that it is too deep (>15 ft.) to connect to a surface water pond.

M.5.2.1.2   Is Ground Water Quality Suitable for Aquatic Life?

The quality of surface water is critical to the support of aquatic life and is affected by (1) the chemical and physical conditions that exist in the pond, (2) runoff from exposed soil, and (3) condensation/entrapment of contaminants from the air (e.g., pollutants, acid rain, etc). Recommended standards for surface waters have been proposed [Viessman and Hammer, 1985] and are listed in Table M.5.

Table M.5   EPA Standards for Surface Waters to Support Freshwater Aquatic Life

Component	Recommended Limits
Dissolved oxygen	5 mg/L (minimum)
Suspended solids	0.90 × (transmission from seasonally established norm)
Fecal coliform bacteria	14 per 100 mL (shellfish)
pH	6.5–9.0
Oil and grease	$0.01 \times LC_{50}$*
Elemental phosphorus	0.0001 mg/L
Phosphate	1.0 mg/L
Chlorine	0.01 mg/L
Ammonia	0.2 mg/L

* LC_{50} represents the concentration that kills 50% of the test specimens.

The concentration of dissolved oxygen in surface water is affected by the biochemical oxygen demand (BOD) of the ecosystem. Sedimentation of suspended solids can cause a buildup of organic matter in sediments. These materials undergo metabolic degradation by aerobic soil microorganisms with the concomitant depletion of dissolved oxygen. Other contaminants, such as dissolved ammonia, can contribute to oxygen depletion by nitrification. Ammonia is toxic to fish and other aquatic animals. Acute toxicity occurs to warm-water species at ammonia levels of 0.4 mg/L.

The presence of coliform bacteria is sometimes indicative of other, more virulent pathogens in surface water and should be considered when fish or other aquatic animals are produced for human consumption.

If the quality of the ground water (and hence the pond) lies outside of the acceptable standards for aquatic life, the aquatic pathway should be removed from the resident farmer scenario.

Documentation to be Submitted to NRC

Ground Water Unsuitable for Aquatic Life: USGS or independent consultant report showing that ground water quality is poorer than the standards listed for this use.

M.5.2.1.3   Is Ground Water Quality Suitable for Agriculture?

The quality of ground water for agricultural uses varies depending on the type of agribusiness or agricultural enterprises conducted at the site. For example, ground water with infiltrated fertilizers and herbicides can be very beneficial to crop land through irrigation, but can have an adverse effect on the health and productivity of livestock and poultry. Based on extensive studies by the USDA, recommended limits for chemicals in drinking water for livestock and poultry have been published [http://www.montana. edu/wwwpb/ag/baudr146.html, and http://www.cahe.nmsu. edu/pubs/_m/m-112.html]. Table M.6 identifies common contaminants in ground water and the recommended maximum concentrations for consumption by livestock and poultry.

Table M.6   Recommended Limits for Components in Drinking Water for Livestock and Poultry

Component	Maximum Concentration (mg/L)
Aluminum	5
Arsenic	0.02
Boron	5
Cadmium	0.05
Chromium	1
Cobalt	0.5
Copper	2
Fluoride	2
Iron	5
Lead	0.05–0.10
Mercury	0.01
Nitrate + Nitrite	100
Nitrite	10
Selenium	0.05–0.10
Vanadium	0.1
Zinc	25
(Mg,Na) sulfates	5000
Alkalinity	2000

In additional to acute and chronic toxicity from the elements in Table M.6, high concentrations of dissolved solids in drinking water can lead to various degrees of mineral toxicity in animals. Most minerals and dissolved solids found in water provide nutritional benefits when present within limited concentration ranges (e.g., selenium). At high concentrations, however, common minerals can lead to acute or chronic effects that impact the quality of animal products and overall productivity.

The salinity, or total dissolved solids, should be a consideration when evaluating ground water for animal consumption. Although 10,000 mg/L is acceptable under some conditions, the health, and ultimately the productivity, of animals is affected to various degrees by the salinity.

Table M.7 provides a breakdown of conditions that have been observed and documented in livestock and poultry for various concentrations of dissolved solids in drinking water.

Table M.7   Effects of Salinity of Drinking Water on Livestock

Salinity Level Limits for Drinking Water	Conditions
Less than 1,000 mg/L	Excellent for all classes of livestock and poultry
1,000–3,000 mg/L	Temporary mild diarrhea in livestock and poultry
3,000–5,000 mg/L	Satisfactory for livestock. Increased morbidity contributes to poor growth in poultry.
5,000–7,000 mg/L	Marginal quality for livestock. Not suitable for poultry and pregnant and lactating animals.
7,000–10,000 mg/L	Considerable risk for pregnant and lactating animals
Above 10,000 mg/L	Unacceptable

If the quality of the ground water is less than what is considered acceptable for irrigation, the irrigation pathway should be removed from the resident farmer scenario.

If the quality of the ground water is less than what is considered acceptable as a drinking source for farm animals, that pathway should be removed from the resident farmer scenario.

Documentation to be Submitted to NRC

Ground Water Unsuitable for Agriculture: USGS or independent consultant report showing that ground water quality is poorer than the standards listed for this use.

M.5.2.1.4   Is Ground Water Suitable for Drinking Water?

This question can be addressed by comparing the quality of the ground water with EPA drinking water standards. National Primary Drinking Water Regulations, 40 CFR Part 141 defines regulations for public water systems in the United States. Primary drinking water standards specify approval limits for microorganisms, including bacteria and viruses, specific inorganic and organic chemicals, radionuclides, and turbidity while secondary standards identified in 40 CFR Part 143, National Secondary Drinking Water Regulations, recommend limits on benign contaminants and define physical characteristics that address aesthetics of drinking water (e.g., color and odor).

Tables M.8 to M.11 specify the Maximum Contaminant Levels (MCLs) of contaminants in drinking water delivered to any user of a public water system. The contaminants are distinguished as (1) inorganic chemicals, (2) organic chemicals, (3) radionuclides, and (4) microorganisms. Although turbidity is a measured physical parameter, it is included with microorganisms because turbid water is generally associated with microorganisms or provides a medium for microbial growth.

Table M.12 specifies recommended secondary standards for drinking water. Although the secondary standards are not regulated, they serve as a guide for water quality and may, in some instances, be regulated at the State or local level.

Documentation to be Submitted to NRC

Ground Water Not Potable: USGS or independent consultant report that shows that ground water quality is poorer than the standards listed for this use.

M.5.2.2   Topography and Soil

M.5.2.2.1   Is Soil Suitable for Agriculture?

Soil performs several functions related to plant growth. It forms a media in which roots penetrate, thereby providing a source of stability and nourishment. Nourishment can be provided by the nutrients available in the soil, by fertilizers, or by soil amendments.

Table M.8   National Primary Drinking Water Regulations for Inorganic Chemicals

Contaminant	Maximum Contaminant Level
Antimony	0.006 mg/L
Arsenic	0.05 mg/L
Asbestos (<10μm)	7×10^6 fibers/L
Barium	2 mg/L
Beryllium	0.004 mg/L
Cadmium	0.005 mg/L
Chromium	0.1 mg/L
Copper	1.3 mg/L
Cyanide	0.2 mg/L
Fluoride	4.0 mg/L
Lead	0.015 mg/L
Mercury	0.002 mg/L
Nitrate	10 mg/L
Nitrite	1 mg/L
Selenium	0.05 mg/L
Thallium	0.002 mg/L

Table M.9   National Primary Drinking Water Regulations for Radionculides

Contaminant	Maximum Allowable Concentration
Beta particles and photon emitters	4 mrem/yr to whole body or organ
Gross alpha particle activity	15 pCi/L
Uranium	30 μg/L
Radium-226 + Radium-228	5 pCi/L

Table M.10   National Primary Drinking Water Regulations for Microorganisms

Contaminant	Maximum Allowable Concentration
Giardia lamblia	99.9% killed/inactivated
Heterotrophic plate count	<500 bacterial colonies per mill
Legionella	No limit (if *Giardia* and viruses are controlled)
Total Coliforms (including fecal coliform and *E. Coli*)	5%
Turbidity	5 NTU
Viruses (enteric)	99.99% killed/inactivated

Table M.11 National Primary Drinking Water Regulations for Organic Chemicals

Contaminant	MCL	Contaminant	MCL
Acrylamide	0.05% dosed at 1 mg/L	Epichlorohydrin	0.01% dosed at 20 mg/L
Alachlor	0.002 mg/L	Ethylbenzene	0.7 mg/L
Atrazine	0.003 mg/L	Ethylene dibromide	0.00005 mg/L
Benzene	0.005 mg/L	Glyphosate	0.7 mg/L
Benzo(a)pyrene	0.0002 mg/L	Heptachlor	0.0004 mg/L
Carbofuran	0.04 mg/L	Heptachlor epoxide	0.0002 mg/L
Carbon tetrachloride	0.005 mg/L	Hexachlorobenzene	0.001 mg/L
Chlordane	0.002 mg/L	Hexachloroch-clopentadiene	0.05 mg/L
Chlorobenzene	0.1 mg/L	Lindane	0.0002 mg/L
2,4-Dichlorophenoxyacetic acid (2,4-D)	0.07 mg/L	Methoxychlor	0.04 mg/L
Dalapon	0.2 mg/L	Osamyl	0.2 mg/L
1,2-Dibromo-3-chloropropane (DBCP)	0.0002 mg/L	Polychlorinated biphenyls (PCBs)	0.005 mg/L
o-Dichlorobenzene	0.6 mg/L	Pentachlorophenol	0.001 mg/L
p-Dichlorobenzene	0.075 mg/L	Picloram	0.5 mg/L
1,2-Dichloroethane	0.005 mg/L	Simazine	0.004 mg/L
1,1-Dichloroethylene	0.007 mg/L	Styrene	0.1 mg/L
cis-1,2-Dichloroethylene	0.07 mg/L	Tetrachloroethylene	0.005 mg/L
trans-1,2-Dichloroethylene	0.1 mg/L	Toluene	1 mg/L
Dichloromethane	0.005 mg/L	Trihalomethanes	0.10 mg/L
1,2-Dichloropropane	0.005 mg/L	Toxaphene	0.003 mg/L
Di(2-ethylhexyl)adipate	0.4 mg/L	Silvex	0.05 mg/L
Di(2-ethylhexyl)phthalate	0.006 mg/L	1,2,4-Trichlorobenzene	0.07 mg/L
Dinoseb	0.007 mg/L	1,1,1-Trichloroethane	0.2 mg/L
Dioxin (2,3,7,8-TCDD)	3×10^{-8} mg/L	1,1,2-Trichloroethane	0.005 mg/L
Diquat	0.02 mg/L	Trichloroethylene	0.005 mg/L
Endothall	0.1 mg/L	Vinyl chloride	0.002 mg/L
Endrin	0.002 mg/L	Xylenes (total)	10 mg/L

Table M.12 National Secondary Drinking Water Regulations

Contaminant	Secondary Standard
Aluminum	0.05–0.2 mg/L
Chloride	250 mg/L
Color	15 (color units)
Copper	1.0 mg/L
Corrosivity	noncorrosive
Fluoride	2.0 mg/L
Foaming Agents	0.5 mg/L
Iron	0.3 mg/L
Manganese	0.05 mg/L
Odor	2 threshold odor number
pH	6.5–8.5
Silver	0.10 mg/L
Sulfate	250 mg/L
Total Dissolved Solids	500 mg/L
Zinc	5 mg/L

With suitable fertilizers or soil amendments, plants can readily be grown in "soil free" materials, such as mineral sand, gravel, perlite, pumice, crushed bricks, or glass wool. Consequently, the absence of soil in the traditional sense at a site does not eliminate plant ingestion as a pathway. Because soilless gardening requires more management than traditional gardening methods, it is more likely to be used for growing vegetables and herbs than for the production of commodity items such as grains or livestock fodder [Nicholls, 1997].

Agriculture could be excluded from a scenario if the site is an outcropping of bedrock without appreciable soil, or debris that could serve to anchor plants.

Areas consisting of made land, where there is abundant debris and cobbles with little or no soil, would not lend themselves to mechanized agriculture in short-term scenarios. In the absence of mechanized agriculture, commodity food items and fodder are not likely crops. However, it would be difficult to exclude vegetable gardens from scenarios at such sites. In addition, it would be difficult to justify exclusion of livestock forage from scenarios for such sites.

Agriculture pathways could be eliminated in short-term scenarios if the soil is outright toxic or inhospitable to plants. As examples, (1) no agriculture is apt to occur on the bed of a dry salt lake, and (2) crops are not apt to be grown in made land that contains such a high percentage of concrete materials that extraordinary efforts would be required to maintain the soil pH in a range that is tolerated by plants.

If it can be documented that the soil at this site would not support the resident farmer's agricultural efforts, this pathway should be eliminated or modified.

Documentation to be Submitted to NRC

Soil Unsuitable for Agriculture: NRCS (SCS) or independent consultant report that shows quality of soil is poorer than the standards listed for this use.

M.5.2.2.2   Is Topography Suitable for Agriculture?

In the past few hundred years, the Dutch built dikes and converted shoals into productive farmlands. Today, explosives and earth-moving equipment can easily change features of the landscape, making them suitable for agricultural or residential use. Consequently, locality or accessability may form a basis for eliminating certain agricultural pathways from scenarios in the next century, but not for a period of 1,000 years.

Ignoring the fact that topography may change with time as a result of civil engineering projects, there are probable limits to the types of terrain where mechanized agriculture can be used. Tractors may likely always be unstable on slopes, so there may probably always be a practical limit on the slopes that can be put under mechanized agriculture. In the absence of mechanized agriculture, persons are more likely to practice gardening than to grow commodity food items. They are also more likely to allow livestock to forage than to grow fodder crops.

While there is no predictable maximum safe slope that tractors may traverse without the danger of rollover, operating a tractor on a 30 degree (2 to 1) slope is so hazardous that the average member of the critical group is not likely to attempt it.

If the topography at the site is too steep or too erratic to support the type of farming expected within the resident farmer scenario, the agricultural pathway should be removed or modified in accordance with this finding. There may also be aspects of the topography that would limit farming or other specific activities at the site.

Documentation to be Submitted to NRC

Topography Unsuitable for Agriculture: USGS or similar topographic map, hand-drawn map, or description that provides enough detail to illustrate the topography that limits farming at this site.

M.6   Summary

The process presented in this document is an extension of the Decommissioning Framework. It uses a logical step-by-step procedure for introducing site-specific information to develop alternate scenarios by eliminating pathways from the default resident farmer scenario. As the process schematic leads the licensee through the steps that are required to remove pathways, iterative dose assessments assure that no more information than is necessary should be assembled and analyzed for this purpose. Once the TEDE to an average member of the critical group drops below 0.25 mSv/y (25 mrem/y), the process is completed and the licensee may proceed to license termination. Following the initial dose assessment and each of the iterative dose assessments, sensitivity analyses help the licensee focus on the introduction of evidence that can rule out those pathways that are responsible for the high dose.

Physical and cultural information are introduced to answer a series of questions about the site. The future use of the land may be key to what assumptions the licensee can make about the starting scenario. Information on current land use, past land use, and a history of land use changes can be used to determine the probable future use of the land. If the future land use can reasonably be predicted to be either urban or industrial, the resident farmer scenario can be bypassed, allowing the licensee to concentrate on these two land uses.

The residential farmer scenario is meant to be applied to sites with land and water residual radioactivity and the building occupancy scenario is to be applied to sites with contaminated structures. If a resident farmer scenario is assumed, the most important aspect of the physical nature of the site is the nature and availability of water. The answers to each of four critical questions about water at the site can be used to determine if major pathways can be removed from the scenario. If ground water is not available, all of the pathways that rely on ground water as a key component can be removed: irrigation, aquatic, and drinking. If ground water is not suitable for aquatic life, the aquatic pathway can be removed. If ground water is not suitable for agriculture, irrigation and drinking water pathways can be removed. If the water is not potable, the drinking water pathway can be removed. Detailed discussion is presented to help the licensee answer these questions, to understand the standards that would have to be met for this pathway to be ruled out, and to identify the documentation that would have to be presented to the NRC.

M.7   References

Bonano, E.J., Hora, S.C., Keeney, R.L., and D. Winterfeldt, 1990, *Elicitation and Use of Expert Judgement in Performance Assessment for High-Level Radioactive Waste Repositories*, NUREG/CR–411 (SAND89–1821), U.S. Nuclear Regulatory Commission, Washington, DC; Sandia National Laboratories, Albuquerque, NM.

Cochran, J.R., 1996, "A Proposed Alternative Approach for Protection of Inadvertent Human Intruders from Buried Department of Energy Low Level Radioactive Wastes," *Proceedings from Waste Management '96*, Session 59, Waste Management Symposia, Inc., Tucson, AZ.

Cranwell, R.M., Campbell, J.E., Helton, J.C., Iman, R.L., Longsine, D.E., Ortiz, N.R., Runkle, G.E., and Shortencarier, M.J., 1987. *Risk Methodology for Geologic Disposal of Radioactive Waste: Final Report*, NUREG/CR–2452 (SAND81–2573), U.S. Nuclear Regulatory Commission, Washington, DC; Sandia National Laboratories, Albuquerque, NM.

Cranwell, R.M., Guzowski, R.V., Campbell, J.E., and Ortiz, N.R., 1990. *Risk Methodology for Geologic Disposal of Radioactive Waste: Scenario Selection Procedure*, NUREG/CR–1667 (SAND80–1429), U.S. Nuclear Regulatory Commission, Washington DC; Sandia National Laboratories, Albuquerque, NM.

Department of Energy, 1994. *Decommissioning Handbook,* DOE/EM–0142P, U.S. DOE, Office of Environmental Restoration.

Department of Energy, 1998. *Effects of Future Land Use Assumptions On Environmental Restoration Decision Making.* DOE/EH–0413/980, U.S. DOE, Office of Environmental Policy and Assistance.

EPA. 1996. *Exposure Factors Handbook.* EPA/600/P–95. Office of Research and Development, Environmental Protection Agency, Washington, DC. (Current draft not citable.)

Federal Register, Vol. 64, February 22, 1999.

Guzowski, R.V. 1991. *Evaluation of Applicability of Probability Techniques to Determining the Probability of Occurance of Potentially Disruptive Intrusive Events at the Waste Isolation Pilot Plant,* SAND90–7100, Sandia National Laboratories, Albuquerque, NM.

Haaker, R., T. Brown, and D. Updegraff. Comparison of the Models and Assumptions used in the DandD 1.0, RESRAD 5.61, and RESRAD–Build 1.50 Computer Codes with Respect to the Residential Farmer and Industrial Occupant Scenarios Provided in NUREG/CR–5512. Draft Report for Comment, October 1999. [NOTE: this report is also SAND99–2147.]

Hora, S.C., D. von Winterfeldt and K.M. Trauth, 1991. *Expert Judgement on Inadvertent Human Intrusion into the Waste Isolation Pilot Plant.* SAND90–3063, Sandia National Laboratories, Albuquerque, NM.

IAEA, 1982. "General Models and Parameters for Assessing the Environmental Transfer of Radionuclides from Routine Releases," Vienna: IAEA; Safety Series No. 57.

IAEA, 1995. "Criteria for Clean-Up of Contaminated Areas," Consultants Meeting Draft Report, December 4–8.

Kennedy, Jr., W.E., and D.L. Strenge. 1992. "Residual Radioactive Contamination from Decommissioning: Technical Basis for Translating Contamination Levels to Annual Total Effective Dose Equivalent," NUREG/CR–5512, Volume 1, U.S. Nuclear Regulatory Commission, Washington, DC.

Kozak, M.W., N.E. Olague, R.R. Rao and J.T. McCord, 1993. *Evaluation of a Performance Assessment Methodology for Low-Level Radioactive Waste Disposal Facilities: Evaluation of Modeling Approaches*, NUREG/CR–5927 Volume 1, U.S. Nuclear Regulatory Commission, Washington, DC.

Leigh, C.D., B.M. Thompson, J.E. Campbell, D.E. Longsine and R.A Kennedy, 1993. *User's Guide for GENII–S,* SAND91–056, Sandia National Laboratories, Albuquerque, NM.

Locke, P.A., M.C. Carney, Tran, N.L., T.A. Burke, 1998. *Chemical and Radiation Enviornmental Risk Management: Foundations, Common Themes, Similarities and Differences,* Workshop Proceedings, Anapolis MD.

NEA, 1992. *Systematic Approaches to Scenario Development*, Nuclear Energy Agency, Organization for Economic Co-Operation and Development, Paris.

National Research Council. 1995. *Technical Basis for Yucca Mountain Standards*, National Academy Press, Washington, DC.

Nicholls, R.E., 1997. Beginning Hydroponics Soilless Gardening, Running, Running Press, Philadelphia.

NRC, 1977. Regulatory Guide 1.109 Calculation of Annual Doses to Man from Routine Releases of Reactor Effluents for the Purpose of Evaluating Compliance with 10 CFR Part 50, Appendix I.

NRC, 1981a, "Proposed Rules for Licensing Requirements for Land Disposal of Radioactive Waste," *Federal Register*, Vol. 46, No. 142, pp. 38081–38015.

NRC, 1981b, *Draft Environment Impact Statement on 10 CFR Part 61 "Licensing Requirements for Land Disposal of Radioactive Waste"*, NUREG–0782, U.S. Nuclear Regulatory Commission, Office of Nuclear Material Safety and Safeguards, Washington, DC.

NRC, 1982, *Final Environment Impact Statement on 10 CFR Part 61 "Licensing Requirements for Land Disposal of Radioactive Waste,"* NUREG–0945, U.S. Nuclear Regulatory Commission, Office of Nuclear Material Safety and Safeguards, Washington, DC.

NRC, 1990, "Components of an Overall Performance Assessment Methodology," NUREG/CR-5256, U.S. Nuclear Regulatory Commission, Office of Nuclear Material Safety and Safeguards, Washington, DC.

NRC, 1993. 10 CFR Part 20. U.S. Nuclear Regulatory Commission, "Standards for Protection against Radiation," U.S. Code of Federal Regulations.

NRC. 1998. *Decision Methods for Dose Assessment to Comply With Radiological Criteria for License Termination*. NUREG–1549, U.S. Nuclear Regulatory Commission, Office of Nuclear Regulatory Research, Washington, DC.

Nuclear Remediation Week, July 31, 1995.

Sara, M.N., *Standard Handbook for Solid and Hazardous Waste Facility Assessment,* Lewis Publishers, Boca Raton Florida.

Trauth, K.M., S.C. Hora and R.V. Guzowski, 1993. *Expert Judgement on Markers to Deter Inadvertent Human Intrusion into the Waste Isolation Pilot Plant.* SAND92–1382, Sandia National Laboratories, Albuquerque, NM.

Viessman, Jr., Warren; Hammer, Mark J. Water Supply and Pollution Control. Fourth Edition. New York: Harper and Row; 1985.

Wernig, M.A., A.M. Tomasi, F.A. Duran, and C.D. Updegraff. Residual Radioactive Contamination from Decommissioning, User's Manual, Draft Report, May 1999.

Wood, D.E., Curl, R.U., Buhl, T.E., Cook, J.R., Dolenc, M.R., Kocher, DC, Napier, B.A., Owens, K.W., Regnier, E.P., Roles, G.W., Seitz, R.R., Thorne, D.J., and Wood, M.I., 1992, *Performance Assessment Task Team Draft Progress Report*, DOE/LLW–157 Revision 0, Idaho National Engineering Laboratory, Idaho Falls Idaho, Prepared for the U.S. Department of Energy.

Wood, D.E., Curl, R.U., Armstrong, D.R., Cook, J.R., Dolenc, M.R., Kocher, DC, Owens, K.W., Regnier, E.P., Roles, G.W., Seitz, R.R., and Wood, M.I., 1994, *Performance Assessment Task Team Progress Report*, DOE/LLW–157 Revision 1, Idaho National Engineering Laboratory, Idaho Falls Idaho, Prepared for the U.S. Department of Energy.

Appendix N

ALARA Analyses

In order to terminate a license, a licensee should demonstrate that the dose criteria in Subpart E have been met, and should demonstrate whether it is feasible to further reduce the levels of residual radioactivity to levels below those necessary to meet the dose criteria (i.e., to levels that are ALARA). This section describes methods acceptable to NRC staff for determining when it is feasible to further reduce the concentrations of residual radioactivity to below the concentrations necessary to meet the dose criteria. This section does not apply to, nor replace guidance for, operational ALARA programs. This guidance does involve the same principle as the operational ALARA guidance:

> "'Reasonably achievable' is judged by considering the state of technology and the economics of improvements in relation to all the benefits from these improvements. (However, a comprehensive consideration of risks and benefits will include risks from nonradiological hazards. An action taken to reduce radiation risks should not result in a significantly larger risk from other hazards.) NRC Regulatory Guide 8.8, Revision 3 (1978)." [Quotes in original.]

For ALARA during decommissioning, all licensees should use typical good-practice efforts such as floor and wall washing, removal of readily removable radioactivity in buildings or in soil areas, and other good housekeeping practices. In addition, licensees should provide a description in the FSSR of how these practices were employed to achieve the final activity levels.

In light of the conservatism in the building surface and surface soil generic screening levels developed by NRC, NRC staff presumes, absent information to the contrary, that licensees who remediate building surfaces or soil to the generic screening levels do not need to provide analyses to demonstrate that these screening levels are ALARA. In addition, if residual radioactivity cannot be detected, it may be assumed that it has been reduced to levels that are ALARA. Therefore, the licensee may not need to conduct an explicit analysis to meet the ALARA requirement.

Areas that have been released under then-existing requirements would not have to be reevaluated under 10 CFR 20.1401(c). According to 10 CFR 20.1401(c), NRC would require additional cleanup following license termination only if it determines, based on new information, that the criteria of Subpart E were not met and that residual radioactivity remaining at the site could result in significant threat to public health and safety. Because ALARA represents an optimization technique below dose criteria, it is not considered reasonable to reopen consideration of a previously released area, where radioactive materials were handled that meets the appropriate dose criterion.

In general, a method for determining whether levels of residual radioactivity are ALARA would have the following characteristics.

- **The method is simple**. The method for most licensee applications should be simple, because the effort needed for very sophisticated models cannot generally be justified. In an ALARA analysis of a remediation action, the primary benefit (i.e., the collective radiation dose that may actually be averted in the future) is uncertain because of future land uses, the number of people that may actually occupy a site, and the types of exposure scenarios are all uncertain. These uncertainties mean that the future collective dose cannot be known with precision. Because of the inherent limitation on the ability to precisely determine the future collective dose at a particular site, it is not useful to perform a complex analysis when a simple analysis can be appropriate. Licensees may use more complex or site-specific analyses if more appropriate for their specific situations (e.g., restricted release analyses, situations that include a number of unquantifiable benefits and costs).

- **The method is not biased and uses appropriate dose modeling to relate concentrations to dose**. The determination of ALARA should not be biased. This is different from demonstrating compliance with a dose limit. The analyses for dose assessments and surveys for compliance with the dose criteria described in this volume include a reasonably conservative bias for demonstrating compliance. Unlike a demonstration of compliance, an ALARA analysis is an optimization technique that seeks the proper balance between costs and benefits below the dose limit. To achieve a proper balance, each factor in the ALARA analysis should be determined with as little bias as possible. If the ALARA analysis were intentionally biased, it would likely cause a misallocation of resources and could deprive society of the benefits from other uses of the resources. Thus, the ALARA analysis should provide an unbiased analysis of the remediation action, which can both avert future dose (a benefit to society) and cost money (a potential detriment because it can deprive future generations of the return on the investment of this money). Sections N.1.1 and N.1.2 discuss the methods that should be used in estimating benefits and detriments, or costs, including scenarios, models, and parameters for relating concentration to dose at a site. The Office of Management and Budget guidance to Federal agencies that implements the President's Executive Order 12866 "Regulatory Planning and Review," in Title 3 of the 1993 Compilation of the U.S. Code of Federal Regulations, January 1, 1994 (page 638), provides guidance on balancing benefits and detriments for analyzing the potential benefits of Federal regulations (Office of Management and Budget, "Economic Analysis of Federal Regulations under Executive Order 12866," January 11, 1996).

- **The method is usable as a planning tool for remediation**. Before starting a remediation action, the licensee should be able to determine generally what concentration of residual radioactivity would require a remediation action to meet the ALARA requirement. It would be inefficient if the licensee could not tell whether the area would pass the ALARA test until after the remediation. Establishing ALARA post-remediation would also likely result in it being less likely for a licensee to remediate below the dose limit(s) because of the additional manpower start-up costs associated with performing additional remediation.

- **As much as possible, the method uses the results of surveys conducted for other purposes**. The demonstration that the ALARA requirement has been met should not require surveys beyond those already performed for other purposes, such as the characterization survey and the FSS. It would be inefficient (and unnecessary) to collect additional sets of measurements to demonstrate that remediation actions were taken wherever appropriate to meet the ALARA requirement if measurements undertaken for other purposes could be used.

N.1   Benefits and Costs for ALARA Analyses

Subpart E contains specific requirements for a demonstration that residual radioactivity has been reduced to a level that is ALARA (10 CFR 20.1402, 20.1403(a), 20.1403(e), and 20.1404(a)(3)). A simplified method for demonstrating compliance with the ALARA requirement is described below. Licensees may use more complex or site-specific analyses if more appropriate for their specific situation. In general, more complex analyses should follow the general concepts presented here. Evaluation of more complex analyses should be handled on a case-by-case basis and early involvement of the appropriate regulatory agencies and members of the public is suggested.

Sometimes it is very difficult or impossible to place a monetary value on an impact. A best effort should be made to assign a monetary value to the impact, because there may be no other way to compare benefits to costs. However, there may be situations for which a credible monetary value cannot be developed. In these situations, a qualitative treatment may be the most appropriate. Qualitative analyses should be evaluated on their merits on a case-by-case basis.

The simplified method presented here is to estimate when a remediation action is cost-effective using generalized estimates for the remedial action. If the desired beneficial effects (benefits) from the remediation action are greater than the undesirable effects or "costs" of the action, the remediation action being evaluated is cost-effective and should be performed. Conversely, if the benefits are less than the costs, the levels of residual radioactivity are already ALARA without taking the remediation action. An example of various benefits and costs are listed in Table N.1. Other than Collective Dose Averted, the additional benefits listed are generally only important in comparisons between alternatives that address whether restricted release can be pursued by the licensee. The value of any benefit or cost can be negative in some cases.

Table N.1   Possible Benefits and Costs Related to Decommissioning

Possible Benefits	Possible Costs
Collective Dose Averted	Remediation Costs
Regulatory Costs Avoided	Additional Occupational/Public Dose
Changes in Land Values	Occupational Nonradiological Risks
Esthetics	Transportation Direct Costs and Implied Risks
Reduction in Public Opposition	Environmental Impacts
	Loss of Economic Use of Site/Facility

In order to compare the benefits and costs of a remediation action, it is necessary to use a comparable unit of measure. The unit of measure used here is the dollar; if possible, then all benefits and costs are given a monetary value. Benefits and costs can be calculated as described in Sections N.1.1 and N.1.2.

The method should be applied during remediation planning, prior to the start of remediation, but after some or all of the characterization work is done. The method should be used only to determine whether and where particular remediation actions should be taken to meet the ALARA requirement.

If the licensee has already decided to perform a remediation action, there is no need to analyze whether the action was necessary to meet the ALARA requirement. The analysis described in this section is needed only to justify *not* taking a remediation action. For example, if a licensee plans to wash room surfaces (either to meet the dose limit or as a good-practice procedure), there is no need to analyze whether the remediation action of washing is necessary to meet the ALARA requirement.

N.1.1   Calculation of Benefits

Collective Dose Averted

In the simplest form of the analysis, the only benefit estimated from a reduction in the level of residual radioactivity is the monetary value of the collective averted dose to future occupants of the site. The collective averted dose should be based on the same exposure scenario(s) used for the compliance calculations. Additional considerations related to ground water residual radioactivity are discussed in Section N.1.6.

The benefit from collective averted dose, B_{AD}, is calculated by determining the present worth of the future collective averted dose and multiplying it by a factor to convert the dose to monetary value:

$$B_{AD} = \$2000 \times PW(AD_{collective}) \qquad (N–1)$$

where B_{AD} =   benefit from an averted dose for a remediation action, in current U.S. dollars

\$2000   =   value in dollars of a person-rem averted (see NUREG/BR–0058, "Regulatory Analysis Guidelines of the U.S. Nuclear Regulatory Commission," Revision 2, November 1995)

$PW(AD_{collective})$ =   present worth of a future collective averted dose

An acceptable value for a collective dose is \$2000 per person-rem averted, discounted for a dose averted in the future. See Section 4.3.3 of "Regulatory Analysis Guidelines of the U.S. Nuclear Regulatory Commission," NUREG/BR–0058, Revision 2, November 1995. For doses averted within the first 100 years, a discount rate of 7 % should be used. For doses averted beyond 100 years, a 3 % discount rate should be used.

The present worth of the future collective averted dose can be estimated from Equation N–2, for relatively simple situations:

$$PW(AD_{collective}) = P_D \times A \times 0.025 \times F \times \frac{Conc}{DCGL_W} \times \frac{1 - e^{-(r+\lambda)N}}{r+\lambda} \qquad (N–2)$$

where P_D =   population density for the critical group scenario in people/m^2;

A =   area being evaluated in square meters (m^2)

0.025   =   annual dose to an average member of the critical group from residual radioactivity at the Derived Concentration Guideline Level ($DCGL_W$) concentration in rem/y;

F =   effectiveness, or fraction of the residual radioactivity removed by the remediation action;

$Conc$ =   average concentration of residual radioactivity in the area being evaluated in units of activity per unit area for buildings or activity per unit volume for soils;

$DCGL_W$ =   derived concentration guideline equivalent to the average concentration of residual radioactivity that would give a dose of 0.25 mSv/y (25 mrem/y) to the average member of the critical group, in the same units as *"Conc"*;

r =   monetary discount rate in units per year;

λ =   radiological decay constant for the radionuclide in units per year; and

N =   number of years over which the collective dose will be calculated.

The present worth of the benefit calculated by Equation N–2, above, assumes that the peak dose occurs in the first year. This is almost always true for the building occupancy scenario, but not always true for the residential scenario where the peak dose can occur in later years. When the peak dose occurs in later years, Equation N–2 would overestimate the benefit. The licensee may perform a more exact calculation that avoids this overestimation of the benefit of remediation by calculating the dose during each year of the evaluation period and then calculating the present worth of each year's dose. A detailed derivation of Equation N–2 and some of the other equations are in Section N.5.

The $DCGL_W$ used should be the same as the $DCGL_W$ used to show compliance with the 0.25 mSv/y (25 mrem/y) dose limit. The population density, P_D, should be based on the dose scenario used to demonstrate compliance with the dose limit. Thus, for buildings, the licensee should estimate P_D for the building occupancy scenario. For soil, the P_D should be based on the residential scenario. The factor at the far right of the equation, which includes the exponential terms, accounts for both the present worth of the monetary value and radiological decay.

If more than one radionuclide is present, the total benefit from a collective averted dose, B_{AD}, is the sum of the collective averted dose for each radionuclide. When multiple radionuclides have a fixed concentration (i.e., secular equilibrium), residual radioactivity below the dose criteria is normally demonstrated by measuring one radionuclide and comparing its concentration to a

$DCGL_W$ that has been calculated to account for the dose from the other radionuclides. In this case, the adjusted $DCGL_W$ may be used with the concentration of the radionuclide being measured. The other case is where the ratio of the radionuclide concentrations is not fixed and varies from location to location within a survey unit; this benefit is the sum of the collective averted dose from each.

Regulatory Costs Avoided

This benefit usually manifests in ALARA analyses of restricted release versus unrestricted release decommissioning goals. By releasing the site with no restrictions, the licensee may avoid the various costs associated with restricted release. These can include (1) additional licensing fees to develop an Environmental Impact Statement, (2) financial assurance related to both the decommissioning fund [10 CFR 20.1403(c)] and the site restrictions [10 CFR 20.1403(d)(1)(ii)], (3) costs (including NRC-related) associated with public meetings or the community review committee [10 CFR 20.1403(d)(2)], and (4) future liability. When evaluating the ability of a licensee's proposal for restricted release according to 10 CFR 20.1403(a), avoiding these costs should be included in the benefits of the unrestricted release decommissioning alternative. These should not be included as costs related to the restricted release (see Section N.1.2).

Changes in Land Values

The licensee should account for any expected change in the value of the site or facility caused by the different decommissioning options. This may be difficult to quantify.

Esthetics/Reduction in Public Opposition

These can be very difficult to quantify. The licensee may wish to evaluate the effect of the available decommissioning options with respect to the overall esthetics (including the decommissioning activities themselves) of the site and surrounding area. Another factor the licensee may wish to consider is the potential reduction in opposition, if there is any, to the decommissioning activities/goal the license is attempting to propose.

N.1.2   Calculation of Costs

The licensee should evaluate the costs of the selected alternative remediation actions being evaluated. When performing a fairly simple evaluation, the costs generally include the monetary costs of: (1) the remediation action being evaluated, (2) transportation and disposal of the waste generated by the action, (3) workplace accidents that occur because of the remediation action, (4) traffic fatalities resulting from transporting the waste generated by the action, (5) doses received by workers performing the remediation action, and (6) doses to the public from excavation, transport, and disposal of the waste. Other costs that are appropriate for the specific case may also be included.

The total cost, $Cost_T$, which is balanced against the benefits, has several components.

$$Cost_T = Cost_R + Cost_{WD} + Cost_{ACC} + Cost_{TF} + Cost_{WDose} + Cost_{PDose} + Cost_{other} \quad \text{(N–3)}$$

where $Cost_R$ =   monetary cost of the remediation action (may include "mobilization" costs);

$Cost_{WD}$ =   monetary cost for transport and disposal of the waste generated by the action;

$Cost_{ACC}$ =   monetary cost of worker accidents during the remediation action;

$Cost_{TF}$ =   monetary cost of traffic fatalities during transporting of the waste;

$Cost_{WDose}$ =   monetary cost of dose received by workers performing the remediation action and transporting waste to the disposal facility;

$Cost_{PDose}$ =   monetary cost of the dose to the public from excavation, transport, and disposal of the waste; and

$Cost_{other}$ =   other costs as appropriate for the particular situation.

All the costs described below do not necessarily have to be calculated. For example, if one or two of the costs can be shown to be in excess of the benefit and if none of the other costs are negative, the remediation action will be unnecessary. However, some of these costs may in fact be negative (i.e., the alternative may cost less than the preferred option) in some comparisons between alternate decommissioning options, and thus, all costs may need to be evaluated.

Remedial Action Costs

Calculation of the incremental remedial action costs include the standard manpower and mechanical costs. The licensee can account for any additional licensing fees from NRC (e.g., if the option to meet the ALARA goal requires another year of remediation). Lower concentrations may change sampling/survey requirements. Increased survey costs can be considered in the remedial action but note that this is the incremental costs of surveying below the dose limit. Survey costs related to evaluating compliance at the dose limit are not part of the ALARA analysis.

Transport and Disposal of the Waste

The cost of waste transport and disposal, $Cost_{WD}$, may be evaluated according to Equation N–4.

$$Cost_{WD} = V_A \times Cost_V \quad \text{(N–4)}$$

where V_A =   volume of waste produced, remediated in units of m³; and

$Cost_V$ =   cost of waste disposal per unit volume, including transportation cost, in units of \$/m³.

Nonradiological Risks

The cost of nonradiological workplace accidents, $Cost_{ACC}$, may be evaluated using Equation N–5.

$$Cost_{ACC} = \$3{,}000{,}000 \times F_W \times T_A \qquad (N–5)$$

where $\$3{,}000{,}000$ = monetary value of a fatality equivalent to \$2000/person-rem (see pages 11–12 of "Reassessment of NRC's Dollar per Person-Rem Conversion Factor Policy," NUREG–1530, December 1995);
F_W = workplace fatality rate in fatalities/hour worked; and
T_A = worker time required for remediation in units of worker-hours.

Transportation Risks

The cost of traffic fatalities incurred during the transportation of waste, $Cost_{TF}$, may be calculated similar to Equation N–6.

$$Cost_{TF} = \$3{,}000{,}000 \times \left(\frac{V_A}{V_{SHIP}}\right) \times F_T \times D_T \qquad (N–6)$$

where V_A = volume of waste produced in units of m³;
F_T = fatality rate per truck-kilometer traveled in units of fatalities/truck-km;
D_T = distance traveled in km; and
V_{SHIP} = volume of a truck shipment in m³.

The actual parameters should depend on the site's planned method of waste transport. Some facilities may consider a mix of trucking and rail transport to get the waste to the disposal site. In these cases, the cost would be equivalent to the total fatalities likely from the rail transport and the limited trucking, not just the trucking alone.

Worker Dose Estimates

The cost of the remediation worker dose, $Cost_{WDose}$, can be calculated as shown in Equation N–7.

$$Cost_{WDose} = \$2000 \times D_R \times T \qquad (N–7)$$

where D_R = total effective dose equivalent (TEDE) rate to remediation workers in units of rems/hr; and
T = time worked (site labor) to remediate the area in units of person-hour.

The cost of worker dose usually should not be discounted because the dose is all incurred close to the time of license termination.

Loss of Economic Use of Property

A cost in the "other" category could include the fair market rental value or economic use for the site during the time the additional remediation work is being performed. These costs are usually associated with locations such as laboratories, hospital rooms, and industrial sites, etc. This cost may be added to the costs in Equation N–3.

Environmental Impacts

Another cost that could fall into the other category would be a remediation action that may damage an ecologically valuable area or cause some other adverse environmental impact. These impacts should be included as costs of the remediation action.

Default Parameters

For performing these calculations, acceptable values for some of the parameters are shown in Table N.2.

Table N.2   Acceptable Parameter Values for Use in ALARA Analyses

Parameter	Value	Reference and comments
Workplace accident fatality rate, F_W	4.2 x 10^{-8}/hr	NUREG–1496, "Final Generic Environmental Impact Statement in Support of Rulemaking on Radiological Criteria for License Termination of NRC-Licensed Nuclear Facilities," and NUREG–1496, July 1997, Volume 2, Appendix B, Table A.1
Transportation fatal accident rate, F_T	Trucks: 3.8 x 10^{-8}/km	NUREG–1496, Volume 2, Appendix B, Table A.1
Dollars/person-rem	$2000	NUREG/BR–0058
Monetary discount rate, r	0.07/y for the first 100 years and 0.03/y thereafter, or 0.07 for buildings and 0.03 for soil	NUREG/BR–0058
Number of years of exposure, N	Buildings: 70 years Soil: 1000 years	NUREG–1496, Volume 2, Appendix B, Table A.1
Population density, P_D	Building: 0.09 person/m^2 Land: 0.0004 person/m^2	NUREG–1496, Volume 2, Appendix B, Table A.1
Excavation, monitoring, packaging, and handling soil	Soil: 1.62 person-hours/m^3 of soil	NUREG–1496, Volume 2, Appendix B, Table A.1
Waste shipment volume, V_{SHIP}	Truck: 13.6 m^3/shipment	NUREG–1496, Volume 2, Appendix B, Table A.1

N.1.3   Residual Radioactivity Levels that are ALARA

The residual radioactivity level that is ALARA is the concentration, *Conc,* at which the benefit from removal equals the cost of removal. If the total cost, *Cost$_T$,* is set equal to the present worth of the collective dose averted in Equation N–2, the ratio of the concentration, *Conc,* to the $DCGL_W$ can be determined (derivation shown in Section N.5).

$$\frac{Conc}{DCGL_W} = \frac{Cost_T}{\$2000 \times P_D \times 0.025 \times F \times A} \times \frac{r + \lambda}{1 - e^{-(r + \lambda)N}} \qquad \textbf{(N–8)}$$

All the terms in Equation N–8 are as defined previously.

Since P_D, N, and r are constants that have generic values for all locations on the site, the licensee only needs to determine the total cost, $Cost_T$, and the effectiveness, F, for a specific remediation action. If the concentration at a location exceeds $Conc$, it may be cost effective to remediate the location by a method whose total cost is $Cost_T$. Note that the concentration, $Conc$, that is ALARA can be higher or lower (more or less stringent) than the $DCGL_W$, although licensees should meet the $DCGL_W$.

N.1.4 Examples of Calculations

Example 1: Washing Building Surfaces

This example considers a building with cesium-137 residual radioactivity ($\lambda = 0.023$/y). The remediation action to be considered is washing a floor of 100 m^2 area. The licensee estimates that this may cost \$400 and may remove 20 % ($F = 0.2$) of the residual radioactivity. For buildings, generic parameters are: $P_D = 0.09$ person/m^2, $r = 0.07$/y, and $N = 70$ years. Using these values in Equation N–8:

$$\frac{Conc}{DCGL_W} = \frac{\$400}{\$2000 \times 0.2 \times 0.025 \times 0.09 \times 100 \ m^2} \times \frac{0.07 + 0.023}{1 - e^{-(0.07 + 0.023)70}} \quad \text{(N–9)}$$

$$\frac{Conc}{DCGL_W} = 0.41 \quad \text{(N–10)}$$

To meet the ALARA requirement, the floor should be washed if the average concentration exceeds about 41 % of the $DCGL_W$. This is more stringent than the dose limit. This calculation shows that washing building surfaces is often necessary to meet the ALARA requirement. If the surfaces may be washed, there is no need for the licensee to perform the ALARA evaluation or to submit the evaluation to NRC. If the licensee decided not to wash the building surfaces, the licensee could submit this evaluation and demonstrate in the FSS that all surfaces have a concentration below 41 % of the $DCGL_W$.

Example 2: Scabbling Concrete in a Building

This example is the same as above except that it evaluates use of a scabbling tool that removes the top one-eighth of an inch of concrete. The licensee estimates the total cost of the scabbling may be \$5000 for the 100 m^2 floor and estimates that it may remove all the residual radioactivity so that $F = 1$. Using these values in Equation N–8 gives:

$$\frac{Conc}{DCGL_W} = \frac{\$5000}{\$2000 \times 1 \times 0.025 \times 0.09 \times 100 \ m^2} \times \frac{0.07 + 0.023}{1 - e^{-(0.07 + 0.023)70}} \quad (N-11)$$

$$\frac{Conc}{DCGL_W} = 0.97 \quad (N-12)$$

The licensee could decide to scabble depending on the concentrations present. In lieu of scabbling, the licensee could provide this analysis and demonstrate that the floor concentration is less than 0.97 $DCGL_W$.

Example 3: Removing Surface Soil

In this example, soil with an area of 1000 m^2 is found to contain radium-226 ($\lambda = 0.000433$/y) residual radioactivity to a depth of 15 cm (6 in). The licensee estimates that the cost of removing the soil ($F = 1$) may be \$100,000. For soil, the generic parameters are $P_D = 0.0004$ person/m^2, $r = 0.03$/y, and $N = 1000$ y. Using these values in Equation N–8 gives:

$$\frac{Conc}{DCGL_W} = \frac{\$100,000}{\$2000 \times 1 \times 0.025 \times 0.0004 \times 1000 \ m^2} \times \frac{0.03 + 0.000433}{1 - e^{-(0.03 + 0.000433)1000}} \quad (N-13)$$

$$\frac{Conc}{DCGL_W} = 152 \quad (N-14)$$

Thus, meeting the dose limit would be limiting by a considerable margin. Based on these results, it would rarely be necessary to ship soil to a waste disposal facility to meet the ALARA requirement. The licensee could use this evaluation to justify not removing soil.

The advantage of the approach shown in these examples is that it allows the licensee to estimate a concentration at which a remediation action may be cost-effective prior to starting remediation and prior to planning the FSS. Thus, it is a useful planning tool that lets the licensee determine which remediation actions may be needed to meet the ALARA requirement.

N.1.5   When Mathematical Analyses Are Not Necessary

In certain circumstances, the results of an ALARA analysis are known on a generic basis and an analysis is not necessary. For residual radioactivity in soil at sites that may have unrestricted release, generic analyses (see NUREG–1496, the examples in Sections 1.4, and other similar examples) show that shipping soil to a low-level waste disposal facility is unlikely to be cost effective for unrestricted release, largely because of the high costs of waste disposal. Therefore shipping soil to a low-level waste disposal facility generally does not have to be evaluated for unrestricted release. In addition, licensees who have remediated surface soil and surfaces to the NRC default screening criteria have remediated soil such that it meets the unrestricted use criteria in 10 CFR 20.1402, or if no residual radioactivity distinguishable from background, may be left at the site would not be required to demonstrate that these levels are ALARA.

Removal of loose residual radioactivity from buildings is almost always cost-effective except when very small quantities of radioactivity are involved. Therefore, loose residual radioactivity normally should be removed, and if it is removed, the analysis would not be needed.

N.1.6 Additional Considerations for Residual Radioactivity in Ground Water

The method described above is adequate for most situations and has minimal cost for analyses. However, other factors, as described below, should be included if the site may be released if it has residual radioactivity from site operations in ground water.

If there is residual radioactivity from site operations in ground water, it may be necessary to calculate the collective dose from consumption of the ground water. Default or generic ground-water models typically assume that potable aquifers have small volumes and cannot supply large populations. When this is the case, dose calculations for the site critical group may adequately represent the collective dose from ground water. However, when site-specific ground water modeling is used, and the residual radioactivity is diluted in an aquifer of large volume and there is also an "existing population deriving its drinking water from a downstream supply using a downstream supply" (see page 39075 of "Radiological Criteria for License Termination," Final Rule, *Federal Register,* Volume 62, 62 FR 39058, July 21, 1997), the collective dose for that population should be included in the ALARA calculation. The possibility of reducing the collective dose by remediation should be one of the items evaluated as one of the benefits, even if remediation would not affect the critical group's doses significantly. Another consideration for ground-water residual radioactivity would be the reduction of any potential costs incurred by other entities, such as a public water supply utility, to meet the requirements of the Safe Water Drinking Act, if the licensee's residual radioactivity levels would potentially lead to concentrations at the wellhead that would exceed the U.S. Environmental Protection Agency's Maximum Contaminant Levels.

N.2 Determination of "Net Public or Environmental Harm"

Subpart E, 10 CFR 20.1403(a) and 10 CFR 20.1403(e)(2)(i) address circumstances in which a licensee would be required to demonstrate that further remediation would cause net public or environmental harm. The calculation to demonstrate net public or environmental harm is a special case of the general ALARA calculation described above that compares the benefits in dose reduction to the cost of doses, injuries, and fatalities incurred. The calculation does not consider the monetary costs for performing further remediation, $Cost_R$, or the costs of waste disposal, $Cost_{WD}$. Thus, if the benefit from averted dose B_{AD} is less than the sum of the costs of workplace accidents, $Cost_{ACC}$, the costs of transportation fatalities, $Cost_{TF}$, the costs of remediation worker dose, $Cost_{WDose}$, and the costs of any environmental degradation, $Cost_{ED}$, then there is net public or environmental harm. Thus, there is net public or environmental harm if:

$$Net \ harm \ if \quad B_{AD} < Cost_{ACC} + Cost_{TR} + Cost_{WDose} + Cost_{ED} \qquad (N\text{--}15)$$

In some cases it may be very difficult to assign a credible monetary value to environmental degradation. For example, environmental harm could be caused by an action such as remediation of a wetlands area. There may be no way to assign a monetary value to this action. In these cases it is acceptable to use qualitative arguments, which should be evaluated on a case-by-case basis.

N.3   Demonstration of "Not Technically Achievable"

Subpart E, 10 CFR 20.1403(e)(2)(i) addresses circumstances in which a licensee would be required to demonstrate that further reductions in residual radioactivity are not technically achievable. Remediation of residual radioactivity is almost always technically achievable even if not economically feasible. This provision allows for special cases that may not be foreseeable; thus, specific guidance on this provision cannot be provided. Instead, NRC staff will evaluate licensee submittals on a case-by-case basis.

N.4   Demonstration of "Prohibitively Expensive"

Subpart E, 10 CFR 20.1403(e)(2) addresses circumstances in which a licensee would be required to demonstrate that further reductions in residual radioactivity would be prohibitively expensive. This can be demonstrated by an analysis like the ALARA analysis described above, but using a value of $20,000 per person-rem when calculating the value of the averted dose. This value reflects NRC's statement in the final rule on radiological criteria for license termination that NRC considers it is appropriate that a remediation would be prohibitively expensive if the cost to avert dose were an order of magnitude more expensive than the cost recommended by NRC for an ALARA analysis (see page 39071 of "Radiological Criteria for License Termination," Final Rule, *Federal Register,* Volume 62, 62 FR 39058, July 21, 1997). However, NRC also stated that "...a lower factor may be appropriate in specific situations when the licensee could become financially incapable of carrying out decommissioning safely." Thus, values lower than $20,000 per person-rem may be used when remediation actions based on $20,000 per person-rem could cause the licensee to become financially incapable of safely carrying out the decommissioning.

N.5　Derivation of Main Equations to Calculate ALARA Concentrations

The ALARA analysis compares the monetary value of the desirable effects (benefits) of a remediation action (e.g., the monetary benefit of collective averted dose) with the monetary value of the undesirable effects (e.g., the costs of waste disposal). If the benefits of a remediation action would exceed the costs, the remediation action should be taken to meet the ALARA requirement.

The primary benefit from a remediation action is the collective dose averted in the future, i.e., the sum over time of the annual doses received by the exposed population. Assume:

$$\textit{If benefits > costs, the remediation action should be taken.} \qquad \text{(N–16)}$$

1. A site has an area with residual radioactivity at concentration, *Conc*.

2. The concentration equivalent to 0.25 mSv/y (25 mrem/y) ($DCGL_W$) for the site has been determined (for soil or for building surfaces, as appropriate).

3. The residual radioactivity at a site has been adequately characterized so that the effectiveness of a remediation action can be estimated in terms of the fraction F of the residual radioactivity that the action may remove.

4. The peak dose rate occurs at time 0 and decreases thereafter by radiological decay.

The derived concentration guideline ($DCGL_W$) is the concentration of residual radioactivity that would result in a TEDE to an average member of the critical group of 0.25 mSv/y (25 mrem/y). Therefore, the annual dose D to the average member of the critical group from residual radioactivity at concentration *Conc* is:

$$D = 0.025 \ rem/yr \times \frac{Conc}{DCGL_W} \qquad \text{(N–17)}$$

If a remediation action would remove a fraction, F, of the residual radioactivity present, the annual averted dose to the individual, $AD_{individual}$, is

$$AD_{individual} \ (rem/yr/person) = F \times 0.025 \ rem/yr \times \frac{Conc}{DCGL_W} \qquad \text{(N-18)}$$

The annual collective averted dose, $AD_{collective}$, can be calculated by multiplying the individual averted dose, $AD_{individual}$, by the number of people expected to occupy the area, A, containing the residual radioactivity. The number of people in the area containing the residual radioactivity is the area, A, times the population density, P_D, for the site.

Thus:

$$AD_{collective} = F \times 0.025 \ rem/yr \times \frac{Conc}{DCGL_W} \times A \times P_D \qquad \text{(N–19)}$$

The annual monetary benefit rate at time 0, B_0, from the averted collective dose in dollars per year can be calculated by multiplying the annual collective averted dose, $AD_{collective}$, by \$2000/person-rem (\$200,000/person-sievert):

$$B_0 = \$2000 \times F \times 0.025 \ rem/yr \ \frac{Conc}{DCGL_W} \times A \times P_D \qquad \text{(N-20)}$$

The total monetary benefit of averted doses can be calculated by integrating the annual benefit over the exposure time in years, considering both the present worth of future benefits and radiological decay. It is the policy of the Office of Management and Budget (OMB) and NRC to use the present worth of both benefits and costs that occur in the future.

The equation for the present worth, P_{WB}, of a series of constant future annual benefits, B (in dollars per year), for N years at a monetary discount rate of r (per year) using continuous compounding is:

$$PW_B = B \times \frac{e^{rN} - 1}{r \ e^{rN}} \qquad \text{(N–21)}$$

The continuous compounding form of the present worth equation is used because it permits an easy formulation that includes radiological decay. If the annual benefit rate, B, is not constant but is decreasing from an original rate, B_0, because of radiological decay, the radiological decay rate acts like an additional discount rate that can be added to the monetary discount rate of decrease so that the present worth of the annual benefits P_{WB} becomes:

$$PW_B = B_0 \times \frac{e^{(r+\lambda)N} - 1}{(r+\lambda) \, e^{(r+\lambda)N}}$$

(N–22)

Dividing the numerator and denominator of the right hand term by e(r + λ)N yields:

$$PW_B = B_0 \times \frac{1 - e^{-(r+\lambda)N}}{r+\lambda}$$

(N–23)

As $N \rightarrow \infty$, Equation N–23 has the limit:

$$PW_B = B_0 \times \frac{1}{r+\lambda}$$

(N–24)

When the discount rate, r, is zero and the radiological decay rate is very small so that $r + \lambda \rightarrow 0$, and Equation N–23 has the limit:

$$PW_B = B_0 \times N$$

(N–25)

The total benefit from the collective averted dose, B_{total}, is the present worth of the annual benefits. B_{total} can be calculated by combining Equations N–20 and N–23:

$$B_{total} = \$2000 \times F \times 0.025 \times \frac{Conc}{DCGL_W} \times A \times P_D \times \frac{1 - e^{-(r+\lambda)N}}{r+\lambda}$$

(N–26)

Now consider the total cost of a remediation action, $Cost_T$. The costs included in $Cost_T$ are (1) the direct cost of the remediation action itself, $Cost_{RA}$, (2) the cost of waste disposal including its shipping cost, $Cost_{WD}$, (3) the monetary costs of workplace accidents during the remediation, $Cost_{ACC}$, (4) the monetary costs of transportation accidents during the shipping of waste, $Cost_{TF}$, (5) the monetary value of the dose that remediation workers receive, $Cost_{WD}$, and (6) other costs as appropriate for the specific site, $Cost_{other}$. Thus,

$$Cost_T = Cost_R + Cost_{WD} + Cost_{ACC} + Cost_{TF} + Cost_{WDose} + Cost_{other} \qquad \text{(N–27)}$$

What is of interest in this derivation is the concentration, $Conc$, at which the benefit, B_{total}, equals the total cost, $Cost_T$. Thus, in Equation N–26, $Cost_T$ can be substituted for B_{total}, and then Equation N–26 can be solved for the concentration, $Conc$, relative to the $DCGL_W$, as in Equation N–28.

$$\frac{Conc}{DCGL_W} = \frac{Cost_T}{\$2000 \times F \times 0.025 \times P_D \times A} \times \frac{r + \lambda}{1 - e^{-(r + \lambda)N}} \qquad \text{(N–28)}$$

Equation N–28 can be used to determine the concentration in an area for which a remediation action should be taken to meet the ALARA criterion. The equation appears complicated, but can be solved in a few minutes with a hand-held calculator, and it only has to be done once for each type of remediation action at a site. P_D, N, and r are constants. Generic values for P_D and N are given in Section N.1.2, or may be determined on a site-specific basis. Values for r are given in NUREG/BR–0058, Revision 2, and OMB policy (OMB 1996). The only site-specific information that the licensee needs is the total cost, $Cost_T$, and the effectiveness, F, for each remediation action being evaluated.

Appendix O

Lessons Learned and
Questions and Answers to
Clarify License Termination Guidance
and Plans

This appendix provides information to clarify existing guidance associated with the License Termination Rule (LTR). Specifically, it provides (a) questions and answers; (b) lessons learned by NRC staff during the review of recently submitted DPs and LTPs; and (c) lessons learned from decommissioning final status survey inprocess inspections and confirmatory surveys. Some of the information provided in this appendix was originally developed for reactor licensees. However, much of this information is also applicable to other types of facilities.

O.1 Nuclear Energy Institute Questions and Answers to Clarify License Termination Guidance

O.1.1 Introduction

As discussed in the June 1, 2001, public workshop on NRC's Decommissioning Guidance Consolidation Project (i.e., this NUREG report series), the Nuclear Energy Institute (NEI) and NRC staff identified an approach to clarify existing guidance associated with the License Termination Rule (10 CFR Part 20, Subpart E), in concert with the guidance consolidation project. Under this approach, NEI's License Termination Task Force (Task Force) generated questions (Qs) associated with decommissioning issues that are common to the industry. The Task Force also proposed answers (As) to the questions and submitted the Questions and Answers (Q&As) to NRC staff for review. NRC staff reviewed the Q&As and the supporting technical bases and provided comments to NEI on September 28, 2001. An open meeting was held between NRC, NEI, and industry representatives on December 4, 2001, to discuss each Q&A and the technical issues to ensure that the questions were properly asked and answered and were supported by a defensible technical basis. NRC staff and NEI further developed the Q&As so that they adequately reflect NRC regulations and guidance and include a sound technical basis. Nothing in this set of Q&As modifies or negates the guidance presented in NUREG–1700, "Standard Review Plan for Evaluating Nuclear Power Reactor License Terminations Plans"; Regulatory Guide 1.179 , "Standard Format and Content of License Termination Plans for Nuclear Power Reactors"; and the Multi-Agency Radiation Survey and Site Investigation Manual (MARSSIM, NUREG–1575). It should be noted that when using the guidance provided in the responses to the questions in the Q&A section of this volume in preparing a DP or LTP, the licensee remains responsible for compliance with the LTR, its implementation, and providing the staff with the information necessary to prepare the Safety Evaluation Report and Environmental Assessment.

Seven Q&As found acceptable by NRC staff are provided below.

Question 1: Development of Radionuclide Profiles for Reactor Facilities

In support of the MARSSIM process, radionuclide distribution profiles are necessary to ensure that survey and analysis techniques are appropriate and that dose assessments properly consider all the radionuclides that may be present. During the process of developing initial radionuclide profiles for characterizing commercial light water reactor sites and facilities, which radionuclides are considered, and what resources and methodologies are appropriate?

Answer to Question 1

A unique radionuclide profile must be developed for each of the major types of materials expected to remain onsite after remediation. A commercial light-water power reactor facility will likely require profiles for contaminated soil or sediments, surface contaminated materials, and activated materials. The licensee must consider that activation products in steels and concretes vary with the constituents and operational history. Concrete will also differ between facilities because of different trace elements. While one generic list cannot be developed that would be applicable to all power reactor licensees and types of contaminated materials, once radioactive decay has been considered to the time when final status surveys (FSSes) will be conducted, a set of radionuclides may be developed for surface contamination and for activated materials. The profiles listed below in Table O.1 are not meant to be all-inclusive, and other radionuclides should be added, as necessary, based on site-specific considerations.

Table O.1   Example Radionuclide Profile

Contamination Suite		Activation Suite
H-3	Sb-125	H-3
C-14	Cs-134	C-14
Mn-54	Cs-137	Fe-55
Fe-55	Eu-152	Ni-63
Co-57	Eu-154	Co-60
Co-60	Ce-144	Cs-134
Ni-59	Pu-238	Cs-137
Ni-63	Pu-239/240	Eu-152
Sr-90	Pu-241	Eu-154
Nb-94	Am-241	Eu-155
Tc-99	Cm-243/244	Mn-54, Ni-59, Zn-65

The licensee should confirm, by using characterization surveys and historical assessments, that the radionuclide lists developed are applicable to the facility and appropriate for each medium. Technical considerations and limitations are discussed in: NUREG/CR–3474, "Long-Lived Activation Products in Reactor Materials"; NUREG–0130, "Technology, Safety and Cost of Decommissioning"; and NUREG/CR–4289, "Residual Radionuclide Contamination Within and Around Commercial Nuclear Power Plants." Characterization surveys conducted according to NUREG–1575, "MARSSIM," provide information on the important radionuclides that must be considered. The licensee may also use (a) radionuclide distributions developed for waste classification, to demonstrate compliance with requirements of 10 CFR Part 61, and (b) analyses such as ORIGEN computer code runs, to help determine which radionuclides to consider. It is important to recognize the limitations of such methods as they apply to the MARSSIM process. The licensee should also consider historical fuel performance, operational history, and time since shutdown. It is incumbent on the licensee to ensure that the list of radionuclides for each material type is developed according to NRC guidance (such as that in MARSSIM) and using good laboratory practices.

Question 2: Radionuclide Deselection

When developing derived concentration guideline levels (DCGLs) for the FSS, which radionuclides can be deselected from further consideration?

Answer to Question 2

Guidance in Section 3.3 of this volume states, "Once a licensee has demonstrated that radionuclides or exposure pathways are insignificant, then (a) the dose from the insignificant radionuclides and pathways must be accounted for in demonstrating compliance, but (b) the radionuclides and pathways may be eliminated from further detailed evaluations." Therefore, during characterization of a facility, if a profile contains radionuclides that collectively contribute less than 10 % of the dose criterion, those nuclides may be deselected from the list. Since DCGLs are developed to equate to the radiological criteria for license termination (0.25 mSv/y (25 mrem/y)) TEDE to the average member of the critical group and ALARA, for unrestricted release in 10 CFR 20.1402), those radionuclides that collectively contribute less than 0.025 mSv/y (2.5 mrem/y) may be considered insignificant, given all appropriate exposure scenarios and pathways are considered. It is incumbent on the licensee to have adequate characterization data to support and document the determination that some radionuclides may be deselected from further detailed consideration in planning the FSSes. Radionuclides that are undetected may also be considered insignificant, as long as the MDCs are sufficient to conclude that the dose contribution is less than 10 % of the dose criterion (i.e., with the assumption that the radionuclides are present at the MDC concentrations). In addition, licensees should note that they are required to comply with the applicable dose criteria in 10 CFR Part 20, Subpart E, and thus the dose contribution from the insignificant radionuclides must be accounted for in demonstrating compliance with the dose criteria.

Question 3: Embedded and Buried Piping Characterization

What are acceptable methods to characterize embedded piping and buried piping?

Answer to Question 3

Several methods have been used to characterize the residual activity within embedded pipe, and these methods can be used for buried piping, as well. By definition, "embedded piping" is piping (e.g., part of a plant system) that is found in buildings and encased in concrete floors and walls, while "buried piping" is piping (e.g., culvert) that is buried in soils. To be found acceptable, the methods should each address the following nine issues:

- radionuclides of interest and chosen surrogate,

- levels and distribution of contamination,

- internal surface condition of the piping,

- internal residues and sediments and their radiation attenuation properties,

- removable and fixed surface contamination,

- instrument sensitivity and related scan and fixed minimum detectable concentrations,

- piping geometry and presence of internally inaccessible areas/sections,

- instrument calibration, and

- data quality objectives (DQOs).

An industry study (Cline, J. E., "Embedded Pipe Dose Calculation Method," Electric Power Research Institute Report No. 1000951, November, 2000) evaluated several techniques for measuring the radiological contamination on the inside of embedded pipe. Measurement techniques included pipe crawlers, gamma-ray scanners, dose rate measurements with dose-to-curie computations, scraping samples with radiochemical analyses, and smear samples with radiochemical analyses. A brief description of these methods is provided below.

- The pipe crawler uses a beta sensitive detection system that is inserted into the pipe with a cable. Spacers keep the detectors at a fixed distance from the pipe wall. Measurements can be made at various points or as a continuous scan within the pipe to provide a profile of the extent and distribution of the contamination. Scaling factors based on a laboratory radiochemistry analysis of the deposited material can be applied to the measurements to provide radionuclide quantities in the pipe.

- The gamma-ray scanner uses a calibrated, collimated high-purity germanium or sodium iodide spectrometer to make external measurements on the pipe. This gamma-ray scanning yields an average concentration over the length of the pipe within the field of view of the detector. The sensitivity of this method may be limited by the thickness of the piping itself and concrete between the pipe and the detector. Some radionuclide identification is possible and scaling factors can be applied as discussed above for the pipe crawler.

- The dose rate measurements are also made on the external surface of the walls or floors containing the embedded pipe using a sensitive gamma detector capable of reading in the roentgen per hour range. The dose rate readings may be used directly in determining compliance with the dose criteria or used to make dose-to-curie conversions based on other measurements providing radionuclide identification.

- Radionuclide identification for the contamination in the pipe may be accomplished by smear or scraping samples and radiochemical analysis. The industry report compared radionuclide ratios determined by smears and by scrapings with those found by etching the surface of the pipe. The report concluded that either of these techniques yields radionuclide mixes that are representative of the average total deposits. Each approach is useful in specific applications and multiple methods might be used in complex facilities like power plants. Each method also has limitations and uncertainties that must be addressed.

Other useful information on embedded pipe characterization may be found in sources such as the DOE Innovative Technology Reports and case studies published in open literature.

Regardless of the source of the information, it is incumbent on the licensee to develop and document a comprehensive approach to embedded pipe and buried piping characterization that accounts for limitations and uncertainties, taking into account MARSSIM guidance in developing the related DQOs. It should also specifically address each of the critical issues in the bulleted list above.

Question 4: Development of Site-Specific Distribution Coefficient Values for Soil or Concrete

What is an acceptable approach for the development of input distribution coefficient (K_d) values for soil or concrete when using site-specific dose modeling codes?

Answer to Question 4

K_d values for input into site-specific dose modeling codes may be determined by the following:

Use sensitivity analyses, which include an appropriate range of K_d values, to identify the importance of the K_d to the dose assessment and how the change in K_d impacts the dose (i.e., how dose changes as K_d increases or decreases). The range of K_d values that bound the sensitivity analysis may be obtained from (a) the literature, (b) the default distribution in DandD, or (c) the default distribution in the probabilistic code of RESRAD (please refer to the "Basis" section that follows).

Using the results of the sensitivity analysis, choose a conservative K_d value, depending on how it affects the dose (e.g., if higher K_d values result in the larger dose, an input K_d value should be selected from the upper quartile of the distribution, or if lower K_d values result in the larger dose, an input K_d value should be selected from the lower quartile of the distribution). For those isotopes where the K_d does not have a significant impact on the dose assessment (i.e., K_d is not a sensitive parameter), the median value within the range is an acceptable input parameter.

If the licensee feels that the K_d value is overly conservative, the licensee is encouraged to perform a site-specific K_d determination, so that the dose assessment reflects true site conditions.

Basis

The licensee is encouraged to use sensitivity analyses to identify the importance of the K_d parameter on the resulting dose either (a) to demonstrate that a specific value used in the analysis is conservative or (b) to identify whether site-specific data should be obtained (if the licensee feels K_d is overly conservative). The sensitivity analysis should encompass an appropriate range of K_d values. As noted above, the input range for the sensitivity analysis may be obtained from literature, DandD default distribution, or RESRAD probabilistic default distribution.

Literature

It is noted that K_d values commonly reported in the literature may vary by as much as six orders of magnitude for a specific radionuclide. Generally, no single set of ancillary parameters, such as pH and soil texture, is universally appropriate in all cases for determining appropriate K_d values. Although K_d values are intended to represent adsorption, in most cases, they are an aggregate parameter representing a myriad of processes. Given the above, the proper selection of a range of K_d values, for either soils or concrete, from the literature will require judicious selection.

DandD

The use of the default K_d values from DandD Version 1.0 outside of the scope of DandD may not be justified, since the single set of default parameters derived for DandD was developed assuming a specific set of exposure pathways and a specific source term. Any single parameter value taken from the default set of parameters outside of the context of the given exposure scenario, source term, and other parameters will have no meaning in terms of the original prescribed probability; therefore there is no basis to conclude that any default K_d value will give a conservative result. However, the *distribution* of K_d values, used in DandD (which can be found in NUREG/CR–5512, Volume 3, "Residual Radioactive Contamination from Decommissioning—Parameter Analysis," Table 6.86), can be used as the range of K_d values for the sensitivity analysis.

RESRAD

RESRAD default parameter values (including K_d values) should not be used. The defaults were included in the code primarily as place holders that enable the code to be run; it was assumed that site-specific values would be developed. However, it is appropriate to use the default parameter distribution, developed for RESRAD Version 6.0, as the range for use in the sensitivity analysis.

After performing sensitivity analysis with the appropriate K_d ranges, the K_d value at the upper or lower quartile of the distribution, resulting in the highest derived dose, is an acceptable value to use in into the dose code, and no further justification is required. For those K_d values that are overly conservative, a site-specific K_d value may be determined by the direct measurement of site samples. Appropriate techniques for K_d determination include American Society for Testing and Materials (ASTM) and U.S. Environmental Protection Agency (EPA) methods 9–83, "Distribution Ratios by the Short-Term Batch Method"; ASTM D 4646–87, "24-h Batch-Type Measurement of Contaminant Sorption by Soils and Sediments"; and "Understanding Variation in Partition Coefficient, K_d Values," Volumes I and II, EPA 402–R–99–004A, available at http://www.epa.gov/radiation/technology/partition.htm#voli.

Question 5: Demonstrating Appropriate Selection of Survey Instrumentation by Illustrative Example

Using appropriate illustrative examples in the license termination plan (LTP), is it acceptable to define (a) the data quality objectives (DQO) process and (b) the acceptance criteria for demonstrating that radiation survey instrumentation, selected for use in the FSS, is sufficiently sensitive for a given derived concentration guideline level (DCGL) and expected survey conditions?

Answer to Question 5

Yes, it is acceptable to define the DQO process and acceptance criteria using examples that demonstrate the appropriate selection of radiation survey instrumentation for the expected types of FSS surface conditions and radionuclides forming the basis of the DCGL.

For example, the selection of instrumentation may be grouped by category of surfaces with similar features and expected instrument responses over these surfaces. For each of the defined categories of survey instrumentation and methods presented in the LTP (e.g., soil scanning, surface scanning and surface fixed measurements), the licensee should provide the derivation of scan and fixed minimum detectable concentrations (MDCs). The derivation of the MDCs must take into account instrument efficiencies (surface and detector), scan rates and distances over surfaces, surveyor efficiency, and minimum detectable count rate, using the guidance in MARSSIM and NUREG–1507, "Minimum Detectable Concentrations with Typical Radiation Survey Instruments for Various Contaminants and Field Conditions."

Instruments, other than those provided as examples in the LTP, may be used for the FSS as long as the process approved in the LTP is used to show that the substitute instrument has equal or better performance. If a licensee were to use new technologies (e.g., *in situ* gamma spectroscopy) or different instrumentation than those that were considered at the time of the LTP submittal, the new technology or instrumentation must be shown to perform with sensitivities that allow detection of residual radioactivity at an appropriate fraction of the DCGL and corresponding investigation levels. In addition, the new technology or instrumentation must be at least as efficient as examples of survey instrumentation provided in the LTP. A licensee should also demonstrate and document that conducting the FSS by this new method also will meet all related DQOs in demonstrating that survey units meet the site-established DCGLs.

Question 6: Characterization of Items to be Removed Prior to License Termination

Is the collection of additional characterization data, beyond that available from periodic radiation protection surveys, required in the license termination plan (LTP) for structures, components, and soils that will be removed from the facility prior to license termination?

Answer to Question 6

No. In general, radiological data obtained during characterization surveys are used to determine the radiological status of the site, including facilities, buildings, surface and subsurface soils, and surface and ground water. In turn, this information is used to support the planning and design of final status surveys (FSS). In addition to providing the basis of the design of FSS, characterization surveys are used to support the following:

- Identification of remaining site dismantlement activities

- Development of new (or revisions to existing) remediation plans and procedures

- Revisions to decommissioning costs and trust fund

- Identification of environmental aspects not previously considered

- Revisions to the Environmental Report

Since the license termination process is only concerned with the status of facilities after the completion of all remediation activities, radioactivity associated with structures, components, and soils that will be removed from the facility and appropriately disposed of elsewhere, is not an issue as it cannot contribute to public dose controlled under 10 CFR 20.1402 – "Radiological Criteria for Unrestricted Use." Therefore, additional characterization data need not be collected.

Question 7: Characterization for Initial Classification of Class 1 Areas

Is characterization data required to support initial classification of Class 1 areas?

Answer to Question 7

Areas classified as Class 1 do not require characterization data to support that classification.

Note that characterization data are needed to support decommissioning activities for all areas including:

- Determination of radionuclide distribution profiles and identification of surrogate radionuclides

- Dose modeling and development of derived concentration guideline levels

- Final status survey design and instrument selection

- Structuring the data quality objectives

- Assessment of spatial variability of radioactive contaminants on building surfaces and in surface and subsurface soils

- Assessment of whether ground water is impacted, using the results of the surface and subsurface soil characterization surveys

- Initially defining and changing the boundaries of Class 1 survey units with bordering and adjacent survey units

- Reclassification of survey units (using guidance in MARSSIM and Section A.1 of Appendix A of this volume).

O.2   Lessons Learned Related to Recently Submitted Decommissioning Plans and License Termination Plans

O.2.1   Introduction

Since the implementation of the LTR, NRC staff has reviewed several DPs and LTPs. As a result of these reviews, the NRC staff learned several lessons, the details of which are discussed in the Regulatory Issue Summary (RIS 2002–02), "Lessons Learned Related to Recently Submitted Decommissioning Plans and License Termination Plans." The information in this section is taken directly from the RIS and is provided to help materials and reactor licensees develop more complete DPs and LTPs, as appropriate. There has been some minor editing of the RIS especially to provide the appropriate reference to other sections in this volume.

O.2.2   Lessons Learned

The issues concerning lessons learned include the following 10 lessons.

Lesson 1: Communications

Early and frequent consultations between NRC staff and licensees are encouraged during the planning and scoping phase supporting the preparation of the DPs or LTPs. In this context, a licensee may schedule a meeting with the NRC license reviewer assigned to the site to discuss the planning and content of the DP or LTP. The discussions would address (among other topics) past and current licensed operations; types and quantities of radioactive materials used or stored; activities (current or past) that may have an impact on decommissioning operations; decommissioning goals (restricted versus unrestricted license termination); basis for cleanup criteria and development of site-specific derived concentration guideline levels (DCGLs), or commitment to use NRC default DCGLs; potential impact on public health and safety or the environment; funding plan and financial assurance; and the minimum information required to be contained in the DP or LTP. Regarding the aforementioned topics, licensees are encouraged to review the three volumes of this NUREG report. The principal purpose of this NUREG report is to provide guidance on review of DPs. However, the guidance in this NUREG report supplements that in NUREG-1700, "Standard Review Plan for Evaluating Nuclear Power Reactor License Terminations Plans," in such areas as site characterization, dose modeling, final radiation survey, and institutional controls. This NUREG report provides a structure, using various sections, with which to provide information for staff review. Each section addresses very specific elements of the decommissioning process and related data and information needs. Given that this NUREG report presents the information in a generic context, it is the responsibility of the licensee to go over each section and determine which technical elements or regulatory requirements apply to the facility. Appendix D of Volume 1 of this NUREG report provides a checklist ("DP Evaluation Checklist") to facilitate this process. Given that the checklist is a brief summary of the material presented in each section, it is recommended that each section be reviewed to gain a full understanding of the requirements as the checklist is being prepared.

Before meeting with the NRC staff, a licensee is encouraged to prepare a checklist that identifies technical elements that are applicable (based on a preliminary review); areas that require clarifications from the NRC staff before decisions can be made as to their applicability to the site or facility; and the scope and level of technical details addressing technical elements and regulatory requirements. In addition, the licensee may wish to make a brief presentation describing the past and current use of the facility and the most current radiological status. During the meeting, the NRC staff and licensee representative would go over each item of the checklist and address specific questions. NRC staff would present an overview of its review process, including discussions of the time line and major milestones. The end product of the meeting is a marked-up checklist that defines the technical elements and regulatory requirements to be covered in the DP or LTP submittal. The NRC staff expects that this process will result in a better understanding of the type of information to be included in either document and to familiarize the licensee with the process that NRC staff will use to evaluate the information contained in the DP or LTP. This approach is expected to minimize the need for requests for additional information, reduce the number of iterations and submittals, and expedite NRC staff's technical review.

Lesson 2: Groundwater

Operational environmental monitoring of ground water, although adequate for its intended purpose, may not be adequate for site characterization and to support dose assessments. To support site characterization and dose assessments, information supplied by licensees may need to address the types and movement of radioactive contamination in ground water at the facility, as well as the extent of this contamination." The actual number, location, and design of monitoring wells depend on the size of the contaminated area, the type and extent of contamination, the background quality, hydrogeologic system, and the objectives of the monitoring program. For example, if the only objective of monitoring is to indicate the presence of ground water contamination, relatively few downgradient and upgradient monitoring wells are needed. In contrast, if the objective is to develop a detailed characterization of the distribution of constituents within a complex aquifer as the design basis for a corrective action program, a large number of suitably designed and installed monitoring wells may be necessary. Power reactors normally have ground water monitoring programs as part of their radiological environmental monitoring programs (REMPs). Although data derived from a REMP may provide useful information, the data still tend to be insufficient to allow the staff to fully understand the types and the movement of radioactive material contamination in groundwater at the facility, as well as the extent of this contamination. Therefore, a licensee may need to gather additional data to address this lack of understanding.

Lesson 3: Data Quality Objectives

In developing the final survey design, the licensee should identify all appropriate data quality objectives (DQOs) in planning and designing the final status survey plan (FSSP). The process of identifying the applicable DQOs ensures that the survey plan requirements, survey results, and data evaluation are of sufficient quality, quantity, and robustness to support the decision on whether cleanup criteria have been met using statistical tests. In brief, the major elements of the DQO process include the following:

- a clear statement of the problem (i.e., a full understanding of the radiological status of the facility and extent and magnitude of the contamination);

- the identification of all related decision statements and alternative actions, including selection of the most appropriate scenario for the site and objectives (i.e., How will compliance be demonstrated?);

- the identification of the information needed to support the decision-making process, such as radionuclide distributions and concentrations, methods used to obtain the data, etc.;

- the definition of the site physical, temporal, and spatial boundaries for all environmental media and structures, including reference areas, that will be covered by the decision process and modeling;

- the development of a decision rule in defining action levels (e.g., DCGL–Wilcox rank (DCGL$_W$), DCGL–elevated measurement comparison (DCGL$_{EMC}$), minimum detectable concentrations (MDCs)), grid size and layout; statistical tests, and hypothesis;

- specification of limits for Type I and II decision errors in support of the null hypothesis and impacts on sample size and use of prospective and retrospective power curves; and

- optimization of the data collection process and updating the design of the survey plan, while meeting all DQOs.

In purpose and scope, the DQO process can include a flexible approach in planning and conducting surveys and for assessing whether survey results support the conclusion that release criteria have been met. The DQO process can be an iterative process that continually reviews and integrates, as needed, new information in the design of the final survey plan and decision-making. Finally, the selection and optimization of DQOs will facilitate the later evaluation of survey results and decision-making processes during the data quality assessment phase. The NRC staff has observed that licensees have had difficulties in developing DQOs and have not taken full advantage of the DQO process, especially the optimization step. Experience has shown that the process is often rigidly structured by relying too much on characterization data and not being readily open to the possibility of incorporating new information as it becomes available. This approach makes the implementation of any changes difficult and is an inefficient use of resources, since it imposes time delays when determining how to implement any changes.

Lesson 4: Inspections

In-process inspections are more efficient than one-time confirmatory surveys. In one case, the confirmatory survey was conducted after the licensee had completed most of the FSS and many of the staff supporting the final survey were no longer available to address questions and issues that were discovered while conducting the confirmatory survey. Simply put, the confirmatory survey was conducted too late in the process.

The in-process approach has allowed the licensee and NRC to take side-by-side measurements, compare instrument readings and sensitivity, and address survey issues early in the process rather than at the end of the process. The in-process approach has resulted in significant savings in cost, assured a more accurate survey, and helped the licensee in maintaining the release schedule.

Lesson 5: Flexibility

Continued communications between NRC staff and the licensee during the NRC staff's review is to help ensure that the licensee is able to take full advantage of the inherent flexibility in MARSSIM and the three volumes of this NUREG report. In reviewing DPs and LTPs, the NRC staff has observed that licensees are often boxing their approaches into rigid structures and formats, thereby locking out any operational flexibility in implementing MARSSIM and negating cost savings. This approach may reflect, in part, the interpretation of NRC guidance as regulatory requirements. However, it is possible to meet NRC requirements, while instilling operational flexibility into the overall decommissioning process. For example, large waste volumes alone do not necessarily make a remediation project a complex one, assuming that adequate resources are available to accommodate the higher disposal cost. What makes a decommissioning project complex includes such considerations as groundwater contamination; the presence of hard-to-detect and transuranic radionuclides (TRU); heterogeneous distributions of contaminants; the presence of mixed waste; onsite disposal using engineered features; and reliance on institutional controls to maintain doses within NRC limits under restricted-release scenarios, among others. Even under such conditions, there still is an opportunity to simplify the process, maximize operational flexibility, and benefit from economies of scale.

Another example involves how final surveys are structured and designed around survey units, in recognition that some sites may have literally hundreds of survey units, with licensees perceiving that NRC staff needs to approve the FSS design of each one. The NRC staff expects that licensees should group survey units into a manageable number of categories, taking into account the types of buildings, rooms, areas, built-in equipment, and other specific features. This approach is expected to provide the means to identify and address survey unit features and design requirements that are specific for each category, while treating all other common aspects of the survey design in a generic and systematic manner. The NRC staff suggests that the descriptions identify and address, as is applicable, specific survey design requirements, data quality objectives, sampling methodology, applicable plans and procedures, quality assurance requirements, and data analysis and interpretation for each category. This approach will relieve the NRC staff of having to review and approve each survey design package, before its implementation, and will expedite the final phases of the remediation work, while leaving the development and implementation of each final survey design package subject to periodic regional inspection. Finally, in structuring the final status survey report, licensees should identify and summarize the specific characteristics of each survey unit and discuss their relevance in the analysis of all survey results and interpretation supporting the conclusion that each survey unit meets the cleanup criteria.

Lesson 6: Modeling Issues

The derivation of DCGLs should include the assumptions and justification for parameters used, and justification for how these DCGLs will be applied to various survey units onsite. DCGLs will be captured by license condition as part of the LTP approval process, and will require NRC staff approval for changes to the approved DCGLs.

- Area Factors
  Area factors are needed in the FSS to determine the required scan MDCs and to develop $DCGL_{EMC}$ values that are needed to identify small areas that may need further investigation. However, area factors are typically not provided for residual radioactivity on building surfaces. The primary reason for this is that such factors cannot be calculated by using the DandD computer code. Therefore, when screening DCGL values are used, which were derived from DandD, an alternative approach must be used to calculate area factors for residual radioactivity on building surfaces.

  One approach that has been successfully used is to develop the area factors by using the RESRAD-BUILD computer code and adjusting these derived area factors to account for the fact that RESRAD-BUILD typically gives less conservative dose estimates. With this approach, the screening DCGL values are converted into the appropriate concentration unit for RESRAD-BUILD (i.e., from "disintegrations per minute per 100 square centimeters" to "pico-curie per square meter"). Area factors calculated by RESRAD-BUILD can then be adjusted by the ratio of the dose from RESRAD-BUILD to 25 milli-roentgen equivalent man per year (i.e., the equivalent dose from DandD).

- Volumetric Contamination
  Licensees often have volumetric contamination (e.g., contamination below the surface) in the containment structure from activation products. Because the contamination occurs within a building structure, some licensees have assumed that it is appropriate to use DCGL values developed for building surface contamination for these areas, without additional justification regarding the appropriateness of their use. However, DCGL values developed for building surface contamination may not be appropriate for areas with volumetric contamination, because the potential future exposure routes may be different, especially if the structure is later torn down.

  It is advisable for licensees to develop specific DCGL values, for volumetric contamination, which consider the potential routes of exposure for residual radioactivity in the material if the structure is eventually torn down. As an alternative, licensees can demonstrate that the DCGL values developed for surface contamination will bound the possible effects from exposures for other configurations of the building structure.

- Model Results
  Licensees using RESRAD, DandD, or other computer codes to generate DCGL values or perform dose analyses often do not include the printout from these codes as part of the decommissioning submittal. This information is typically omitted because the output results tend to be voluminous. However, without this information it is difficult for NRC staff to

undertake confirmatory analyses (if needed) or to complete the review of the licensee's analyses.

It is advisable for licensees to provide output results from any analyses used to develop DCGL values or used to perform dose analyses. If the output results do not provide an echo of the inputs used in the analyses, it may be necessary to also provide copies of the input files.

- Nondispersion Versus Mass Balance Models
  In using the RESRAD computer code to develop DCGL values or to perform dose analyses, licensees often use a nondispersion model for evaluating the groundwater pathways. This model is commonly used because it is the default in RESRAD, and therefore will be used unless specifically changed. However, the nondispersion model makes certain assumptions about the location of the future hypothetical well and will generally give lower estimated doses than the mass balance model (if the ground water is an important pathway).

  It is advisable for licensees to either use the more conservative mass balance models or provide justifications for using nondispersion models. Specific guidance on justification for using the nondispersion model can be obtained from Section I.4.3.2.1.2 from Appendix I of this volume.

- Parameters
  Licensees often use a combination of default and site-related parameters in their analyses to develop DCGL values or in dose analyses. In many cases, little or no justification is provided for the reason for using the specific parameter values used in the analysis. This can lead to uncertainties in assessing the appropriateness of the DCGL values or calculated dose in demonstrating compliance with the standard.

  Given the large number of parameters that may have to be justified in an analysis to develop DCGL values or a dose analysis, Appendix I, Section I.6, of this volume discusses an approach for focusing on those parameters most important to the results. This approach entails classifying parameters as either behavioral, metabolic, or physical, as defined in NUREG/CR-5512, Volume 3. Licensees may use default values for behavioral and metabolic (primarily those prescribed for DandD) as long as the values are consistent with the generic definition of the average member of the critical group, and the screening scenarios are used. Site-specific physical parameter values should be used and justified. The level of justification needed is dependent on the significance of the parameter to the results. The relative significance of parameters to the results can be determined through a sensitivity analysis. In the sensitivity analysis, the default statistical distributions provided in RESRAD 6.0 and RESRAD-BUILD 3.0 should be used, supplemented with what is known about the site (note: default distributions should not be used as a substitute for known information). Known parameter values should be treated as a constant in the sensitivity analysis. The relative significance of the various parameters can be determined based on the ranks listed in the regression and correlation results in the uncertainty report. The default surface contamination values for alpha-emitting radionuclides are rather low, and in some cases below the detection limit. This results from a conservative resuspension factor (RF) used in the DandD code. Therefore, the licensee may wish to consider using a more realistic RF value for site-specific analyses.

Lesson 7: Decommissioning Cost Estimate

There needs to be a clear relationship between the planned decommissioning activities and the associated cost estimate. At the license termination stage, the NRC staff must make decisions on the proposed actions described in the LTP. The NRC staff typically considers (a) the licensee's plan for assuring sufficient funds will be available for final site release; (b) radiation release criteria for license termination; and (c) the adequacy of the final survey required to verify that the site release criteria have been met. 10 CFR 50.82(a)(9)(ii)(F) requires the licensee to provide, in part, an updated site-specific decommissioning cost estimate. If little decommissioning has been completed, and inflation and disposal costs have not changed, the cost estimate required by 10 CFR 50.82(a)(8)(iii) may be acceptable. NRC staff is not requiring the licensee to submit any contractual documents or agreements that exist between the licensee and the decommissioning contractor, and the cost estimate should not be impacted by the election of the licensee to decommission the facility, or contract to decommission the facility. However, for NRC staff to be able to make a finding that sufficient funding is available to complete decommissioning, the updated cost estimate of the remaining site dismantlement activities, and the remediation plan that outlines how the decommissioning will be conducted, must correlate. The updated cost estimate should be based on the remaining activities and the plans on how the actions will be completed. The updated site-specific cost estimate must address the remaining activities necessary to complete decommissioning, to assure sufficient funds are available, because the financial assurance instrument required under 10 CFR 50.75 must be funded to the amount of the cost estimate, and during decommissioning, the licensee has been allowed to withdraw the funds set aside for decommissioning.

Lesson 8: Records

Old records may be inadequate or inaccurate for the purpose of developing either the historical site assessment (HSA) or site characterization. The NRC staff suggests that these records not be relied on as the sole source of information for the HSA or site characterization. Interviews with current and former staff and contractors play an essential role in formulating the HSA. Experience has shown that old records and results of operational surveys and post-shutdown scoping surveys have been submitted as substitutes for characterization surveys. For example, the results of operational surveys may represent radiological status, describing conditions over a limited time span, or may have been conducted to address specific events (i.e., post-spill cleanup assessment). In a few instances, the results of personnel interviews and information, which can only be considered as anecdotal, have been presented in the HSA. In fact, it could not be determined whether this information was part of an unbroken chronological history of the site or contained time gaps for which operational milestones or occurrences were missing. Although NRC staff encourages licensees to review old records and conduct personnel interviews (past and current employees and key contractors), there is a need to present this information in its proper context and qualify its usefulness and how it might be supplemented (e.g., via additional data searches or characterization surveys). To achieve the purpose of the HSA, a complete history of the residual contamination is needed. Given their importance, the NRC staff suggests that characterization surveys be developed only after the licensee has conducted a thorough evaluation of the information collected during the site historical assessment.

Based on the review of several DPs and LTPs, the NRC staff has found that licensees have generally done extensive characterizations of facilities slated for decommissioning. A review of selected characterization files (in support of decommissioning and turnover surveys) revealed that a wealth of information is indeed available, but that it is not conveyed or presented clearly in DPs and LTPs. The information NRC staff seeks can be drawn from existing characterization records or supplemental analysis of existing samples, thereby avoiding the need to conduct additional surveys and to send workers into radiation areas — all while minimizing costs. The type of information that is needed to support the preparation of DPs and LTPs focuses primarily on residual levels of contamination remaining on building surfaces or in soils (surface and subsurface), after the remediation work has been completed. The characterization of elevated contamination levels typically found in radiation areas is of no concern in addressing the design of FSSes, since these areas are contaminated at levels that obviously exceed any realistic $DCGL_W$. NRC staff is seeking a better presentation, and perhaps evaluation, of existing data supporting specific DQO elements and justification for the approach proposed in developing survey designs. In most instances, it is not a question of generating more data — rather, it is a question of making use of all existing data. There may be some exceptions where additional characterizations might be warranted. Such exceptions might apply to the characterization of subsurface soils, ground water, and TRU, since these may present unique challenges, but can be resolved without unnecessary radiation exposures.

Lesson 9: Environmental Reviews

Environmental assessments should address nonradiological impacts of the proposed action. In accordance with the provisions of the National Environmental Policy Act (i.e., Public Law 91-190) all agencies of the Federal Government are required to assess the environmental impact of any major Federal action that may significantly affect the quality of the human environment. As part of NRC's approval of either a DP or an LTP, NRC staff is required to determine if that approval is a Federal action. Therefore, the impacts on the human environment associated with NRC approving either a DP or an LTP must be assessed. Further, this assessment must include both radiological and non-radiological impacts. Although most licensees normally provide sufficient information for the NRC staff to assess the radiological impacts on the human environment, some licensees have not provided sufficient information related to current site-specific non-radiological impacts.

Because actions associated with NRC's approval of a DP are different than those associated with NRC's approval of an LTP, the information required to assess the impacts on the human environment are different. That is, when NRC approves a DP, NRC is approving the licensee performing the activities necessary to remediate radiological contamination at a site.

Therefore, a DP should include information addressing non-radiological impacts on the human environment associated with these proposed activities. Non-radiological impacts include, but are not limited to the following: land use; water quality; transportation; air quality; ecological, historical, and cultural resources; hazardous material/waste; noise; visual/scenic quality; socioeconomics; and public and occupational health. However, under the provisions of 10 CFR 50.82, most if not all activities necessary to complete site remediation can be completed under the provision of 10 CFR 50.59. Therefore, these activities will not require prior NRC approval. Consequently, unless certain site-specific issues exist, NRC, when it approves an LTP, is approving only (a) the adequacy of the decommissioning funding plan to assure that sufficient funding is available to complete the remaining radiological remediation activities, (b) the radiation-release criteria for license termination, and (c) the adequacy of the design of the final survey to verify that the release criteria have been met.

Lesson 10: Characterization Surveys and Classification of Survey Units

The NRC staff recommends that submittal of the DP or LTP occur only after sufficient site characterization has occurred. The NRC staff suggests that the DP or LTP provide sufficient information demonstrating the characterization of the radiological conditions of site structures, facilities, surface and subsurface soils, and groundwater. The NRC staff has observed that some DPs and LTPs have been submitted with incomplete or inadequate characterizations of radiological conditions. A review of such DPs or LTPs has shown that the lack of information makes it difficult to agree with the rationale justifying the proposed classification of survey units. The NRC staff suggests that the following issues related to the use of characterization survey results and classification of survey units be considered when developing either a DP or an LTP:

- *Use of operational, post-shutdown scoping, or turnover surveys as characterization surveys* — Characterization surveys are the most comprehensive of all surveys, yield the most information, provide the basis to design the FSSP, and are used for dose modeling as well. Characterization surveys are conducted to determine the current extent and magnitude, and variability (as surface and depth profiles) of the contamination, and radionuclide distributions and concentrations. Characterization survey results are used to guide remediation efforts, provide information with which to update waste volume and cost estimates, and develop DCGLs. Given their importance, the NRC staff recommends that characterization surveys be developed only after the licensee has conducted a thorough evaluation of the information collected during the HSA, and the results of operational surveys and post-shutdown scoping surveys. Accordingly, it is not appropriate to use the results of past operational and post-shutdown scoping surveys as substitutes for characterization surveys conducted using the guidance of MARSSIM. For example, the results of operational surveys may represent radiological status describing conditions over a brief operational time span or may have been conducted to address specific occurrences (i.e., post-spill cleanup assessment). Moreover, the results of both operational and post-shutdown scoping surveys may be of limited use unless it can be shown that data quality, instrument calibration methods, and detection sensitivities (fixed and scan measurements) for the anticipated radionuclide mix are comparable to those defined for the characterization surveys based on MARSSIM guidance. These limitations also apply to turnover surveys conducted after the completion of remediation. In all three instances, this approach is also a departure from the MARSSIM methodology in that it defeats the statistical basis intended to confirm that survey units meet the release criteria. As is noted in MARSSIM (Section 5.5.2.5), "Measurement locations based on professional judgement violate the assumption of unbiased measurements used to develop the statistical test described in Chapter 8" (of MARSSIM). If a licensee were to use turnover survey data for part of the final survey, statistical samples and/or measurements may need to be identified in addition to the turnover survey data. Also, the samples and/or measurements should be collected or made in compliance with MARSSIM guidance (i.e., random start and systematic sampling/measurements using an established grid) or other survey methods found acceptable to NRC staff.

- *Reclassification of Survey Units* — It may not always be appropriate to simply separate out an area of elevated activity, from a Class 2 or Class 3 survey unit, as an individual Class 1 survey unit since the initial basis for evaluating a Class 2 or 3 survey unit is based on specific criteria [i.e., 10 to 100 % scan coverage for Class 2, and judgement (typically <10 %) for Class 3 survey units]. Similarly, licensees should provide the basis in delineating Class 3 survey units as buffer zones around Class 1 and 2 survey units and areas with insufficient justification to be classified as non-impacted. If survey results were to reveal elevated levels of contamination in an arbitrarily selected portion of a Class 2 or 3 survey unit, then the classification of the entire survey unit should be deemed suspect and re-evaluated, using MARSSIM guidance. In this context, the NRC staff suggests first, that there should be considerations of the assumptions made as to how the survey unit was initially classified, most likely or known causes of contamination, and the possibility that other similarly contaminated areas within the original survey unit might have gone undetected. The NRC staff also suggests that a DP or LTP address these considerations and describe the method, consistent with MARSSIM, that will be used if a survey unit or portion of a survey unit must be upgraded to a higher classification level. In general, increasing the coverage of the scan is less expensive than finding areas of elevated contamination levels later in the process. Finding areas with elevated levels of contamination later in the process will require the conduct of additional surveys, lead to delays in reconsidering the initial classification of the survey unit, and will lead to additional regulatory scrutiny. The NRC staff recognizes, in many instances, that DPs or LTPs are submitted at a time when some characterization work is still ongoing and that supplemental data may lead to the reclassification of some survey units. Accordingly, a DP or LTP should include the flexibility to accommodate changes in the classification of survey units as more characterization data are obtained and evaluated.

- *Completeness of Characterization Survey Design and Results* — In some submittals, the NRC staff has noted that contamination results for plant structures, systems, and components; surface and subsurface soils; and groundwater are at times incomplete. For example, the review of data characterizing such areas or media has revealed that only limited information is being provided about the presence of TRU (e.g., plutonium-239, americium-241) and hard-to-detect radionuclides (e.g., hydrogen-3, carbon-14, nickel-63). In other instances, the data fail to provide sufficient information in determining the fraction of surface radioactivity that is fixed and removable. Similar shortcomings were noted for removable alpha and beta radioactivity found in embedded piping, usually contained in residues, sediments, and internal film coatings. Although reporting histories of fuel cladding failures, some plants have not provided information on the presence of TRU in plant systems and at effluent discharge points. The characterization of neutron activation products in concrete and rebar is often limited in scope, and the presentation of the results fails to address the significance of the reported radionuclide concentrations and their applicability to other areas of the plant. In summarizing characterization results, there are instances when both the average and maximum surface beta activity results are below the stated MDCs. Such results are misleading since it is not clear if the stated MDCs are representative of all areas within a survey unit or whether there might be multiple MDCs that could be unique to distinct areas within each survey unit. Such results imply that the variability may apply to all areas within a survey unit, when perhaps the variability of the contamination might be multi-modal if it were evaluated by separate and smaller areas. This problem, in part, is attributed to how the data are edited for summarization. In other instances, licensees have proposed radiological results characterizing radionuclide distributions and concentrations using smears/wipes, air filters, and debris, with no rationale as to the relevance of the information. It should be noted that characterization survey results provide the most important information (i.e., the basis to design the FSSP; define radionuclide distributions and concentrations; identify hard-to-detect radionuclides and develop surrogate ratios; define survey area classifications; and assign the sigma characterizing the variability of the contamination (a key parameter in determining the number of samples in survey units)). Accordingly, the planning and execution of any characterization surveys should be conducted in a manner that will generate technically defensible results with which to design the FSSP.

Lesson 11: Embedded Piping

Nuclear power reactors and other types of nuclear facilities contain embedded piping that may become radiologically contaminated as a result of licensed operations. The NRC staff suggests that DPs and LTPs include a discussion on the methodology for conducting surveys of embedded piping planned to be left behind. The NRC staff suggests that sufficient justification for the assumptions considered in the computer modeling and dose analysis for embedded piping be described in the basis. Also, the NRC staff suggests that copies of relevant computer code printouts be included for NRC staff evaluation.

One approach that has been approved for surveys of embedded piping is to establish a separate site-specific dose criterion for external penetrating gamma radiation emitted from the internal surface of embedded piping present in structures (e.g., walls, floors, ceilings) which are also in the same survey unit. In this approach, the predominant radionuclide of concern from a dose perspective (e.g., cobalt-60) is determined by isotopic analysis of scale or residue samples collected within such piping during the licensee's radiological characterization program. The dose criterion should be based on bounding conditions developed from characterization data, computer modeling using a radiation shielding computer code, and a detailed dose analysis of the exposure scenario. In the model, grit blasting of the internal surface of embedded piping may need to be considered to assess (a) any gains from the removal of loose surface activity and

(b) whether the application of grout to immobilize and encapsulate fixed residual surface contamination would reduce radiation exposures.

It is important to describe the mechanism in which the dose contribution from the embedded piping and the non-embedded piping portion in a given survey unit is evaluated, when the dose to either component is determined to be equal to/or greater than the respective established dose limit, to ensure that the entire survey unit does not exceed the release criteria. Further, the NRC staff recommends that licensees discuss how adequate scan and static investigation levels will be implemented and further evaluated, as needed, in the FSS. It is also advisable that radiation detectors used for embedded piping surveys be properly calibrated for this specific geometry [including the use of National Institute of Standards and Technology (NIST) traceable radiation source(s)], which are appropriate for types, energies, and residual concentrations expected in the FSS.

Lesson 12: MDCs

The decommissioning process typically involves sites with multiple radionuclides present at the time the FSS is conducted. Although individual radionuclides and their respective $DCGL_W$ values and initial-scan MDCs for the principal radionuclides of concern have been identified, DPs and LTPs should describe the methodology and basis on which to implement a scan MDC to account for a mixture of radionuclides that may remain in a given survey area/unit. The NRC staff recommends that parameter values such as source (ε_s) and instrument (ε_i) efficiencies, surveyor efficiency (ρ), and performance criteria (d'), which determine the scan MDC, be evaluated before implementation; also, changes in the default parameter values (e.g., $\rho = 0.5$, $d' = 1.38$) need to be clearly justified in the DP or LTP.

In MARSSIM, decisions are made on selecting appropriate detection sensitivities or MDCs for radiological survey and laboratory instruments in the DQO process. Static MDCs within 10 to 50 % of the $DCGL_W$ of the individual radionuclide are often readily achievable; however, the scan MDC involves a larger number of arbitrary assumptions and decisions. The NRC staff generally considers the ε_s values described in International Organization for Standardization (ISO) 7503–1 and ISO 7503–3 guidance for alpha- and beta-emitters to be acceptable estimates, absent site-specific information, for surface contamination detectors in the final status survey design. The NRC staff suggests that, in situations where surface contamination measurements are planned on irregular and uneven surfaces such as scabbled concrete and embedded piping, licensees determine an appropriate site-specific s value(s). Further, the NRC staff recommends that the methodology and basis for the ε_s value(s) be provided for NRC review.

When multiple radionuclides are present in the survey area/unit, application of an ε_i value, the use of a representative, conservative, or beta-weighted average energy for the anticipated radionuclide mixture, has been acceptable to NRC staff.

Because the estimated–scan MDCs for open land areas (soils) (Table 6.7 of MARSSIM) are premised on certain decisions and assumptions involving human factors and survey techniques, detector characteristics and performance, and computer modeling, it is advisable that licensees validate (e.g., *a posteriori*–scan MDC) the *a priori*–scan MDC used for design goals, as information is collected and assessed, so that an actual-scan MDC can be calculated for implementation in the FSS, for demonstration of compliance.

O.3 Lessons Learned During Decommissioning Final Status Survey In-Process Inspections and Confirmatory Surveys

O.3.1 Introduction

Confirmatory surveys and in-process inspections conducted at various NRC materials and reactor licensee facilities undergoing decontamination and decommissioning have identified a number of issues with implementation of FSSes in accordance with current guidance. The issues identified are related to the following categories: instrumentation, procedural, and survey planning and data evaluation. Each issue is discussed together with the potential problems that may result and recommended solutions. It is important to note that this is not an all-encompassing discussion of identified issues, rather the discussions represent more recent and pervasive technical deficiencies.

O.3.2 Instrumentation

The issues discussed in the following sections relate to the selection and operation of either radiation detectors or instruments selected for radiological surveys.

O.3.2.1 Temperature Effects on Gas Proportional Detectors

Industry guidance for the calibration of instrument/detector combinations used for assessing residual radioactive material contamination levels requires calibrations be performed in a manner that simulates the environmental and set up conditions under which the equipment will be used (ANSI 1997 and NCRP 1991). Recent evaluations of surface activity discrepancies when evaluating comparative licensee and confirmatory survey measurements found a systematic under response in the licensee's reported activity levels during cold weather periods. In these cases, alpha plus beta or beta-only scans or measurements were being conducted using gas proportional detectors.

Normal practice is to conduct calibrations in a laboratory setting. When these instruments are distributed for use, the conditions may change once the user is out in the field. Temperature variations on the order of 30 to 40° F over the course of a work day are common. Past investigations have determined that the optimal operating voltage plateau, established during calibration, shifts while the instrument is being used under varying temperature conditions. For example, a controlled experiment showed for a specific brand and model of detector and instrument that a detector voltage plateau calibration at 70° F provided a plateau ranging from approximately 1700 to 1760 volts (optimal setting of 1725 to 1750 volts — the mid-point of the plateau). For the same detector calibrated at 20° F, the voltage plateau ranged from 1775 to 1875 volts, with an optimal setting of 1825 volts. The above data show that the voltage set point for the 70° F calibration is actually below the knee of the 20° F calibration. This shift — found to begin around 40° F — results in a detector calibrated at room temperature, then operated in a cold environment, to under respond to the source of radiation.

There are solutions to this issue to ensure calibration is conducted to match any expected temperature extremes. Voltage plateaus may be performed at multiple temperatures when possible. The proper voltage set point may then be selected to match conditions. Otherwise, to minimize the effects of the plateau shift during cold weather operation, select an operating voltage that is closer to 3/4 of plateau maximum. As conditions change, the high voltage may be adjusted to bring the detector back to within established operational parameters and verified through appropriate procedures. For hot weather, the opposite effect is probable and the operating voltage may have to be reduced to avoid a shift into the continuous discharge region of the plateau.

O.3.2.2 Instrument Count Rate Plateaus

Some types of data logging instrumentation selected for FSSes have audio response limitations that impact the surveyors ability, under certain conditions, to discern the presence of elevated activity. These instruments have a preset audio response that plateaus once the count rate reaches 4500 counts per minute (cpm). This condition does not impact alpha contamination assessment and normally does not interfere with assessing beta surface activity when ambient gamma backgrounds are at typical environmental levels. On the other hand, when these instruments are used for conducting gamma scans using NaI scintillation detectors, complicating factors occur. Typical background gamma levels range from approximately 2,500 to 12,000 cpm when using the more common NaI detector crystal sizes. It can be immediately seen that the background may saturate the audio capability of the instrument making it impossible for the surveyor to rely on increases in audio response to identify locations of elevated direct gamma radiation.

There are essentially three solutions to this problem, all of which result in either additional complicating factors that must be addressed or potential further project costs. One such approach that has been implemented is the use of an alarm set point action level — rather than relying on the audio response — that roughly correlates a specified count rate to a concentration in soil. The difficulty encountered with this approach occurs when the scan MDC calculations prescribed in MARSSIM are adapted. This is further discussed in Section O.3.3.1 of this appendix. Furthermore, the use of such an action level should build-in adequate conservatism — and corresponding confidence level — to account for the statistical variance normally seen in the data that are used to generate the count rate to pCi/g relationship. Another option is to use the audio divide feature of the instrument to bring the audio response below a suitable fraction of the plateau. When doing so, the survey planner will need to ensure that the reduced audible background count rate is factored into the MDC calculations. Lastly, consideration may be given to using a different instrument or smaller NaI crystal size (to lower the background) for gamma scanning.

O.3.2.3   Miscellaneous Instrument Issues

Other instrument issues that have been noted include static (disconnection from a continuous gas supply) operation of gas proportional detectors, long detector to instrument cables, and altitude effects on the calibration of gas proportional detectors. When gas proportional detectors are operated in a static mode, there will be some gas leakage from the detector. As the gas supply decreases, the detector efficiency degrades accordingly. The rate of gas leakage greatly varies among detectors, particularly once the factory-installed face and gasket are removed for maintenance. The rate of leakage has been observed to range from minutes to days. Past field observations of FSSes and comparative measurements have found that these detectors may have had only a partial purge, resulting in the underestimation of surface activity levels. Therefore, procedures should specify that when surveying in a static mode, the operational parameters should be checked regularly through either a background or source check. If the detector falls below established parameters, repurging the detector would be required prior to continuing surveys. Operation at the alpha plus beta voltages more readily allows the surveyor to distinguish a drop in efficiency caused by gas leakage as the background levels — generally in the 200 to 500 cpm range for hand-held detectors — will noticeably decrease. However with the 0 to 5 cpm alpha voltage backgrounds of most hand-held gas proportional detectors, a decrease in efficiency will not be immediately observable and therefore will necessitate a regular operational source check to validate performance.

Section O.3.2.1of this appendix discusses the importance of calibrating instruments under the same environmental and set up conditions in which they will be used. Two additional factors that have occurred are differences in detector performance that result from significant changes in altitude between the calibration and use point and when long cables are used. A gas proportional detector calibrated at 1000 feet above mean sea level will under respond when operated at higher altitudes. Therefore, this impact must be addressed by either calibrating at the site where the equipment will be used or otherwise adjusting the electronics once the equipment is received at the site to ensure operational parameters are correct. Similarly, there have been cases where the original 5- or 6-foot cable that an instrument/detector combination was calibrated with is replaced with a longer cable to permit access of the detector to difficult to reach places. The longer cable may increase the electrical impedance and again result in an under-response. The instrument/detector combinations should then have separate calibrations performed, both with the standard and long cables.

O.3.3   Procedural

The issues discussed in this section were identified either as a result of observation of FSSes during in-process inspections or following the review of licensee procedures.

O.3.3.1 Alarm Set-Points and the MARSSIM Scanning MDC Calculation

There have been a number of instances where FSS procedures have implemented the use of various detectors coupled to data logging instruments. These instruments in several cases were set to alarm at a pre-determined count rate action level that is calculated to correspond to the $DCGL_W$, rather than relying on the surveyor listening to the audible response. Although this may be an acceptable practice, with the provision of an adequate technical basis, the MARSSIM scan MDC equations are no longer appropriate. The reason for this position is that the derivation of the scan MDC equations are based on signal detection theory. That is, how a human observer theoretically processes the audible input and then makes decisions. Refer also to Section O.3.3.3 for related discussions.

Furthermore, where a human may continually adjust to varying backgrounds, an alarm set point is normally established as a multiple of a static background. However, once an alarm (or MDC) is set using a static background, the electronics are not capable of discriminating when lower or higher background areas are encountered. As a result, any significant changes in background levels would necessitate a re-evaluation of the basis of the MDC in determining whether the new MDC still meets the related data quality objectives. In the case of operating in a lower background area, the instrument may not alarm when required.

The following example illustrates this point: Background is established at 10,000 cpm. The action level is determined to correlate to 5,000 net cpm above the selected background, or 15,000 gross cpm. However, backgrounds fluctuate between 8,000 and 11,000 cpm in the survey unit, dependent upon surface types. While surveying in an area where the background is 8,000 cpm, residual contamination contributing an additional net count rate of 5,000 cpm would fail to activate the alarm (13,000 gross cpm) — increasing the false negative rate. Conversely, when operating in an area where background is higher than the set point background, one would expect a higher false positive rate.

The use of an alarm set point therefore requires a number of considerations for calculating the scan MDC, procedures for addressing varying backgrounds, and specific investigation requirements for when an alarm occurs (i.e., second stage scanning and soil sampling).

O.3.3.2 Gamma Fixed Point Measurement in Place of Surface Scanning

Confirmatory surveys conducted in Class 1 soil survey units at several sites have identified small areas of residual gamma-emitting contamination that when evaluated, exceeded the $DCGL_{EMC}$. A root cause analysis was performed and determined that the site procedures required systematically spaced, fixed point gamma measurements rather than prescribing surface scanning over 100 % of the survey unit area in accordance with MARSSIM.

Experience has shown that for characterization surveys, where contamination may be more distributed, systematic fixed point gamma measurement can be useful for identifying large areas requiring investigation. However, once an area is remediated, contamination generally becomes more isolated and heterogeneously distributed. Scanning surveys are designed to specifically address this condition where small areas of elevated activity may be present that would go undetected by systematically-spaced measurements. Therefore, to resolve this issue, surface scans should be performed over, not only Class 1, but all survey units following the MARSSIM recommendations for coverage.

O.3.3.3 Not Listening to Audio Response While Conducting Surface Scans

A significant number of facilities assessed during decontamination and decommissioning do not require the surveyor to listen to the instrument audio response while conducting radiological surface scans. Rather, the analog meter is visually observed, an instrument alarm is set to notify the surveyor when to pause and investigate, a peak trap mode (the maximum observed count rate value is stored in the instrument memory) is used and the data are reviewed for anomalies post-survey, or a second person — rather than the individual using the detector — listens to the instrument audio.

Each of these techniques have inherent deficiencies that impact one's ability to identify locations of residual contamination. The instrument alarm comments were detailed in Section O.3.3.1 of this appendix, with one additional comment provided here. That is, it has been previously observed during in-process inspections that a only a single alarm may occur when multiple hot spots were known to be present or, if a peak trap mode is used to assess scan data, only the maximum value is available for review. With both of these approaches, information on the presence of multiple areas of elevated direct radiation is not available to the surveyor.

Reviews of procedures and direct observations of FSS field scanning techniques identified a unique variation of this issue. In these cases, a surveyor separate from the one performing the survey listens to the audio output of the instrument. The previous discussion of the applicability of the MARSSIM recommended scan MDC calculation also should be considered when using the dual surveyor approach. A second key component that should be addressed is the impact on second stage scanning. That is, the mechanism for when the surveyor moving the detector is caused to stop the detector and investigate an increase in the count rate.

Lastly, there are two less common methods that have been used during FSS scans. The first is reliance on visually observing fluctuations in the instrument readout — either needle deflections or digital readout. Again, this is contrary to the MARSSIM scan MDC paradigm and this method is significantly less sensitive than the audio output due to instrument smoothing functions built into the readout, may require a greater degree of vigilance, and also may result in additional safety concerns. The second method is the use of ratemeter-scalers capable of counting alpha and beta interactions simultaneously. These instruments provide a different tone for alpha or beta counts. Although in one specific case, the surveyors were listening to the audio response, it was found during confirmatory surveys that a significant quantity of alpha contamination had not been identified. The most probable cause was the difficulty in discerning the low alpha activity guideline over the higher beta background. In other words, the beta activity count rates overwhelmed the surveyors ability to audibly detect the low alpha count rates that required further investigation. It is therefore recommended that separate alpha and beta scans be performed.

For any of these cases, the surveyors should listen, using head phones — especially in high noise environments — to the audio output. The use of the other techniques described above do not adhere to the MARSSIM guidance and therefore may require preparation of a technical basis that details the approach, calculated scan MDCs for the specific approach, procedures for second stage scanning and investigation requirements.

O.3.3.4 Instrument Calibration for Assessing Surface Activity Using ISO 7503-1

The implementation of the instrument calibration guidance for assessing alpha and beta surface activity recommended in ISO 7503-1 (ISO 1988) and adapted into the MARSSIM is not always consistently applied. This issue was identified while reviewing either license termination plans or specific licensee calibration procedures. The ISO 7503-1 guidance more accurately accounts for surface conditions encountered at decommissioning sites — typically rough, dirty, or porous — and emission energy of the radionuclides of concern. Without the proper application of the ISO 7503–1 guidance, surface activity levels for alpha and low-energy beta-emitting contaminants will be significantly underestimated. The guidance recommends a total efficiency that is the product of two components — an instrument efficiency (ϵ_i) and a source efficiency (ϵ_s).

The most commonly encountered calibration findings have identified the use of a 4π total efficiency instead of the ISO 7503-1 and MARSSIM–adapted 2π instrument efficiency which is then modified to address surface conditions ($\epsilon_i \times \epsilon_s$). As an example, if technicium-99 — a low energy beta emitter — were the contaminant of concern at a site, an expected laboratory derived 4π efficiency for a Geiger-Mueller (GM) detector would be approximately 0.17. However, this efficiency is overly optimistic because of the expected attenuating surfaces that will be measured in the field. The comparative ISO 7503-1 derived technicium-99 efficiency would be approximately 0.05 (0.20 for the $\epsilon_i \times$ 0.25 for the ϵ_s) for the same GM detector. The resultant surface activity would be underestimated by a factor of almost 70 % using the 4π efficiency versus the two component ($\epsilon_i \times \epsilon_s$) efficiency. A related issue identified that also results in an underestimation of residual contamination was the application of an ϵ_s for alpha calibrations of 0.5 — the correct default value is 0.25 for alpha emitters and low energy (< 400 keV maximum energy) beta emitters.

In general it has been seen that licensees have adequately accounted for mixtures of varying energy beta emitters, hard-to-detect radionuclides, and unusual surface configurations such as corrugated metal in determining total efficiencies.

O.3.3.5 Performing Alpha Rather than Beta Surface Activity Measurements for Natural Thorium Surface Contamination

There have been several instances where residual natural thorium surface contamination was assessed by performing only alpha activity measurements. Natural thorium emits both alpha and beta radiations, therefore, either alpha or beta activity may be measured for determining the residual activity of the thorium contaminant. However, beta measurements provide a more accurate evaluation of thorium contamination on structural surfaces due to the problems inherent in measuring alpha contamination on rough, porous, and/or dirty surfaces. For the thorium series in secular equilibrium, for each beta emission there are approximately 1.5 alpha emissions — a beta to alpha ratio of 0.67. At one site, both alpha and beta surface activity measurements were performed during confirmatory surveys at the same location and the results compared. The data clearly showed the significant and widely varying alpha attenuation with beta to alpha ratios ranging from 3 to 280 — much greater than the theoretical ratio of 0.67. This provides further evidence that alpha activity is difficult to measure on surfaces that are typically encountered during radiological surveys and when possible, beta measurements should be performed. Alternatively, the alpha efficiency should be empirically reduced to account for the attenuation.

Uranium contamination on surfaces presents similar challenges as natural thorium when planning for the type of surface activity assessments that will be performed. As with thorium, the uranium series also emits both alpha and beta radiations. The specific alpha to beta ratio for the type of uranium (natural, natural processed, enriched or depleted) should be determined. Dependent upon the uranium isotopic abundances, these alpha to beta ratios can range from approximately 1:1.6 for depleted uranium, up to 20:1 for highly enriched uranium. This information is necessary for selecting which emission to measure, calculating an appropriate efficiency, and quantifying surface activity.

Several sites where natural thorium was the contaminant measured the alpha component rather than the beta component due to high ambient gamma background levels. This approach was followed because the high ambient gamma background present at the site resulted in a static beta surface activity measurement MDC that exceeded the thorium surface activity guideline. Alpha measurements were therefore selected to demonstrate compliance. However, the significant alpha attenuation was not accounted in the detector calibration. As a result, the reported alpha surface activity significantly underestimated the residual thorium contamination.

The impact of the high ambient background can be readily resolved by revising procedures and adapting one of two methods. If a given approach must rely solely on alpha measurements to assess residual thorium (or uranium) activity, alpha calibrations should then be conducted in accordance with the ISO 7503-1 and MARSSIM guidance. That is, the total efficiency of the detectors should be modified to account for the significant source attenuation. A second approach, would be to conduct beta activity measurements corrected for the high ambient gamma background. This is accomplished by performing both shielded (using a sufficiently thick Plexiglas™ shield) and unshielded measurements both in the survey unit and a suitable reference area. The surface activity:

$$\text{dpm} \,/\, 100 \text{ cm}^2 = \frac{\text{Net count rate } (N)}{\varepsilon_{tot} \times \text{Geometry} \times \text{other modifying factors}} \qquad \textbf{(O–1)}$$

is calculated with correction for the gamma component. The net count rate used in the numerator of the surface activity equation is acquired as follows:

$$N = (R_{u,su} - R_{s,su}) - R_{rm} \qquad \textbf{(O–2)}$$

where N = net counts
$R_{u,su}$ = unshielded survey unit count rate
$R_{s,su}$ = shielded survey unit count rate
R_{rm} = reference material background count rate (ambient background subtracted out)

$$R_{rm} = R_u - R_s \qquad \textbf{(O–3)}$$

where R_u = unshielded (gross) on background reference material and
R_s = shielded, background count rate.

Example: Beta activity measurements are required on a survey unit concrete floor. There are high ambient gamma levels in the survey unit due to contaminated sub-floor soils. A non-impacted concrete floor in another part of the facility is identified for background reference measurements. The count times are for one minute, the ϵ_{tot} is 0.20, and the geometry factor is 1.26. The following background reference material data for the concrete floor are obtained: $R_u = 400$ cpm; $R_s = 300$ cpm (the gamma component of the background). $R_{rm} = 400$ cpm $-$ 300 cpm $= 100$ cpm.

The following survey unit concrete floor data are obtained: $R_{u,\,su} = 1000$ cpm; $R_{s,\,su} = 500$ cpm. Therefore, $N = (1000 \text{ cpm} - 500 \text{ cpm}) - 100 \text{ cpm} = 400$ cpm. When this value and the previously provided count time, geometry, and total efficiency are substituted into the surface activity equation, the reported surface activity result equals approximately 1,600 dpm/100 cm^2.

O.3.4   Survey Planning and Data Evaluation

The following sections describe issues encountered that are related to survey planning input parameters and subsequent post-survey data evaluation.

O.3.4.1   Contaminant Variability Ratio: Difference Across a Site

There have been several instances where a limited number of soil samples were used to determine a site-wide ratio between various contaminants. A surrogate contaminant was then to be measured and the ratio used to account for the remaining site contaminants. In one case, the sampling procedure did not take into account the actual site spatial contaminant distribution. Instead, a limited sample data set from one area of the site was relied upon to prepare the radionuclide ratios. A review of site data collected during earlier scoping surveys clearly demonstrated that the ratio varied among the radionuclides of concern, dependent upon which area of the site the sample represented. When the varying ratios were analyzed, it was determined that the site-specific surrogate ratio that had been developed would significantly underestimate the inferred radionuclide concentrations for portions of the site.

This issue can be readily avoided provided representative samples are collected in such a manner that the ratio developed accurately represents both spatial, and in some cases, depth variability. Furthermore, it may not be reasonable to select a single ratio for application across a site. Rather, it may be necessary to develop multiple ratios and specifically identify sites areas where each ratio will apply. In other cases, the ratio may vary to the extent that no consistent ratio can be inferred, meaning the surrogate approach would not be an option and radionuclide-specific measurements are then required. Additionally, the ratio is typically verified for a percentage of the FSS samples. This is especially true in remediated areas where the decontamination may alter the ratio through either physical or chemical processes.

O.3.4.2   Unity Rule Not Used with Multiple Contaminants

Recent reviews of FSS data packages have identified a critical oversight with demonstrating compliance with the release criteria at some sites with multiple contaminants. What has occurred is that each individual radionuclide is compared with the respective $DCGL_W$ and a conclusion reached as to the acceptability of a survey unit for release. However, an additional requirement is to apply the unity rule (also known as "sum of fractions") to the data to ensure that the basic dose limit is met. This is based on the $DCGL_W$ for each radionuclide equating to the dose limit for release of the site. Due to the additive nature of the dose from each radionuclide, the total residual activity must be proportionality reduced to ensure the sum of each radionuclide divided by its $DCGL_W$ does not exceed one (unity). Application of the unity rule is detailed in Section 2.7 of this volume. Licensees should ensure that when multiple radionuclides of concern are present that the unity rule is applied both in data evaluation and in the initial survey planning phase.

O.3.4.3   Survey Unit Misclassification

Evaluations of licensee survey unit designations and confirmatory surveys have identified inconsistencies with recommendations on survey unit classification; primarily involving contaminated Class 2 survey units. That is, contamination in excess of the $DCGL_W$ that has been found during past confirmatory surveys within Class 2 survey units. As expected, the contamination was usually identified in that portion of the survey unit bordering adjacent Class 1 areas. The simplest solution for the observed occurrences would have been for the licensee to have extended the size of the Class 1 survey units to include adjacent regions. In one case, the contamination was found on the wall portion of the interface between the Class 1 floor and Class 2 wall.

O.3.4.4   Demonstrating Compliance with Hot Spots Present in a Survey Unit

There have been isolated instances where reviews of FSS data packages or confirmatory survey findings identified survey units where the $DCGL_W$ was statistically satisfied, but hot spots were not fully addressed. When hot spots remain in a survey unit, MARSSIM recommends additional data assessment to ensure compliance with the basic dose limit. The first recommendation is that each hot spot be evaluated against the $DCGL_{EMC}$, relative to hot spot size and allowable concentration within the hot spot area. Generally, for hot spots documented in FSS packages, this recommendation is addressed adequately. A component for demonstrating compliance that has been overlooked is showing that the combination of residual hot spot contamination in addition to any uniformly distributed activity is less than the basic dose limit. MARSSIM, Section 8.5.2, provides the equation and narrative guidance for implementation and documentation in survey units where this condition exists.

O.3.5   Implications of Contamination Identified During Confirmatory Surveys

The question is frequently asked: What should be done when contamination is identified during the confirmatory process? This question is directly pertinent because many of the lessons learned presented here have been identified as the root cause for missed contamination. There is no single answer to the question, as each situation is unique. The data quality objective (DQO) process should be followed to establish remedies to the given situation. For confirmatory survey results contrary to the FSS reports for the site, the NRC staff and the licensee should determine what is the magnitude of the finding (number of anomalies identified, size of the anomalies, classification of the area where they were identified) and the proposed remedy. Anomalies that are identified should be evaluated for compliance with the $DCGL_W$ and $DCGL_{EMC}$ and a determination made if the area affected is acceptable relative to size and concentration, has the licensee previously documented and adequately addressed the anomalies, and are they within the bounds of survey unit classification? For multiple anomalies, determine the root cause and reevaluate DQOs. Also consider what percent of a site was subjected to confirmatory surveys. If a small percentage was investigated and multiple areas of residual contamination were identified, the confidence level that the FSS procedures were adequate and that the remaining site areas are acceptable would be low and may necessitate further licensee activities to remedy the data gaps. On the other hand, when a large percentage is confirmed and few anomalies are found, the confidence interval increases significantly and further activities to provide added assurance of guideline compliance may be minimal.

O.3.6 References

American National Standards Institute (ANSI). ANSI N323A–1997, "Radiation Protection Instrumentation Test and Calibration, Portable Survey Instruments." ANSI: New York, New York. 1997.

American Society for Testing and Materials (ASTM). ASTM D4646–87, "24-h Batch-Type Measurement of Contaminant Sorption by Soils and Sediments." ASTM: West Conshohocken, Pennsylvania. 1987.

—————. ASTM D4319–93 (2001), "Standard Test Method for Distribution Ratios by the Short-Term Batch Method." ASTM: West Conshohocken, Pennsylvania. 2001.

Cline, J. E. "Embedded Pipe Dose Calculation Method." Electric Power Research Institute Report No. 1000951. EPRI: Palo Alto, California. November 2000.

Environmental Protection Agency (U.S.) (EPA), Office of Air and Radiation. EPA 402–R–99–004A, Vols. I and II. "Understanding Variation in Partition Coefficient, K_d Values." EPA: Washington, DC. August 1999.

International Organization for Standardization (ISO). ISO–7503–1, "Evaluation of Surface Contamination – Part 1: Beta Emitters and Alpha Emitters (first edition)." ISO: Geneva, Switzerland. 1988.

—————. ISO–7503–3, "Evaluation of Surface Contamination – Part 3: Isomeric Transition and Electron Capture Emitters, Low Energy Beta-Emitters." ISO: Geneva, Switzerland. 1996.

National Council on Radiation Protection and Measurements (NCRP). NCRP Report No. 112, "Calibration of Survey Instruments Used in Radiation Protection for the Assessment of Ionizing Radiation Fields and Radioactive Surface Contamination." NCRP Publications: Bethesda, Maryland. 1991.

Nuclear Regulatory Commission (U.S.) (NRC). NUREG–0130, "Technology, Safety and Cost of Decommissioning." NRC: Washington, DC. June 1978.

—————. NUREG–1507, "Minimum Detectable Concentrations with Typical Radiation Survey Instruments for Various Contaminants and Field Conditions." NRC: Washington, DC. June 1998.

—————. NUREG–1575, Rev. 1, "Multi-Agency Radiation Survey and Site Investigation Manual (MARSSIM)." EPA402–R–97–016, Rev 1, DOE/EH–0624, Rev. 1. U.S. Department of Defense, U.S. Department of Energy, U.S. Environmental Protection Agency, and NRC: Washington, DC. August 2000.

— — — — —. NUREG–1700, "Standard Review Plan for Evaluating Nuclear Power Reactor License Terminations Plans." NRC: Washington, DC. April 2000.

— — — — —. NUREG/CR–3474, "Long-Lived Activation Products in Reactor Materials." NRC: Washington, DC. August 1984.

— — — — —. NUREG/CR–4289, "Residual Radionuclide Contamination Within and Around Commercial Nuclear Power Plants." NRC: Washington, DC. February 1986.

— — — — —. NUREG/CR–5512, Vol. 3, "Residual Radioactive Contamination From Decommissioning, Parameter Analysis, Draft Report for Comment." NRC: Washington, DC. October 1999.

— — — — —. Regulatory Guide 1.179, "Standard Format and Content of License Termination Plans for Nuclear Power Reactors." NRC: Washington, DC. January 1999.

— — — — —. Regulatory Issue Summary (RIS) 2002–02, "Lessons Learned Related to Recently Submitted Decommissioning Plans and License Termination Plans." NRC: Washington, DC. January 2002.

Appendix P

Example of a Graded Approach for Erosion Protection Covers

The purpose of this appendix is to provide a detailed example of the application of a graded approach to the design and implementation of engineered barriers for erosion protection.

P.1   Graded Approach for Erosion Protection Covers

The graded approach can be readily used for the design of an engineered barrier, specifically a soil or rock cover that is provided for erosion control. This approach provides significant flexibility and can be adapted for design of covers at a wide variety of sites with a wide range of waste inventories.

This section describes how the criteria provided in NUREG-1623, "Design of Erosion Protection Covers for Long-Term Stabilization," can be adapted to specifically implement erosion protection guidance for lower and higher risk decommissioning sites. This NUREG was chosen because the staff has used its suggested criteria to review and approve erosion protection designs at approximately 40 different uranium mill tailings impoundments.

Erosion control covers for uranium mill tailings impoundments were chosen because they provide examples of design and construction of robust engineered barriers for long-term protection of radioactive materials. Under the Uranium Mill Tailings Radiation Control Act (UMTRCA), 40 CFR Part 192, and NRC's implementing regulations in 10 CFR Part 40, Appendix A, erosion control covers are required to be designed to remain effective for up to 1000 years without reliance on ongoing active maintenance. Although UMTRCA also requires State or DOE ownership and long-term care of the uranium mill tailings sites, including maintenance of the covers as needed, the covers are designed to function independently of maintenance. Therefore, the monitoring/surveillance and maintenance provided by DOE can be thought of as a backup to the robust design, or a defense in depth approach to long-term protection. Over 25 years of experience is available, including NRC's guidance and technical basis for design of robust erosion control covers and DOE's construction of these covers and monitoring of their performance. This program offers an approach and lessons learned for one kind of robust engineered barrier that could have some application to design of other types of robust barriers. Although the graded approach to engineered barriers in this decommissioning guidance offers greater flexibility to select appropriate options for engineered barriers (including reliance on maintenance), the UMTRCA covers offer an excellent example of a robust erosion control barrier to maintain stability for sites with a long-term hazard.

Robust engineered barriers for erosion control should be designed, but a graded approach to the design should be taken with respect to selection of the design flood, evaluation of rock durability, and selection of appropriate design factors. Each of these three areas is described below.

Selection of Design Flood

One of the phenomena most likely to affect long-term stability is surface water erosion. To mitigate the potential effects of surface water erosion, the staff considers that it is very important to select an appropriate rainfall event on which to base the erosion protection designs. Further, the staff considers that the selection of a design flood event should not be based on the extrapolation of limited historical flood data, due to the unknown level of accuracy associated with such an extrapolation. The Probable Maximum Precipitation (PMP) is computed by deterministic methods (rather than statistical methods) and is based on site-specific hydrometeorological characteristics. The PMP has been defined as the most severe reasonably possible rainfall event that could occur as a result of a combination of the most severe meteorological conditions occurring over a watershed. No recurrence interval is normally assigned to the PMP; however, the staff has concluded that the probability of such an event being equaled or exceeded during a 1000-year stability period is very low. Accordingly, the PMP is considered by the NRC staff to provide an acceptable design basis.

The Probable Maximum Flood (PMF) is based on the occurrence of the PMP and is considered to represent the most severe flood that can reasonably be expected to occur over a particular drainage basin. There is no probability assigned to the PMF, but the staff will generally not accept the use of statistically-derived floods when the analysis time period significantly exceeds recorded data, due to the unreliable extrapolation of flood records based on short-term data. Additional discussion of use of the PMP and PMF may be found in NUREG-1623.

Rock Durability

Rock durability is defined as the ability of material to withstand the forces of weathering. Primary factors that affect rock durability are (1) chemical reactions with water, (2) saturation time, (3) temperature of the water, (4) scour by sediments, (5) windblown scour, (6) wetting and drying, and (7) freezing and thawing.

For rock to remain effective to control erosion, the rock size selected and emplaced should not be reduced in size by weathering processes. Therefore, if the rock size used for the cover does not diminish over the 1000-year compliance time period, its ability to control future erosion will be sustained. However, uncertainties exist with estimating future rock durability. For example, quantitative studies of weathering rates of different rock types and minerals are limited, in general, and not expected to be available for specific rock sources that might be selected by licensees. As a result, three evaluations of rock durability should be conducted to provide multiple and complimentary lines of evidence and greater confidence in the future durability of the rock source selected. These evaluations are: (1) rock durability testing and scoring; (2) absence of adverse minerals and heterogeneities; and (3) evidence of resistence to weathering. Each of these evaluations are described below.

Rock durability testing and scoring

In general, rock durability testing is performed using standard test procedures, such as those developed by the American Society for Testing and Materials (ASTM). The ASTM publishes and updates an Annual Book of ASTM Standards, and rock durability testing should be performed using these standardized test methods.

NUREG-1623 provides a procedure for determining the acceptability of a rock source, which was developed by NRC staff and used for selecting durable rock for erosion covers for uranium mill tailing sites. This procedure provides a consistent and quantitative way to evaluate rock sources at NRC licensed sites using standard parameters that are good indicators of rock durability. Test samples of the selected rock source should be representative of the specific rock expected to be used. The number and location of the samples should be determined based on the variability of the rock source, such as texture and mineralogy that could affect the individual test results. The test procedure generally includes the following:

- Test results of four parameters (specific gravity, absorption, sodium sulfate, and Los Angeles abrasion) from representative samples are scored on a scale of 0–10.

- The score is multiplied by a weighting factor, which focuses the scoring on those tests that are the most applicable for the particular rock type being tested.

- The weighted scores are totaled, divided by the maximum possible score, and multiplied by 100 to determine the rating.

- The rock quality scores are then compared to the criteria which are measures of acceptability.

After these tests are conducted, an overall rock quality score is determined. Rock scores of 80% or greater indicates high quality rock that can be used for most applications. Rock scores between 65%–80% indicate less durable rock that can also be used for most applications. Rock scoring less than 65% cannot be used for critical areas such as diversion ditches or poorly drained toes and aprons. Rock scoring between 50%–65% can be used in non-critical areas such as well drained side slopes provided it is over-sized. Rock scoring less than 50% is not recommended for use in any application. Additional discussion of specific tests and the scoring procedures are found in NUREG-1623.

Absence of adverse minerals and heterogeneities

Results should be provided of petrographic analyses of the selected rock source and available published data that supports the absence of adverse minerals that could cause rapid degradation the rock, such as clays, olivine, or calcite cement. If adverse minerals are present, evaluate the potential effect on future weathering of the selected rock.

Particular emphasis should be placed on selecting rocks that do not have appreciable clay content or do not contain minerals that could rapidly weather to clays. The staff examined the causes of rock durability failures in typical applications such as placement on dam slopes or stream channels. In most cases, it was determined that durability failures were caused by the presence of expanding clay-lattice minerals, which, when exposed to moisture or freeze thaw cycles, caused the rapid deterioration of the rock.

Licensees should identify either the absence or presence of heterogeneities that could adversely affect the selected rock source such as clay or shale partings, interbeds, fractures, alteration zones, or vein deposits. Such heterogeneities can be zones of water flow and associated chemical alteration or zones of breakage due to freeze thaw. Licensees should characterize heterogeneities present and discuss the ability to avoid these adverse heterogeneities when removing the selected rock for use. Evaluations should also be included of how heterogeneities could impact achieving the acceptable size of the selected rock. Breakage along thin interbeds or fractures during rock quarrying, transport, or emplacement could result in reduction of the rock size that is necessary for erosion protection.

Evidence of resistance to weathering

Direct evidence from the selected rock source should be provided, such as minerals that are resistant to weathering, resistant cements, and regional or local geomorphic evidence of slow weathering of the selected rock that outcrops in other locations. Examples include rounded boulders or thin weathering rinds. Weathering rind thickness and alteration of minerals and rock properties from exposures of the weathered selected rock source could provide insights on the extent and nature of future weathering of the selected rock source from fresh quarry exposures. Identify and describe any available studies of weathering rates of the selected rock source.

Indirect evidence can also be used from other locations where the general rock type is similar to the selected rock source. For example, evidence of durability from a diabase igneous rock found in Europe could be used to provide insights on a diabase rock source in Maryland because the general mineralogy of diabase is similar, regardless of the location. This approach allows the use of datable natural or archaeological/historical rock sites (called analog sites), that could provide general evidence on rock weathering rates or time periods during which rock types have remained resistant to weathering. For example, as mentioned in Section P.3.3, numerous datable archaeological sites, such as Stonehenge (constructed about 4,000 years ago of diabase and silica cemented sandstone), Hadrian's Wall (constructed by the Romans over 2000 years ago of primarily diabase), and numerous buildings, monuments, and megoliths in Europe, could be used to demonstrate that these rock types have been resistant to weathering over time periods that exceed the 1000-year period of regulatory compliance for a robust erosion cover. Historical evidence can also provide useful insights on the durability of certain rock types. One example is the comparison of dated Civil War photographs of diabase outcrops in Devils's Den at the Gettysburg Battle Field Park to present-day conditions of the same outcrop. Such a comparison indicates that this diabase has been resistant to weathering for about 150 years. Similarly, dated grave markers or historical buildings made from the selected rock source or similar rock type can also provide evidence of resistance to weathering for 100–200 years. Appendix A of NUREG/CR-2642 provides additional information on rock weathering, durability, and examples of analogs that provide insights on general weathering rates of various rock types.

As previously noted, natural rock sites, such as age-dated glacial erratics could also be used to show that the particular rock type has been resistant to weathering for over 10,000 years. Examples that illustrate this approach include age-dated diabase erratics from New York and North Dakota that are over 10,000 years old and river scour features in diabase that have been preserved for about 20,000 years in Pennsylvania.

For each of the above evaluations, the licensee should provide the appropriate supporting data, information, and references. Also identify the location of the selected rock source along with regional and local rock source descriptions.

Selection of Appropriate Design Factors

In the selection of appropriate input parameters for calculating erosion protection size and thickness, it is important to choose values that reflect the degree of risk and the importance of the rock layer, as it contributes to overall stability. However, the selection of many input parameters to various models can sometimes be subjective and will need to be based largely on engineering judgment. Where there are large ranges in values, or where a parameter cannot be well-defined, or is not well-known, it has been the general policy of the NRC staff to accept the use of reasonable ranges and distributions of input parameters. For well-known or accepted parameters with narrow empirical distributions or very narrow ranges, expected values should be used as appropriate. For less well-known parameters, such as those based on little empirical data or with broad distributions, conservative values should be chosen from within the observed distributions or estimated range. In any case, there should a reasonable and defensible technical basis for the choice of a design basis event or design criteria, and the staff will accept values that can be justified. Otherwise, reasonably conservative values will be needed.

P.2   Erosion Protection Cover Analysis Process

At a lower risk site, where engineered barriers are needed to meet applicable requirements for only about 100 years and there is a lower hazard level should institutional controls and maintenance fail [up to the public dose limit of 1.0 mSv/y (100 mrem/y)], the principal design basis and goal is to assure the relative stability of the contaminated material by providing a cover design that maintains control of the material. Control of the material is achieved by providing a relatively robust design that prevents offsite movement (e.g., erosion by natural forces) of the material. The rock erosion protection barrier could protect a second layer of material, for example one that might be a radiation shielding barrier or an infiltration barrier depending on the radionuclides at the site and natural processes important to achieving compliance with the dose criteria.

The design should be able to survive the occurrence of relatively rare events, and the erosion protection should be sufficiently robust to remain effective for about 100 years. Using the guidance and rationale contained in NUREG-1623 for a 100-year stability period, the barriers should be designed to resist a flood equivalent to either the regional historic flood of record or about half of the PMF, whichever is greater. The licensee should provide rainfall, flood, and erosion analyses that justify the design. A design that meets the suggested flooding and erosion protection criteria of NUREG-1623 is acceptable. The rock itself should be sufficiently durable to remain effective for at least 100 years by obtaining a rock quality score of at least 65. The computations and selection of input parameters should rely on reasonable and justified estimates.

For a higher risk site, with long-lived radionuclides or where failure of institutional controls and maintenance could result in a higher hazard of 1.0–5.0 mSv/y (100–500 mrem/y), the principal design basis and goal is to assure the control and stability of the contaminated material by providing a robust design that will remain effective for a period of 1000 years or more by preventing erosion by natural forces. These covers should be designed to maintain control and stability with no reliance on active maintenance. However, monitoring and maintenance will be conducted as a backup to provide defense-in-depth.

The staff could approve an engineered barrier design that is effective and maintains control of the material for a period exceeding 1000 years. Using the guidance and rationale contained in NUREG-1623, the barriers should be designed to resist severe localized rainfall events and large floods on nearby streams. The design rainfall event should be the PMP, and the design flood should be the PMF. A design that meets the suggested flooding and erosion protection criteria of NUREG-1623 is acceptable. The rock quality score should be at least 85, and selection of input parameters to various models should account for the unknowns associated with a very long stability period and the high-risk site.

For sites like this, if erosion is a significant issue and there are some uncertainties associated with the magnitude of this erosion, the staff will approve a design that would likely incorporate: (1) covers designed to resist erosion for a stability period exceeding 1000 years; (2) a long-term surveillance program that monitors the magnitude and rate of erosion; and (3) sufficient funding for the surveillance, repair, and replacement of some of the erosion protection. The staff will work closely with the expected long-term custodian to determine the amount of funding needed.

It is important to reiterate that the requirements of 10 CFR Part 40 are very prescriptive and may have precluded the use of many types of erosion protection designs. The staff considers that the design criteria suggested in NUREG-1623 may be used at decommissioning sites using approaches that were not necessarily used in uranium mill tailings applications. For example, nearly all tailings sites were designed with disposal cell side slopes of about 1 vertical (V) on 5 horizontal (H). Based on the stability of erosion protection placed on much steeper slopes of stream channels, levees, and/or dam slopes, there is no reason why slopes steeper than 1V on 5H could not be used for the side slopes of disposal cells. The criteria in NUREG-1623 were not developed for use on specific slopes and may be adapted for steeper side slopes, as necessary. Minor changes to construction specifications, emphasizing careful rock placement on steeper slopes, may be the only added design consideration.

Table P.1 summarizes the application of the graded approach for the design of erosion protection for the sites discussed above.

Table P.1   Summary of the Graded Approach for the Design of Erosion Protection Systems

Level of Risk	Flood Design Basis	Rock Durability Score	Confidence in Selection of Input Parameters
Lower	½ PMP / ½ PMF	> 65	Reasonable
Higher	PMP/ PMF	> 85	Very High

P.3   Technical Basis for Design and Performance of Erosion Protection Covers

NRC staff considers that the use of the guidance presented in this document and in NUREG-1623 will result in designs that provide acceptable long-term erosion protection at decommissioning sites. These documents contain design criteria that incorporate a strong technical basis, including: (1) use of NRC experience and lessons learned at various sites to develop and implement erosion protection guidance, (2) use of appropriately conservative design bases and computational procedures for erosion protection designs; and (3) an extensive archaeological and natural basis for long-term stability.

P.3.1   NRC Experience and Lessons Learned

The NRC staff has over 20 years of experience in the design and review of erosion protection covers. This experience includes activities associated with (1) development of design guidance for uranium mill tailings reclamation, including use of technical studies sponsored by NRC to address specific design problems; (2) review of reclamation plans and decommissioning plans at about 50 sites, including review of specific problem sites where challenging erosion problems were addressed; and (3) review and inspections of construction problems and deficiencies. The staff has applied this experience in its review of uranium mill reclamation plans and decommissioning plans on numerous occasions.

Guidance Development

Following passage of the UMTRCA in 1978, implementing regulations were developed in 40 CFR 192 and 10 CFR 40, Appendix A. These regulations established design standards to be met at uranium mill sites for Title I (inactive sites) and Title II (licensed active sites), as required by UMTRCA. Specifically, the design standard for long-term stability was established to be 1000 years to the extent reasonably achievable, or in any case at least 200 years. At that time, there was very little experience associated with providing designs that would remain effective for such long periods of time. To address this problem, the NRC staff worked closely with DOE and various contractors in the 1980s to establish design criteria and guidance for long-term stabilization. Several joint meetings were held with DOE and DOE contractors to formulate guidance and design procedures that would be used to address the long-term stability of erosion protection covers. These procedures were published in various technical and construction documents that were developed by DOE, DOE contractors, and NRC staff.

During the development of design guidance, it was recognized that some of the design procedures normally used for design of erosion protection were not necessarily appropriate for long-term stabilization or for conditions that would be encountered at various sites. To address such issues, the NRC sponsored extensive technical studies by independent contractors. These studies were conducted in the 1990s at Colorado State University and included: prototype flume studies to determine rock sizing procedures for overland flows; flume studies to determine rock sizing and volume requirements for aprons and toes of slopes; gully studies to determine rates, magnitude, and location of gully development on reclaimed slopes; and rock durability studies to address rock weathering rates and tests needed to assure adequate rock quality. The results of the studies were published in nationally-recognized journals and were peer-reviewed by many experts in the field of erosion protection design. One example of a publication that has been widely used includes detailed guidance for design of riprap for flood flows that would be expected down the side slope of a disposal cell (Abt and Johnson, 1991). Results from this and other publications are included in NUREG-1623.

NRC Staff Review of Challenging Sites

Vegetated soil covers, rock covers, and composite covers (soil and rock) were used extensively in the reclamation and stabilization of all uranium mill tailings sites. In general, most of these sites presented no significant erosion problems, and the DOE and several NRC licensees had a great deal of success in designing and constructing these covers. However, several sites presented challenging erosion problems. Table P.2 provides examples of several significant erosion problems that were encountered and solved with engineered barriers, using design criteria contained in NUREG-1623.

Table P.2   Examples of Engineered Barriers Used to Solve Erosion Problems

Site	Erosion Problem	Solution to Problem
Maybell UMTRA Site	Downstream gullies that could scour to a depth of about 20 feet could encroach on the disposal cell. Unstable local base levels could cause further increase in the potential for gullying.	DOE provided extensive rock aprons to check erosion advance to the disposal cell, using criteria for scour depth and rock sizing provided in NUREG-1623.
Grand Junction UMTRA Site	Deep gullies with relatively large drainage areas existed in the approximate center of the proposed disposal cell. Hydraulically-steep slopes were present near the toe of the cell.	DOE provided diversion channels and channel outlets with very large riprap to safely convey flows around the disposal cell, using criteria found in NUREG-1623.
Rifle (Estes Gulch) UMTRA Site	The disposal cell was excavated in a steep gully located in the center of the site, requiring diversion of flood flows.	An upstream diversion channel was constructed to convey flows around the cell into a different drainage basin.
Atlas Title II Site	The disposal cell is immediately adjacent to the Colorado River, with a drainage area of thousands of square miles. There is a potential for the river to erode and migrate towards the cell, where high river velocities could impinge on the cell.	The licensee provided a design to resist the erosional forces associated with river migration, proposing to construct a large rock apron in accordance with the suggested criteria of NUREG-1623. (It should be noted that a decision was later made to completely move the entire tailings pile.)

It can be seen that a large amount of experience was gained at various sites where erosion was found to be a significant problem. The staff analyzed and became familiar with various design solutions that could be used to mitigate significant erosion problems. Based on this experience, the staff considers that the erosion protection criteria suggested in NUREG-1623, combined with the graded regulatory approach discussed in this example, will provide a significant degree of flexibility in solving difficult erosion problems at complex sites.

NRC Staff Inspections and Reviews of Construction Deficiencies

For the last 15–20 years, it has been routine NRC staff practice to conduct inspections at nearly all sites that have been reclaimed in the Title I and Title II programs. The inspections included evaluations of soil cover and rock cover placement during construction and final closeout inspections of the completed work.

During these inspections, the staff noted that adequate placement of rock riprap layers was difficult for many contractors to achieve. There were numerous instances where: rock layers were not placed to correct and consistent design thicknesses; in-place riprap layers did not have correct gradations and varied from tested samples; and areas of segregation existed where rock sizes were much smaller than required over large areas of the disposal cell. In many cases, the staff required corrective actions to be taken. Based on these problems encountered in staff reviews and its experience with corrective actions, the staff developed guidance for proper rock placement. The staff's construction experience was used to develop suggested quality control procedures for use by contractors. The suggested guidance may be found in Appendix D of NUREG-1623.

At the present time, the staff also routinely accompanies the DOE and/or DOE contractors on annual inspections at some sites where DOE now has custody and licensed responsibilities. These inspections have generally shown that covers are performing rather well and regulatory requirements continue to be met in all cases.

The staff also has experience with rock durability problems at uranium mill tailings reclamation sites. These problems occurred, for example, when rock sources consisting of minerals susceptible to weathering (such as olivine basalts) or sources that had undergone hydrothermal alteration were used. In addition, the staff reviewed information regarding 149 case histories (Esmiol, 1964) associated with rock durability problems at facilities constructed by other Federal agencies (such as Bureau of Reclamation dam sites). Using the lessons learned from both uranium mill sites and other Federal sites, the rock durability criteria suggested in NUREG-1623 were developed.

P.3.2   Use of Design Bases Appropriate for Long Stability Periods

As discussed in previous sections, the staff considers that erosion protection designs that meet the suggested criteria and guidance in this document and NUREG-1623 will provide adequate protection against extreme erosion events that could occur over the period of regulatory interest. The guidance reflects staff review and construction experience, past practices with regard to selection of design bases, and good engineering practices. The staff considers that appropriate guidance is provided with regard to: (1) selection of conservative rainfall and flooding events that reflect the long stability periods needed to meet regulatory requirements; (2) selection of parameters for determining flood discharges that account for the uncertainties associated with flood calculations; (3) computation of flood discharges using appropriate and/or conservative methods; (4) computation of appropriate flood levels and flood forces associated with the design discharge; (5) use of widely-accepted, state-of-the-art, and standardized methods for determining erosion protection sizes and thicknesses; (6) selection of a rock type for the riprap layer that will be durable and maintain its size and ability to provide protection for a long period of time; and (7) placement of riprap layers in accordance with accepted engineering practice and in accordance with appropriate testing and quality assurance controls.

P.3.3   Archaeological and Natural Bases for Long-Term Stability

A strong archaeological basis exists to demonstrate the long-term stability of erosion protection materials. NUREG/CR-2642 presents substantial information to demonstrate the long-term survivability of various rock structures. NUREG-1623 provides information on long-term weathering rates, based on observations of rock petroglyphs that could be dated to a period of nearly 1000 years before present. Further, as discussed in Section 3.5.3 of this document, observations of Native American burial mounds in West Virginia, Ohio, Illinois, and Louisiana serve to illustrate the survivability of man-made earthen structures for long periods of time (1000–5500 years).

Numerous other archaeological sites, such as Stonehenge (constructed about 4,000 years ago of diabase and silica cemented sandstone), Hadrian's Wall (constructed by the Romans over 2000 years ago of primarily diabase), and numerous buildings in Europe demonstrate that certain rock types used in these structures have been resistant to weathering over time periods greater than the 1000-year period of regulatory compliance. In addition to archeological and historical evidence, natural rock sites can offer further insights that certain rock types and sources have been resistant to weathering. For example, dating of glacial erratics (rocks transported by glaciers) and preservation of river scour features in diabase results from high flow during glacial melting demonstrate that these rocks have been resistant to weathering and preserved for over 10,000 years.

The natural and archaeological insights noted above serve to demonstrate that engineered structures and construction materials can survive for very long periods of time. The intent of this guidance is to develop procedures for improving long-term stability by further enhancing design concepts where structures have generally remained intact for many years.

P.4   Degradation Mechanisms for Erosion Control Covers

The erosion control cover at a typical decommissioning site could consist of either a rock layer or a soil layer and underlying rock layer. One of the most likely degradation mechanisms would be gully erosion. To account for that process, the design consideration that should be analyzed (and is considered by the staff to be the most likely) is the formation of a gully in the top soil cover, caused by surface erosion, flow concentrations, and/or the uprooting of large trees. The erosion should be assumed to continue and be deep enough to expose the rock layer, and thus the rock layer would need to be designed to resist further erosion and down-cutting of the gully. The licensee should design the soil cover, as described above, to be stable for rainfall events and runoff as large as the PMP/PMF. Further, the rock layer should be designed as a separate backup system for the soil cover and should also be designed for the PMP/PMF occurring in that gully, with flow concentrations produced by the growth of a drainage network to that gully. Further, the rock should meet the durability criteria suggested in NUREG-1623, with particular emphasis placed on the petrographic examination that indicates that no clay minerals are present (see Table 3.2).

References*

Abt, S. R. and T. L. Johnson, "Riprap Design for Overtopping Flow," American Society of Civil Engineers, Journal of Hydraulic Engineering, Vol. 117, No. 8, August, 1991.

Esmiol, E.E., "Rock as Upstream Slope Protection for Earth Dams—149 Case Histories," Report No. DD-3, U.S. Department of the Interior, Bureau of Reclamation, September 1967.

* References not listed here are found in Table 3.2.

NRC FORM 335
(9-2004)
NRCMD 3.7

U.S. NUCLEAR REGULATORY COMMISSION

BIBLIOGRAPHIC DATA SHEET

(See instructions on the reverse)

1. REPORT NUMBER
(Assigned by NRC, Add Vol., Supp., Rev., and Addendum Numbers, if any.)

NUREG-1757, Vol. 2, Rev. 1

2. TITLE AND SUBTITLE

Consolidated Decommissioning Guidance:
Characterization, Survey, and Determination of Radiological Criteria

Final Report

3. DATE REPORT PUBLISHED

MONTH	YEAR
September	2006

4. FIN OR GRANT NUMBER

5. AUTHOR(S)

D.W. Schmidt, K.L. Banovac, J.T. Buckley, D.W. Esh, R.L. Johnson, J.J. Kottan, C.A. McKenney, T.G. McLaughlin, and S. Schneider

6. TYPE OF REPORT

Technical

7. PERIOD COVERED *(Inclusive Dates)*

8. PERFORMING ORGANIZATION - NAME AND ADDRESS *(If NRC, provide Division, Office or Region, U.S. Nuclear Regulatory Commission, and mailing address; if contractor, provide name and mailing address.)*

Division of Waste Management and Environmental Protection
Office of Nuclear Material Safety and Safeguards
U.S. Nuclear Regulatory Commission
Washington, DC 20555-0001

9. SPONSORING ORGANIZATION - NAME AND ADDRESS *(If NRC, type "Same as above"; if contractor, provide NRC Division, Office or Region, U.S. Nuclear Regulatory Commission, and mailing address.)*

Same as above

10. SUPPLEMENTARY NOTES

D.W. Schmidt and K.L. Banovac, NRC Project Managers

11. ABSTRACT *(200 words or less)*

As part of its redesign of the materials license program, the U.S. Nuclear Regulatory Commission (NRC), Office of Nuclear Material Safety and Safeguards has consolidated and updated numerous decommissioning guidance documents into a three-volume NUREG.

Volume 2 of the NUREG series, entitled, "Consolidated Decommissioning Guidance: Characterization, Survey, and Determination of Radiological Criteria," provides guidance on compliance with the radiological criteria for license termination (License Termination Rule (LTR)) in 10 CFR Part 20, Subpart E. This guidance takes a risk-informed, performance-based approach to the demonstration of compliance. The approaches to license termination described in this guidance will help to identify the information (subject matter and level of detail) needed to terminate a license by considering the specific circumstances of the wide range of NRC licensees. Licensees should use this guidance in preparing decommissioning plans, license termination plans, final status surveys, and other technical decommissioning reports for NRC submittal. NRC staff will use the guidance in reviewing these documents and related license amendment requests.

Volume 2 is applicable to all licensees subject to the LTR.

12. KEY WORDS/DESCRIPTORS *(List words or phrases that will assist researchers in locating the report.)*

consolidated
decommissioning
guidance
risk-informed
license termination
LTR

13. AVAILABILITY STATEMENT

unlimited

14. SECURITY CLASSIFICATION

(This Page)

unclassified

(This Report)

unclassified

15. NUMBER OF PAGES

16. PRICE